A Legacy in Arms

THE WESTERN LEGACIES SERIES

Published in Cooperation with the
NATIONAL COWBOY & WESTERN HERITAGE MUSEUM

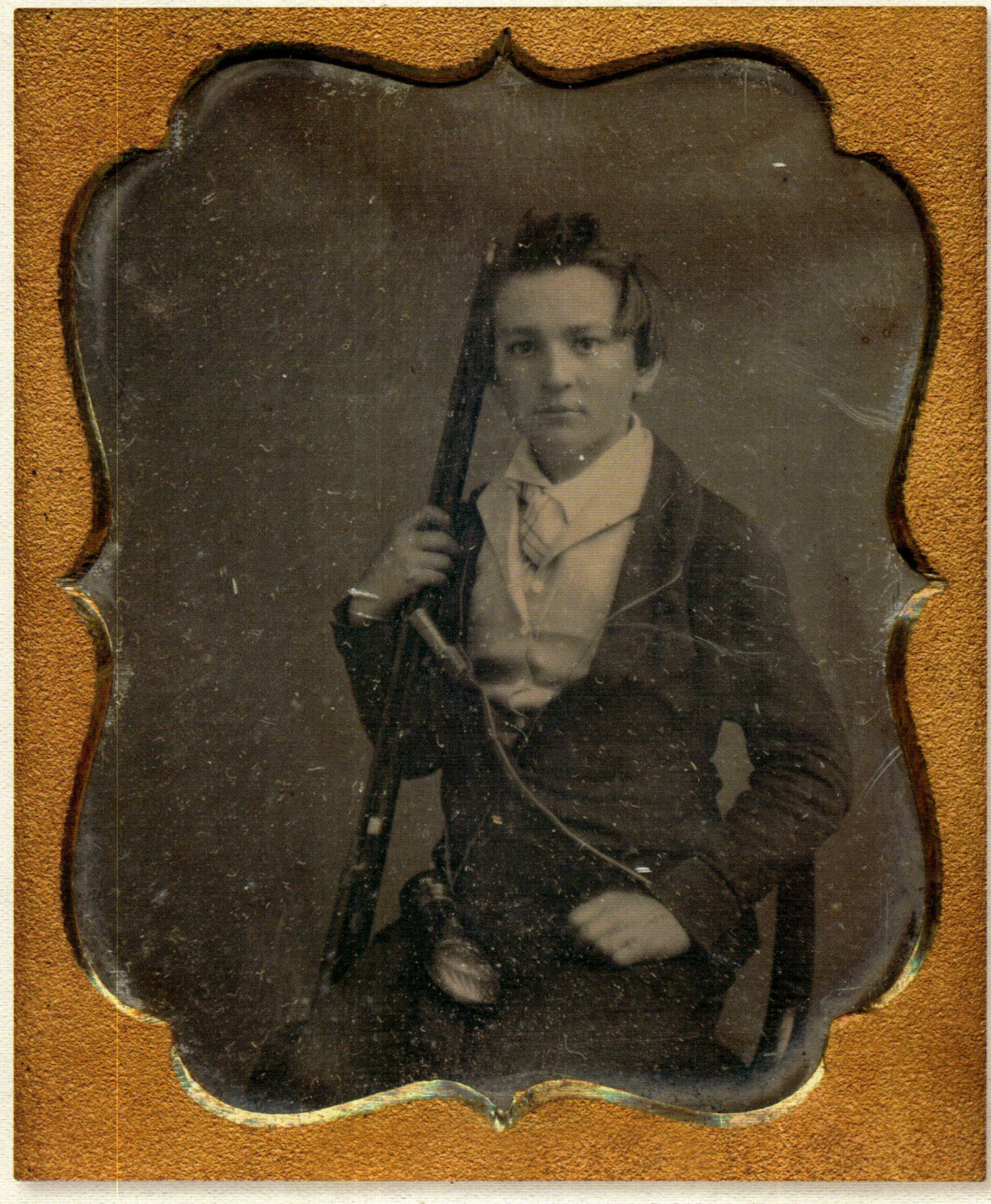

Jesse Whitehurst, photographer. [Untitled.] Sixth-plate daguerreotype, Boston or New York (?), circa 1845–1850. Dressed for a quite-formal portrait, this ardent young huntsman poses with his rather substantial, half-stocked percussion sporting rifle at his side, his sheath knife at his chest, and his powder flask in his lap.
Dickinson Research Center, National Cowboy & Western Heritage Museum, 2003.081

A LEGACY IN ARMS

American Firearm Manufacture, Design, and Artistry, 1800–1900

By ***Richard C. Rattenbury***

Foreword by ***R. L. Wilson***

Collection Photography by ***Ed Muno***

UNIVERSITY OF OKLAHOMA PRESS : NORMAN

Also by Richard C. Rattenbury

Arena Legacy: The Heritage of American Rodeo

Hunting the American West: The Pursuit of Big Game for Life, Profit and Sport, 1800–1900

A Western Legacy: The National Cowboy & Western Heritage Museum (contributor)

The Art of American Arms Makers: Marketing Guns, Ammunition and Western Adventure

Packing Iron: Gunleather of the Frontier West

The Browning Connection: Patent Prototypes in the Winchester Museum

LIBRARY OF CONGRESS CATALOGING-IN-PUBLICATION DATA
Rattenbury, Richard.
A legacy in arms : American firearm manufacture, design, and artistry, 1800–1900 / by Richard C. Rattenbury ; foreword by R. L. Wilson ; collection photography by Ed Muno.

pages cm. — (Western legacies series ; volume 10)
Includes bibliographical references and index.

ISBN 978-0-8061-4477-1 (hardcover : alk. paper)

1. Firearms—United States—History—19th century—Catalogs.
2. Firearms—United States—Design and construction—Catalogs.
3. Firearms industry and trade—United States—History—19th century—Catalogs. I. Title.

UD383.R42 2014
683.40973'09034—dc23

2013049567

The paper in this book meets the guidelines for permanence and durability of the Committee on Production Guidelines for Book Longevity of the Council on Library Resources, Inc. ∞

1 2 3 4 5 6 7 8 9 10

A Legacy in Arms: American Firearm Manufacture, Design, and Artistry, 1800–1900 is Volume 10 in The Western Legacies Series.

To
Norm Flayderman
antiquarian, author, mentor, and friend
1928–2013

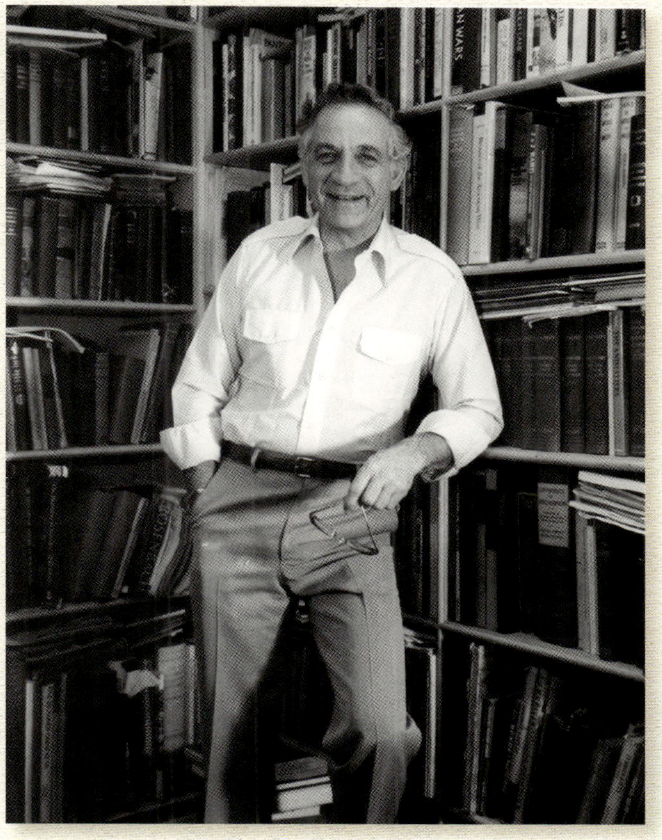

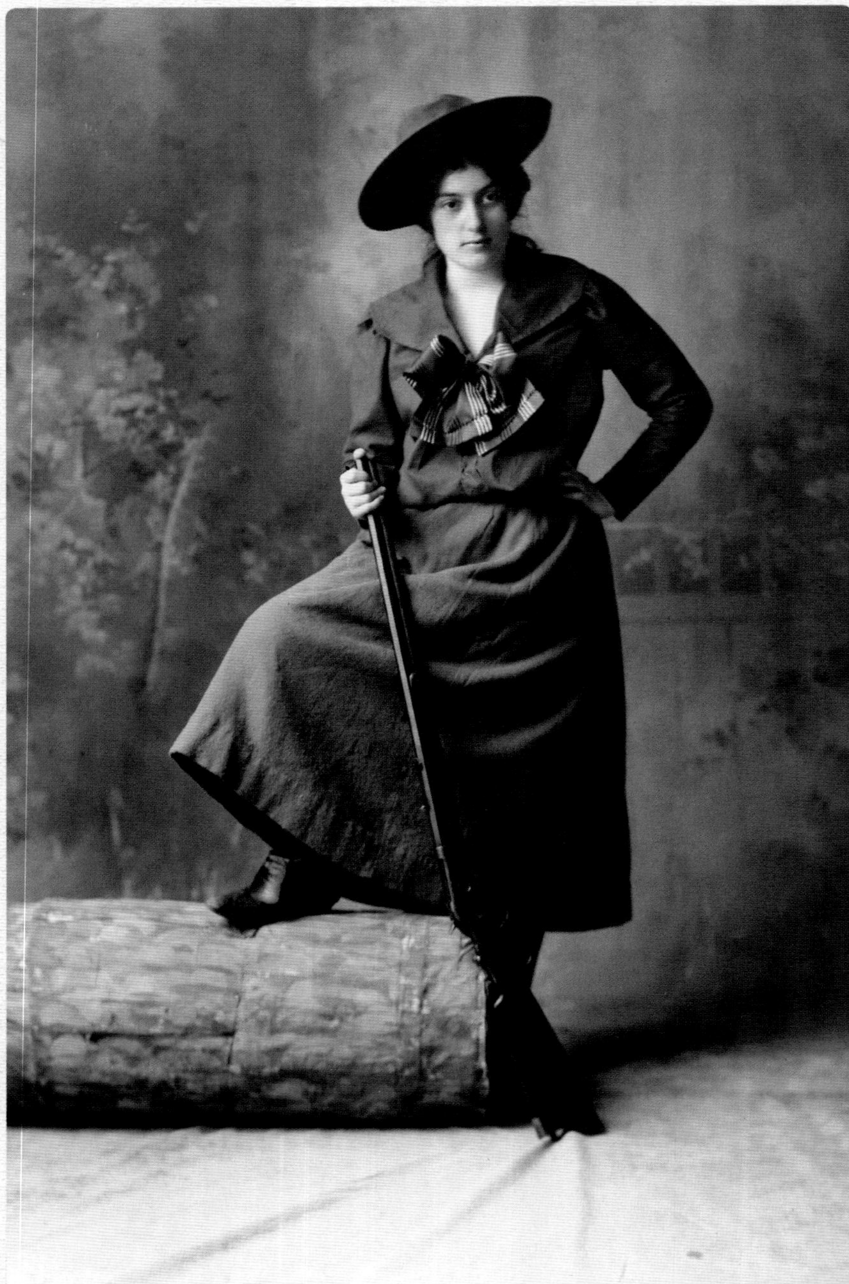

Photographer unknown. [Untitled.] Sepia print, America, circa 1905–1910. With her standard-grade Winchester Model 1892 lever-action, repeating magazine rifle in hand, this latter-day huntress in wide-brimmed hat, middy blouse, and skirt strikes a rather risqué pose for the time.

Dickinson Research Center, National Cowboy & Western Heritage Museum, 2004.045

*As to the species of exercise, I advise the gun.
While this gives a moderate exercise to the body, it gives boldness, enterprise, and independence to the mind.*

THOMAS JEFFERSON

Arms, women, and books should be studied every day.

NINETEENTH-CENTURY
RUSSIAN MAXIM

CONTENTS

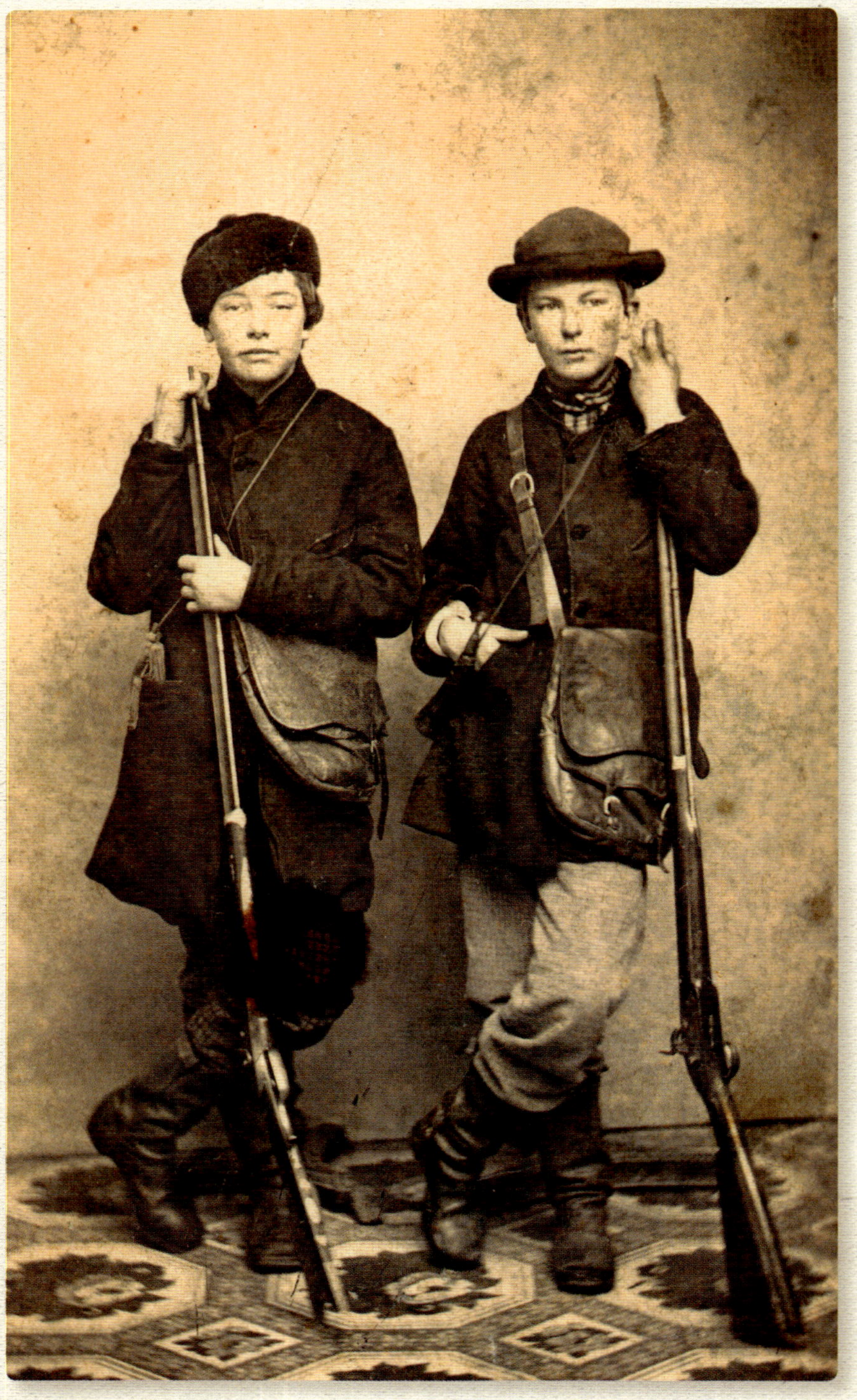

Photographer unknown. [Untitled.] Carte de visite, America, circa 1865–1870. Striking a jaunty pose, these lads wear leather game bags and hold half-stocked percussion sporting rifles. Rifle at left—fitted with double set triggers—appears to be a comparatively fancy specimen for a youth.

Dickinson Research Center, National Cowboy & Western Heritage Museum, 2003.196

FOREWORD

R. L. Wilson

WHEN SIR HAROLD EVANS wrote his landmark book *They Made America—From the Steam Engine to the Search Engine: Two Centuries of Innovators,* the four hundred inventors and entrepreneurs featured in that title were reviewed by WGBH (Boston) for a four-hour television documentary series for broadcast on PBS. Only sixteen of those select four hundred were chosen for that broadcast, and of that final cut, two were gun makers: Eli Whitney and Samuel Colt. This fact by itself is revealing of the significance of firearms in the history of American invention, design, manufacturing, and innovation.

That distinctive triumph, termed the "American System of Manufactures," the creation primarily of Yankee inventors and "mechanics" in our nation's early-nineteenth-century firearms trade, proved to be a pivotal influence in American industrial history. It took the country from the diverse cottage industries of 1800 to the concentrated plateau of factory enterprises rivaling any in the world in 1900.

Crucial to that amazing evolution were the arms makers themselves—from the singular Kentucky rifle artisans of 1800 to the big-four manufacturers of 1900 (Colt, Remington, Smith & Wesson, and Winchester). The story of that transition involves art and craftsmanship, invention, technology, machinery, parts interchangeability and mass production, manifest destiny, and myriad and extraordinary examples of energetic and progress-fueled ingenuity.

In many ways those nineteenth-century achievements in arms manufacture have been paralleled by advancements in the late twentieth and early twenty-first centuries through the magic of computer and digital technology. As the latter period had its Steve Jobs, so the earlier had its Samuel Colt. Innovative, dominating, and inspiring in his achievements, Colonel Colt, the first industrial tycoon, may be seen as the prototype and precursor to Steve Jobs—frequently termed the Thomas Edison and Henry Ford of the digital age. It was Samuel Colt who, more than anyone else, was the precursor of Edison and Ford.

Under the colonel and his principal associate, E. K. Root, Colt's Patent Fire Arms Mfg. Co. (incorporated in 1855) became a training ground for inventors and manufacturers. Indeed, Colt intended on fomenting such creativity by establishing in Hartford his own institution of technology that would have preceded MIT by several years. Both Sam Colt and Steve Jobs were geniuses of incredible energy and innovation, and both left the world they were instrumental in creating years before they should have—Colt at 47 and Jobs at 56. The cultures from which they departed had been changed dramatically, each having shepherded in the future and impacted the larger world beyond any of their contemporary pioneers.

These creators and entrepreneurs were visionaries with a capital *V*. Colt and Jobs also designed innovative products that were of aesthetic merit—form following function in graceful, often elegant ways. And they achieved international success as masters, too, of branding and marketing. Each represented the most advanced technology of their day. Both employed teams that operated at the cutting edge of imagination, intelligence, scale, and dedication. Both became legends in their own time.

Richard Rattenbury's *A Legacy in Arms* captures the essence of the supercharged universe of nineteenth-century American gun making. Through informative narrative text and captions, revealing imagery, and supporting notes and bibliography, little of this vital story is left without discovery.

The focus of the collection brings into full dimension the fascinating world of Samuel Colt, the Remington family, Horace Smith and Daniel Wesson, Christian Sharps, John Marlin, and Oliver Winchester—and their contemporary designers, makers, and dreamers. And in this world must also be acknowledged the fascinating characters whose lives and callings depended on the reliability of their weapons—from intrepid frontiersmen and hunters the likes of Theodore Roosevelt to famed and flamboyant shootists such as Wild Bill Hickok, Buffalo Bill Cody, and so many more—icons and legends all.

Richard Rattenbury is perfectly qualified to research and author this captivating work. For years he has served as curator of history with the National Cowboy & Western Heritage Museum in Oklahoma City, and he formerly held curatorial positions at the Panhandle-Plains Historical Museum in Canyon, Texas, and at the Winchester Museum at the Buffalo Bill Historical Center in Cody, Wyoming.

Over a multidecade career, the author has placed on public view four permanent exhibition galleries, as well as some twenty-five temporary exhibitions treating firearms and a myriad of themes in western history, material culture, and graphic art. Rattenbury has also authored several books and exhibition catalogs, as well as many articles treating firearms history and technology. All of these pursuits have contributed to a well-balanced perception and understanding of America and its fascinating tangible heritage.

Rattenbury's insightful new publication will contribute significantly to understanding our innovative past and will prove an inspiration to the vast audience fascinated by nineteenth-century American history, firearms, and achievements in inventing, manufacturing, marketing, artistry, and craftsmanship. The result is a magnet to the reader—*A Legacy in Arms* is a vivid tribute to the era when America came of age.

PREFACE

THE NATIONAL COWBOY & WESTERN HERITAGE MUSEUM holds a significant assemblage of American firearms that spans all the nineteenth century and reflects both civilian and military provenance. Though hardly rivaling the broad, prominent, and specialized collections at venues such as the Cody Firearms Museum, the National Firearms Museum, and the Springfield Armory Museum, the institution's collection nonetheless encompasses a representative sampling notable for its overall quality, its historical merit, its famed marques, and its several significant rarities. It is, certainly, an arms collection worthy of broader public appreciation.

Thus, in furthering its essential educational mandate and continuing its association with the University of Oklahoma Press, the museum is pleased to bring this work to publication as volume 10 in the Western Legacies Series. As with previous collection-oriented publications in the series (*A Western Legacy: The National Cowboy & Western Heritage Museum*, vol. 1, 2005; *Luis Ortega's Rawhide Artistry: Braiding in the California Tradition*, vol. 7, 2010; and *Arena Legacy: The Heritage of American Rodeo*, vol. 8, 2010), this work seeks to highlight the more significant pieces among the several hundred available through a melding of historical narrative, full-color imagery, and detailed object description.

The museum's arms collection represents an accretion of pieces over some forty years, but it is notable for two particularly significant acquisitions in 1991 and 2001. The first, the Joe Ruiz Grandee Collection, embodies a broad assemblage of more than 4,000 items of nineteenth-century frontier material culture and includes some 450 firearms and related accouterments of both military and civilian derivation. The second, the A. Max Weitzenhoffer

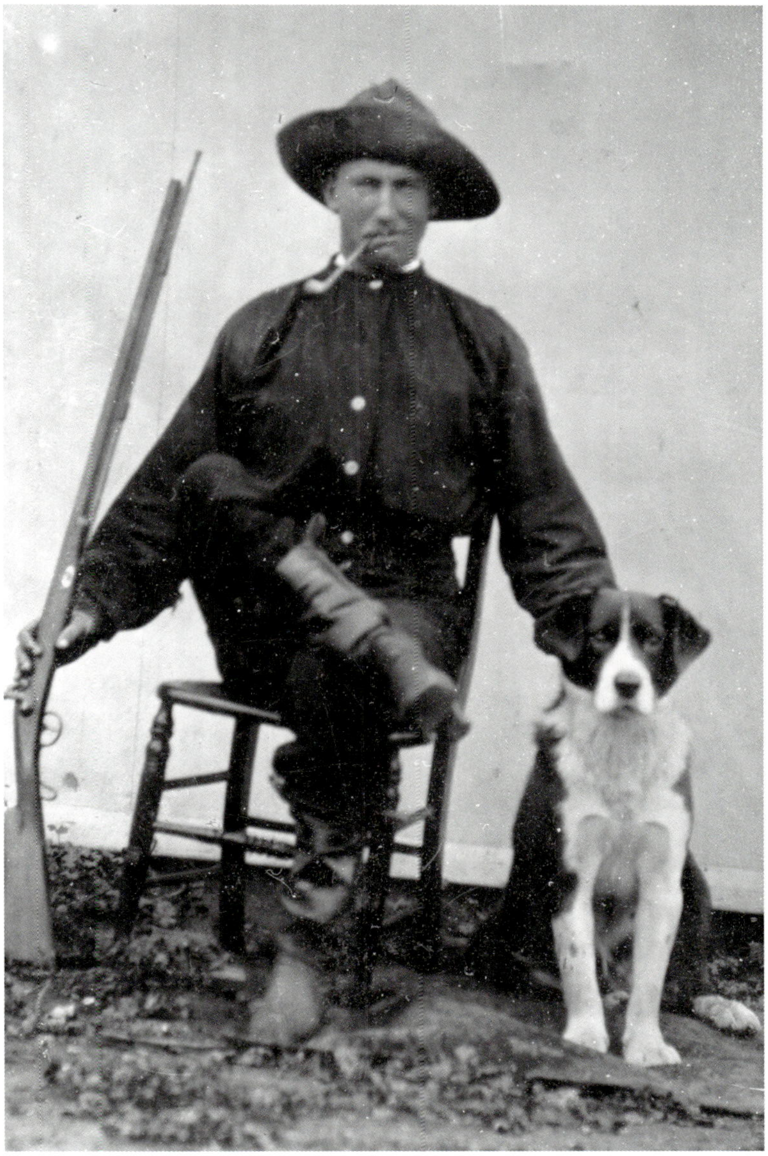

Photographer unknown. [Untitled.] Sixth-plate tintype, America, circa 1865–1870. Posed with his faithful canine, this rustic, pipe-smoking hunter conspicuously exhibits his half-stocked percussion sporting rifle—finished with rather unusual checkering at wrist and extended, handhold trigger guard strap.

Dickinson Research Center, National Cowboy & Western Heritage Museum, 2003.268

Collection, consists of a focused holding of approximately 100 arms encompassing many of the most notable names in American gun making during the second half of the nineteenth century—Colt, Remington, Smith & Wesson, Sharps, Marlin, and Winchester. The latter holding is notable for its consistently fine condition and its several important and sought-after rarities—among them, a Colt Double Rifle, a Winchester Model 1873 "One of One Thousand" rifle, and many engraved specimens by the leading artisans of the mid- to late nineteenth century.

Over the past thirty years, in excess of a dozen illustrated books treating the history of firearms in the frontier West have seen publication. Among the very best of these for historical research are Louis A. Garavaglia and Charles G. Worman's *Firearms of the American West, 1803–1865* and *Firearms of the American West, 1866–1894* (University of New Mexico Press, 1984–1985). Certainly, the most notable for the presentation of superlative color imagery of rare and historically significant frontier arms is R. L. Wilson's *The Peacemakers: Arms and Adventure in the American West* (Random House, 1992). Although the museum's weapons holdings dovetail nicely with this oft-reiterated frontier-oriented interpretation, it is thought that—rather than plough the same well-worn furrow yet again—a different narrative approach might serve as well to place the collection in a meaningful historical context.

Here, the material presented is treated from the perspectives of its manufacturing and technological evolution, its marque production names, and its aesthetic qualities. In particular, a treatment of the instrumental contributions of antebellum firearms manufacturers to the broader process of domestic industrialization provides general readers with a greater appreciation of the place of guns in American culture. Some due attention, nonetheless, also is rendered to the frontier milieu, as that turbulent environment certainly served as a dramatic stage—and an essential proving ground—for virtually all of the arms discussed and illustrated herein. In that vein, numerous historical images drawn from the holdings of the museum's Dickinson Research Center also are included to provide something of a human dimension to the narrative. I have exercised my authorial (and curatorial) prerogative to select firearms that are, with but few exceptions, of commercial manufacture and typically for civilian use. The virtual exclusion of military arms and their attendant accouterments is made in favor of a less cumbersome narrative and, perhaps, in anticipation of another collections publication in future.

Conceived for a general audience, this work draws almost entirely on secondary sources and makes no assertion of presenting particularly new historical material (except, perhaps, in its interpretation). Readers wishing to delve further into the development of American arms making should consult the sources cited in the notes and bibliography—many of which draw on primary source materials. Though perhaps repetitive, the inclusion of firearm production years and totals in chapters 3 through 8 serves to emphasize the far-reaching effects of mechanization and mass production over the second half of the nineteenth century. Such statistics also may provide some notion of the comparative utility of various firearm makes and models, as well as underscoring the successive impact of design improvements and design departures within a given product line.

The author hopes that this publication will help readers develop a broader familiarity with the museum's rich holdings, will foster a fuller understanding of the place of firearms within traditional American manufactures, and will engender a greater appreciation of their mechanical, historical, and aesthetic merit.

In bringing a selection of the firearms collections of the National Cowboy & Western Heritage Museum to publication, I am particularly indebted to my fellow staff members for their unstinting cooperation and professionalism. Registrar Melissa Owens and assistant registrar Tobie Cunningham provided indispensable collection documentation, while exhibits preparator Hal Prestwood—with the able assistance of Jim Cardoni and Russ Leonard—furnished signal aid in accessing the many items on exhibition. At the museum's Donald C. & Elizabeth M. Dickinson Research Center, director Gerrianne Schaad and librarian Karen Spilman proved invaluable in locating research and graphic materials, while digital archivist Laura Heller very cordially scanned scores of transparencies, processed scores of new digital images, and copied dozens of historical images from the center's holdings. My sincere thanks, as well, go to curatorial colleagues Anne Morand and Don Reeves for acting as sounding boards during the writing and for their early and helpful readings of the manuscript.

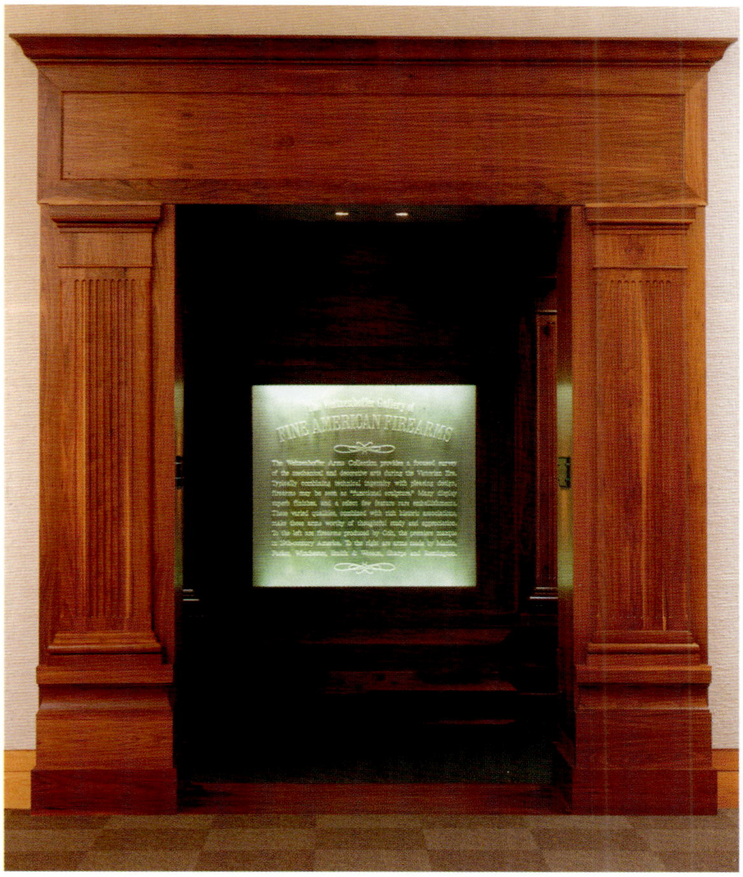

Entrance to the Weitzenhoffer Gallery of Fine American Firearms at the National Cowboy & Western Heritage Museum, Oklahoma City, Oklahoma.

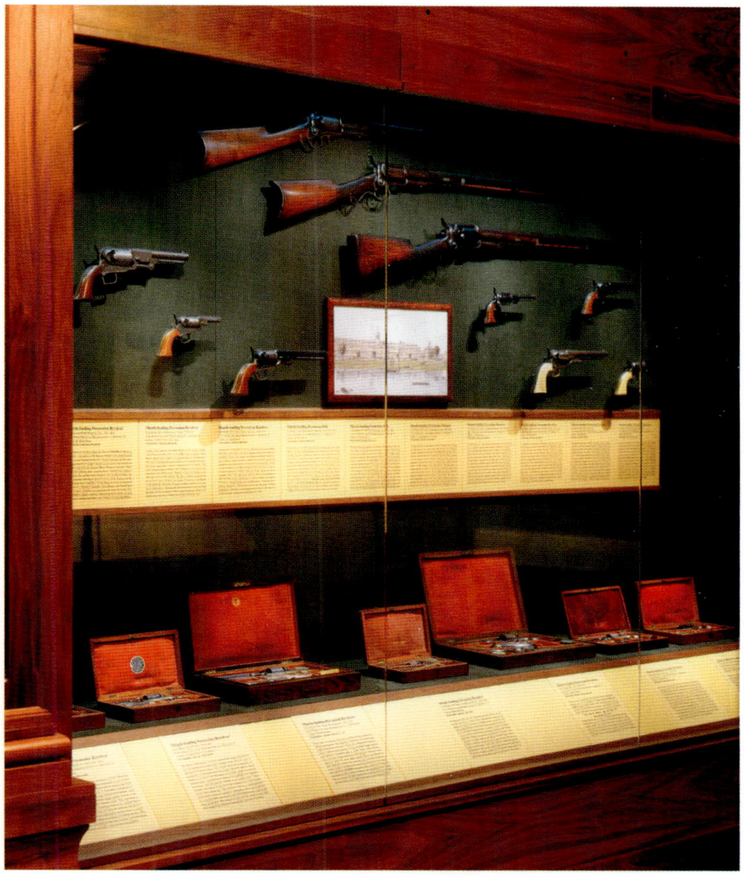

Colt percussion revolving-cylinder arms exhibited in the Weitzenhoffer Gallery of Fine American Firearms at the National Cowboy & Western Heritage Museum, Oklahoma City, Oklahoma.

Special appreciation is extended to my longtime friend and former museum curator Ed Muno, who once again demonstrated his generosity and many photographic talents in producing the vast majority of the color images of the collections. Thanks as well to the following auction houses for the provision of images of recently acquired firearms: Amoskeag Auction Company, Manchester, New Hampshire; James D. Julia Auctioneers, Fairfield, Maine; and Greg Martin Auctions, San Francisco, California. And I wish to express my particular gratitude to Mr. S. P. Fjestad, publisher of Blue Book Publications, Inc., for permission to reproduce the Colt cylinder scenes appearing in part 3.

My sincere thanks are extended to those who graciously read and critiqued the manuscript, thus improving the whole and saving me from several embarrassments: Dr. Merritt Roe Smith, Cutten Professor of the History of Technology at Massachusetts Institute of Technology; Mr. R. L. Wilson, widely published firearms historian and engraving authority; and Mr. Norm Flayderman, noted American arms authority, antiquarian, and author. To R. L. Wilson, as well, goes my additional gratitude for his generosity in agreeing to author the foreword.

At the University of Oklahoma Press, particular appreciation is due director B. Byron Price and editor in chief Charles E. Rankin for adopting the project and seeing it to publication. Also at the press, the way was smoothed by assistant director and production manager Emmy Ezzell, by managing editor Steven Baker, by graphic designer Julie Rushing, and by "image wrangler" Anna María Rodríguez. My thanks to one and all.

Last, but certainly not least, the National Cowboy & Western Heritage Museum and I are especially grateful to Mr. Joe Ruiz Grandee for generously supporting the production of this Western Legacies Series volume.

__Richard C. Rattenbury,__ Curator of History
National Cowboy & Western Heritage Museum
Oklahoma City, Oklahoma

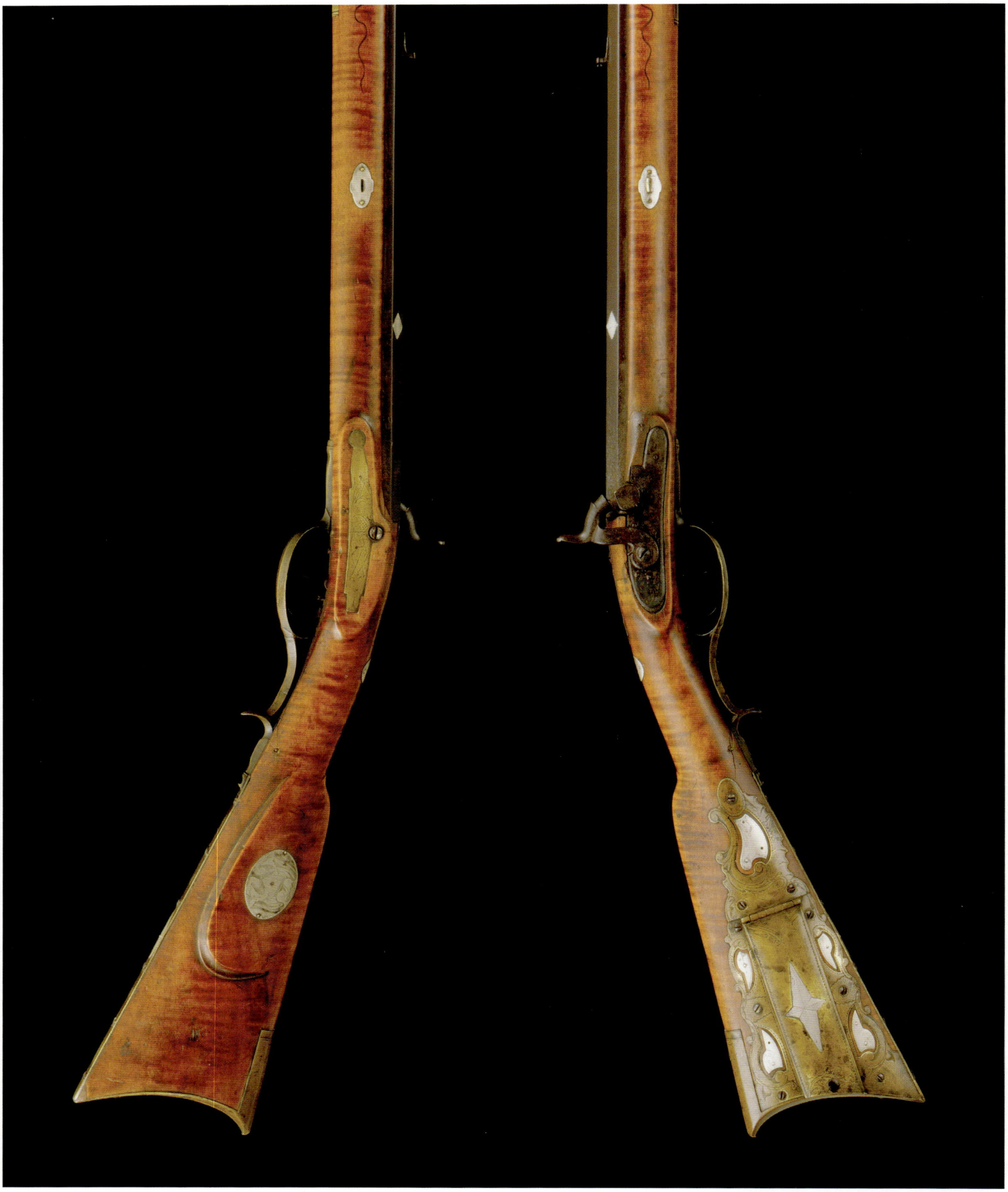

Transitional Kentucky–Plains rifle by Philip Creamer, St. Clair Co., Illinois, circa 1820–1825. For more information, see page 6.

Part I

From Craftsman to Industrial Mechanic

Evolution in American Arms

Firearms have played an integral role in American history and culture ever since the first European explorers and colonists set foot on the continent. Used for both community protection and the procurement of sustenance, the matchlock and various flintlock guns introduced into the New World represented perhaps the most sophisticated technology of the time. On the isolated colonial frontier, however, the intermittent importation of arms (and gunpowder) from Europe proved inadequate to secure settlement and advance expansion. Thus, even before 1650, a small number of military armorers and civilian gunsmiths already operated in the American colonies out of simple necessity. If not the first, one Eltwood (or Eltweed) Pomeroy, who set up shop in the Boston area in 1630 and established a dynasty of New England gunsmiths, certainly ranked among the earliest. He, his descendants, and a handful of other "artificers" laid the foundation of a nascent arms-making industry in North America.[1]

The emergence of firearms manufacture in the colonies manifested itself perhaps most notably among the German and Swiss emigrant gunsmiths of eastern Pennsylvania, who, from about the 1720s, commenced developing the first indigenous American firearm—the now-famed Pennsylvania, or Kentucky, rifle. Though made with traditional Old World methods and skills, this new weapon manifested distinctive characteristics and qualities fostered in response to the challenges of a frontier environment and colonial economy. Reaching its technical and aesthetic zenith around 1800, the Kentucky rifle—and its later regional interpretations—best epitomized the work of the individual American gunsmith in the century between 1750 and 1850. These rifles, as well as other domestic firearms of the period, represented the products of the singular artisan. Bereft of the specialized crafts and guilds that underpinned the gun-making trade in Europe, the gunsmith in Colonial America had no recourse to a cadre of skilled tradesmen for the finished components that might refine his product or increase his output. The early American gunsmith had, by circumstance, to perform as the "master of all trades," be it in metal or wood, and each of his creations represented a unique statement of his craft.

Although the demand for firearms in the United States increased apace with the growth in population and territory during the early 1800s, the necessary increase in skilled labor for the gun-making trade failed to meet the need. Ever ingenious and entrepreneurial, American arms makers gradually turned to the adoption of increasingly sophisticated, labor-saving machinery to augment their production. By the 1850s, firearms manufacturing in the United States largely had shifted its focus from the scattered shops of individual gunsmiths to the machine-based factories clustered along the Connecticut River valley in New England. There, in a response to economic necessity and government entreaties for greater uniformity in both firearms and their constituent parts, a fertile confluence of innovative gun designers and manufacturers, ingenious mechanics, and imaginative industrial entrepreneurs gradually had developed the principles, machinery, and practices of mass production and uniform (sometimes interchangeable) parts manufacture. Even by 1860, what already was widely heralded as the "American System of Manufactures" had truly transformed domestic arms production—and it ultimately would change all manner of industrial production down to the present day.[2]

Just as the personnel and methods of gun making evolved over the first sixty years of the nineteenth century, so too did the technology and design of the firearms themselves evolve. For example, ignition systems (the means by which a weapon was discharged) advanced from the long-serving but fussy flintlock to the more reliable percussion lock to the very convenient and ultimately entirely reliable self-contained metallic cartridge. Similarly, even prior to 1850, the means of charging, or loading, guns had progressed from the often-tedious and time-consuming muzzle-loading method to the far more rapid and altogether convenient breech-loading method. As well, successive developments in the firepower of domestic weaponry proved particularly dramatic, advancing from the ubiquitous single-shot arms of 1800, through the multibarreled and revolving-cylinder repeating firearms of the 1830s and 1840s, on to the repeating magazine guns of the 1860s and following. Indeed, the evolution in the design and technology of firearms advanced so rapidly over the nineteenth century that, by the first decade of the twentieth century, semiautomatic, or "self-loading," handguns and longarms had entered the commercial market in substantial number and variety.[3]

In varying degree, all of these evolutionary advances in firearms design, production, and function impacted the concurrent advance of frontier exploration and ultimate settlement over the course of the nineteenth century. The gradual improvement of ignition systems, for example, certainly bolstered the resolve of many a frontiersman for whom a rifle's reliability might prove crucial to survival. The greater facility of loading at the breech instead of at the muzzle allowed an increased rate of fire at a ratio of four or five to one, while the ever-growing availability of repeating arms of revolving-cylinder and magazine design provided isolated frontier parties or individuals with a decided advantage in their occasional confrontations with dangerous game or hostile adversaries. Steady progress in applying machinery to arms manufacture also guaranteed sufficient output to meet a steadily rising demand from Americans flooding into the West.

Equally important on the other side of the historical ledger, America's advancing frontier provided numerous domestic firearms manufacturers with an ever-expanding marketplace and a sometimes-crucial testing arena for their more innovative creations. The latter condition proved true, for example, of the early Colt-Paterson revolving pistol, which demonstrated its potential for mounted combat on the unsettled plains of Texas in the 1840s. As cultural historian William Hosley observed in profiling Colonel Samuel Colt and his deadly arm during the antebellum era, "Guns, technology, and the campaign of western expansion were overlapping layers of the same progressive tendency. Each fed and enabled the other."[4]

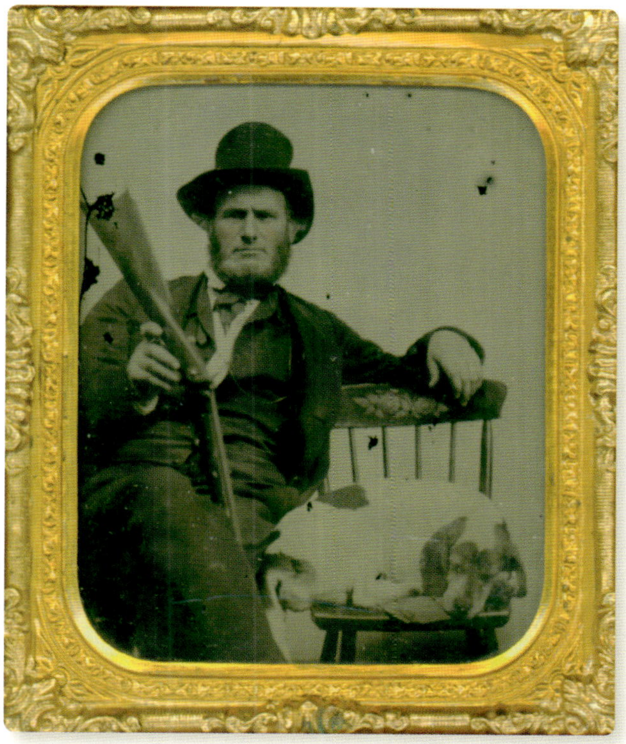

Photographer unknown. [Untitled.] Sixth-plate ambrotype, America, circa 1855–1860. By the 1850s, recreational hunting had become a popular pastime with Americans of varying economic and social station. This rather casual gent poses with his recumbent hound and a quite utilitarian percussion sporting rifle.
Dickinson Research Center, National Cowboy & Western Heritage Museum, 2005.079

Chapter 1

FIREARMS OF THE ARTISAN GUNSMITH

THROUGHOUT THE EIGHTEENTH CENTURY and into the first decades of the nineteenth century, firearms manufacture in America was the domain of individual gunsmiths. Such men might have been familiar with one or another of the particular gun-making trades practiced in Europe, but there appears never to have been a formal, segregated system of guilds or trades among domestic gunsmiths comparable to that operating in the Old World. There, one found specialization in the various processes of barrel making, lock making, stock making, and mount making. Ancillary decorative skills, such as wood carving and inlaying, and metal engraving and inlaying, called as well for specially trained artisans. And the end product, of course, demanded the close attention of keen-eyed and deft-handed fitters and assemblers—men who brought the constituent parts of the firearm to a functional whole that met certain standards of quality.[1]

Given the isolation and comparatively primitive economy prevalent in the colonies, the early American gunsmith had to rely for the most part on his own inherent inspiration, ingenuity, and talent in mastering virtually all of these skills for himself. Out of necessity, he performed as barrel maker, locksmith, stock maker, brass caster, filer, fitter, assembler, and, if one of the more talented smiths, wood-carver and metal engraver. That at least some domestic smiths operated as all-around hands is illustrated in the 1771 probate inventory of Pennsylvania gunsmith Matthius Roessor. For barrel fabrication, his shop included a forge, a barrel anvil, a boring mill, a rifling bench, and associated "britchen tools"—taps and dies used to thread the barrel breech and breech plug. For fashioning gunstocks, the shop held a compass saw, frame saws, draw knives, various rasps, and other unidentified "stocking tools"—likely hand augers, braces, and chisels. As well, it included "casting tools" (likely ladles and molds) and sand for the making of brass buttplates, trigger guards, sideplates, and the other gun mountings, or "furniture," typical of the time. And among dozens of small metalworking tools, Roessor's inventory included files, punches, screw plates, vise clamps, and a "tumbler tool"—all suggesting the repair, if not the fabrication, of gunlocks and trigger mechanisms.[2]

As with Roessor, present in any self-reliant gunsmith's shop were one or two bench vises and a full battery of two dozen or more files of various useful patterns (flat, round, half-round, triangular, round-edged) that ranged in gradation from coarse to fine. Lacking the sophisticated, power-driven machine tools of the next century, the colonial smith relied on such hand files for virtually all phases of metal shaping and finishing, particularly for gunlock fabrication and the fine work of final fitting for assembly. Skill in metal filing remained a crucial gunsmithing talent for another century and longer. In early 1850s Texas, for example, young apprentice J. C. Petmecky—in an initial trial of his aptitude for the gunsmith trade—undertook the impossible task of filing a perfectly square cube from a chunk of scrap iron.[3]

Although Matthius Roessor very likely had apprentices in his shop from time to time, and perhaps took on journeymen assistants as well, his inventory certainly suggests that he worked as an accomplished individual gunsmith capable of creating an arm "lock, stock, and barrel." Like most of his mid-eighteenth-century peers, he probably fashioned arms almost entirely with hand tools and by his own motive power—the only "machines" on site were his boring mill and rifling bench, and these required his exertion via hand crank or foot treadle. So equipped, he probably could turn out a well-finished rifle in two to three weeks, a production capacity equating to only twenty to twenty-five pieces per year. For colonial riflesmiths such as Roessor, arms making demanded the mastery of tools and materials, attention to detail, and full measures of ingenuity and patience. To fabricate firearms that handled well, that looked well (if not elegant), and that functioned dependably and shot true required specialized skills, hard work, and no little pride of craft on the part of the devoted artisan.[4]

By the last decade of the eighteenth century, a number of water-powered, barrel-boring and barrel-grinding mills were in operation in rural Pennsylvania and elsewhere, and a few of these evidently adopted trip-hammers for barrel welding as well. At much the same time, trade and commerce with Europe improved to the extent that good-quality gun barrels, finished locks, trigger assemblies, and brass mountings could be reasonably imported from gun-making centers in England, Belgium, and Germany.[5] Thus, the talented domestic gunsmith, freed from much of the fundamental labor, turned his attention largely to barrel rifling, stock making, gun assembly, and the niceties of fitting, sighting, and embellishment. Continually moving west with frontier expansion, many individual smiths labored in this manner well into the mid-nineteenth century, perpetuating traditional gun-making skills in an already burgeoning industrial age.

The comparatively rapid dispersion of such skills from among the riflesmiths of eastern Pennsylvania appeared first in Maryland, Virginia, and western Pennsylvania during the latter decades of the eighteenth century. By the early nineteenth century, isolated frontier gunsmiths worked in the Ohio Valley country and even farther westward. Talented Maryland riflesmith Philip Creamer, for example, resettled in 1805 on the Illinois side of the Mississippi River near Saint Louis, where he supplied dozens of plain "riffles" to frontier outfitters Bryan & Morrison of Cahokia and Kaskaskia for the Indian trade and, later, for the fur companies penetrating into the upper Missouri country.[6] (As can be seen below, and in chapter 10, Creamer also fabricated deluxe, custom rifles for individual frontiersmen desirous of a personal arm of finer quality and adornment.)

By the 1830s a few well-known Pennsylvania gunsmiths had established "manufactories" to "mass-produce" arms for the frontier trade. Prominent among them were J. J. Henry's Boulton Gun Works near Nazareth, and Henry Leman's Conestoga Rifle Works of Lancaster. Most such factories utilized water (and later steam) power to drive their rudimentary forging and drilling machinery, and they no doubt introduced some division of labor in barrel making, stock making, and gun assembly among their employees to rationalize and increase production. In contrast to the individual gunsmith, such manufactories might turn out several hundred arms per year. Like the individual gunsmiths, some of these nascent factories moved along

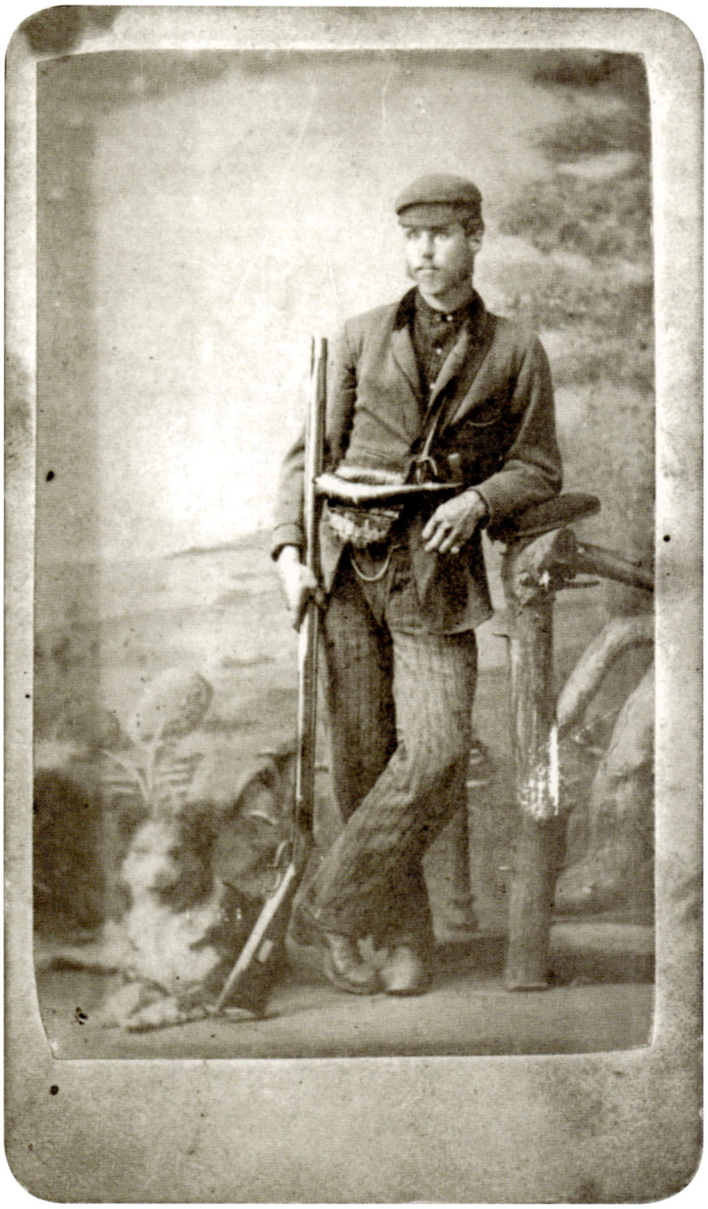

W. A. Smith, photographer. [Untitled.] Carte de visite, Newark, Ohio, circa 1870. This unknown, midwestern huntsman strikes a studio pose with his attentive canine. He holds a muzzle-loading percussion, full-stocked, Kentucky-pattern rifle and is accoutered with hunting bag and powder horn.

Dickinson Research Center, National Cowboy & Western Heritage Museum, RC2007.022

with the swelling tide of settlement, such as Bown & Tetley's Enterprise Gun Works, which commenced manufacturing at Pittsburgh in 1848. All of these small-factory operations, however, aimed at a larger market than the individual, local or regional customer. They catered instead to an ever-growing number of frontier trading companies, to individual firearms dealers located even farther west, or to the burgeoning emigrant trade.[7]

At the opening of the nineteenth century, the only domestically produced firearm that could be considered distinctly American in its characteristics was the Pennsylvania rifle—better known today as the Kentucky rifle. The end product of necessity and inspiration, this flintlock rifle's design sprang from the melding of traits drawn from two dissimilar shoulder arms brought to America from Europe—the long, graceful, lightweight, smoothbore fowling gun from England; and the short, bulky, large-caliber, *jäger* (hunting) rifle from central Europe. Believed to have first started to be developed around Lancaster, Pennsylvania, in the 1720s, this new arm gradually evolved until, by about 1770, a recognizably different weapon had taken shape both in its form and in its function. With its graceful lines, often-handsome brass furniture, and deadly accuracy, the Kentucky rifle represented a superb weapon and a significant expression of American folk culture.[8]

Borrowing from the jäger arm, the muzzle-loading Kentucky's barrel incorporated spiral grooves, or rifling, cut the length of its interior bore. Upon discharge, this caused the tight-fitting projectile (typically a round lead ball seated in a greased linen patch) to spin around its own axis and, thus, stabilize in flight. Rifling proved fundamental to

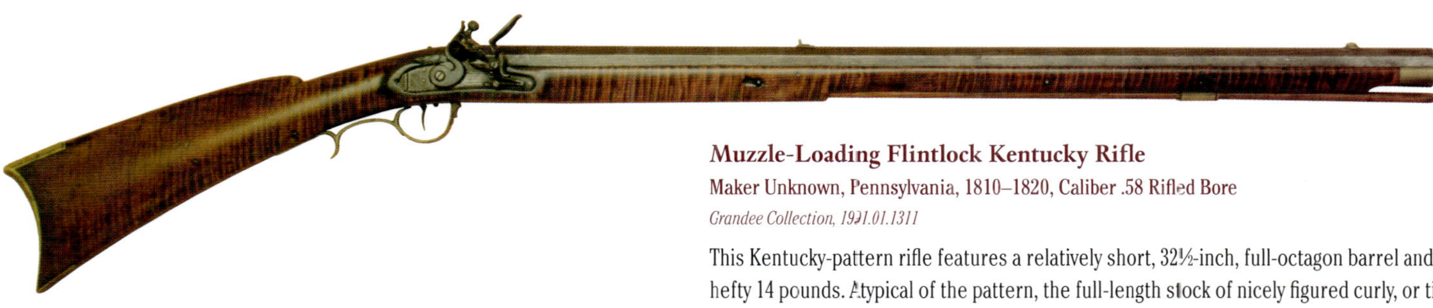

Muzzle-Loading Flintlock Kentucky Rifle

Maker Unknown, Pennsylvania, 1810–1820, Caliber .58 Rifled Bore

Grandee Collection, 1991.01.1311

This Kentucky-pattern rifle features a relatively short, 32½-inch, full-octagon barrel and weighs a hefty 14 pounds. Atypical of the pattern, the full-length stock of nicely figured curly, or tiger-stripe, maple is finished without the classic patchbox in the butt. Unusually short for the era, the rifle would have proven ideal for horseback travel, while its substantial caliber would have been adequate even for buffalo and grizzly bear. Like the following piece, this rifle is representative of eastern-made longarms carried into the trans-Mississippi frontier by trappers, explorers, hunters, and naturalists during the opening decades of the nineteenth century.

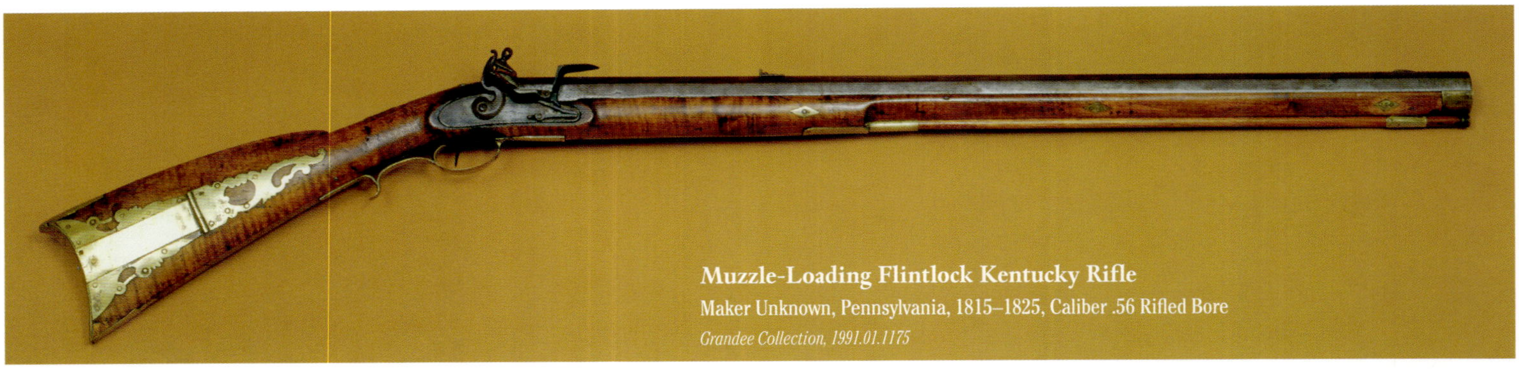

Muzzle-Loading Flintlock Kentucky Rifle
Maker Unknown, Pennsylvania, 1815–1825, Caliber .56 Rifled Bore
Grandee Collection, 1991.01.1175

Of middling quality, this Kentucky rifle features a full stock of lightly figured, tiger-stripe maple and a fairly ornate, four-piece patchbox of brass. The prominent drop of the gun's buttstock and its low-arched, or "Roman-nosed," comb line suggest the influence of the Bethlehem school of Pennsylvania riflesmiths. Originally of flintlock ignition, the piece was converted to percussion ignition during its period of practical use, and it has since been reconverted to flintlock. The 32-inch octagonal barrel is unusually short for the type and probably was cut down for horseback use. Among western frontiersmen, even such handy, large-bored, eastern-made rifles gradually lost favor to the more robust Mountain, or Plains-pattern, rifle during the 1830s and 1840s.

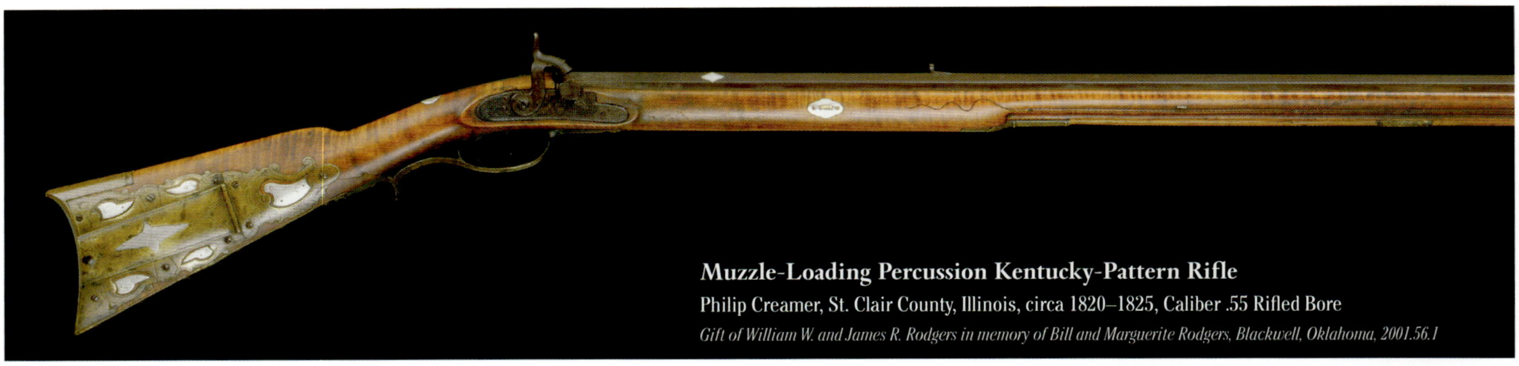

Muzzle-Loading Percussion Kentucky-Pattern Rifle
Philip Creamer, St. Clair County, Illinois, circa 1820–1825, Caliber .55 Rifled Bore
Gift of William W. and James R. Rodgers in memory of Bill and Marguerite Rodgers, Blackwell, Oklahoma, 2001.56.1

Of Kentucky rifle pattern and originally made with flintlock ignition, this transitional piece was a precursor of the Mountain, or Plains-pattern, rifle favored in the Far West from the mid-1820s into the 1860s. Creamer's talents as a frontier riflesmith are evidenced in the finely crafted full stock of figured maple, which displays vestiges of the classic, "Golden Age" Kentucky rifle. Indicative of the gunsmith's technical skills, the piece also incorporates a unique patchbox-cover release at the rear of the trigger guard. The rifle's 42-inch octagonal barrel might possibly have been created by Creamer's one-time apprentice—later-famed mountain man Jim Bridger. Such a fine weapon would have been prized by a frontiersman of some means, such as an agent of the American Fur Company. (See chapter 10 for detail on the ornamentation of this piece.)

the Kentucky's efficacy in an often-challenging wilderness environment, for it made the weapon—typically outfitted with low, open sights and double set triggers—markedly more accurate than the contemporary and ubiquitous smoothbore musket. While the common musket might strike a man-sized target at 50 yards with moderate regularity, the rifle could dependably hit the head of such a target at 150 yards or more. British Major George Hanger, who interviewed several American backwoods riflemen during the Revolutionary War (1775–83), later recorded that his subjects consistently advised him that "an expert marksman, provided he can draw a good and true sight . . . can hit the head of a man at 200 yards."[9]

Other adoptions and modifications of the jäger rifle's attributes also contributed to the Kentucky rifle's function and proportion. The jäger's short, thirty- to thirty-six-inch octagonal barrel was lengthened in the Kentucky rifle to forty-two to forty-eight inches—both to improve accuracy with a longer sight radius and, as was then believed, to better capture the full force of the era's slow-burning black powder. In an effort to conserve lead, the jäger's large bore of .60 to .75 caliber (that is, 60/100 to 75/100 inch in diameter) diminished in the Kentucky to a still-quite-lethal range of .45 to .58 caliber. As well, the lengthening and streamlining of the jäger's chunky full stock created in the Kentucky rifle a remarkably graceful and well-balanced firearm reminiscent of the long English fowling piece. This proved particularly true of the Kentucky's altered buttstock, which lowered the jäger's comb profile and replaced its flat butt with a crescent pattern. The jäger's rudimentary cheekrest and its butt trap with sliding wooden cover were retained and improved in the Kentucky, the latter evolving into the practical patchbox with its handsomely decorated, hinged cover of brass. And instead of the jäger's typical stocking in plain walnut, the Kentucky rifle commonly featured a full-length stock of figured curly or tiger-stripe maple, which created an attractive arm not only in form and line but also in visual appeal.[10]

The gunsmiths of Pennsylvania also fabricated single-shot, muzzle-loading flintlock pistols for the commercial market. Far scarcer than the rifles, and often differing themselves in stock design and finish, these Kentucky-pattern handguns nonetheless shared certain attributes of the type. They typically had full- or half-octagon barrels—oftentimes made from cast brass instead of forged iron—measuring six to ten inches in length and ranging from .40 to .50 caliber. Barrels often featured front and rear sights not unlike the rifle, but they more often were smooth-bored than rifled. Like the rifles, pistols of this pattern typically utilized full-length stocking of curly or tiger-stripe maple having gracefully curved butts and corresponding brass or, sometimes, silver furniture. Other muzzle-loading, single-shot handguns of belt and holster size—in either flint or, later, percussion ignition, and often of a semimartial pattern—accompanied the Kentucky pistol into the frontier.[11]

Although the intrepid American backwoodsman might carry a brace of these pistols on his explorations, it was the Kentucky rifle that served as his principal weapon in advancing and protecting settlement across the forested Appalachian, Allegheny, and Ohio Valley frontiers and into the Northwest Territory during the latter eighteenth century and into the early decades of the nineteenth century. As the prevalence of conflict with foreign and Native adversaries gradually diminished throughout these regions—along with big-game populations of elk and

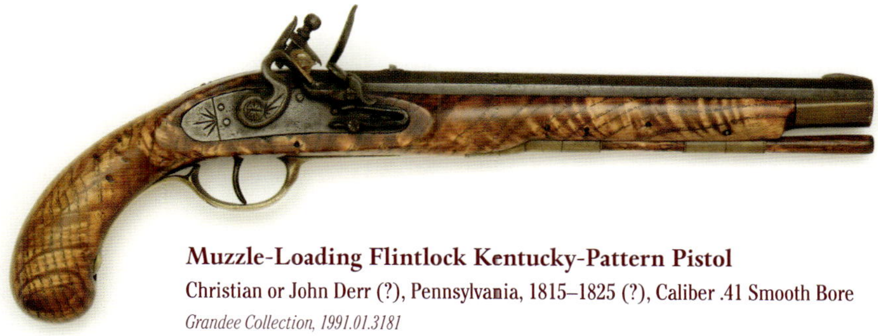

Muzzle-Loading Flintlock Kentucky-Pattern Pistol
Christian or John Derr (?), Pennsylvania, 1815–1825 (?), Caliber .41 Smooth Bore
Grandee Collection, 1991.01.3181

Although the iron barrel of this pistol is inscribed "Derr," the overall craftsmanship and patina of the piece do not reflect the quality and age of a Derr-family work. And though it provides a representative sample of a Kentucky-pattern pistol, it is more than likely a composite assembly, having a restored flintlock and stocking of much later origin. As with many such pistols made in eastern Pennsylvania during the early nineteenth century, this specimen features a finely figured curly, or tiger-stripe, maple stock mounted with typical brass furniture. Usually carried for one's personal protection, the belt-sized, Kentucky-pattern pistol enjoyed some favor among frontier types between 1800 and 1830.

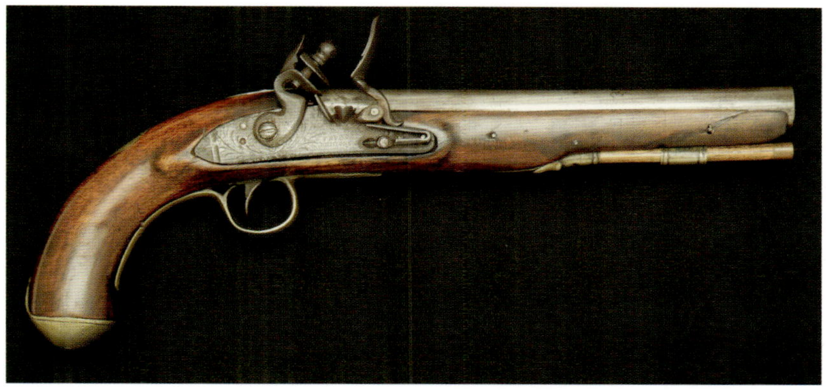

Muzzle-Loading Flintlock Militia or Traveling Pistol
George W. Tryon, Philadelphia, Pennsylvania, 1815–1830, Caliber .54 Smooth Bore
Grandee Collection, 1991.01.3161

Very likely assembled with parts imported from Great Britain, this large pistol incorporates a 9-inch round iron barrel, relatively plain brass furniture, and a sporting-type lock ornamented with simple arabesque engraving. All elements are mounted on a full stock of domestic walnut using pins and screws. Though never manufactured under government contract, pieces like this often are considered secondary martial arms intended for militia service or for officers' private purchase. In fact, the type was sold commercially to all manner of buyers, and an appreciable number probably traveled into the trans-Mississippi West in the belts of intrepid frontiersmen.

buffalo—the Kentucky riflesmiths responded to these changing conditions, gradually reducing the firearm in weight, barrel length, and caliber (often to as little as .36 to .40 caliber). Farther west, however, the Kentucky rifle would encounter challenging new conditions, and it would be found wanting.

As American explorers, trappers, and traders entered the trans-Mississippi West, they encountered an essentially different and very expansive environment occupied by a rich diversity of big-game species and a sometimes-combative Native population. On the Great Plains and in the Rocky Mountains, the Kentucky rifle soon proved inadequate to the vital necessities of long-distance accuracy and, equally if not more important, dependable knockdown power. Its typical powder charge and middling-caliber projectile often failed to sustain velocity and power over the necessarily greater ranges required, the ball tending to "drift," or lose accuracy, on the frequently windy plains. After his initial buffalo-hunting foray on the upper Mississippi River in 1805, for example, explorer Zebulon Pike remarked, "My Rifle was too small a Ball to kill Buffalo. The Ball should not be more than 30 to the pound [.54 caliber]; or an Ounce Ball [.66 caliber] would still be preferable."[12] Additionally, the Kentucky rifle's comparatively light construction frequently failed the rough usage received in the West, while its considerable length also proved unwieldy for easy carriage and manipulation on horseback. A rather sturdier rifle of a relatively larger bore and a somewhat shorter length gradually evolved in response to the demands of the western environment and the frontiersman's requirements—yet another distinctly American arm known as the Mountain rifle (and more commonly referred to today as the Plains rifle).[13]

Although similar in design to the Kentucky rifle, these new arms incorporated markedly heavier barrels measuring only thirty-eight to forty-two inches but ranging in caliber from .54 to .60. Because of its heavier barrel, a rifle of this pattern often weighed twelve to fourteen pounds, and it usually featured a more robust stock—particularly through the wrist area behind the lock where breakage proved most common. What might be recognized as the first Mountain-pattern rifles were, most likely, Kentucky-pattern rifles altered for western use by shortening, reboring, and rerifling the barrel. True Mountain rifles, developed in response to the specific preferences and special requests of the free trappers and frontiersmen, began to appear in the middle 1820s, supplied by western gunsmiths such as Philip Creamer and the famed Hawken brothers of Saint Louis, Missouri. As this new rifle pattern gained adherents, some eastern smiths, notably J. J. Henry's Boulton Gun Works and the Henry Deringer, Jr. and George W. Tryon shops located in Philadelphia, also produced weapons generally conforming to the type.[14]

Certainly among the best-known manufacturers of the Mountain rifle, brothers Jacob and Samuel Hawken set up in partnership in Saint Louis in 1825. Over the next quarter century, they brought the arm to the peak of its technical and design perfection. By the early 1830s, the typical J. & S. Hawken rifle incorporated a heavy, forty-inch octagonal barrel of .53 to .58 caliber with a full stock of plain, dark-

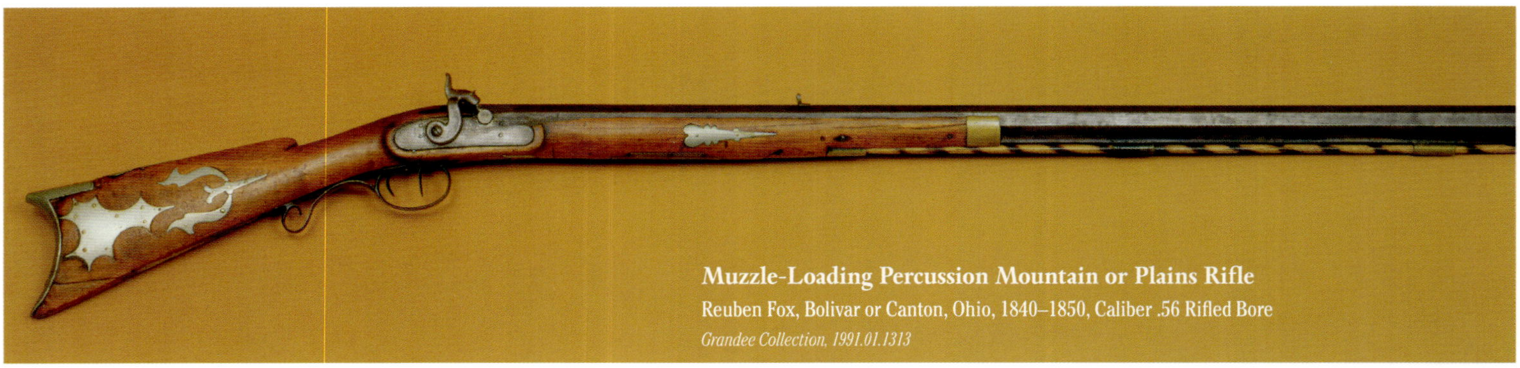

Muzzle-Loading Percussion Mountain or Plains Rifle
Reuben Fox, Bolivar or Canton, Ohio, 1840–1850, Caliber .56 Rifled Bore
Grandee Collection, 1991.01.1313

This is a midwestern interpretation of the celebrated Mountain-pattern rifle that evolved on the Missouri River frontier between 1820 and 1840. The 14-pound piece features a heavy, 40-inch octagonal barrel slightly swamped (enlarged) at the muzzle and fitted with open sights. German silver inlay decorates the plain maple half stock, which is particularly thick at the wrist and incorporates a graceful, scroll-pattern trigger guard—both features of the classic, latter-day Mountain, or Plains-pattern, rifle. Combining fine accuracy with knockdown power, this rifle pattern remained popular in the West well into the 1860s.

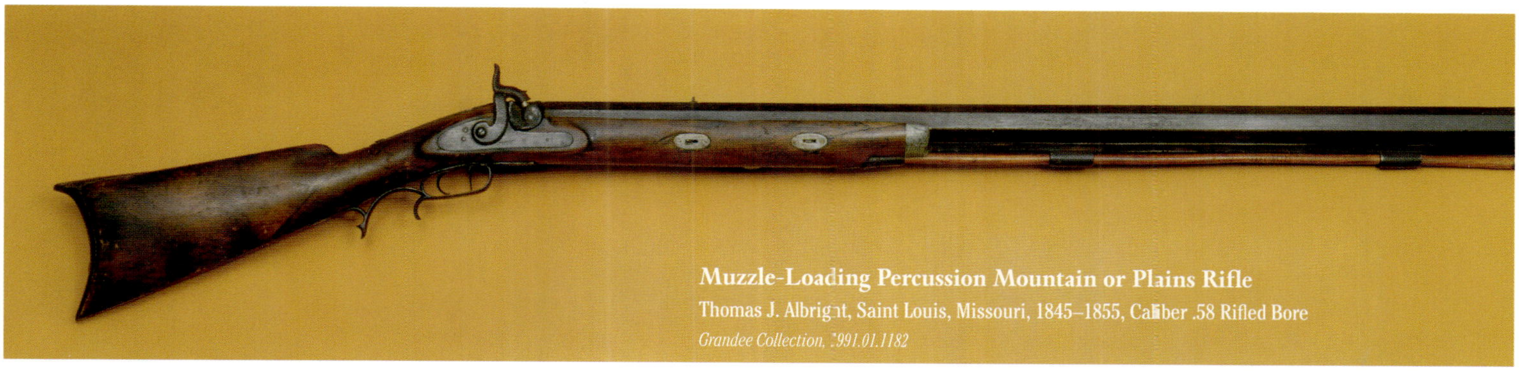

Muzzle-Loading Percussion Mountain or Plains Rifle
Thomas J. Albright, Saint Louis, Missouri, 1845–1855, Caliber .58 Rifled Bore
Grandee Collection, 1991.01.1182

Typical of the latter-generation Mountain, or Plains-pattern, rifle, this 14-pound piece has a heavy, 40-inch octagonal barrel of soft iron fitted with low open sights. The plain walnut half stock with cheekpiece secures the barrel with double crosskeys and is mounted with simple iron furniture. The rifle also features the improved, long-tang, patent breech with snail guard at the nipple, and it is outfitted with ever-popular, double set triggers. T. J. Albright was a widely respected Saint Louis arms importer, dealer, gunsmith, and latter-day competitor of the Hawken brothers' shop—then the most famous maker of Plains-pattern rifles.

stained maple. The patchbox so common on the Kentucky rifle usually now was omitted, while the remaining furniture frequently was of iron instead of brass. This was preferred because iron was stronger and lighter in weight and would not readily glint in the sun. As with the Kentucky rifle, crescent-pattern buttplates, double set triggers, and low, open sights remained relatively standard features.[15]

From the early 1830s onward, the Mountain rifle made by the Hawken brothers and other midwestern smiths appeared with the more dependable percussion lock (discussed more fully below), and it rapidly acquired a reputation in the frontier West for rugged dependability and unrivaled shooting characteristics. The rifle's renown for power and accuracy had much to do with its heavy barrel, which reduced recoil even with excessive powder loads. Equally important to the gun's accuracy was that the barrel was made of comparatively soft iron that reduced internal vibration as the projectile accelerated, and it typically employed slow-twist rifling that would "grab" the projectile and give it rotation even with an extremely heavy powder charge. These qualities were likely what inspired frontiersman William Hamilton to opine in 1842 that "in those days the best rifles used were the Hawkins [*sic*] and they carried three hundred and fifty yards."[16]

The Mountain rifle reached the zenith of its form and function during the middle 1840s. Presumably influenced by the "best-quality" percussion sporting rifles imported from England, the robust frontier rifle now appeared in half-stock configuration, usually with a raised oval cheekpiece on the butt and a graceful, scroll-pattern trigger

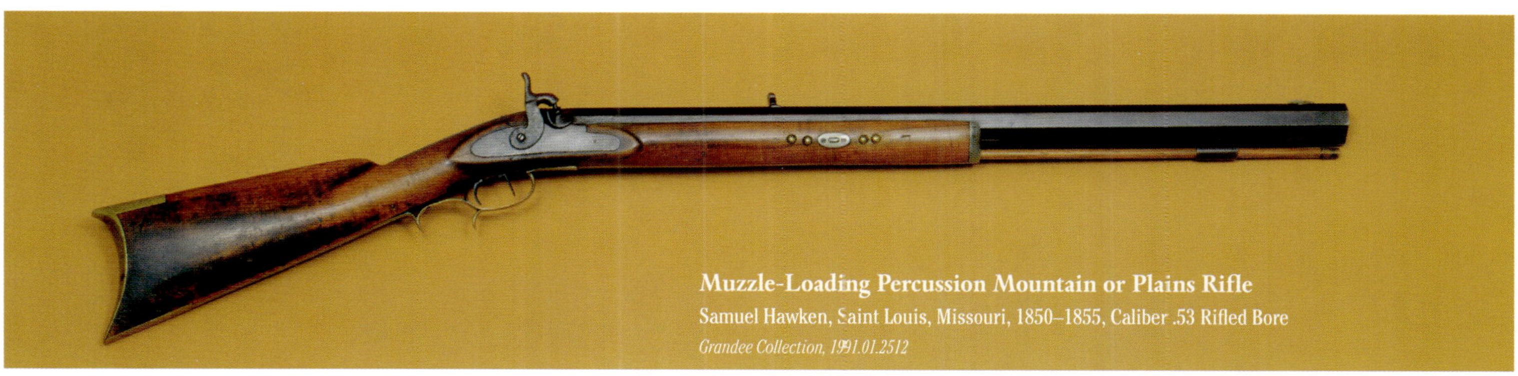

Muzzle-Loading Percussion Mountain or Plains Rifle
Samuel Hawken, Saint Louis, Missouri, 1850–1855, Caliber .53 Rifled Bore
Grandee Collection, 1991.01.2512

A latter-day product of the then-famous Hawken gun shop, this half-stocked Mountain, or Plains-pattern, rifle incorporates a comparatively short, 30-inch octagonal barrel ideal for carriage on horseback, as well as a relatively large bore for bringing down almost any big game. The 12-pound piece, stamped "S. HAWKEN ST. LOUIS" on the top flat of the barrel, features a patent breech with snail guard and extended top tang, double set triggers, and fine open sights. The Hawken name earned a deserved reputation among many trans-Mississippi frontiersmen for unexcelled power, superb accuracy, and rugged dependability.

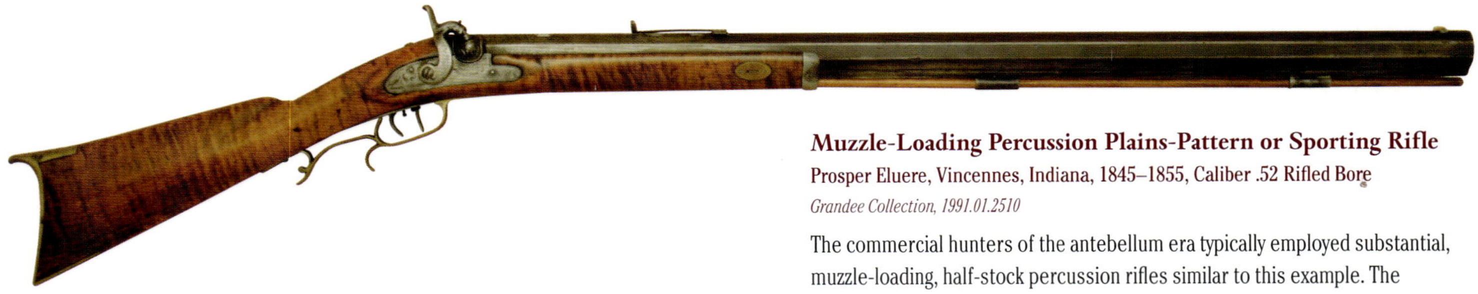

Muzzle-Loading Percussion Plains-Pattern or Sporting Rifle

Prosper Eluere, Vincennes, Indiana, 1845–1855, Caliber .52 Rifled Bore

Grandee Collection, 1991.01.2510

The commercial hunters of the antebellum era typically employed substantial, muzzle-loading, half-stock percussion rifles similar to this example. The 15-pound gun incorporates a heavyweight, 34-inch octagonal barrel, plain open sights, patent breech, and double set triggers—features considered ideal for fine shooting over ranges from 150 to 300 yards. The .52-caliber bore would have been sufficient for deer, elk, sheep, or even buffalo and grizzly bear. This well-crafted and weighty specimen boats a handsomely striped stock of hard maple finished with a pewter tip.

guard. Barrels typically incorporated the improved patent-breech system having a long iron tang reinforcing the wrist portion of the stock, and they averaged thirty-four to forty inches in length and .53 to .55 caliber. Somewhat heavier than in the past, the octagon barrel was now secured to the maple or walnut stock with one or two crosskeys in lieu of pins and had an iron underrib from the forestock cap to the muzzle with two ferrules, or thimbles, to hold the ramrod. Double set triggers and low, knifeblade front and "buckhorn"-pattern rear sights typified these fourteen- to sixteen-pound rifles.[17]

From the free trappers on the upper Missouri River to the traders on the Santa Fe Trail, the classic Mountain, or Plains-pattern, rifle won the devotion of all manner of frontiersmen for its rugged dependability, superb accuracy, and sheer stopping power. And although the classic, Hawken-made Mountain rifle did not achieve its ultimate form until the heyday of the beaver trade had passed, it still was acclaimed and sought after by all types of adventurers in the trans-Mississippi West—from mountain man Jim Bridger and traders Edwin Denig and Charles Bent to explorers Kit Carson and John C. Frémont. Like the Kentucky rifle before it, the Mountain rifle's utility and longevity lay in its adaptation to the frontier conditions in which it originated. Its ultimate practicality can perhaps be discerned in the fact that a few gunsmiths, such as J. P. Gemmer of Saint Louis, Missouri, and J. C. Petmecky of Austin, Texas, continued to produce the type as late as 1870—nearly two decades after more-advanced, breech-loading firearms had become readily available.[18]

Many less-expensive longarms, patterned after both the Kentucky and the Mountain rifle, were manufactured for the far-western trade between 1810 and 1870. Generally of lesser quality, these weapons saw distribution in comparatively large quantities to both Anglo-Americans and Native Americans through various trading companies,

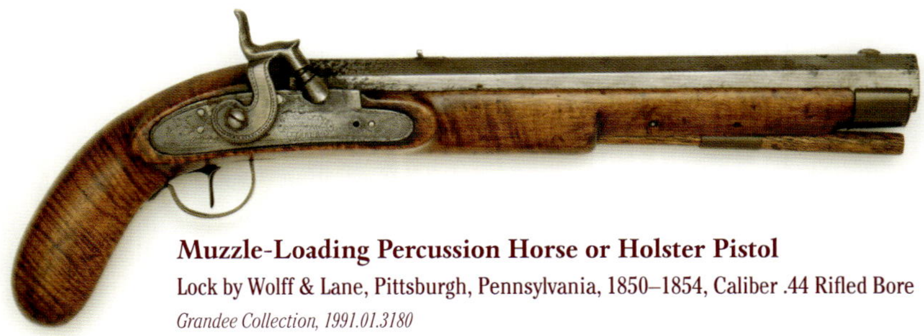

Muzzle-Loading Percussion Horse or Holster Pistol

Lock by Wolff & Lane, Pittsburgh, Pennsylvania, 1850–1854, Caliber .44 Rifled Bore

Grandee Collection, 1991.01.3180

Christian Wolff of Wolff & Lane worked as a lock maker only. This pistol, in total, may have been assembled and sold at their Pittsburgh shop, but the lock is just as likely to have been shipped with dozens of others to a gunsmith farther west, such as one in Saint Louis. Large percussion "horse" pistols, such as this specimen, were carried by antebellum frontiersmen for self-defense or for "running" buffalo on horseback. Percussion locks superseded flintlocks during the 1830s because they proved more reliable in windy or wet conditions and provided faster ignition, or "lock-time," than did flintlock arms.

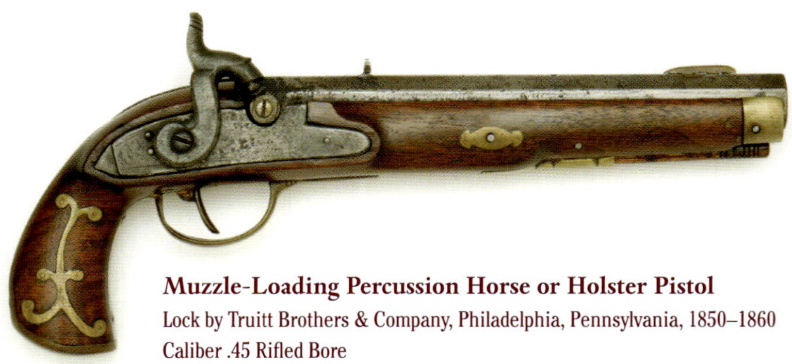

Muzzle-Loading Percussion Horse or Holster Pistol

Lock by Truitt Brothers & Company, Philadelphia, Pennsylvania, 1850–1860
Caliber .45 Rifled Bore

Grandee Collection, 1991.01.0150

Because Truitt Brothers worked only as lock makers, this rather ungraceful percussion pistol probably was made by an anonymous gunsmith closer to the frontier of the time. Large, percussion holster pistols were favored by early frontier sportsmen for hunting, or "running," buffalo on horseback. In his *Narrative of the Texan Santa Fe Expedition* (1844), George Wilkins Kendall observed that "[the sporting method] of hunting buffalo is to sally out after them on horseback, armed with heavy holster pistols, run alongside while under full speed, and shoot from the saddle. Of all the hunting in the world, this is probably the most exciting, at the same time involving a sportsman in no little danger" (78).

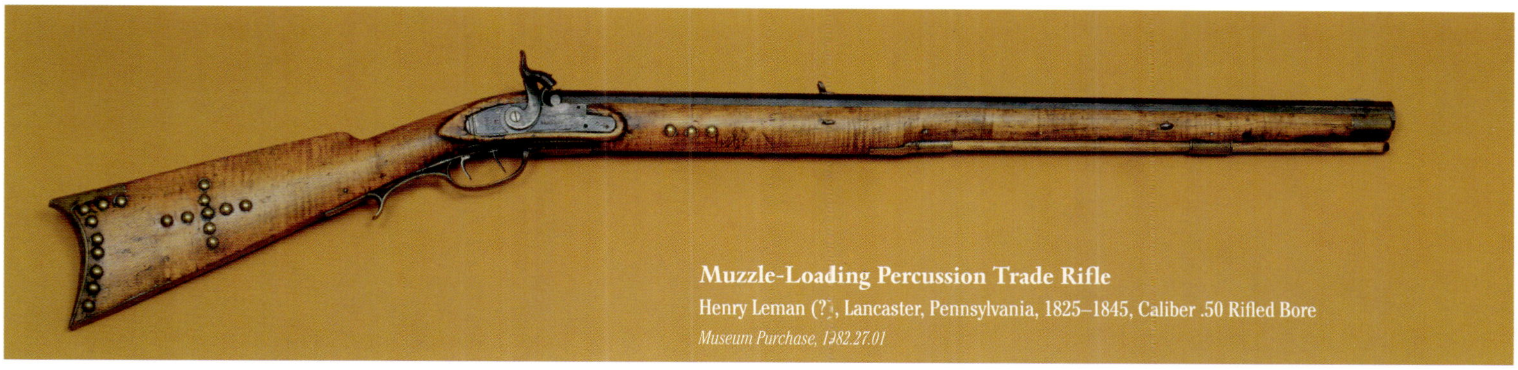

Muzzle-Loading Percussion Trade Rifle
Henry Leman (?), Lancaster, Pennsylvania, 1825–1845, Caliber .50 Rifled Bore
Museum Purchase, 1982.27.01

Exhibiting adaptations common on the fur-trapping frontier, this Lancaster-pattern trade rifle saw several changes and alterations during its active life. It probably was altered from flintlock to percussion ignition before 1840. During the early 1850s, its original 36–40-inch barrel was replaced with a 30-inch octagonal barrel stamped "S. HAWKEN ST. LOUIS" of .50 caliber—likely for easier horseback carriage. Even during the 1860s, the Hawken name remained in high regard among western veterans. Suggesting something of its frontier heritage, the trigger guard of this piece is crudely inscribed "J. Renet / Wind R[iver] 1842."

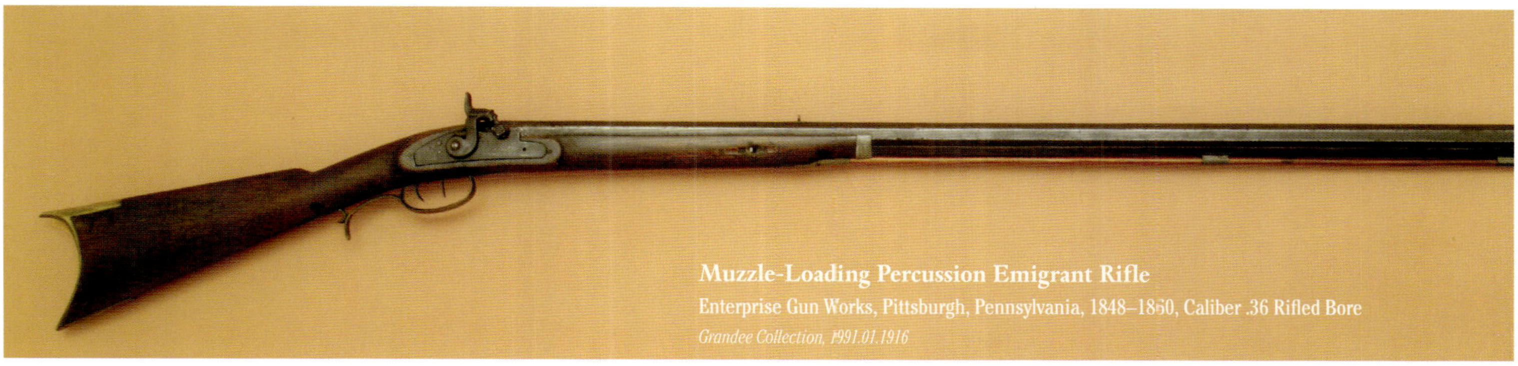

Muzzle-Loading Percussion Emigrant Rifle
Enterprise Gun Works, Pittsburgh, Pennsylvania, 1848–1860, Caliber .36 Rifled Bore
Grandee Collection, 1991.01.1916

Inexpensive percussion-lock rifles such as this were purchased by overland emigrants for hunting and defense along the Oregon and California trails. The comparatively small bore of this specimen helped to conserve limited supplies of powder and lead over the arduous, four- to five-month trip. The 42-inch barrel is stamped "Made Expressly For Child, Pratt & Co. St. Louis, Mo.," identifying a quite famous frontier outfitting house that sold goods to overland travelers at several of the primary jumping-off points along the Missouri River. A rather plain rifle, the piece is fitted with brass furniture and typical double set triggers and open sights.

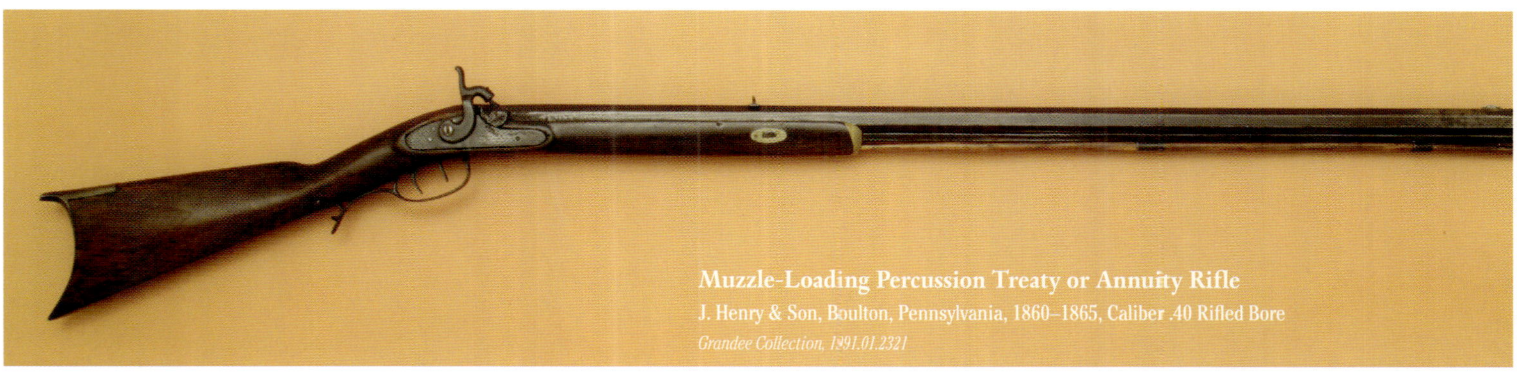

Muzzle-Loading Percussion Treaty or Annuity Rifle
J. Henry & Son, Boulton, Pennsylvania, 1860–1865, Caliber .40 Rifled Bore
Grandee Collection, 1991.01.2321

The U.S. government presented firearms to Native Americans on the frontier for treaties of peace and for land settlements. Stamped "US" on the stock below the trigger guard, this relatively inexpensive, half-stocked rifle was purchased under contract by the United States for just such an exchange. Specimens of this pattern and make are believed by some researchers to have been among the arms distributed to the Cherokees, or another of the "Five Civilized Tribes," earlier removed to Indian Territory (present-day eastern Oklahoma). This well-preserved example features a 37-inch octagonal barrel, open sights, double set triggers, and plain iron mountings.

the U.S. Office of Indian Trade, and the Bureau of Indian Affairs. By the middle 1820s, comparatively inexpensive longarms of plain, conventional Kentucky pattern—known in the West as Lancaster rifles for their initial point of origin in Pennsylvania—enjoyed widespread production. These weapons, with octagon barrels ranging from thirty-six to forty-two inches and .46 to .53 caliber, appeared with either the familiar flintlock ignition or, somewhat later, the more-advanced, percussion-lock ignition. Notable makers included Jacob Dickert, Melchior Fordney, and Henry Leman, all of Lancaster; Henry Deringer, John Krider, and George W. and Edward K. Tryon of Philadelphia; and J. J. Henry's Boulton Gun Works, located in the vicinity of Nazareth. The latter manufactory, later known as J. Henry & Son, also produced Indian trade and annuity rifles under the Indian Intercourse Act of 1834 well into the 1860s. Somewhat reminiscent of the then-established Mountain or Plains pattern, these lighter, muzzle-loading, percussion rifles usually were of half-stock configuration, having either brass or iron mountings and full-octagon barrels of .38 to .45 caliber.[19]

As alluded to in the introduction to part 1, advances in technology and design impacted American firearms manufacturing quite as much as the influences of environment, economy, or individual craftsmanship. Far and away the most influential technological advance of the early nineteenth century involved the development and utilization of percussion ignition as a means of discharging arms. As ultimately refined, the percussion system used an explosive, fulminate-of-mercury compound fixed in small, virtually weatherproof copper caps that fit on a short tube, or "nipple," at the rear of the barrel. When struck by the hammer, the percussion cap sent a flash of flame through the nipple, igniting the main powder charge in the barrel breech and discharging the arm. This proved eminently more practical—both in sureness and in speed—than the flint-and-steel locks that had reigned for the previous two hundred years. First haltingly adopted in the early 1820s, within a decade percussion-ignition locks met with broad acceptance on new guns, and many gunsmiths also did a brisk business converting older, flintlock arms to the cap-lock system.[20]

Most who made regular use of firearms, whether east or west, readily appreciated the merits of percussion-lock ignition—particularly its reliability in wet or windy conditions. In the latter regard, Santa Fe trader Josiah Gregg observed, "Severe winds are very prevalent upon the great western prairies. . . . It is for this reason, as well as on account of the rains, that percussion guns are preferable upon the Prairies, particularly for those who understand their use. The winds are frequently so severe as to sweep away both sparks and priming from a flint lock, and thus render it wholly ineffective." Sportsman Sir William Drummond Stewart, who hunted the West in the 1830s, remarked on the faster lock-time of the percussion system, citing "the difference of the going off of a cap and a flint lock" and noting, "how much easier it would be to shoot at a mark with the more immediate ignition [of the former]." By 1840, then, virtually all individual gun makers and the larger manufactories had phased the percussion-lock ignition system into use on the vast majority of domestic firearms.[21]

Innovative gun makers seeking greater firepower in

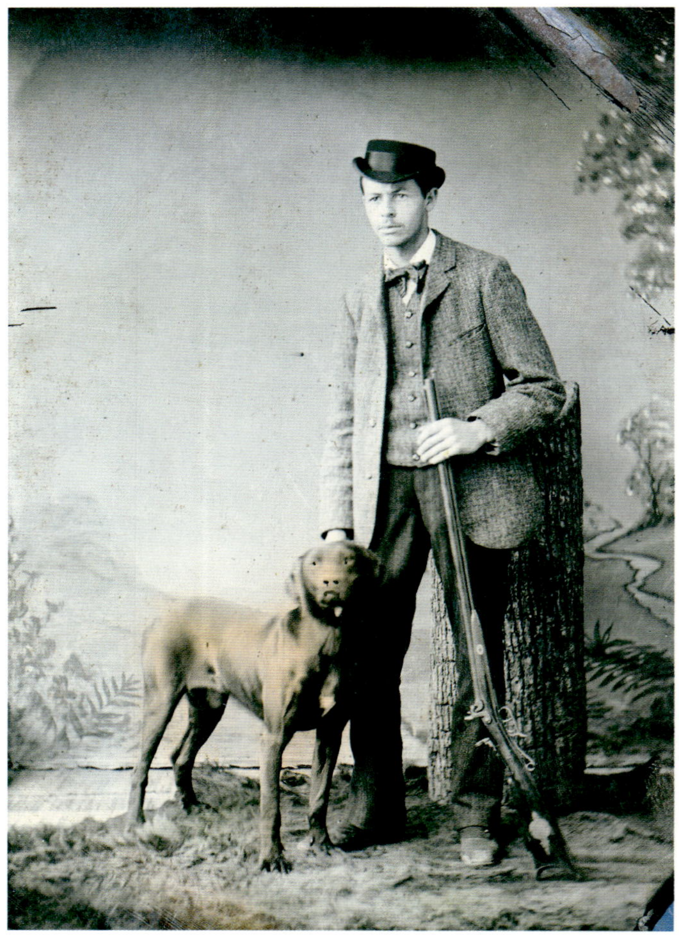

Photographer unknown. [Untitled.] Sixth-plate tintype, America, circa 1880–1885. Reflecting his hunting proclivities, this rather-stern-looking young man poses with his tinted dog and a fancy, muzzle-loading percussion, half-stocked, double-barreled sporting rifle of the classic side-by-side pattern, probably manufactured circa 1845–1855.
Dickinson Research Center, National Cowboy & Western Heritage Museum, 2004.052

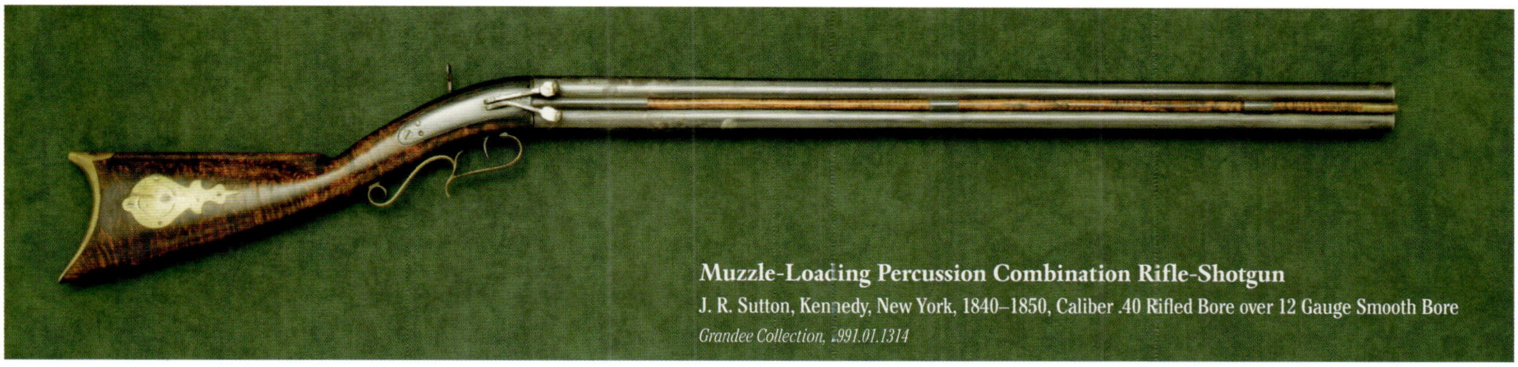

Muzzle-Loading Percussion Combination Rifle-Shotgun
J. R. Sutton, Kennedy, New York, 1840–1850, Caliber .40 Rifled Bore over 12 Gauge Smooth Bore
Grandee Collection, 1991.01.1314

Popular among some sport hunters at midcentury, this percussion, combination firearm served as a dual-purpose field gun for furred or feathered game. The piece incorporates stationary, 29-inch round superposed barrels (rifled bore above, smooth bore beneath) in combination with a back-action, double "mule ear"–hammered lock. A screw-adjustable rear peep sight is mounted forward of the wrist. Of quite an unusual—perhaps ungraceful—silhouette, the burled, bird's-eye maple buttstock is mounted with plain brass furniture.

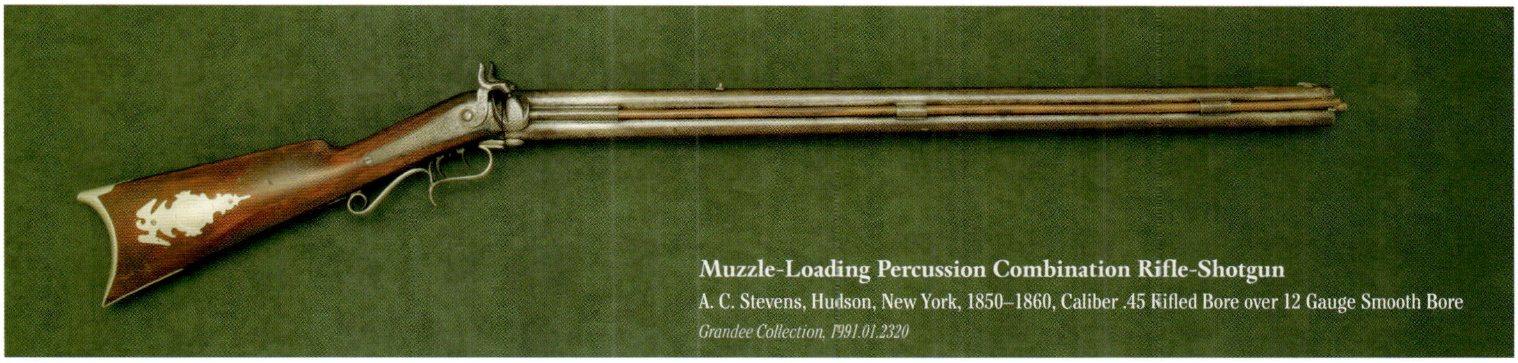

Muzzle-Loading Percussion Combination Rifle-Shotgun
A. C. Stevens, Hudson, New York, 1850–1860, Caliber .45 Rifled Bore over 12 Gauge Smooth Bore
Grandee Collection, 1991.01.2320

Like the previous specimen, this dual-purpose weapon features stationary, superposed barrels for a single rifle ball (above) and a load of shot (below). The piece is discharged with a pair of back-action locks actuated by double triggers. Though providing two quick shots in succession, such arms realized little favor among frontier types because of their heavier weight and lesser accuracy as compared to a single-barreled sporting or Plains-pattern rifle. (The decorative, German silver capbox and other figural ornamentation on this specimen are further illustrated in chapter 10.)

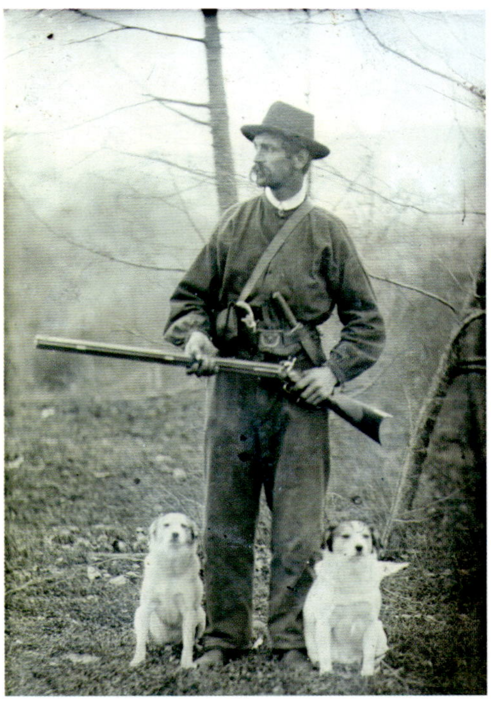

Photographer unknown. [Untitled.] Sixth-plate tintype, America, circa 1865–1870. Accompanied by his attentive dogs, this fellow is armed with a double rifle—or combination rifle-shotgun—with superposed barrels and percussion locks, a Bowie knife, and what appears to be a small-cartridge deringer.
Dickinson Research Center, National Cowboy & Western Heritage Museum, 2002.113

their arms found percussion ignition quite adaptable to multibarreled and breech-loading weapons. A few double rifles of the classic English pattern, having a pair of barrels mounted side by side with double locks, appeared by the latter 1830s. A somewhat more common American pattern employed a pair of over-and-under, or "superposed," barrels. Typically of stationary design, these double guns also utilized a pair of locks. Less common were superposed longarms in which spindle-mounted barrels were rotated manually for discharge by a single lock. Many of these arms combined a rifled barrel for a single patched ball or bullet, and a smoothbore barrel for shot loads. Although the benefit of two quick shots could sometimes be crucial to survival, such multibarreled guns never gained wide favor in the West. Practical frontiersmen found the advantage of increased firepower more than offset by the double gun's

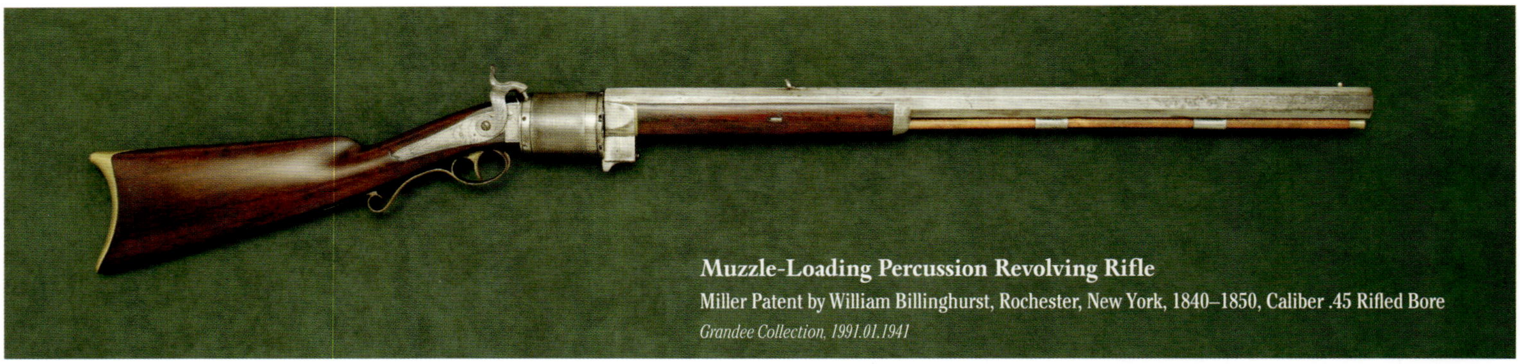

Muzzle-Loading Percussion Revolving Rifle
Miller Patent by William Billinghurst, Rochester, New York, 1840–1850, Caliber .45 Rifled Bore
Grandee Collection, 1991.01.1941

One of the earliest revolving-cylinder arms patented in the United States (1829), this seven-shot, percussion sporting rifle provided its owner with greatly increased firepower. Reliable function could be problematic, however, because the arm used primitive, fulminate pills rather than caps for ignition—a system that required close attention in loading and proved difficult to resupply in remote country. Rapid fire also proved challenging because the cylinder had to be unlocked and rotated by hand for each successive shot—unlike Colt's first Paterson-made revolving rifles, which advanced and locked the cylinder when either the ring lever or the hammer was cocked.

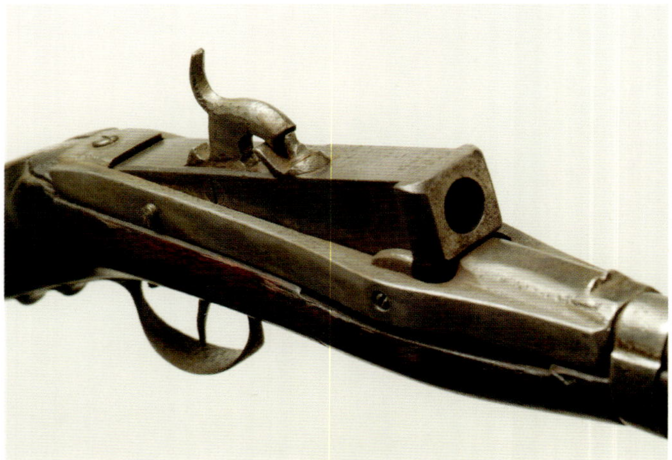

Breech-Loading Percussion Carbine *(detail)*
Simeon North, Middletown, Connecticut, U.S. Model 1833 Hall-North Contract, 1834–1839, Caliber .52 Smooth Bore
Grandee Collection, 1991.01.2305

A significant technological milestone in American arms making, John H. Hall's patented lifting-breech system (1811) was the earliest military weapon design to achieve true interchangeability (circa the mid-1830s). It also was the first U.S.-regulation breech-loading firearm (1819) and the first regulation percussion-ignition firearm (1839) adopted by the U.S. military. Technically still a muzzle-loader, the piece was loaded with a paper-wrapped cartridge at the front of the breechblock, much as the cylinder of a contemporary percussion revolver also was charged.

greater cost, heavier weight, and relative deficits in power and accuracy as compared with the single-barreled rifle. As is observed in the next chapter, however, by the 1840s multibarreled revolving handguns, or pepperboxes, had achieved considerable popularity in the civilian market both east and west.[22]

Other design efforts to obtain increased firepower—in the form of breech-loading and repeating arms—progressed only gradually. A single-shot, breech-loading weapon, John H. Hall's novel lifting-breech system (patented in 1811) saw considerable use by U.S. military forces during the antebellum period, but its rather unconventional mechanism and quite-limited commercial production retarded civilian use. Likewise, the development of repeating weapons by individual smiths progressed only by fits and starts. Beginning in the 1830s, multichambered, revolving-cylinder firearms dominated in this arena. Probably the earliest of these to gain some notoriety was the Miller-patent (1829) revolving rifle featuring a manually rotated, six-shot cylinder using novel pill-lock (percussion) ignition. One such rifle evidently was shipped in the early 1830s to American Fur Company factor Kenneth McKenzie, then at Fort Union on the upper Missouri River.[23]

As suggested above, many notable innovations in arms design sprang from the minds of American craftsmen-gunsmiths and inventors between 1810 and 1840. Many of these, however, demanded close mechanical tolerances and production techniques generally unattainable at the individual artisan's bench. To be practically produced in economical quantities, the relatively complex mechanics of breech-loading and revolving-cylinder firearms required a standard of precision and parts uniformity available only through machine production. Although individual gunsmiths continued making firearms throughout the period surveyed (and beyond to the present day), their traditional methods and output would be almost totally eclipsed in the United States by the 1860s through the ascendance of large, integrated, machine-based factories turning out thousands and tens of thousands of standardized, mass-produced, and technically sophisticated firearms.

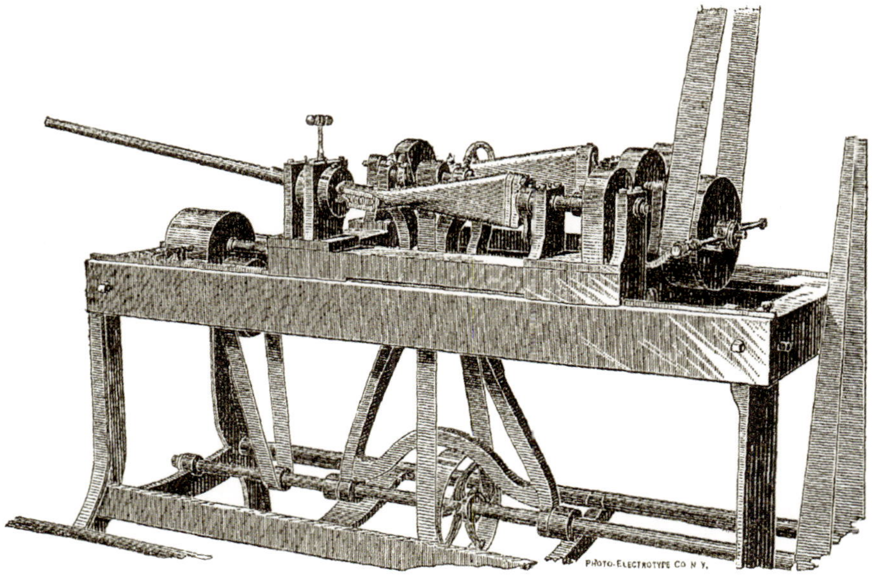

A second-generation gunstock-replicating machine designed by Cyrus Buckland and made by the Ames Manufacturing Company of Chicopee, Massachusetts, circa 1853. Far more refined and compact than the Blanchard machine, this type saw wide use at home and abroad for fifty years.
Drawing from Fitch, "Report on the Manufactures of Interchangeable Mechanism," 1883

Chapter 2

FIREARMS OF THE AMERICAN SYSTEM

BETWEEN 1800 AND ABOUT 1860, a dramatic transformation occurred in the methods and means of firearms production in the United States. The transition from the individual shop of the craftsman-gunsmith to the integrated factory of the specialized machine operator represented a remarkable technological progression in which arms fabrication evolved from *product-oriented* manufacture to *process-oriented* manufacture. In no other domestic industry during the early nineteenth century did the evolving principles and practices of mass production via machinery achieve such signal success or foster such far-reaching influences. Nurtured almost exclusively among an innovative and expanding cadre of arms manufacturers, mechanics, and machine-tool builders, what became known as the "American System of Manufactures" proved to be a crucial catalyst in the nation's industrial revolution. Indeed, this distinctive, rationalized system of mass production in firearms rapidly transferred to all manner of products—from the first mechanical sewing machines and typewriters to contemporary electronic computers and cell phones.[1]

The transformation of domestic arms production sprang from two consequential concepts. The first, largely of government origin, sought the uniformity, and ultimate interchangeability, of a firearm's various constituent parts. In practical terms, the achievement of interchangeable-parts manufacture would contribute both to a standardized weapon's initial production quality and to its much easier repair in the field. The second concept, of both government and civilian origin, sought the mass manufacture of firearms through the application of specialized, labor-saving machinery. Systematically applied, machine production sought the separate manufacture of each component part of an arm in large quantities by a workman typically lacking the skill to create an entire arm by himself. The many iterations of each constituent part would be so alike that any one part would serve in creating a functional arm of a given pattern. The use of specialized machines—rather than specialized men (skilled gunsmiths)—would not only provide the desired parts-uniformity but also dramatically increase a manufacturer's total production and perhaps reduce individual unit cost as well.

The notion of producing "standardized weapons with standardized parts" first arose in the French military in the eighteenth century, and it was first attempted by Frenchman Honoré Blanc in the fabrication of flint gunlocks by the hand filing of parts in conformity to pattern guides or jigs (no machinery yet was involved). While serving as the minister to France in the mid-1780s, Thomas Jefferson became acquainted with the concept of parts uniformity, and he promoted the idea vigorously back in the United States. By 1815, the "uniformity ideal" had become a central tenet of the U.S. Ordnance Department and the national armories at Springfield, Massachusetts, and Harpers Ferry, Virginia. As well, several private arms makers, working with the government through cash-advance contracts, also strived to achieve parts uniformity in their firearms. From there, the ideal of parts uniformity rapidly evolved into the ultimate objective of complete parts interchangeability. Equally significant, individual arms makers and mechanics in both the private and the public (military) sector began applying innovative methods and machinery to the fabrication of various metal and wood firearm components. As a result, much of the fundamental—albeit very incremental—progress in American machine and machine-tool design and production occurred within the nexus of the burgeoning antebellum arms industry.[2]

Private inventor and struggling manufacturer Eli Whitney (1765–1825) of Whitneyville (near New Haven), Connecticut, who won the initial federal arms contract in 1798 for ten thousand muskets, was the first to broach the idea of achieving firearms uniformity via machinery. "One of my primary objects," Whitney wrote, "is to form the tools so the tools themselves shall fashion the work and give to every part its just proportion." As he later explained, his goal was "to substitute correct and effective operations of machinery for that skill of the artist which is . . . not possessed in this country to any considerable extent."[3] For nearly a century and a half, Eli Whitney was credited with achieving interchangeable-parts manufacture with powered machinery. In fact, although he wrote of mechanized gun production and promoted the interchangeability ideal in a government demonstration intended to save his then-failing contract, Whitney actually achieved neither of these objects. And although his enterprise survived into the late nineteenth century, it was never truly a leader in arms manufacturing. Ultimately, Eli Whitney's principal contributions to the nascent American System of arms production lay in espousing some of its most basic principles and in inciting government demands for further progress from the two national armories and the various private contractors.[4]

The first arms maker to achieve substantial progress was Simeon North (1765–1852) of Middletown, Connecticut, who undertook a series of government pistol contracts that included increasingly stringent requirements for interchangeable-parts production. His 1813 contract,

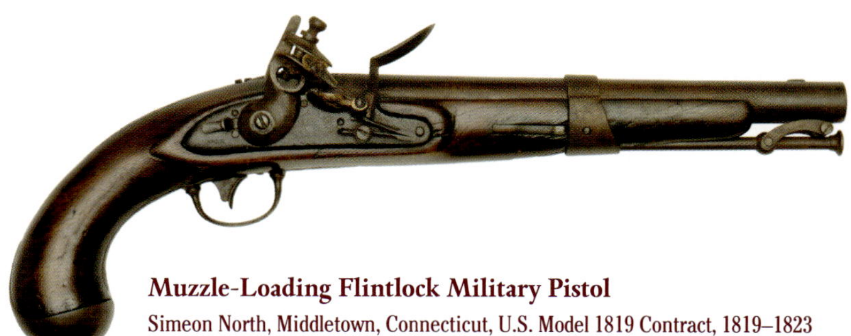

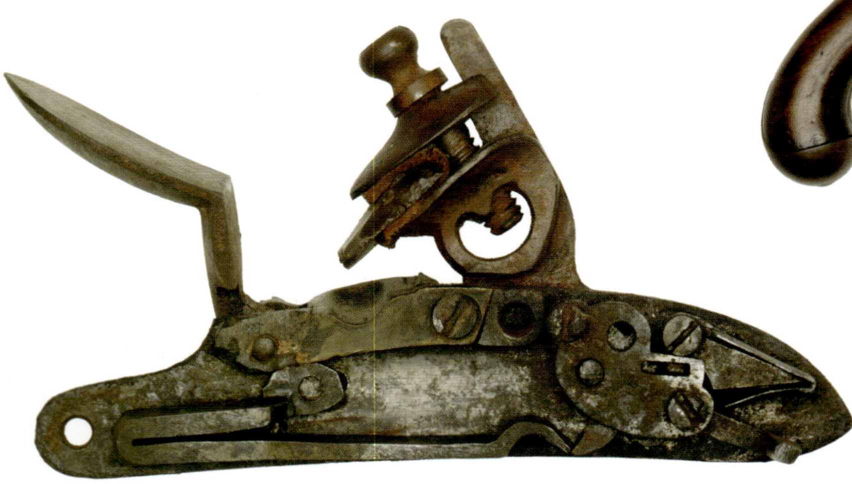

Muzzle-Loading Flintlock Military Pistol
Simeon North, Middletown, Connecticut, U.S. Model 1819 Contract, 1819–1823
Caliber .54 Smooth Bore
Grandee Collection, 1991.01.0146

Military contract manufacturer Simeon North is believed to have achieved interchangeability in the fabrication of his flint gunlocks by the middle 1810s. The interior lockworks of this Model 1819 pistol (left) suggest this was so, as neither the lockplate nor its constituent parts carry any of the lot or batch marks—usually an identical letter, digit, or number of punched dots or scribed hash lines—commonly employed in the assembly of earlier and contemporaneous martial arms and their gunlocks.

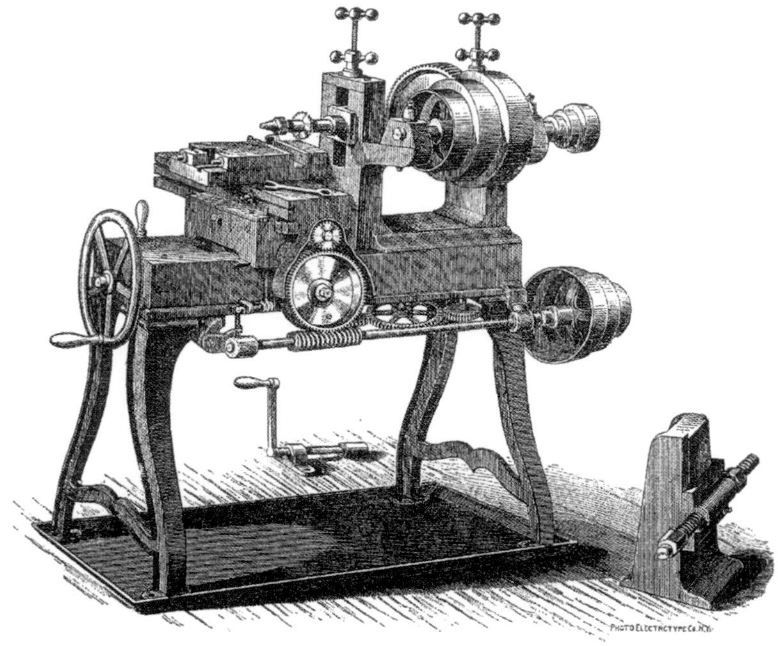

A standard, Lincoln-pattern milling machine designed by F. A. Pratt and manufactured by G. S. Lincoln & Company of Hartford, Connecticut, circa 1860. A simpler milling machine was used by Simeon North at Middletown, Connecticut, prior to 1820. By 1860, such machinery—used in tandem with drop-forging—had largely eliminated laborious metal grinding, chiseling, and filing.
Drawing from Fitch, "Report on the Manufactures of Interchangeable Mechanism," 1883

for example, specified that "the component parts of the pistols are to correspond so exactly that any limb or part of one pistol may be fitted to any other pistol of the twenty thousand."[5] North began fabricating individual gunlocks to a standard, *model lock* that served as a uniform template or gauge—a manufacturing innovation that resulted in his production of fully interchangeable locks around 1816. This, in turn, culminated in the establishment by 1840 of a disciplined inspection regimen among the government armories and contractors in which scores of different pattern, receiving, limit ("go no-go"), and thread gauges of hardened steel were employed to test each and every part of a given arm against a universally accepted standard. Equally, if not more, significant, Simeon North (likely in company with his son) is now believed to have developed and utilized the first plain milling machine in American manufacturing. This marked a true milestone in power-driven, metal-shaping technology that dramatically reduced the necessity of traditional grinding and hand-filing operations and inspired the further creation of special-purpose machines for the labor-saving fabrication of firearm components.[6]

Yet another ingenious Yankee mechanic, Thomas Blanchard (1788–1864) of Millbury, Massachusetts, devised a practical turning machine for shaping the irregular breech end of gun barrels in 1818. Inspired by this success, he went on to devise his greatest invention—a pattern-replicating lathe for shaping gunstocks via tracing and cutting wheels that moved in tandem over the master iron-gunstock pattern and the wooden stock blank, respectively. Replacing "a whole battalion" of individual stock makers using traditional hand tools, this revolutionary machine was quickly adopted by the Springfield Armory and a few private manufactories in New England. By 1830 Blanchard had devised fourteen special-purpose woodworking machines for musket stocks that shaped the blank to profile from butt to muzzle, grooved out the barrel bed, and cut precise inlets or mortises for the lockworks, sideplate, barrel breech and tang, trigger guard, and barrel bands. As important as these special-purpose machines proved in their output alone, Blanchard also employed them in a logical, sequential order that contributed directly to the mass-production principles fundamental to the American System.[7]

Perhaps the man best epitomizing Yankee ingenuity in arms making during the formative years of the American

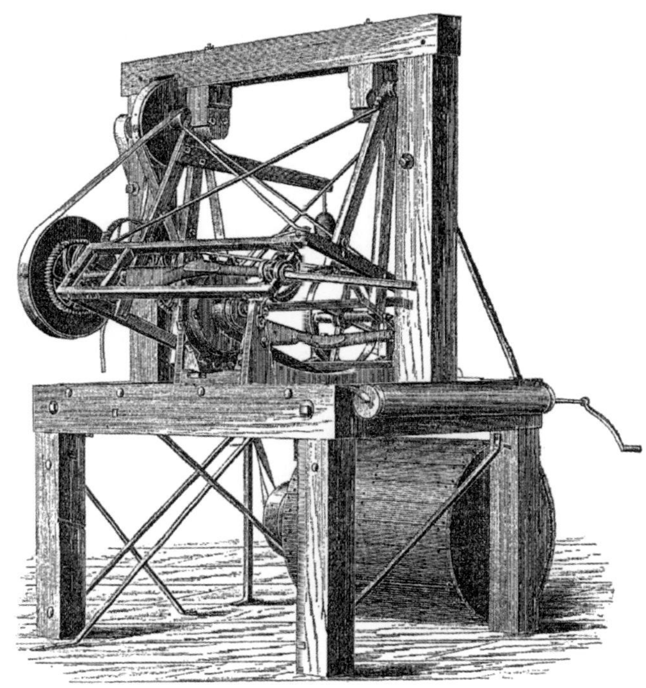

The first-generation gunstock-replicating machine designed by Thomas Blanchard, circa 1818, and installed for operation at the U.S. Springfield Armory, circa 1820. The template guiding wheel and the cutting wheel were both about 18 inches in diameter and moved simultaneously.
Drawing from Fitch, "Report on the Manufactures of Interchangeable Mechanism," 1883

System, Maine native John H. Hall (1781–1841) labored for twenty years to produce his patent breech-loading rifle via machinery and with complete parts interchangeability. Through the Ordnance Department, Hall worked as a semiautonomous contractor from 1820 to 1840 at the Rifle Works in Harpers Ferry, Virginia. There, he designed and utilized an extensive series of self-acting die-forging, milling, and drilling machines in which the various workpieces were exactly placed in locating fixtures having consistent bearing points. In combination with the application of more than sixty different inspection gauges, Hall's system of arms manufacturing was the first to attain true interchangeability in all the weapon's constituent elements. This was conclusively demonstrated when Simeon North, using Hall's methods and gauges, produced Hall-patent breechloaders under subcontract in which the component parts he made in Connecticut interchanged not only among themselves but also with those made by Hall in Virginia.[8]

By the 1850s, with the U.S. Springfield Armory serving as an informal clearinghouse, the regular sharing and transfer of technical information among government armories and contractors, private arms manufacturers, innovative mechanics, and machine-tool builders within the American gun industry had given rise to a growing number of substantial, machine-based firearms factories scattered along or near the Connecticut River valley from central Vermont south to Long Island Sound. In varying degree, all employed the machinery, production principles, and labor practices then virtually unique in either domestic or foreign manufacturing. In Europe, certainly, there was no corresponding advance in arms manufacture over the same period. There, the traditional guild practices of home-based, specialized piecework via hand tools continued to hold sway well into the 1880s and even later.[9]

Indeed, Great Britain's 1855 *Report of the Committee on Machinery of the United States of America* clearly recognized these distinctive Yankee manufacturing traits, observing that

> the private establishments of Colonel Colt, Robbins & Lawrence, and Sharpe's [*sic*] Rifle Company, are all conducted on the thorough manufacturing system, with machinery and special tools applied to the several parts. . . . Besides the machinery and tools . . . there are hundreds of valuable instruments and gauges that are employed in testing the work through all its stages, from the raw material to the finished gun, others for holding the pieces whilst undergoing different operations, such as marking, drilling, screwing, etc., the object of all being to secure thorough identity [uniformity or interchangeability] in all the parts.[10]

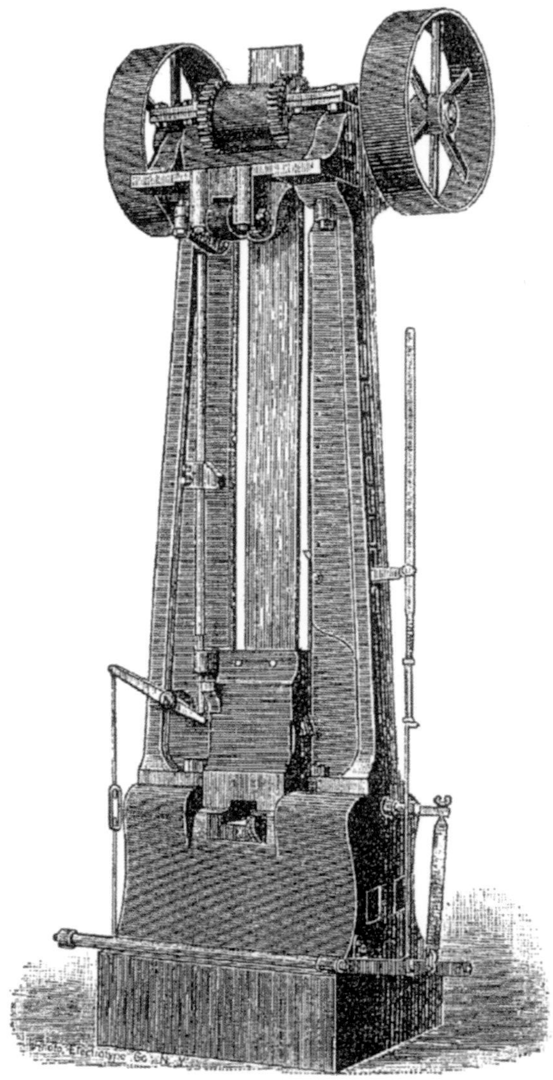

A plank-type, drop-hammer machine for die-forging gun parts, designed by Pratt & Whitney of Hartford, Connecticut, circa 1865. John H. Hall devised a chain-drop, die-forging machine at Harpers Ferry, circa 1821, and E. K. Root designed a crank-drop variation at Colt's circa 1850.
Drawing from Fitch, "Report on the Manufactures of Interchangeable Mechanism," 1883

The Colt and Robbins & Lawrence firms cited by the British committee represented, at the time, the fruition of machine-based gun manufacture in the private sector. Both companies already had brought the American System to world attention at London's Crystal Palace Exhibition in 1851, and both benefited by the export of arms and arms-making technology immediately thereafter. In the United States, literally all of the famous names in American gun making discussed in part 2 adopted the fundamental elements of the interchangeable system, among them producing tens and hundreds of thousands of commercial firearms during the second half of the nineteenth century.

One obvious, yet significant, result of the American System was that firearms no longer represented singular creations, each a unique product of the craftsman-gunsmith's bench. Now, with machine production and precision gauging, the long-sought ability to make "standardized weapons with standardized parts" resulted in what are today recognized as "model" arms within a manufacturer's product line. Firearms of a particular model share certain characteristics and might be differentiated from others merely on the basis of scale and caliber, as between Colt's Model 1860 Army and Model 1861 Navy revolvers. Or they might be differentiated because of major changes or improvements in their internal operating mechanisms, as, for example, between the toggle-link and vertical-bolt breech-locking systems utilized in Winchester's Model 1876 and Model 1886 rifles, respectively. Early on, several firearms manufacturers, such as Remington, Sharps, and Smith & Wesson, identified their various models by marking them with successive patent-issue dates or by the imprint "New Model." By the mid-1870s, the Winchester firm had commenced stamping its different rifle models with introductory-year designations (Model 1873, Model 1876, and so forth), and other companies, such as Marlin, followed suit. The model names today applied by scholars and collectors to commercial arms without original maker designations generally derive from design or technical differentiation among a maker's product line, or from a firm's internal nomenclature or advertising literature.[11]

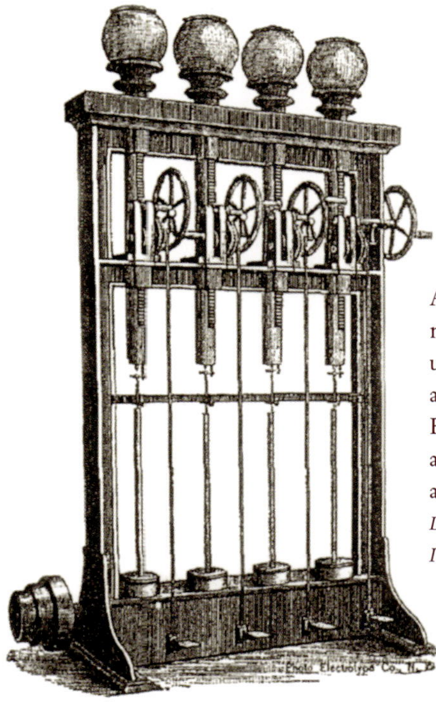

A four-spindle, vertical barrel-drilling machine designed by Frederick Howe and utilized at the Robbins & Lawrence factory at Windsor, Vermont, circa 1852. At Colt's, E. K. Root devised a similar machine at about the same time, having the spindles arranged around a central column.
Drawing from Fitch, "Report on the Manufactures of Interchangeable Mechanism," 1883

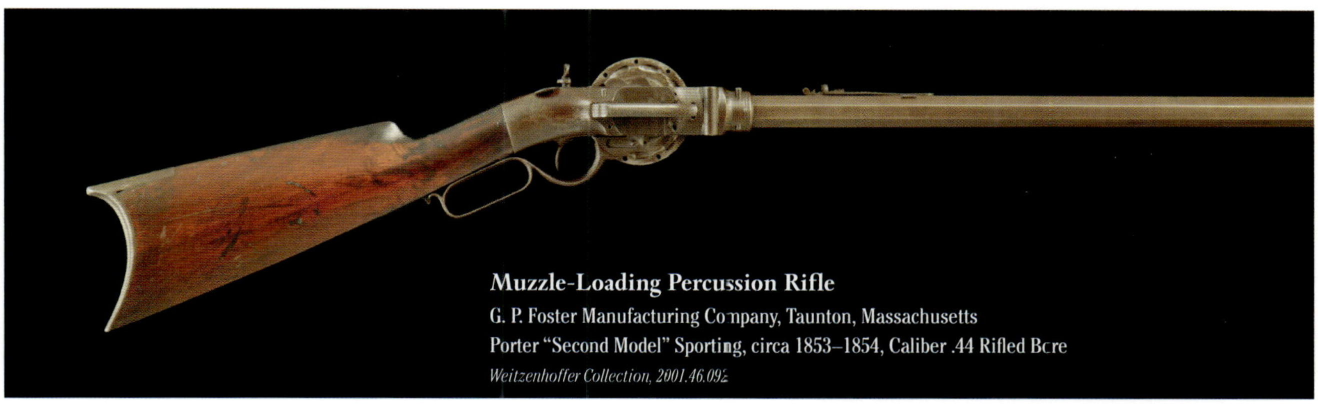

Muzzle-Loading Percussion Rifle
G. P. Foster Manufacturing Company, Taunton, Massachusetts
Porter "Second Model" Sporting, circa 1853–1854, Caliber .44 Rifled Bore
Weitzenhoffer Collection, 2001.46.092

Designed by Parry W. Porter of Memphis, Tennessee, and patented in July 1851, this revolving "turret" rifle epitomizes the rampant innovation in American arms invention and manufacturing during the middle of the nineteenth century. The Porter system incorporated a vertically mounted, nine-shot, revolving turret advanced by a loop-pattern underlever and discharged by a straight, mule-ear side hammer. Such an intricate mechanism required sophisticated machine production, which, at the time, was readily available only among New England's several arms makers. Like many other such specimens, this rifle's relatively complex, automatic percussion capping system later was altered to a manually capped (single-shot) system.

As suggested previously, the adoption of powered machinery and precision gauging not only increased arms production and uniformity but also allowed for the practical manufacturing of more mechanically complex breech-loading and repeating firearms. Even during the brief era of percussion ignition, circa 1830–1870, machinery for die forging, boring, rifling, milling, drilling, broaching, and pattern replicating contributed directly to the relative efficacy of complicated arms systems demanding close working tolerances. Prominent among these during the antebellum era were the Sharps lever-actuated, falling-block, breech-loading, single-shot rifle; the Allen double-action, revolving-barreled, muzzle-loading, repeating "pepperbox" pistol; and the now-famed Colt single-action, multichambered, revolving-cylinder, muzzle-loading, repeating handguns and longarms. These, along with many competitor firearms of the era, saw production in eastern arms factories in which the necessary standards and methods of precision forging, machining, and gauging had already commenced to flourish.

It fell to gunsmith Christian Sharps, during the late 1840s and early 1850s, to perfect the first truly practical breech-loading rifle design—a falling-block system so functional and robust that it later adapted to metallic-cartridge ignition with little modification. At first plagued with gas leakage at the breech, Sharps introduced several improvements prior to the formation of the Sharps Rifle Manufacturing Company in 1851 and the commencement of contract production with the Robbins & Lawrence firm of Windsor, Vermont, then the largest private firearms factory in the country. The Sharps slant-breech sporting rifles (Models 1851, 1852, 1853, and 1855), utilizing percussion pellet or tape primers and firing linen or paper-wrapped cartridges, could be loaded and discharged eight or nine times per minute—about four times the rate of a common muzzle-loading rifle. In 1854, well-known Saint Louis firearms-dealer Horace Dimick enthusiastically advertised that the Sharps could be "loaded on horseback at a smart gallop, or lying down in the grass. . . . Two hundred balls can be shot with this Rifle, four hundred yards, in a small space [target], and in less time, than with any other military or hunting Rifle on this or 't'other side of Jordan.' We will back our judgement with the 'Filthy Lucre' on the above proposition. It is self evident that they are the best guns for persons exposed to the dangers of a frontier life."[12] Approximately 3,750 Sharps slant-breech sporting rifles were produced between 1852 and 1859. Some appreciable number of these, typically in .44 or .52 caliber, no doubt served in the West with a variety of frontier types who agreed with Mr. Dimick on the merits of the weapon. Of the later, straight-breech, percussion Sharps rifles and carbines (New Models 1859, 1863, and 1865), virtually all were of military pattern manufactured for service in the Civil War (1861–1865).[13]

Another breech-loading rifle popular with frontier sportsmen in the 1860s, and following, incorporated Dr. Edward Maynard's tip-up breech design, first patented in 1851 and improved in 1856 with the introduction of his innovative, reloadable cartridge cases. As manufactured by

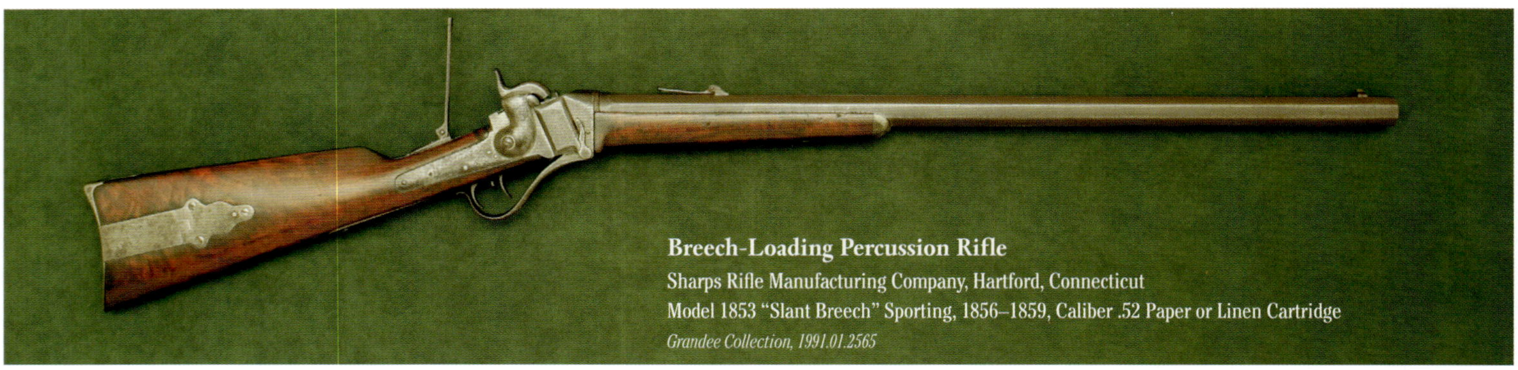

Breech-Loading Percussion Rifle
Sharps Rifle Manufacturing Company, Hartford, Connecticut
Model 1853 "Slant Breech" Sporting, 1856–1859, Caliber .52 Paper or Linen Cartridge
Grandee Collection, 1991.01.2565

Providing much faster loading at the breech, the Sharps rifle met with widespread approval among frontiersmen and sport hunters. While the antebellum hunter with a muzzle-loading rifle might shoot one buffalo from a herd per minute, a hunter with a breechloader such as the Sharps could harvest three or even four. This rifle utilizes the inventor's own pellet priming system for percussion ignition and also features figured walnut stocking with iron patchbox and pewter forestock tip, a special-order or aftermarket folding tang sight, a midweight 26-inch octagonal barrel, and double set triggers

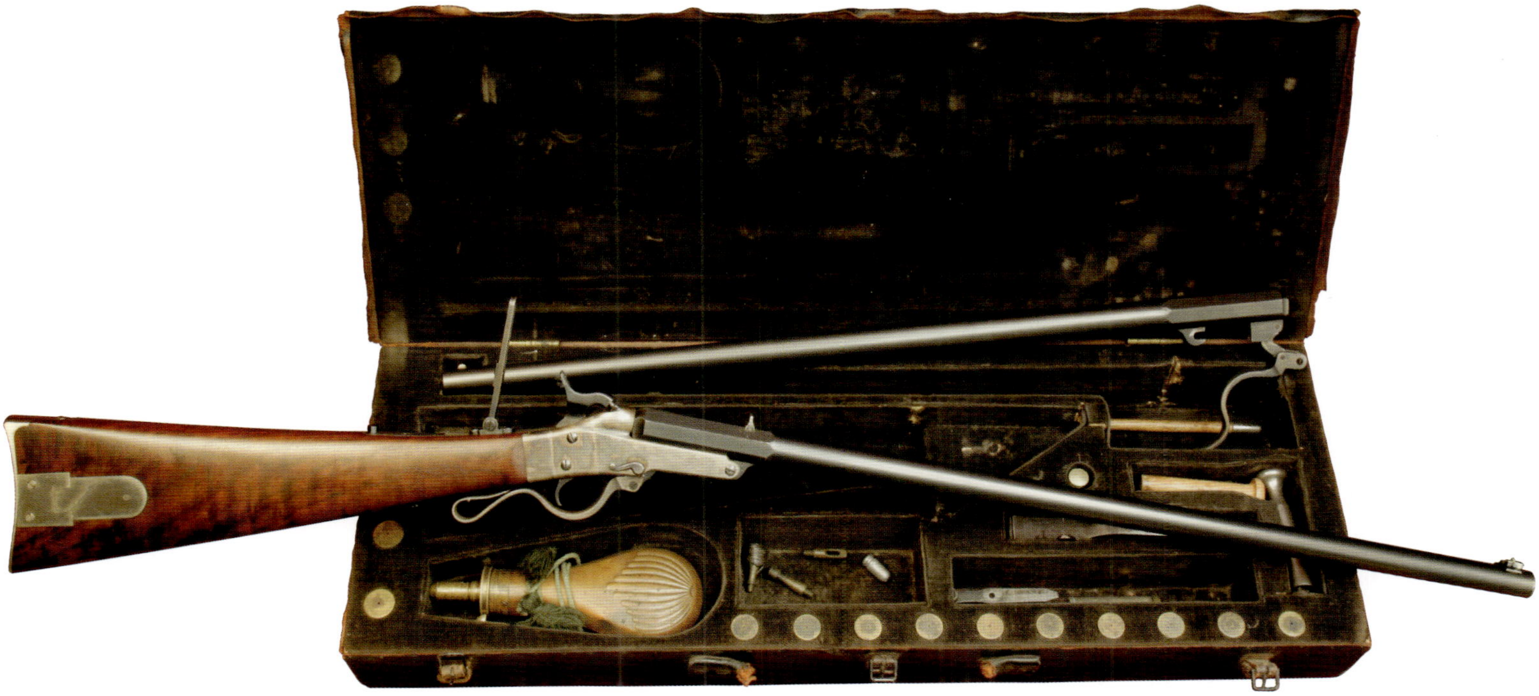

Breech-Loading Percussion Rifle
Massachusetts Arms Company, Chicopee Falls, Massachusetts
Maynard Model 1865 Sporting, circa 1865–1872, Caliber .40 Rifled Bore
Weitzenhoffer Collection, 2001.46.093

The Maynard breech-loading system utilized a simple, yet strong, tip-down action in combination with a well-designed, nonprimed cartridge case (holding the propellant and projectile) that provided a gas-tight breech seal. The weapon's simplicity of design allowed for the quick and easy replacement of barrels—switching to a rifled barrel of different caliber or to a smooth-bored barrel for shot loads. A few Maynard rifles, such as this example, were sold with custom-made casings outfitted with extra barrels, cartridge cases, and loading accessories. This rifle sports a blued barrel, case-hardened frame, and figured walnut stock with German silver patchbox and buttplate.

the Massachusetts Arms Company during ensuing years, this simple yet remarkably practical breech-loading design allowed for the convenient interchange of rifled and smooth-bored barrels. In fact, the well-to-do sportsman could acquire a specially cased Maynard outfitted with a selection of barrels, cartridge cases, and loading accessories.[14] Noted British adventurer Richard Burton, who traveled across the West in 1860, praised the firearm. "Those . . . who can use light weapons," he advised, "should prefer the Maynard breech-loader, with an extra barrel for small shot. . . . The Maynard is the best of breech-loading guns . . . ; a mere carbine in size, it kills at 300 yards, and in the U. S. costs only $40."[15]

During the antebellum years, efforts to increase the firepower of percussion arms focused largely on two design types—revolving-barreled and revolving-cylinder repeaters. Typically made as handguns, arms of the former type were termed "revolving pistols" by their various makers, though today they are widely known as "pepperboxes." Typical American-made pepperbox arms incorporated a cluster of four to six barrels bored in a single cylindrical metal block mounted on a center spindle in the face of the pistol's frame. The entire barrel cluster rotated and successively discharged through manipulation of the pistol's single- or double-action mechanism. With the double-action firing system, the weapon provided surprisingly fast firepower. However, as they typically appeared with smoothbore barrels and without sights, these rather muzzle-heavy pistols proved notoriously inaccurate much beyond fifteen feet. While traveling overland to California in 1861, Samuel Clemens (Mark Twain) wrote of the weapon, "To aim along the turning barrel and hit the thing aimed at was a feat which was probably never done. . . . [A]s one of the stage-drivers afterward said, 'If she didn't get what she went after, she would fetch something else.'"[16] Such pistols, nonetheless, marketed in pocket, belt, and holster sizes ranging from .28 to .36 caliber, enjoyed flourishing sales from the late 1830s well into the 1850s among overland emigrants, California gold seekers, and sundry others desiring close-range firepower in a weapon that was pointed rather than aimed.[17]

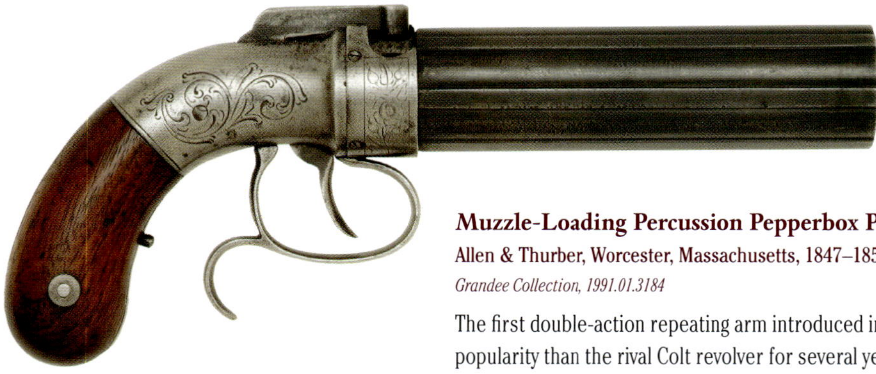

Muzzle-Loading Percussion Pepperbox Pistol

Allen & Thurber, Worcester, Massachusetts, 1847–1854, Caliber .36 Smooth Bore

Grandee Collection, 1991.01.3184

The first double-action repeating arm introduced in America, the Allen pepperbox enjoyed wider popularity than the rival Colt revolver for several years. The revolving barrels were rotated, and the bar hammer cocked, simply by pulling the trigger. Many of these pistols went west with the emigrant trains, particularly to the gold fields of California. As a weapon of self-defense, this dragoon-sized specimen, with its muzzle-heavy cluster of six 6-inch barrels, no doubt provided ample firepower but rather uncertain accuracy beyond 15 feet. It was certainly no match for the contemporary Colt Navy or Dragoon revolvers, which, though using slightly slower single-action mechanisms, were both more powerful and far more accurate.

By far the most influential—and most prolific—manufacturer of pepperbox pistols, Ethan Allen ranked among the first private arms makers to make extensive use of the interchangeable-parts principle in production. Perhaps as important, in 1837 he devised, patented, and subsequently applied his innovative double-action firing mechanism, in which a pull of the trigger raised the hammer and rotated the barrel cluster in unison and then released the hammer for discharge. (In contrast, major revolver manufacturers such as Colt, Remington, and Smith & Wesson stayed almost exclusively with the somewhat slower and simpler single-action mechanism for another forty years.) Allen, operating from Grafton and Worcester, Massachusetts, and Norwich, Connecticut, under the company name Allen & Thurber, produced his patent revolving pistols by the thousands in more than forty variations, far outpacing notable pepperbox competitors such as Blunt & Syms, Marston, and Robbins & Lawrence. During much of the 1840s, in fact, Allen's pepperbox pistols realized greater popularity among many Americans than the soon-to-be-vaunted Colt revolver.[18]

Though repeating, pepperbox handguns enjoyed widespread use for a time, their favor rapidly declined in direct competition with more practical, multichambered revolving-cylinder repeating arms. Incorporating a single, stationary barrel with a rifled bore (and sights), revolvers proved far more accurate at appreciably greater distances than did pepperboxes. As well, their single-action firing mechanisms—in which cocking the hammer released, rotated, and relocked the five- or six-shot cylinder preparatory to firing by pulling the trigger—proved entirely adequate for most practitioners desiring rapid, repetitive fire. Revolving-cylinder handguns proved so practical and efficient, in fact, that they continue to be produced in substantial quantities today among such notable domestic manufacturers as Colt, Ruger, and Smith & Wesson.

Widely recognized as inventor, industrialist, entrepreneur, and superlative self-promoter, Samuel Colt led the way in revolver design and manufacture. He received patent rights for the first truly functional revolving-cylinder repeating-arm system in 1836 and founded his initial gun-making enterprise, the Patent Arms Manufacturing Company of Paterson, New Jersey, in 1837. Paterson-manufactured arms, which included rifles and shotguns as well as revolvers, garnered only limited public acceptance and meager sales owing to their rather delicate mechanisms, frequent malfunctions, and occasional multiple discharges. (Some detractors satirically referred to the ingenious product as "Colt's Patent Wheel of Misfortune.") Coupled with comparatively high unit cost and an inability to obtain substantial government contracts, such deficiencies resulted in the failure of the Paterson firm in 1842. However, Colt's No. 5 Holster Model revolver had proven its merits for superior firepower and adaptability to mounted combat on the frontier with the fledgling Texas Rangers, thereby providing a foundation for future demand with the Mexican-American War (1846–1848). In collaboration with Captain Samuel H. Walker, Colt redesigned his invention

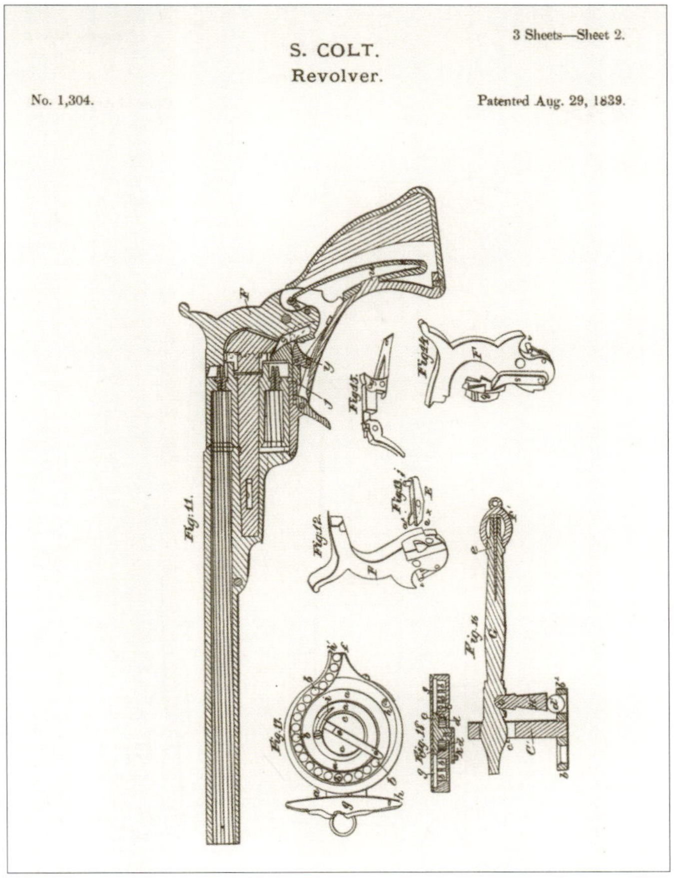

Patent drawing illustrating the internal workings of Colt's No. 5 Paterson Holster revolver and associated loading tools. Note the drop-down trigger mechanism without trigger guard, and the hammer-and-pawl arrangement for advancing the cylinder.
U.S. Patent and Trademark Office, uspto.gov

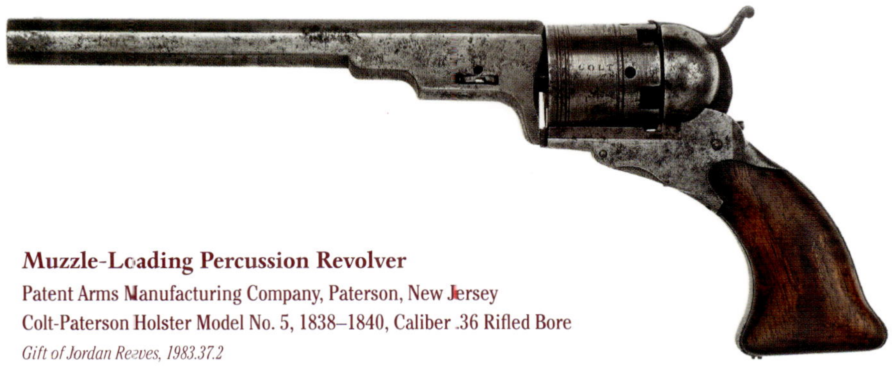

Muzzle-Loading Percussion Revolver
Patent Arms Manufacturing Company, Paterson, New Jersey
Colt-Paterson Holster Model No. 5, 1838–1840, Caliber .36 Rifled Bore
Gift of Jordan Reeves, 1983.37.2

Made famous with the "ranging companies" of the Texas Republic, the five-shot, single-action, Colt-Paterson No. 5, or Holster Model, revolver demonstrated its worth as an effective repeating arm in mounted combat. The gun's firepower proved decisive in several running engagements with Comanche marauders, most notably in the Battle of Wilson's Creek, in which Colonel John Coffee Hays and fourteen Texas Rangers charged and routed more than eighty adversaries. All of the later Walker and Dragoon model Colts (some 21,000 revolvers) carried a cylinder scene commemorating the so-called Hays Big Fight on the Pedernales. Colt himself came to call the Paterson Holster Model revolver the Texas Arm, and present-day collectors generally refer to it as the Texas Model or Texas Paterson. (Also see chapter 3.)

in 1847, creating a more robust and functional revolver that—proved in battle—would put him back in business to stay. With the reissue of his original patent specifications, extending protection out to 1857, regular production started anew in 1848 at Hartford, Connecticut. Later incorporated as the Colt's Patent Fire Arms Manufacturing Company, the resurrected firm had a virtual monopoly on practical revolver production for another decade. (The company's successful 1851 patent-infringement suit against the Massachusetts Arms Company also did much to strengthen its competitive position during the last antebellum decade.)[19]

Within a half-dozen years, Colt's percussion revolvers dominated U.S. military contracts and commandeered much of the civilian market as well. Upon such success, Colt and his trusted engineer-superintendent, Elisha King Root, established the largest, most-advanced private armory in the world in 1855. The principal armory building, measuring

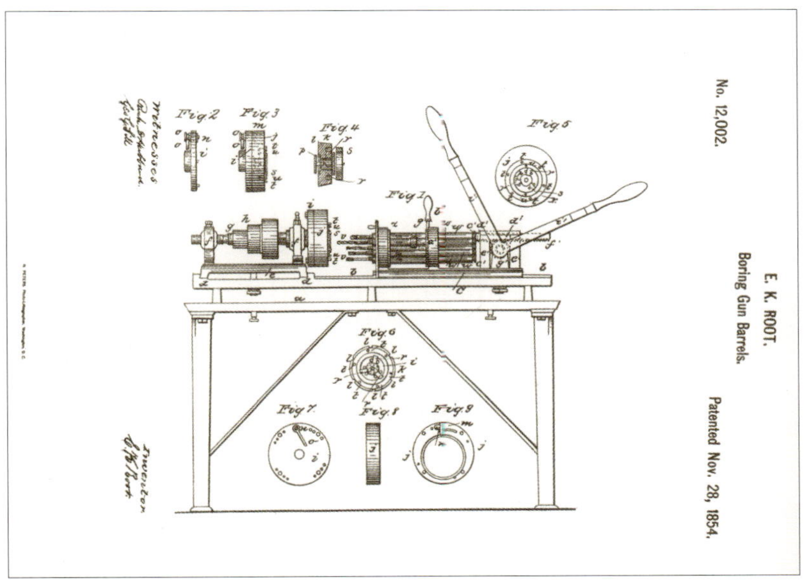

Patent drawing illustrating a multispindle horizontal drilling machine for boring and reaming the chambers of revolver cylinders on an accurate, parallel axis. This was one of several such machines designed by E. K. Root in the mid-1850s to both refine and speed production at the Colt factory.
U.S. Patent and Trademark Office, uspto.gov

sixty by five hundred feet, housed some four hundred special-purpose metal- and woodworking machines—several of them designed by Root and made on the premises. Through a sophisticated system of overhead shafting and belting, all were powered by a seven-ton, 250-horsepower, Corliss walking-beam steam engine. Throughout the facility, workers employed all the specialized machinery and mass-production practices that epitomized the American System of manufacturing, encompassing precision tooling and gauging, meticulous inspection regimens, and the supervision and division of inside-contract labor among a few score trained mechanics and several hundred relatively unskilled machine tenders.[20]

In "A Day at the Armory," a journalist for *United States Magazine* characterized the Colt factory in 1857 as "an immense industrial pile" and provided some notion of the complexity and sophistication of its principal machine room:

> Running through the center is a row of cast-iron columns, sixty in number, to which is attached the shafting and belting for driving the machines. . . . As we enter the door, the first group of machines appears to be exclusively employed in chambering cylinders; the next is turning and shaping them; here another is boring barrels; another group is milling the lock-frames; still another is drilling them; . . . here are the rifling machines, and there the machines for boring rifle barrels; now we come to the jigging machines that mortice out the lock-frames; and thus it goes on all over this hive of physical and mental exertion.

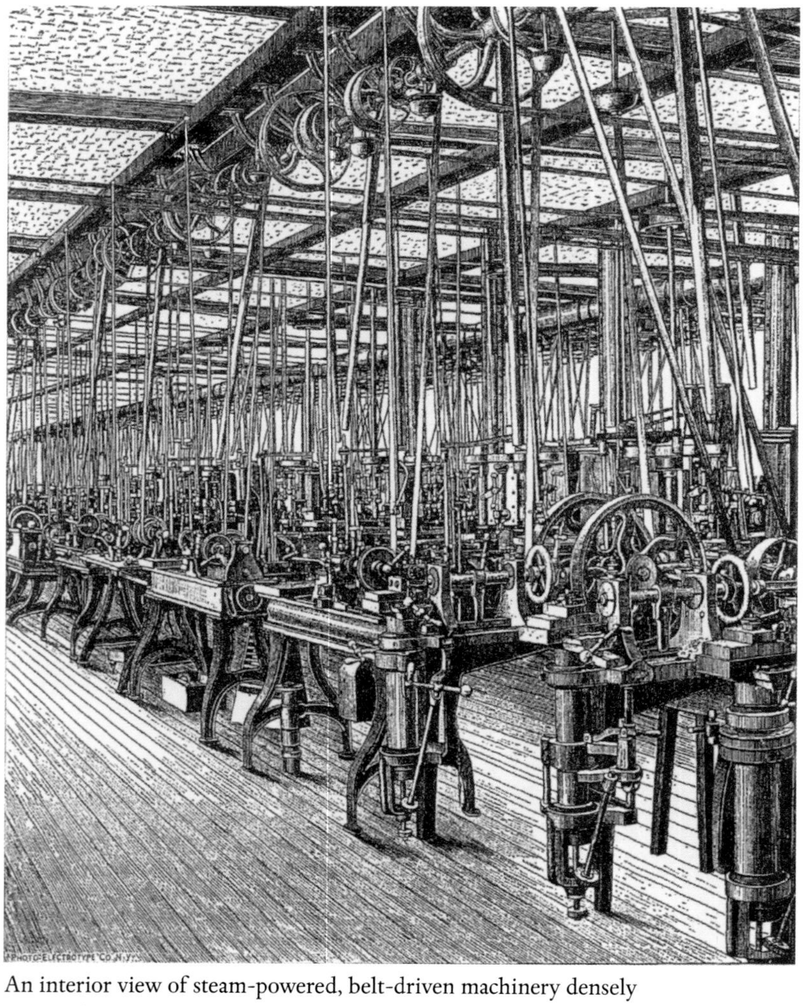

An interior view of steam-powered, belt-driven machinery densely arranged at the Colt factory, circa 1865. In the foreground (from the right) are horizontal slotting machines and small lathes; behind these is a row of E. K. Root–designed vertical-shaft edging or profiling machines.
Drawing from Fitch, "Report on the Manufactures of Interchangeable Mechanism," 1883

During and after the machine processing of various parts, which required from twelve to thirty-six separate operations, all underwent meticulous inspection—the other fundamental element of the American System. To quote the journalist, "The tools [pattern, receiving, and limit gauges] to inspect a cylinder, for example, are fifteen in number, each of which must gauge to a hair. The greatest nicety is observed, and it is absolutely impossible to get a slighted piece of work beyond this point."[21]

The degree to which Colt replaced hand labor with machinery was truly unprecedented. During most of the 1850s, the factory's cost of arms manufacture—excluding raw materials and overhead—was figured at machinery, 80 percent; wages for unskilled machine tenders (of whom there were hundreds), 10 percent; and wages for skilled finishers and assemblers (of whom there were comparatively few), 10 percent. Still, recent appraisals agree that although Colt certainly mechanized arms production as never before, and achieved uniformity in component elements, he never sought to attain truly interchangeable parts by machine processing alone. Hand filing still was required in the initial fitting-up process before the parts were hardened, and again in the final assembly and timing of the weapon's action. Colt (and Root) realized that, at a certain point in the production process, the achievement of absolute parts interchangeability did not correspond with their overriding goal of unfettered mass production. Unlike the U.S. Springfield Armory, strict parts interchangeability did not yet pay for a commercial enterprise unsupported by government largess.[22]

Throughout the 1850s, Colt's revolving-cylinder arms commanded the marketplace for repeating weapons. From

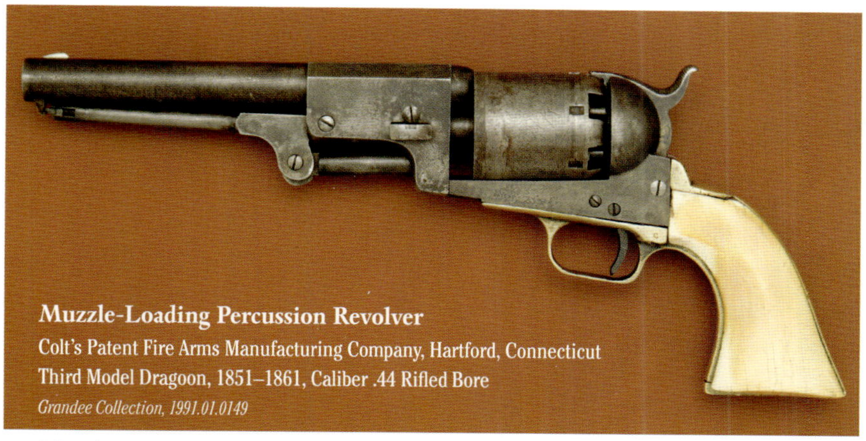

Muzzle-Loading Percussion Revolver
Colt's Patent Fire Arms Manufacturing Company, Hartford, Connecticut
Third Model Dragoon, 1851–1861, Caliber .44 Rifled Bore
Grandee Collection, 1991.01.0149

A favorite among western hunters for "running" buffalo, this large-caliber single-action revolver was powerful enough to kill a bull with one well-placed shot. In 1869, artist-journalist Theodore Davis wrote of the massive handgun in *Harper's Magazine*: "Of the revolvers in use for buffalo hunting, the old style dragoon pistol of the Colt pattern seems the favorite. [It] is heavier and more steady to shoot, and the cylinder [takes] more powder. The Plainsmen hold them in great esteem" (122). This particular model is easily distinguished from earlier variants by its round-backed trigger guard. Approximately 10,500 Third Model Dragoons were made over their ten-year span of production.

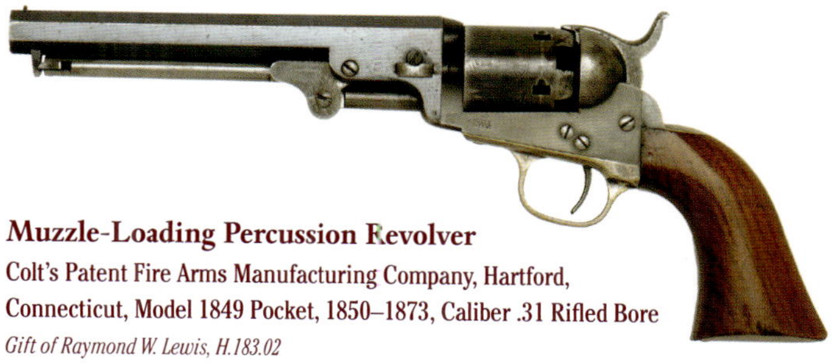

Muzzle-Loading Percussion Revolver
Colt's Patent Fire Arms Manufacturing Company, Hartford, Connecticut, Model 1849 Pocket, 1850–1873, Caliber .31 Rifled Bore
Gift of Raymond W. Lewis, H.183.02

The longest-lived and most popular of Colt's percussion arms, more than 335,000 Model 1849 Pocket revolvers saw production over twenty-three years. This five-shot cylinder, small-trigger-guard variant with a 6-inch octagonal barrel displays standard finish with a case-hardened frame, hammer, and loading lever; blued barrel and cylinder (with roll-died stagecoach holdup scene); silver-plated trigger guard and gripstraps; and one-piece grips of varnished walnut. Providing personal protection with accurate firepower, thousands of these guns accompanied adventurous overlanders and gold seekers into the West. By the mid-1850s, they had largely supplanted the pepperbox pistol.

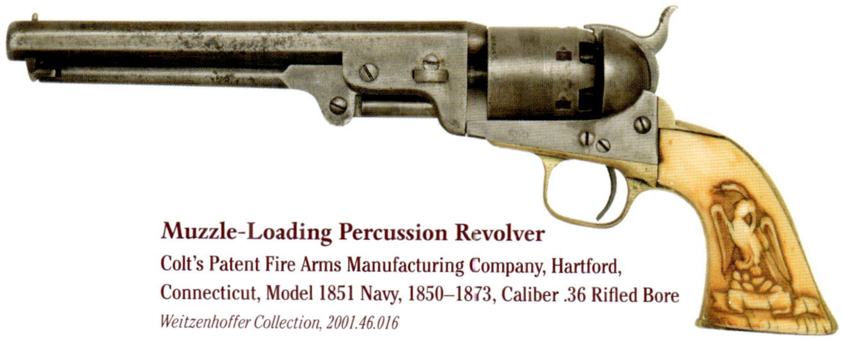

Muzzle-Loading Percussion Revolver
Colt's Patent Fire Arms Manufacturing Company, Hartford, Connecticut, Model 1851 Navy, 1850–1873, Caliber .36 Rifled Bore
Weitzenhoffer Collection, 2001.46.016

Colt's primary medium-caliber handgun throughout the percussion era, the Model 1851 Navy also proved the decided favorite among antebellum westerners—particularly among Texans, as noted by Frederick Law Olmsted. This fourth variation, having the large, round trigger guard of brass, presents standard blued and case-hardened finish with the roll-died naval battle scene on the cylinder. The revolver is fitted with special-order, one-piece grips of raised-carved elephant ivory featuring the Mexican national motif of an eagle with a snake in its beak perched atop a cactus. Over twenty-three years of production, some 215,300 Model 1851s were made at Hartford, with another 42,000 produced at Colt's short-lived London factory.

the large, .44-caliber Dragoon models favored by Indian fighters and sportsmen running buffalo to the diminutive, .31-caliber pocket models carried for personal protection in burgeoning frontier settlements, the Colt revolver reigned supreme in the West. Among the most popular in the antebellum era was the .36-caliber Navy model, of which traveler Frederick Law Olmsted observed, "Of the Colt's [Navy revolver] we cannot speak in too high terms. . . . There are probably in Texas about as many revolvers as male adults, and I doubt if there are one hundred in the state of any other make. . . . A border weapon, so reliable in every sense, would give brute courage to even a dyspeptic tailor."[23] Through the deft marketing of a superior product, Colt handguns had become a ubiquitous facet of American culture by the opening of the Civil War. By the time of Colt's death in 1862, his Hartford enterprise already had made nearly 450,000 arms in a succession of sixteen different models of handguns and longarms for the civilian and military markets alike. (The only near competitor in handgun production at the time was the Remington firm of Ilion, New York.)[24]

The demands of America's Civil War fostered significant advances in arms technology. This proved particularly so in the design of various breech-loading arms and in the development of the self-contained metallic cartridge, which combined the fulminate primer, powder charge, and projectile in a single copper or brass casing that provided a sure gas seal at the breech. The already established principles and methods of standardization and mass production via precision gauging and specialized, sequential machinery found direct application in metallic-cartridge manufacture, as the end product had to consistently meet standard dimensions and ballistic properties. With the

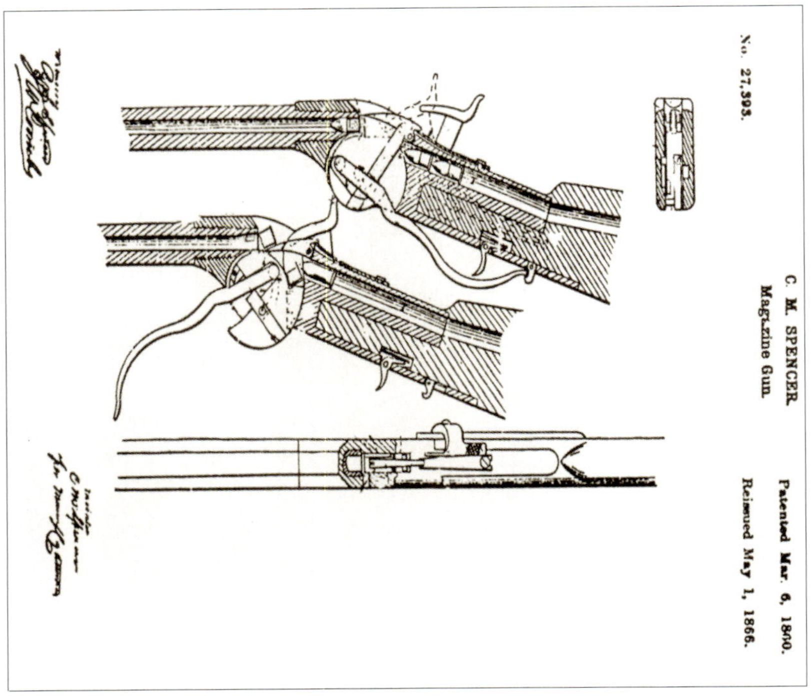

Patent drawing illustrating the internal workings of the Spencer magazine rifle. Note the lever-actuated, rotary breech mechanism and the spring-loaded, tubular cartridge magazine feeding in from the buttstock.
U.S. Patent and Trademark Office, uspto.gov

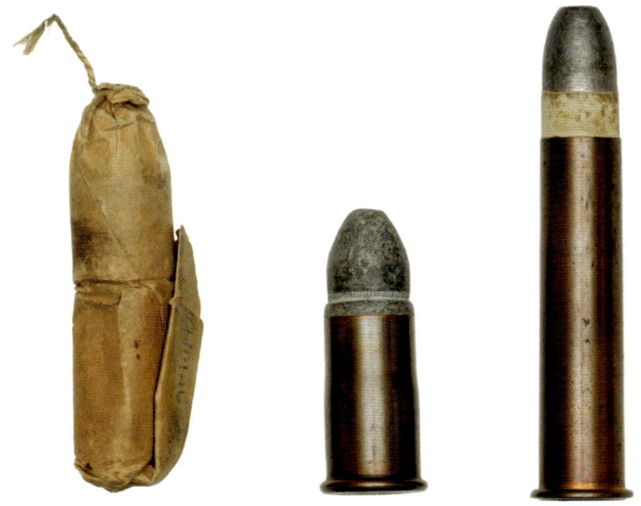

Cartridge Ammunition
Various Manufacturers, 1860–1890
Grandee Collection, Lot 1991.01.3543

These samples illustrate the rapid progression in small-arms ammunition from the loose powder and ball employed in flint and early percussion, muzzle-loading arms to the paper-wrapped and metallic cartridges utilized in muzzle-loading, breech-loading, and repeating firearms. *Left*, a .58 caliber paper cartridge containing a conical bullet and measured charge of black powder requiring a separate fulminate-of-mercury percussion cap for ignition. *Center*, a .56-52 Spencer rimfire, copper-cased cartridge containing a bullet, measured powder charge, and fulminate primer within the rim of the cartridge base. *Right*, a .45-70 Government centerfire, copper-cased cartridge with a paper-patched bullet, measured powder charge, and fulminate primer cap seated in the center of the cartridge base.

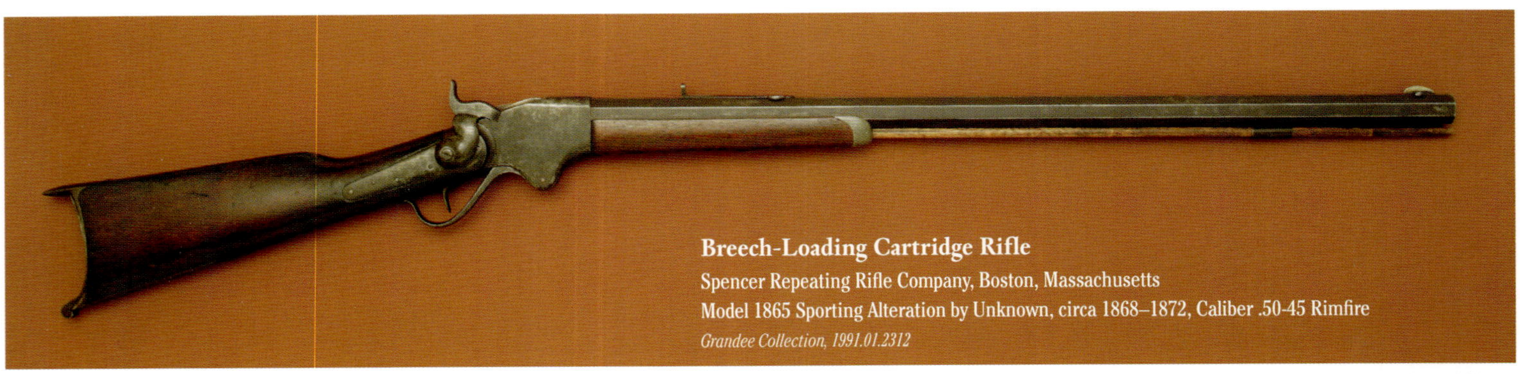

Breech-Loading Cartridge Rifle
Spencer Repeating Rifle Company, Boston, Massachusetts
Model 1865 Sporting Alteration by Unknown, circa 1868–1872, Caliber .50-45 Rimfire
Grandee Collection, 1991.01.2312

A contender with the Henry repeating rifle, the Spencer incorporated a tubular magazine within the buttstock. Hundreds of surplus military Spencers came on the civilian market following the Civil War, and a few—like this specimen—were altered by frontier gunsmiths to sporting arms by fitting the repeating action with a heavy octagonal barrel and new stocking. This example features German silver blade front and adjustable buckhorn rear sights, a fancy pewter forestock cap, and a rather homemade-looking Swiss-pattern buttplate. A few professional hunters used such Spencer repeaters during the early years of the Great Buffalo Hunt, but most favored the stronger, single-shot actions adapted to far more powerful, long-cased centerfire ammunition.

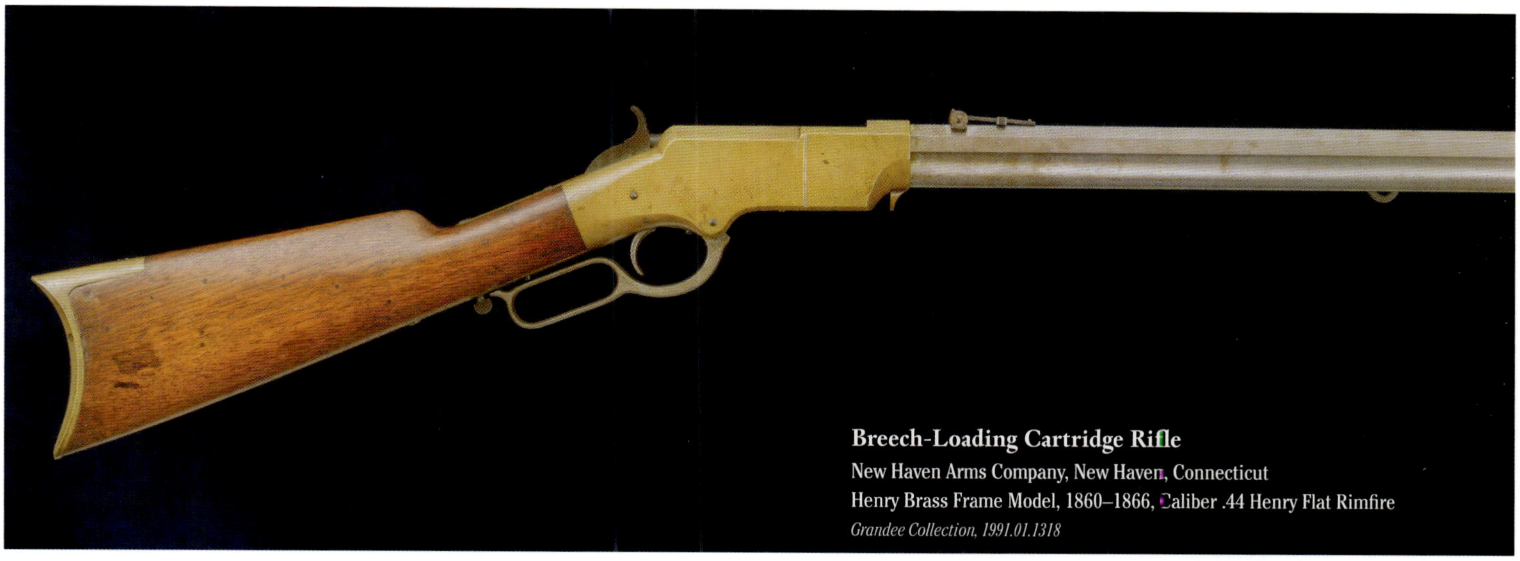

Breech-Loading Cartridge Rifle
New Haven Arms Company, New Haven, Connecticut
Henry Brass Frame Model, 1860–1866, Caliber .44 Henry Flat Rimfire
Grandee Collection, 1991.01.1318

Providing sustained and comparatively reliable firepower, the lever-action Henry repeating rifle incorporated an integral, tubular magazine beneath the barrel that housed up to fifteen metallic cartridges. A direct competitor with the Spencer magazine rifle, the Henry realized considerable commercial success, but the Spencer's production surpassed the Henry's by a ratio of ten to one because of the former's extensive military sales during the Civil War. Nevertheless, the fast-firing Henry garnered sufficient sales and acclaim to underpin a new Winchester enterprise, and it now is recognized as the direct forerunner of that famed line of lever-action repeaters. (The barrel of this piece was draw-filed and buffed sometime in the past to remove corrosion and pitting.)

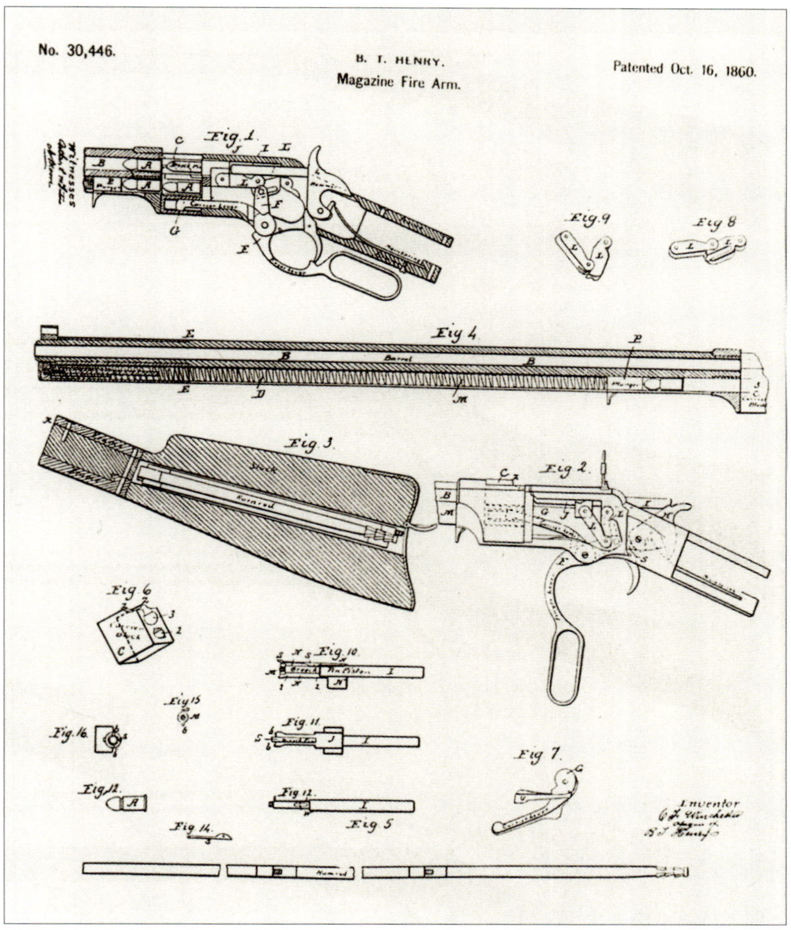

Patent drawing illustrating the internal workings of the Henry magazine rifle. Note the lever-actuated, toggle-link breech mechanism and the integral spring-loaded, tubular cartridge magazine beneath the barrel. (Buttstock recess holds a jointed cleaning rod.)
U.S. Patent and Trademark Office, uspto.gov

relative perfection and mass production of both rimfire and centerfire ammunition by the 1870s, manufacture of the long-traditional muzzle-loading firearm virtually ceased among major arms makers. And the advent of the metallic cartridge did not merely precipitate the widespread adoption of breech-loading firearms; it contributed directly to the practical design of new repeating weapons as well. This was evidenced especially in the design of magazine longarms. Even by 1865, the success of repeating magazine rifles using rimfire cartridges—notably, the lever-actuated Spencer and Henry models—spelled the near demise of revolving-cylinder, percussion repeaters such as Colt's Model 1855 Root rifle. By the latter 1890s, a plethora of metallic-cartridge magazine rifles (as well as a few shotguns) incorporating sophisticated lever-action and slide-action repeating mechanisms had achieved thriving markets with recreational shooters and sport hunters alike.[25] (Although novel bolt-action, repeating magazine rifles of sporting pattern saw production by Remington and Winchester during the 1880s and 1890s, this mechanism type enjoyed little commercial acceptance before the early 1920s, when tens of thousands of American soldiers exposed to the Model 1903 Springfield bolt-action military rifle during World War I returned to civilian pursuits.)

The application of the new metallic-cartridge ammunition to revolving-cylinder handguns also proved rather revolutionary. This involved a now obviously simplistic but nonetheless quite significant adaptation—the introduction of completely bored-through chambers allowing insertion of cartridges at the rear of the cylinder. The rights to this seemingly novel concept, patented by multitalented inventor Rollin White in 1855, were acquired by the gun-making partnership of Smith & Wesson of Springfield, Massachusetts, the following year. Much like Colt's benchmark patent, this significant design detail provided the new partnership with a veritable monopoly in modern revolver manufacture for the succeeding fourteen years. To avoid costly patent-infringement suits, viable competitors such as Colt and Remington had to resort to the conversion of their percussion revolver models through various rather inadequate design modifications. Not until the early 1870s could the domestic rivals compete directly, and by then, Smith & Wesson had established a significant niche in the revolver market both at home and abroad.[26]

The many makes, types, designs, and models of metallic-cartridge arms that proliferated among domestic gun manufacturers from the 1870s onward proved nearly innumerable. Most of the best of these handguns and longarms, which often saw comparatively substantial production and sometimes realized wide acclaim, relied on exclusive patents protecting the particular design features that made them competitive. A representative—though hardly comprehensive—sampling of such firearms produced by several of America's premier manufacturers is presented in part 2.

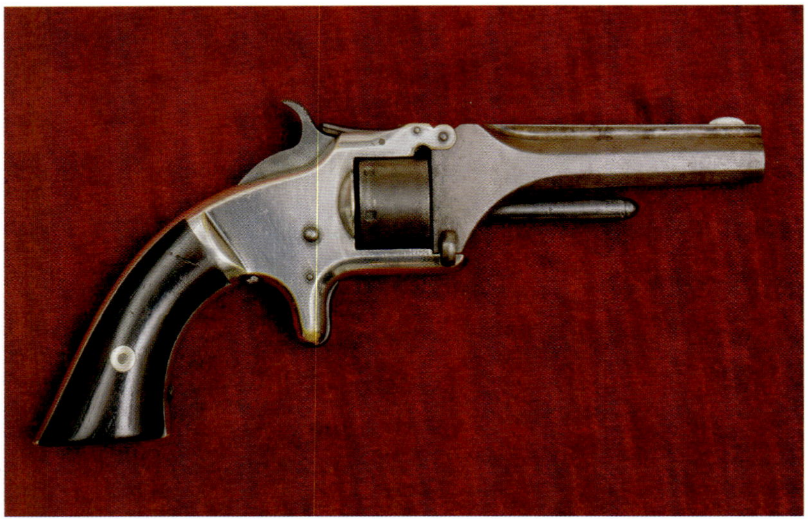

Breech-Loading Cartridge Revolver
Smith & Wesson, Springfield, Massachusetts
Model No. 1 Pocket Second Issue, 1860–1868, Caliber .22 Short Rimfire
Grandee Collection, 1991.01.0153

The smallest of Smith & Wesson's "tip-up" revolvers, the Model No. 1 presented a remarkable advance in firearms technology with its use of self-contained, metallic cartridges in combination with a bored-through, breech-loading cylinder. Although some 117,000 of these pistols were made in only eight years, the diminutive design was deficient for anything beyond close range. In *Roughing It*, budding humorist Samuel Clemens remembered the pistol as a "seven-shooter, which carried a ball like a homeopathic pill, and it required the whole seven to make a dose for an adult" (23).

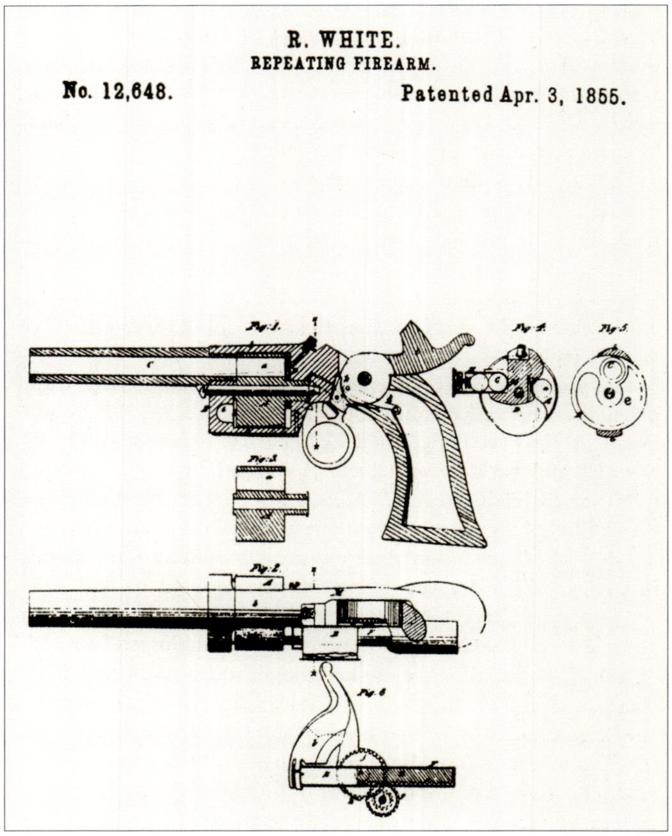

Patent drawing illustrating Rollin White's fundamental claim for revolving cylinder arms having completely bored-through chambers loading self-contained cartridges at the breech.
U.S. Patent and Trademark Office, uspto.gov

Part II

From Colt to Winchester

A Portfolio of Famed American Arms Makers

The establishment of machine-based firearms production in the United States generated remarkable changes in the business of gun making. According to the census report, *Manufactures of the United States in 1860*, the Pennsylvania arms "manufactories" noted in chapter 1 had by 1860 been transformed and all but eliminated as going business concerns. Leman's Conestoga Rifle Works and Bown & Tetley's Enterprise Gun Works—both still concentrating on marketing muzzle-loading, single-shot rifles—had devolved from arms-manufacturing businesses to arms-assembly shops using parts purchased from others. Similarly, the Henry family's famed Boulton Gun Works, though still considered a maker—rather than an assembler—of firearms, retained only six employees and capitalization of only fourteen thousand dollars. In contrast, fully mechanized companies—most of them producing and marketing more complex, patented breech-loading and repeating firearms—flourished in New England. From the middle 1830s onward, private establishments such as Allen & Thurber, Ames Manufacturing, Colt's, Massachusetts Arms, Robbins & Lawrence, and Sharps ascended to totally eclipse the earlier manufactories. In Hartford, Connecticut, for instance, Colt's Patent Fire Arms Manufacturing Company and the Sharps Rifle Manufacturing Company retained about three hundred workers each, and each boasted capitalization in buildings and machinery in excess of a half-million dollars in 1860.[1]

As suggested in part 1, the significance of the individual, small-shop gunsmith markedly declined beneath the rising tide of mass-produced guns that flooded the marketplace from 1850 onward. And this trend ultimately proved true throughout much of American manufacturing. As one twentieth-century historian observed of this industrializing phenomenon, the "concept of standardized and interchangeable parts reversed the normal direction of thought. . . . It remade the economic map of the United States and gave the country a new economic history. It delivered the final

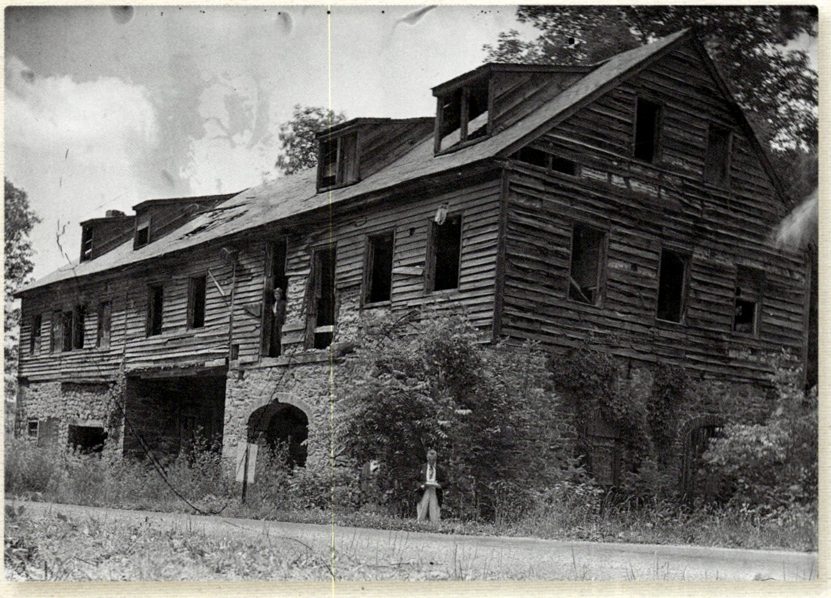

Stanley Jones, photographer. *Henry's Gun Works*. Black-and-white print, WPA Historic American Buildings Survey, 1936. As shown here after having been abandoned for more than fifty years, the derelict Henry family gun manufactory at Boulton, Pennsylvania, epitomizes the decline and ultimate demise of individual and small-factory arms makers owing to the American System's rampant mechanization from the 1850s onward.
Image courtesy Library of Congress Prints and Photographs Division, 136399

blow to handicraft and took manufacture forever out of the home."[2] Though the independent gunsmith never disappeared entirely, many that formerly made complete firearms either went out of business or fell back on the sale and repair of factory-made guns. A few individual smiths with keen design acumen, such as Frank W. Freund, found some prosperity in modifying and improving mass-produced firearms—in Freund's case, the already popular Sharps rifle. John M. Browning, the most ingenious of such independent gunsmiths, created scores of advanced firearm systems that directly impacted the marketing success of major manufacturers Winchester and Colt in the final two decades of the nineteenth century.[3] Such inventive smiths, however, served only to enable and expedite the prevailing trend—the major manufacturers now dominated the commercial production of firearms in the United States. Where the craftsman-gunsmith formerly made arms for individual customers, the mechanized arms factory now made arms for "the market."

The domestic gun-making enterprises that flourished during the latter nineteenth century proved to be highly competitive. All relied on exclusive design patents to establish and maintain a particular demand niche in the growing marketplace, and most sought as well to specialize in—and dominate—particular firearm types. Colt and Smith & Wesson, for example, focused extensively on revolving-cylinder handgun manufacture, while Remington competed, for the most part, in both revolver and single-shot rifle production. The Sharps enterprise devoted itself almost exclusively to the manufacture of single-shot, breech-loading rifles, while the Winchester and Marlin companies both concentrated on the development and production of a succession of improved lever-action, repeating magazine rifles.[4] Samuel Colt had demonstrated particular vigor in protecting his patent rights, especially as manifested in *Colt v. Massachusetts Arms Co.*, and in his attempts to forestall foreign copyists through the selective issuance of *brevete* patent licenses in Europe. Winchester sought market dominance not only through its predecessor patents and its—for a time—exclusive control of John Browning's patent designs but also through the outright elimination of competitor firms by purchase and dissolution. Between 1869 and 1899, Winchester bought out its first rival, the Spencer Repeating Arms Company (then incorporated with the Fogerty Repeating Rifle Company as the American Repeating Arms Company), and followed with the Adirondack Fire Arms Company, the venerable Whitney Arms Company, and the Burgess

Gun Company—one and all manufacturers of repeating longarms.[5]

Colt and Winchester certainly attained the greatest brand recognition among the domestic firearms manufacturers during the last decades of the nineteenth century. This stemmed not only from their business and marketing acumen but also from their products' ubiquity and popularity among the latter-day denizens of the frontier West, where their notoriety achieved almost-mythic proportions. Although there certainly were other legitimate claimants to instrumental roles in the settlement of the West, such as the famed Sharps rifle and the prosaic double-barreled shotgun, the Colt revolver and the Winchester rifle were the arms that ultimately laid contending marketing claims to the title of "the gun that won the West." Today, the firearms of these two companies remain the most treated by historians and scholars, and they also rank as the "blue-chip" marques among contemporary collectors and connoisseurs. Given these historical and popular considerations, the firearms of Colt and Winchester serve here as "bookends" in the following portfolios of domestic makers. Although far from inclusive or exhaustive, the model firearms of each of the commercial manufacturers discussed and illustrated in the following chapters appear, for the most part, in their production chronology from inception through circa 1920.

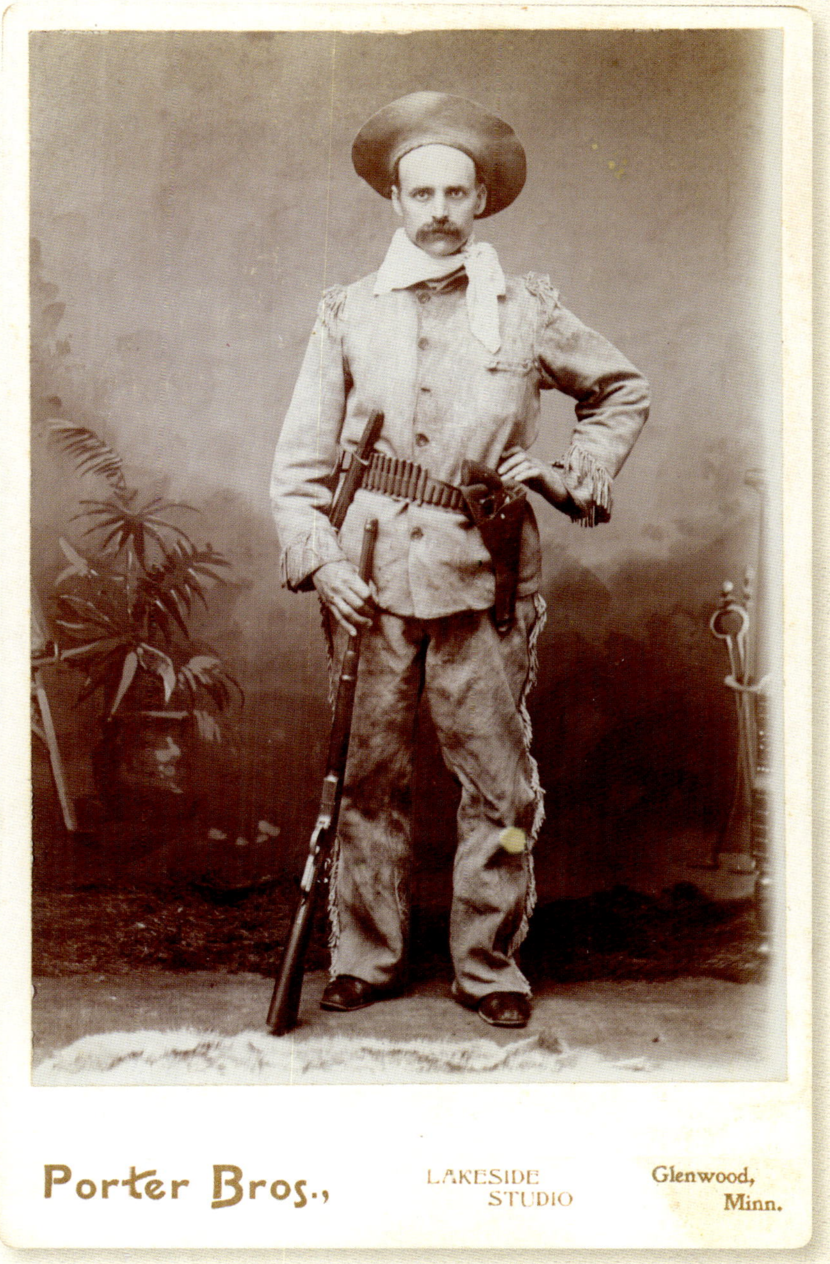

Porter Brothers Lakeside Studio, photographers. [Untitled.] Cabinet card, Glenwood, Minnesota, circa 1880–1890. This frontier type poses with the two marque firearms that later were claimed to have "won the West." The gent holds a Winchester Model 1873 lever-action carbine, while a Colt Model 1873 Single Action Army revolver resides in his open-topped belt holster. (The long-cased cartridges in his belt would fit neither weapon.)

Dickinson Research Center, National Cowboy & Western Heritage Museum, RC2006.038

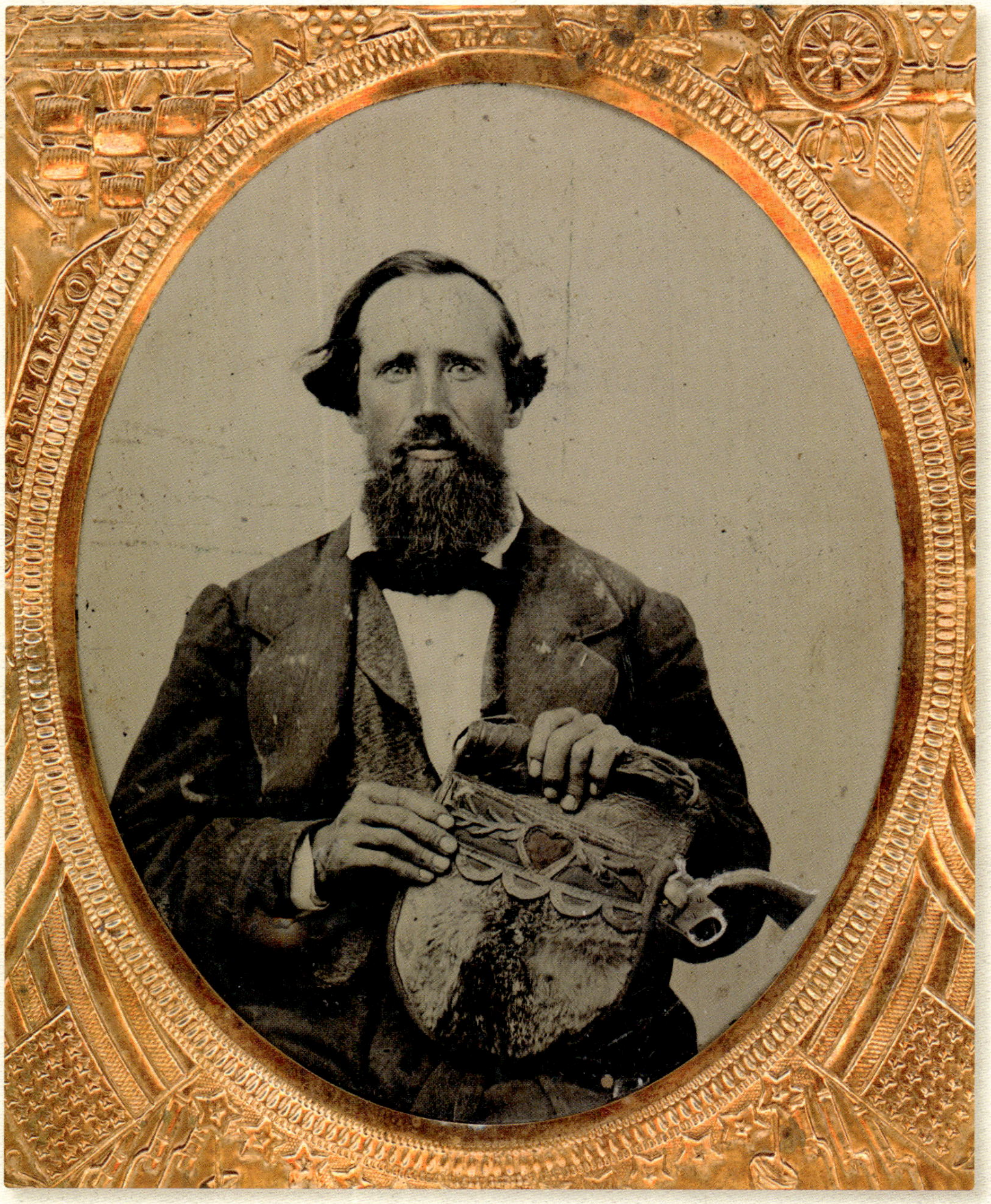

Photographer unknown. [Untitled.] Sixth-plate tintype, America, circa 1860–1865. This formal fellow poses with fancy, Mexican saddle pommel bags, from which a Colt Model 1849 Pocket revolver conspicuously protrudes. Such bags sometimes had holsters sewn beneath their cover flaps. (The stamped-brass, patriotic cover mat on this image is positioned upside down.)

Dickinson Research Center, National Cowboy & Western Heritage Museum, 2002.227

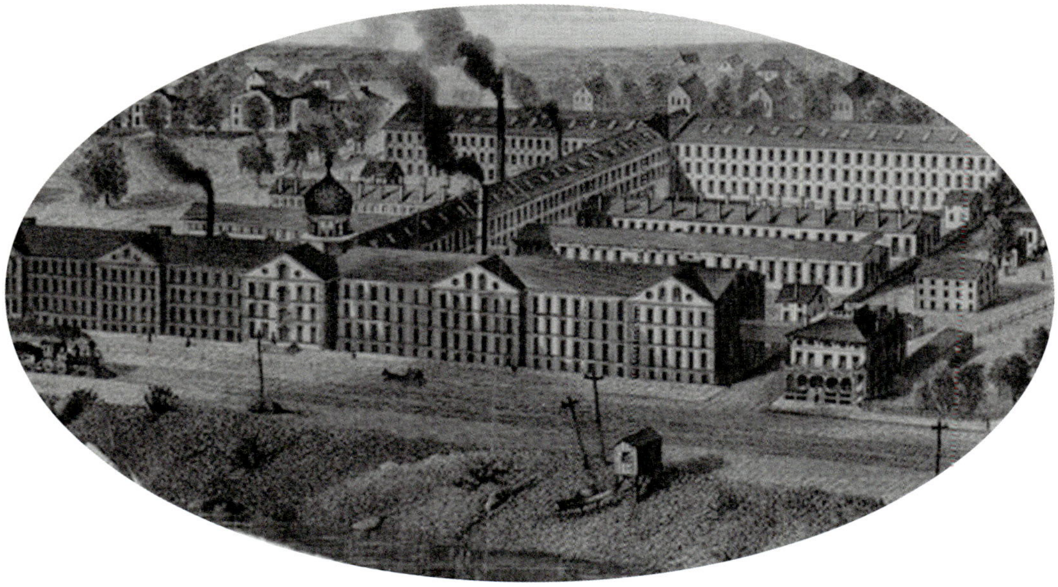

Chapter 3

COLT FIREARMS

SAMUEL COLT (1814–1862) created an industrial empire founded on revolving-cylinder repeating firearms. These innovative weapons had a significant influence on nineteenth-century American history and manufacturing technology. From the early Paterson revolvers that acquired fame with the Texas Rangers; to the Dragoon, Army, and Navy models that helped preserve the Union during the Civil War; to the famed Single Action Army model that "won the West," the Colt firm created many of the arms that aided American expansion, sustained U.S. sovereignty, and defended frontier settlement.[1]

As noted in the previous chapter, although Colt's Paterson enterprise failed in achieving a sustainable market, its arms did establish a reputation for firepower on the frontier. In addition to the notoriety gained in Texas by the No. 5 Holster Model revolver, the Model 1839 Paterson-Colt carbine won several western devotees for its potent firepower. Along the Santa Fe Trail in 1840, for example, Josiah Gregg carried one of these six-shot, .525-caliber arms, which, he observed, "constituted a capacity of defense rarely matched even on the Prairies."[2] The total production of Paterson-Colt firearms (including pocket, belt, and holster revolvers, as well as rifles, carbines, and shotguns) amounted over five years to only some 2,850 handguns and 1,800 longarms.[3] Today, after more than 170 years of attrition, all are considered decidedly rare and desirable.

Colt set up his second gun-making enterprise in 1848 at Hartford, Connecticut, marketing an improved series of percussion revolvers. From the powerful .44-caliber Dragoon to the handy .36-caliber Navy to the diminutive .31-caliber Baby Dragoon and Pocket models, these handguns dominated both military and civilian sales throughout the 1850s—and all but eliminated the once-popular pepperbox pistol. In 1855 Colt incorporated the burgeoning business as Colt's Patent Fire Arms Manufacturing Company; opened a huge new factory in Hartford's South Meadows;

The Colt factory, woodcut print, circa 1860.
Image courtesy private collection

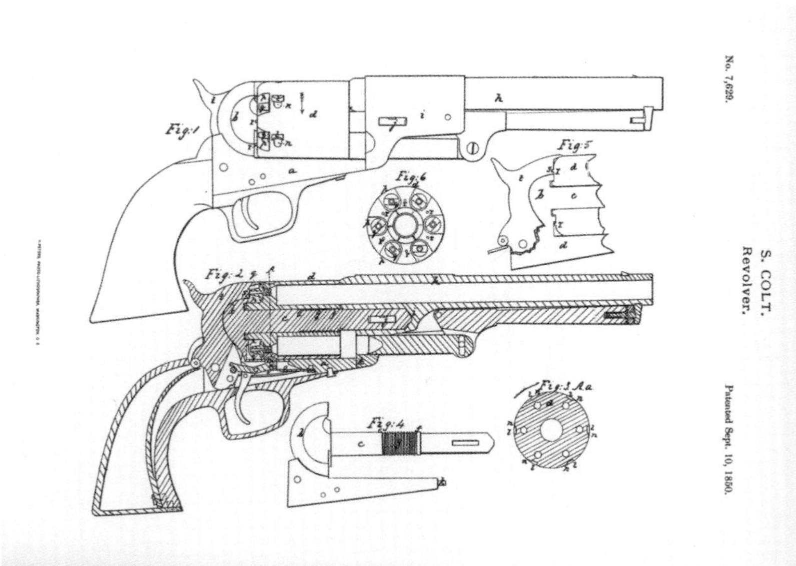

Patent drawing illustrating the internal workings of, and improvements applicable to, the Colt Dragoon, Navy, and Pocket single-action model revolvers.
U.S. Patent and Trademark Office, uspto.gov

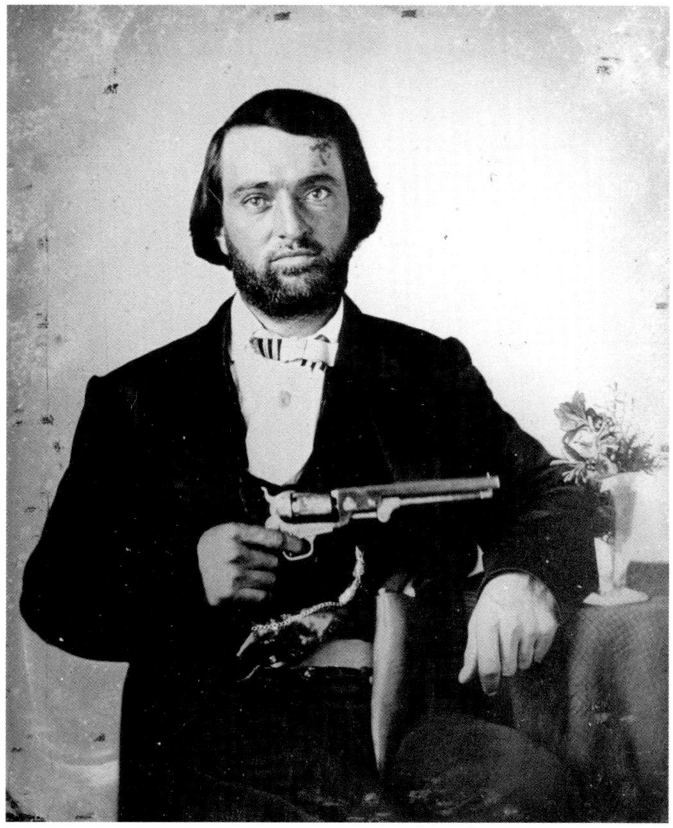

Photographer unknown. [Untitled.] Sixth-plate ambrotype, America, circa 1855–1860. Seated in the studio for a formal portrait, this triple-armed gentleman strikes a common antebellum pose with a Colt Model 1851 Navy revolver prominently displayed across his midsection. (Flowers lend a pleasant—if incongruous—air.) Throughout the 1850s, Colonel Colt's percussion revolvers dominated the domestic market for repeating handguns.
Image courtesy private collection

and introduced a series of novel solid-frame, side-hammer, "Root" revolving-cylinder pocket pistols, rifles, carbines, muskets, and shotguns. Perhaps even more significant to firearms design, in the early 1860s the firm brought out a remodeled series of lightened and streamlined New Model Army, Navy, and Police revolvers that reflected the company's mastery of advanced design and practical metalworking technology. By the time the last of these Colt percussion-model arms were discontinued in 1873, the enterprise had manufactured and marketed more than 900,000 firearms.[4]

Having gained widespread acclaim—and demand—from his presentation of weaponry at London's Great Crystal Palace Exhibition of 1851, Colt opened a London factory in 1853 to serve the British and Continental markets—and perhaps to better thwart foreign patent infringements as well. The facility operated only through 1857, fabricating about 11,000 Model 1849 Pocket and 42,000 Model 1851 Navy revolvers. As well, about 700 Dragoon Model revolvers (known today as the Hartford-English Dragoon) were finished, assembled, proofed, and marketed in London using component parts manufactured in Connecticut. These London-made revolvers usually equaled the quality of their American counterparts, and they attained a rather substantial clientele among the British Empire's far-flung soldiers, adventurers, and colonists. Although the Colt London factory ultimately proved to be technically and financially impractical, the company maintained marketing offices in London into the early decades of the twentieth century.[5]

Anticipating the ascendance of self-contained cartridge ammunition, in 1855 former Colt employee Rollin White received patent rights for revolving-cylinder arms having chambers that were entirely bored through for loading at the breech. Colt evidently ignored the significance of this fundamental design attribute, leaving the nascent Smith & Wesson enterprise to gain control of the patent and allowing it to become Colt's chief rival in revolver manufacture for more than a century. Colt's, thus, was compelled to convert much of its established product line to breech-loading through a variety of alterations devised by talented engineers F. Alexander Thuer, C. B. Richards, and William Mason. Embodying the firm's transition from muzzle-loading percussion arms to breech-loading cartridge arms, these conversions proved a frugal means of consuming otherwise obsolete parts. Many of these arms went to secondary markets in South and Central America and Mexico. Between 1869 and 1880, about 41,000 percussion 1851 Navy, 1860 Army, 1861 Navy, and 1862 Police and [1865] Pocket Navy revolvers were so converted to accept rimfire and centerfire cartridges.[6]

The first Colt firearms actually designed to utilize metallic-cartridge ammunition consisted of a series of small pistols of both single-shot and repeating design. Marketed between 1870 and 1912, these handguns included large-bore deringers (named generically for the pattern popularized by Henry Deringer of Philadelphia in the early 1850s); small-bore, open-topped pocket revolvers; and diminutive, solid-frame revolvers ranging in caliber from .22 to .41. Marketed as personal protection weapons, these Colts acquired a significant clientele among men and women of varying social persuasion, attaining a total production of 286,000 arms over forty-two years. Of satisfactory quality, the entire sequence nevertheless fell victim to the proliferation of inexpensive and often-inferior competitor arms today termed "suicide specials."[7]

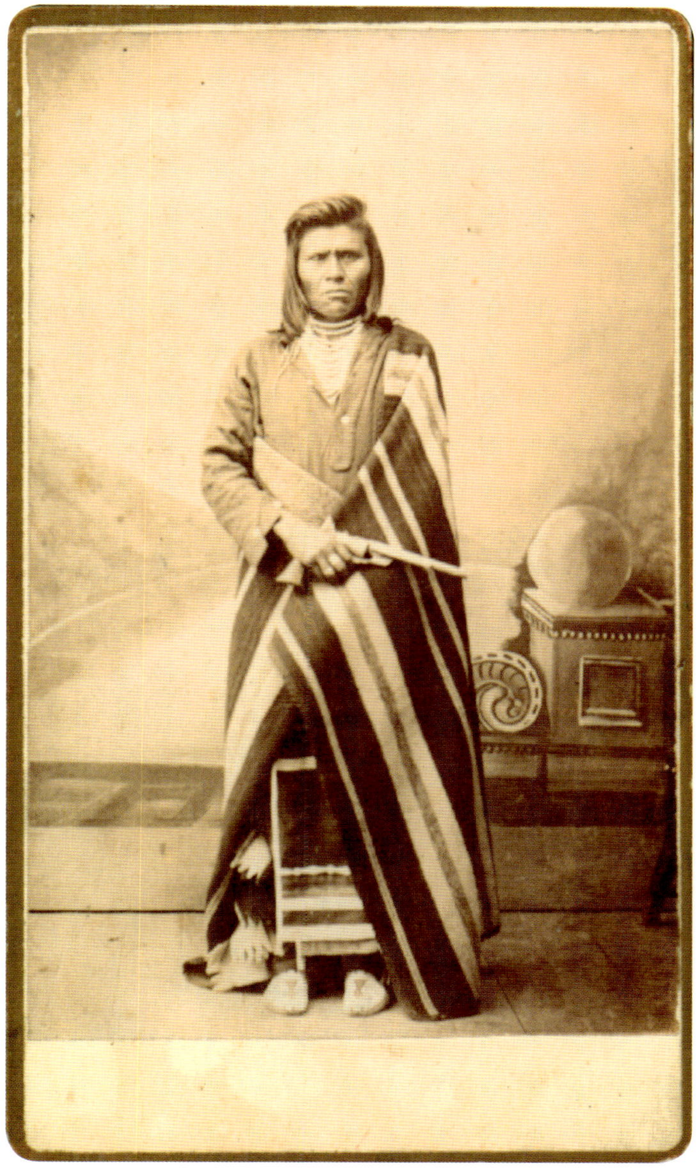

H. Beck, photographer. [*Wolf Head.*] Carte de visite, Winfield, Kansas, circa 1880. Defeated and temporarily deported from his native land, this undaunted and stern-visaged Nez Perce man poses in tribal costume and prominently holds a Colt Model 1860 Army, Richards-conversion revolver.

Dickinson Research Center, National Cowboy & Western Heritage Museum, 2004.015.3

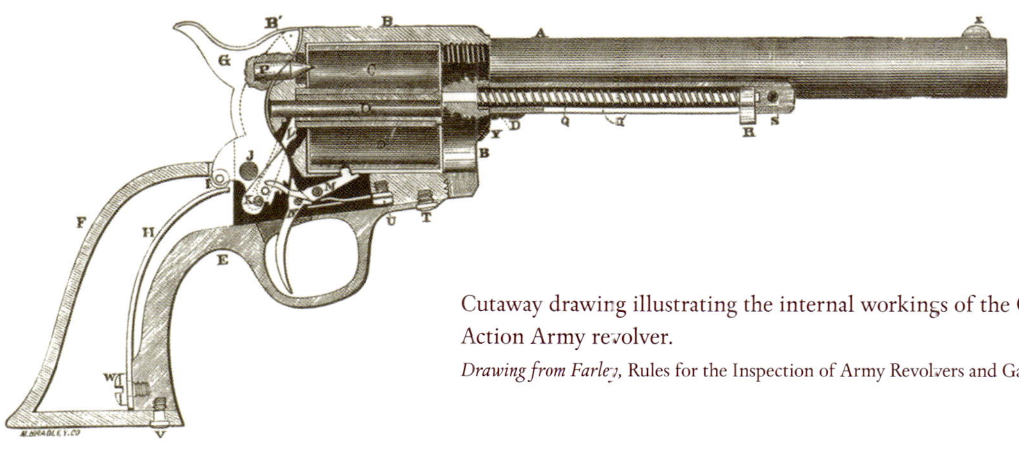

Cutaway drawing illustrating the internal workings of the Colt Single Action Army revolver.

Drawing from Farley, Rules for the Inspection of Army Revolvers and Gatling Guns, *1875*

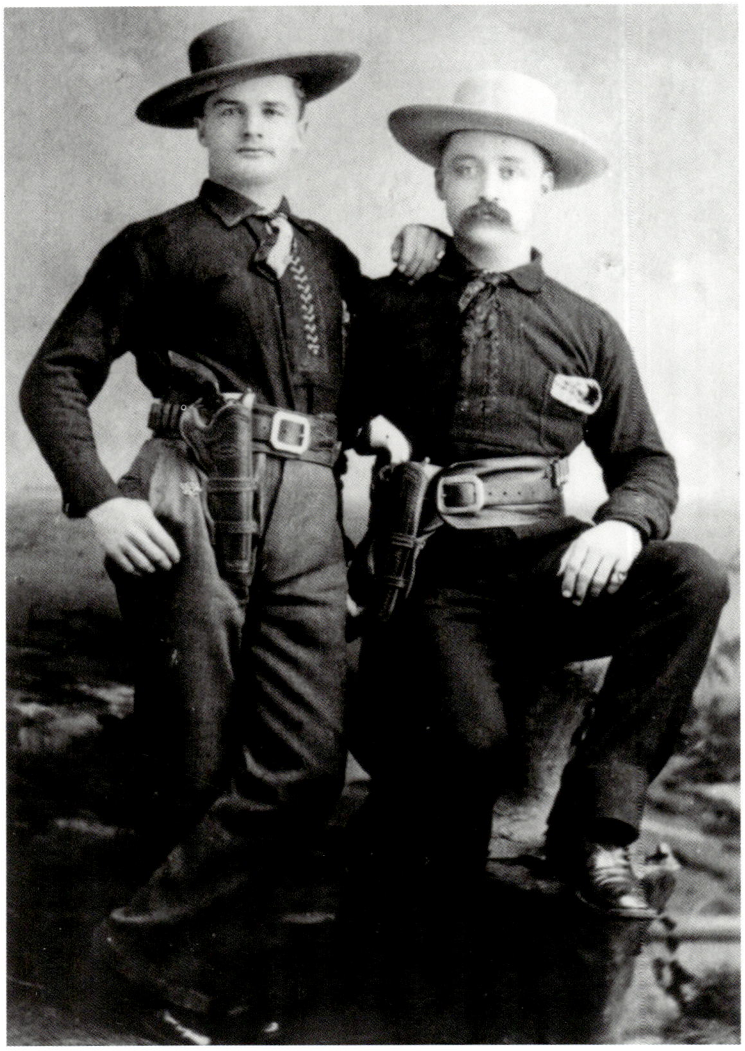

Photographer unknown. [Untitled.] Cabinet card, America, circa 1885–1890. Believed to have posed in Cheyenne, Wyoming Territory, these wonderfully unweathered fellows model their brand-new "cowboy" garb and equipage. Both of their loop-pattern holsters carry Colt Model 1873 Single Action Army revolvers—that at left with checkered, hard rubber grips; that at right with elephant ivory grips possibly adorned with a silver butt cap.
Image courtesy private collection

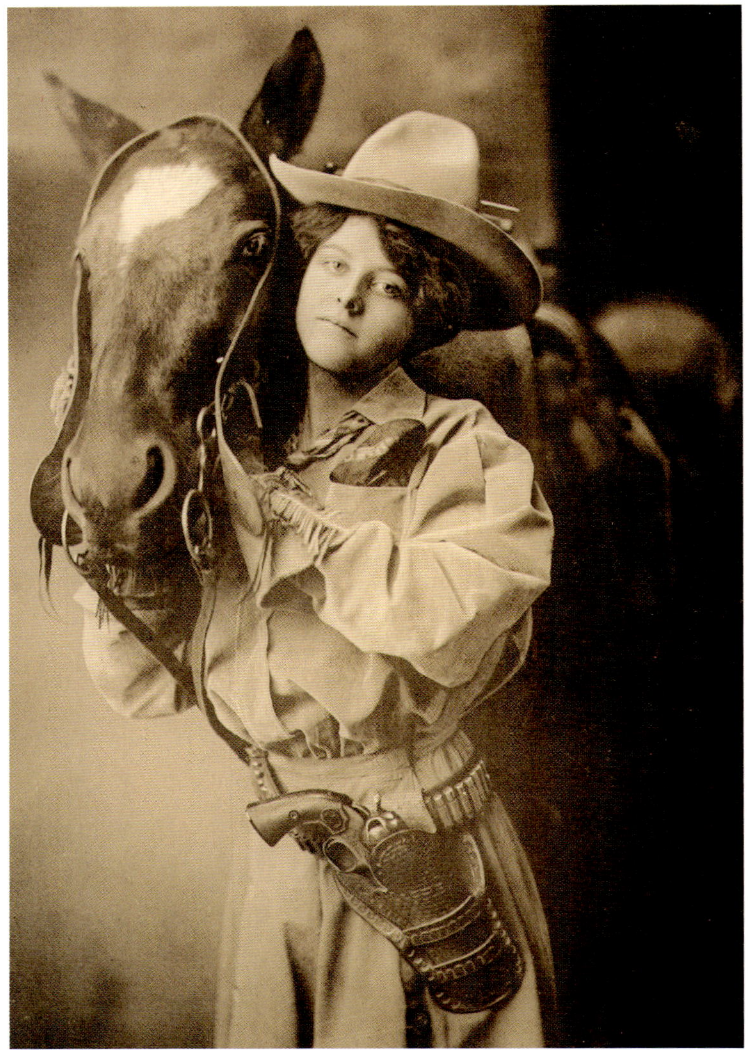

George B. Cornish, photographer. *Belle of the Ranch*. Sepia print, Arkansas City, Kansas, circa 1905. In this photographic interpretation of the iconic "Cow-Boy Girl," the semicomely maiden prominently displays a holstered Colt Bisley Model Single Action Army revolver—identified by its elongated grip profile—at her hip. Her affectionate equine companion appears to be unarmed.
Dickinson Research Center, National Cowboy & Western Heritage Museum, 2003.077

In 1872 Colt manufactured its first revolver expressly for large-caliber, metallic-cartridge ammunition. A transitional arm, the Model 1872 Open Top .44, met with little acclaim because of its percussion-era design. The following year, however, the Hartford company introduced by far the most popular weapon in its storied history—the solid-frame Model 1873 Single Action Army revolver. Known generically as the Peacemaker, Plough-handle, Equalizer, or simply Six-shooter, the basic design (ultimately available in several configurations and in calibers ranging from .22 to .476 Eley) became an icon of the frontier West in both the military and the civilian spheres. In 1878, chambering for the .44-40 centerfire cartridge (that is, a .44-caliber projectile propelled by forty grains of black powder) was introduced, allowing westerners to carry a single type of ammunition for the Colt Single Action Army and the equally popular Winchester Model 1873 carbine or rifle. An English correspondent for *The Field* magazine wrote in 1884, "[It] is the special weapon of the American Cavalry, and of hunters, trappers, cowboys, ranchmen and miners of the frontier, who may be credited with knowing what a revolver should be."[8] (Such popularity continued through much of the twentieth century in Hollywood's celluloid West, where the Colt Single Action Army proved ubiquitous in epics and oaters alike.) Notable Single Action variants included the Sheriff's or Storekeeper's Model, the Flattop Target Model, and the Bisley Model. Production of this now-legendary revolver amounted to some 192,000 arms by 1900 and exceeded 357,000 from its start in 1873 to its first discontinuance in 1940.[9]

From the later 1870s through the 1890s, Colt's also introduced a series of double-action revolvers, in which hammer cocking, cylinder advancement, and discharge were accomplished simply by squeezing the trigger. These included the Model 1877 "Lightning" and Model 1878 "Frontier," both of which gained adherents in the latter-day frontier West and enjoyed production runs of 166,000 and 51,000 to their respective discontinuances in 1909 and 1905. Colt's brought out its first swingout-cylinder revolver design in 1889 and in 1898 introduced its largest-ever solid-frame, double-action, swingout-cylinder handgun—the massive New Service model. Like the Single Action Army, this revolver included a special flattop-target variant and came in a range of potent calibers. Intended largely for military service, more than 356,000 were manufactured from 1898 to the revolver's discontinuance in 1944.[10]

Colt's had discontinued the manufacture of commercial longarms (the 1855 Root models) during the Civil War, but it reentered the market in the late 1870s and 1880s. First came the Model 1878 (hammer) and the Model 1883 (hammerless) double-barrel shotguns, of which some 30,000 saw production. A mere handful of extremely rare double-barreled sporting rifles also were made on the 1878 shotgun frame at the special order of Colt scion Caldwell Colt. (Denver, Colorado, firearms dealer John P. Lower declared his specimen "the Express Dbl Rifle of the Age.")[11] Colt's brought out two innovative repeating rifles in the 1880s—a lever-action devised by prolific designer Andrew Burgess and a slide-action designed by William Elliot (formerly associated with Remington). Although these arms embodied the excellent quality typical of Colt's, they made only moderate inroads into a market dominated by established manufacturers Winchester and Marlin. The Colt-Burgess lever-action rifle dropped from production after two years and less than 6,500 guns—purportedly in response to direct competitive pressure brought by Winchester. Between

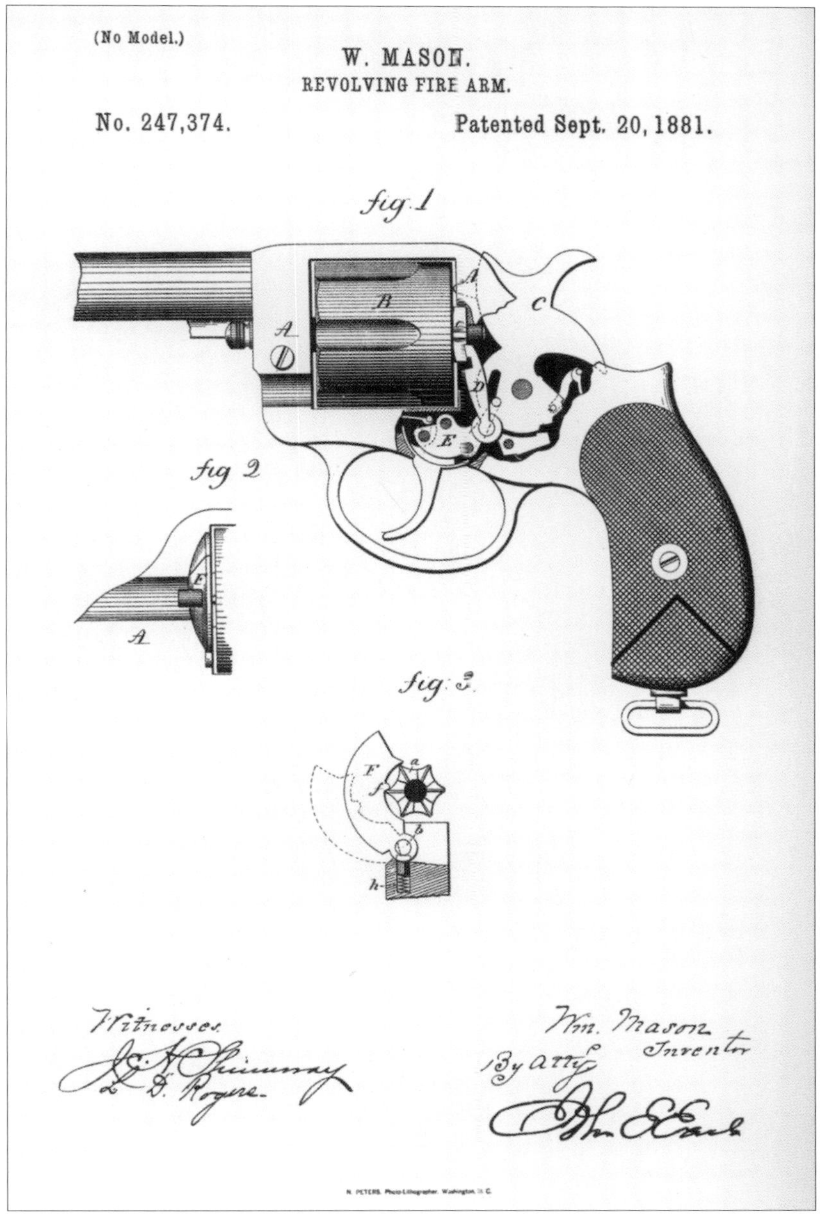

Patent drawing illustrating the internal workings of the Colt Lightning and Frontier double-action revolvers.
U.S. Patent and Trademark Office, uspto.gov

1884 and 1904, however, the fast-firing Colt Lightning slide-action rifle, which was produced in three frame sizes for ammunition ranging from .22 rimfire to .50-95 Express centerfire, saw production of more than 186,000 firearms—the medium- and large-frame variants finding some favor among latter-day western sport hunters.[12]

By the close of the nineteenth century, Colt's Patent Fire Arms Manufacturing Company had produced nearly two million commercial firearms encompassing some forty distinct models over a period of fifty-two years. This was a manufacturing and marketing record approached only by Winchester. Today, the storied firm continues comparatively limited handgun production as the Colt's Manufacturing Company, LLC, in West Hartford, Connecticut.

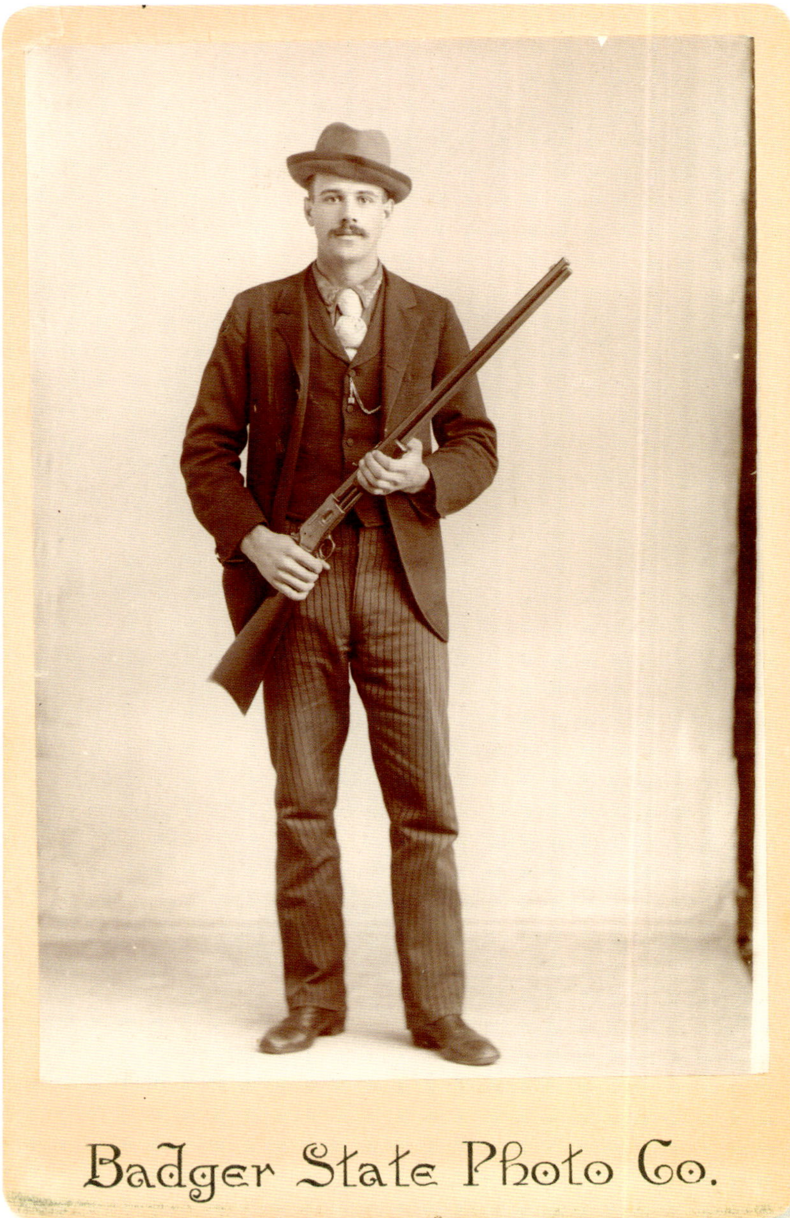

Badger State Photo Company. [Untitled.] Cabinet card, Wisconsin, circa 1900. Certainly dressed for business in tie, vest, and jacket, this young gentleman holds a Colt Lightning medium-frame, slide-action repeating rifle of standard configuration. Because the medium-frame model shared chamberings with the Single Action Army revolver, it enjoyed quite-substantial sales.
Dickinson Research Center, National Cowboy & Western Heritage Museum, 2003.008

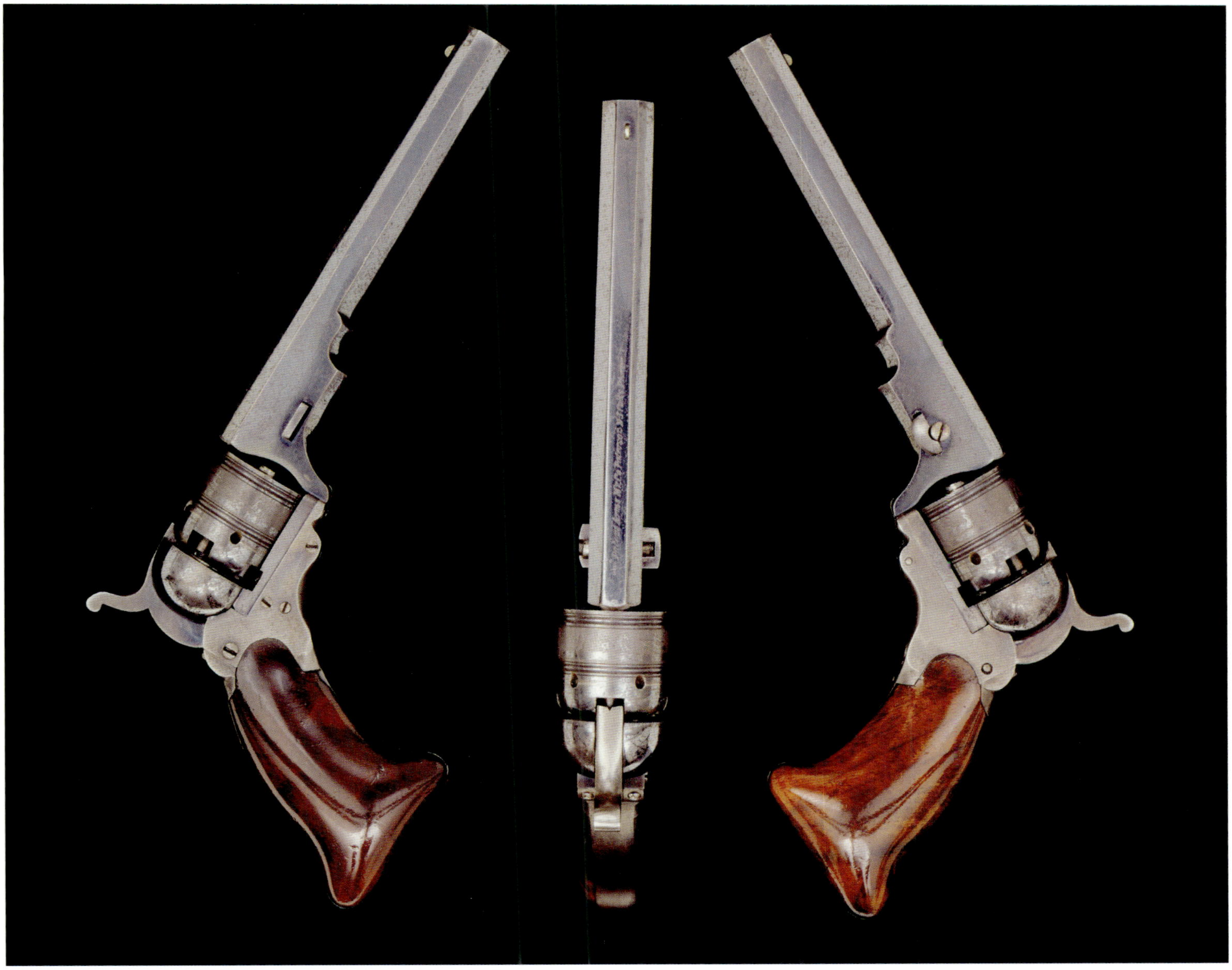

Muzzle-Loading Percussion Revolver
Patent Arms Manufacturing Company, Paterson, New Jersey
Colt-Paterson Belt Model No. 3, 1837–1840 Caliber .31 Rifled Bore
Weitzenhoffer Collection, 2001.46.001

Distinguished by their flared grip profiles, fewer than 500 of the No. 3 Belt Model revolvers saw manufacture. The firearm's five-shot, round-shouldered cylinder bears Waterman L. Ormsby's distinctive, roll-died scene of a dueling centaur and the firm's early, four-horsehead trademark device (see page 170). The 5½-inch octagonal barrel carries the typical marking: "Patent Arms M'g Co. Paterson N-J. Colt's Pt." This example features a blued barrel, cylinder, and recoil shield; case-hardened hammer, frame, and gripstraps; and one-piece grips of varnished walnut. Merging a fair measure of machine-production with considerable handwork, Paterson revolving firearms set a high standard for quality and finish during the late 1830s.

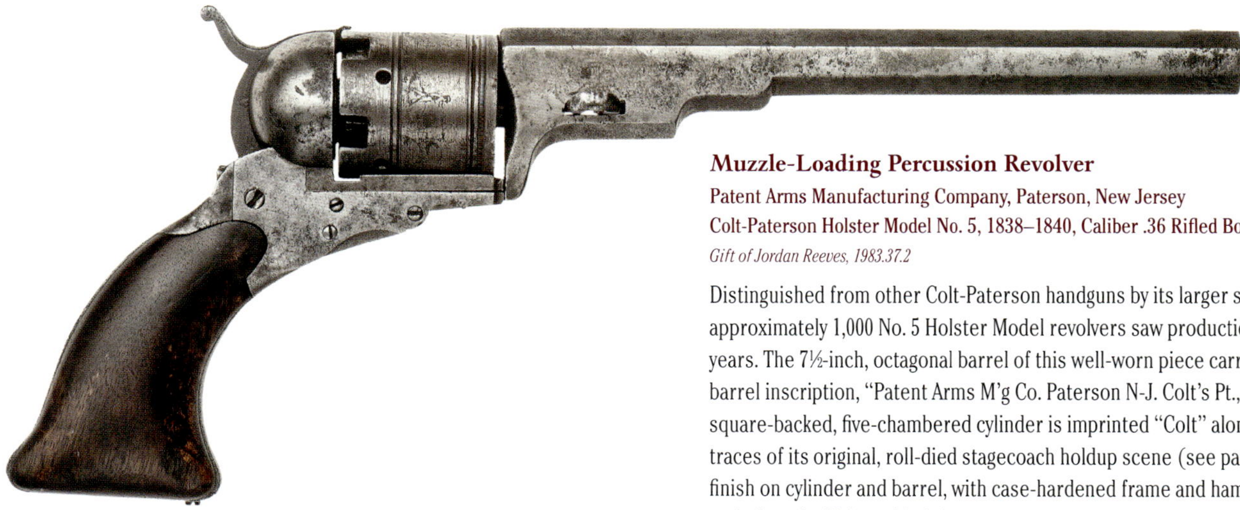

Muzzle-Loading Percussion Revolver
Patent Arms Manufacturing Company, Paterson, New Jersey
Colt-Paterson Holster Model No. 5, 1838–1840, Caliber .36 Rifled Bore
Gift of Jordan Reeves, 1983.37.2

Distinguished from other Colt-Paterson handguns by its larger size and caliber, approximately 1,000 No. 5 Holster Model revolvers saw production over three years. The 7½-inch, octagonal barrel of this well-worn piece carries the typical barrel inscription, "Patent Arms M'g Co. Paterson N-J. Colt's Pt.," while the square-backed, five-chambered cylinder is imprinted "Colt" along with faint traces of its original, roll-died stagecoach holdup scene (see page 170). Blued finish on cylinder and barrel, with case-hardened frame and hammer, was typical on the Holster Model, as were varnished walnut grips of flared profile. Like other single-action Paterson-model revolvers, the folding trigger dropped from the frame when the hammer was cocked.

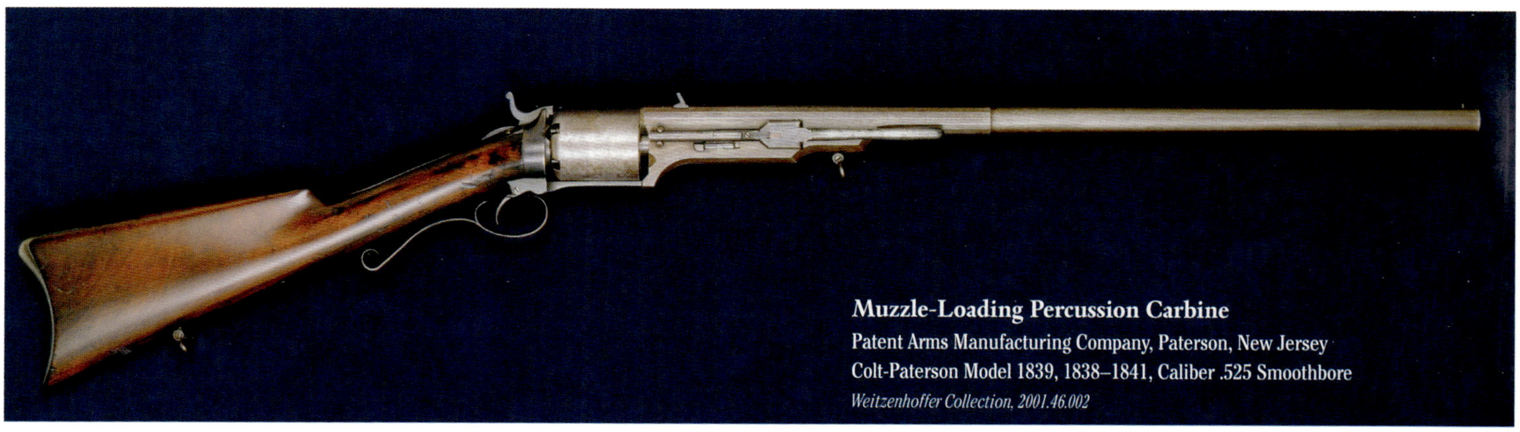

Muzzle-Loading Percussion Carbine
Patent Arms Manufacturing Company, Paterson, New Jersey
Colt-Paterson Model 1839, 1838–1841, Caliber .525 Smoothbore
Weitzenhoffer Collection, 2001.46.002

Certainly the most practical and popular of all the Colt-Paterson longarms, approximately 950 Model 1839 carbines saw production over three years. Fitted with a 24-inch round barrel and attached loading lever, this example includes one of the rare unfinished replacement cylinders made without the usual roll-engraved panel scenes of a lion hunter, a naval battle, and a land battle. The carbine features a browned barrel, a blued frame with the cylinder "in-the-white," and a buttstock of varnished walnut. Artist George Catlin produced a series of six lithographic images promoting Colt arms in the mid-1850s, one of which portrayed him in the West shooting buffalo in rapid succession with his Model 1839 Paterson carbine, which he had affectionately named "Sam Colt."

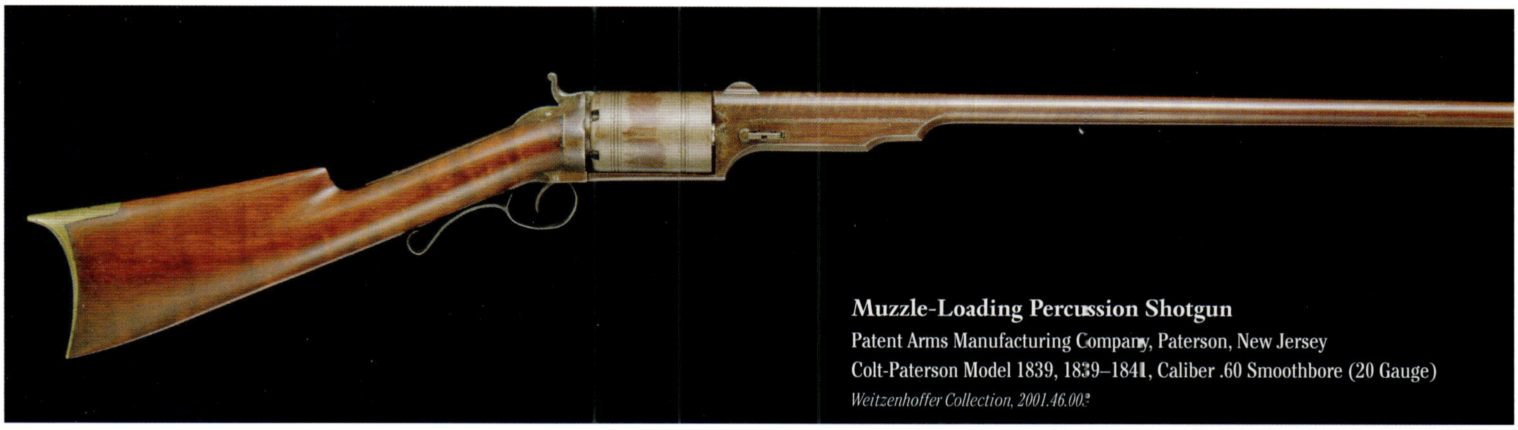

Muzzle-Loading Percussion Shotgun
Patent Arms Manufacturing Company, Paterson, New Jersey
Colt-Paterson Model 1839, 1839–1841, Caliber .60 Smoothbore (20 Gauge)
Weitzenhoffer Collection, 2001.46.003

A comparative rarity in Colt longarm production, only about 225 Model 1839 shotguns were manufactured prior to the demise of the Paterson enterprise. The shoulder arm's massive, six-shot cylinder measures 3½ inches in length and carries ornate, roll-died panel scenes of a running stag, a Native American, and a patriotic American-eagle-and-shield motif (see page 170). The shotgun's 32-inch round barrel is browned with a faux-Damascus finish. The lightly striped, American walnut buttstock is fitted with an unusual rifle-pattern buttplate of brass. The design had some potential as a repeating military musket, but the company's failure in 1842 precluded any U.S. government contracts.

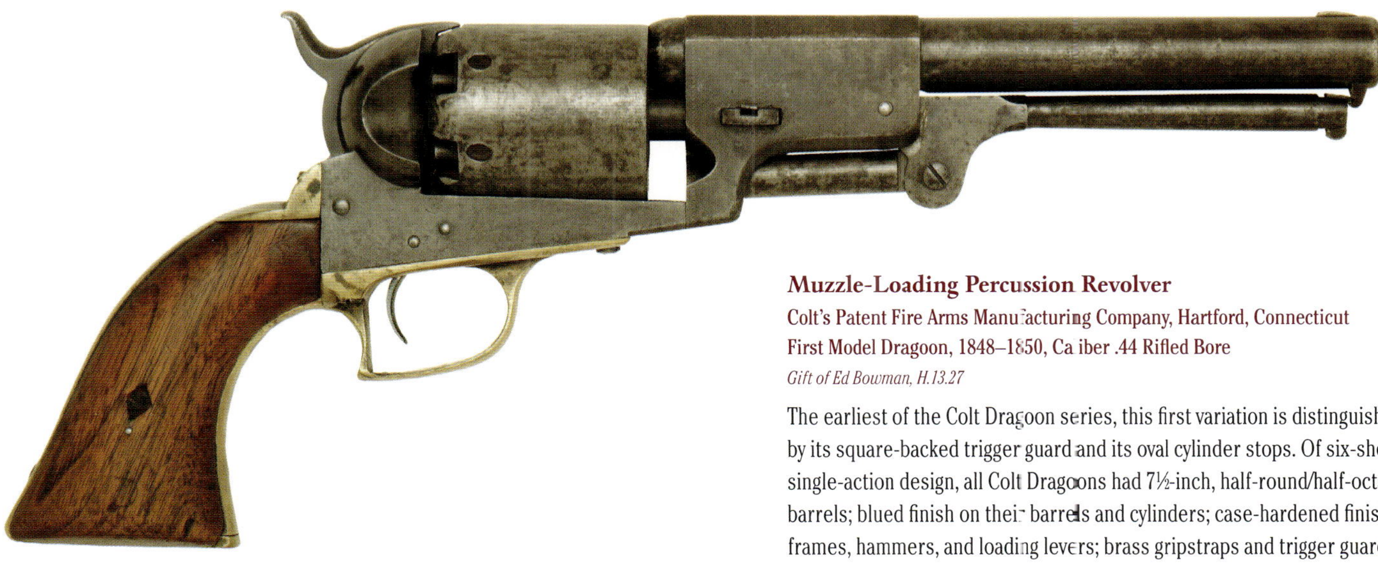

Muzzle-Loading Percussion Revolver
Colt's Patent Fire Arms Manufacturing Company, Hartford, Connecticut
First Model Dragoon, 1848–1850, Caliber .44 Rifled Bore
Gift of Ed Bowman, H.13.27

The earliest of the Colt Dragoon series, this first variation is distinguished by its square-backed trigger guard and its oval cylinder stops. Of six-shot, single-action design, all Colt Dragoons had 7½-inch, half-round/half-octagon barrels; blued finish on their barrels and cylinders; case-hardened finish on frames, hammers, and loading levers; brass gripstraps and trigger guards (silver-plated on civilian production); and one-piece grips of varnished or oiled walnut. Overall surface wear—and the rather flat, period replacement grips—suggests that this military specimen experienced considerable frontier service. About 7,000 of these massive, 4-pound revolvers were produced before the second model variant was introduced.

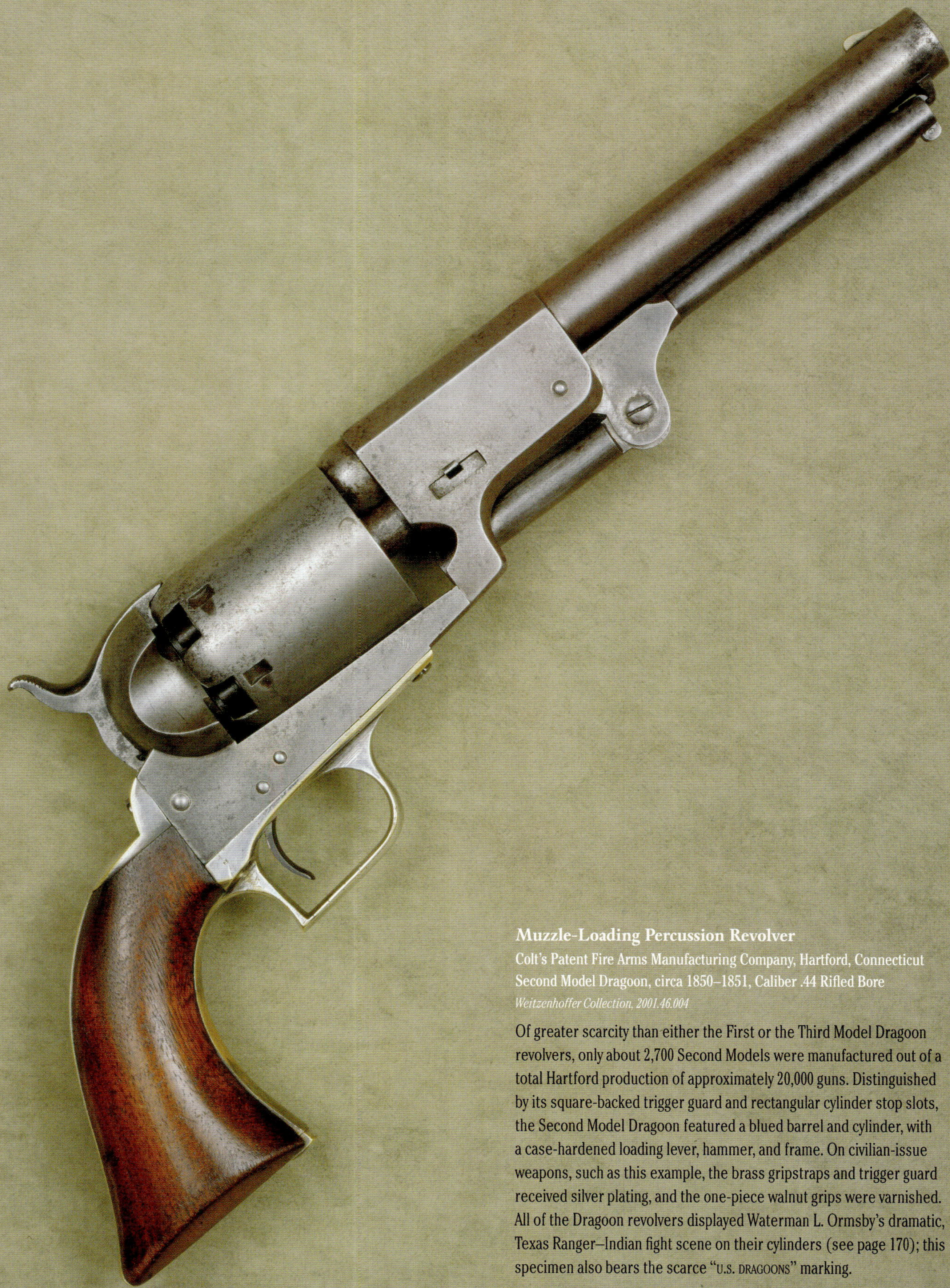

Muzzle-Loading Percussion Revolver

Colt's Patent Fire Arms Manufacturing Company, Hartford, Connecticut
Second Model Dragoon, circa 1850–1851, Caliber .44 Rifled Bore
Weitzenhoffer Collection, 2001.46.004

Of greater scarcity than either the First or the Third Model Dragoon revolvers, only about 2,700 Second Models were manufactured out of a total Hartford production of approximately 20,000 guns. Distinguished by its square-backed trigger guard and rectangular cylinder stop slots, the Second Model Dragoon featured a blued barrel and cylinder, with a case-hardened loading lever, hammer, and frame. On civilian-issue weapons, such as this example, the brass gripstraps and trigger guard received silver plating, and the one-piece walnut grips were varnished. All of the Dragoon revolvers displayed Waterman L. Ormsby's dramatic, Texas Ranger–Indian fight scene on their cylinders (see page 170); this specimen also bears the scarce "U.S. DRAGOONS" marking.

Colt Revolver Cases

Between 1837 and about 1870, Colt percussion revolvers commonly could be purchased with special-order cases. American-made examples typically featured rosewood or mahogany boxes with baize or velvet linings, while those made in London for the British market usually were of oak with baize linings. Single (or, considerably more rare, double) casings had compartmented interiors designed to house the revolver(s) and all the necessary loading accessories—a bullet mold, powder flask, and combination tool, along with percussion cap tin, loose bullets, or packages of combustible cartridges. Regular demand for these custom cases declined following the widespread adoption of metallic-cartridge handguns around 1870, because fewer accessories were required for their use. Still, presentation casings continued to be marketed in limited numbers into the twentieth century, particularly to those in Great Britain who more ardently appreciated long-standing custom and tradition.

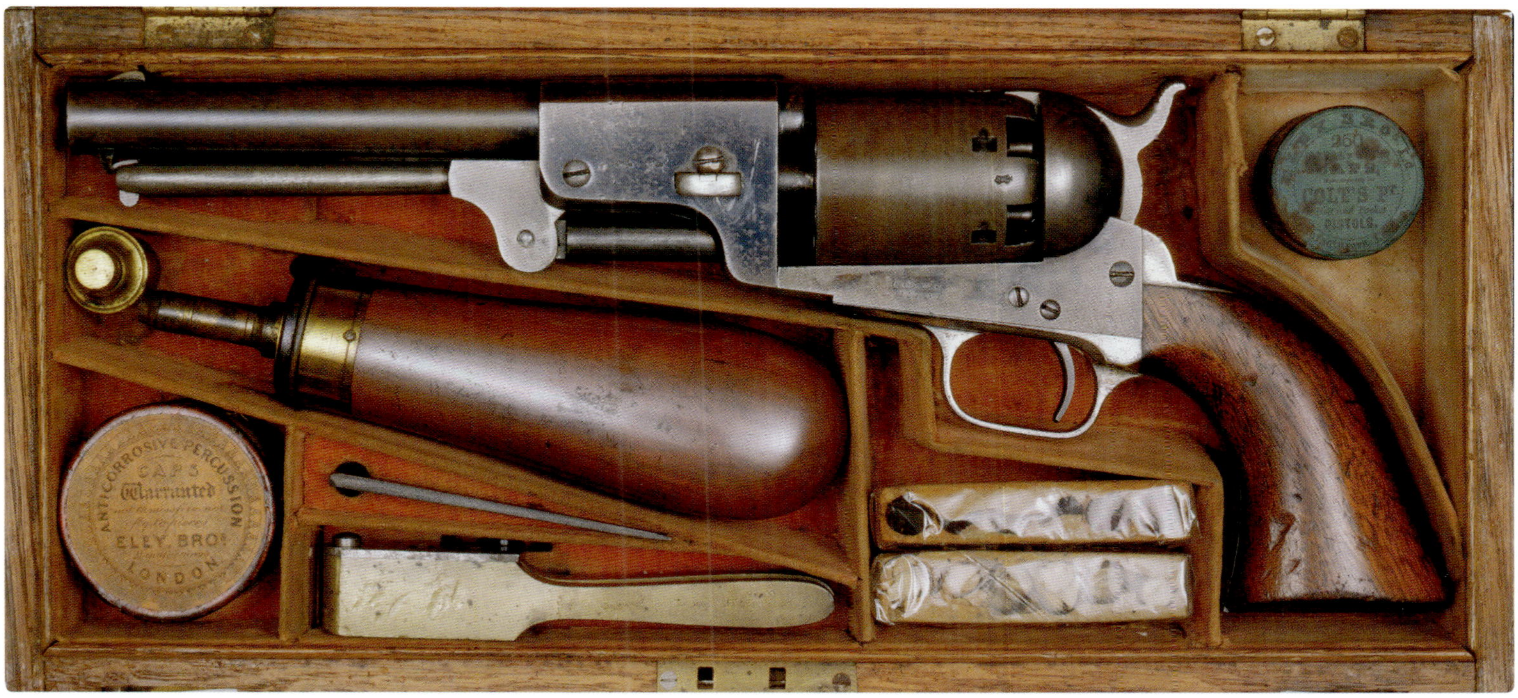

Muzzle-Loading Percussion Revolver
Colt's Patent Fire Arms Manufacturing Company, London, England
Hartford-English Third Model Dragoon, 1853–1854, Caliber .44 Rifled Bore
Weitzenhoffer Collection, 2001.46.005

A relatively scarce model among Colt revolvers, only about 700 Hartford-English Dragoons were produced before Colt's London factory ceased operation in 1857. Distinguished by British proof stamps around the cylinder's breech and on the left barrel lug, and by the hand-engraved "Colts / Patent" marking on the left frame, the components of this specimen actually were made in Hartford but finished, assembled, and proved in London. The piece resides in a compartmented, London oak casing with burgundy velvet lining and is accompanied by a bag-patterned English powder flask, a brass bullet mold, an oil container, two Eley Brothers percussion cap tins, an L-shaped screwdriver/nipple wrench, and two packets of skin cartridges. About 200 of these guns were shipped from Great Britain to the United States in 1861 at the outbreak of the Civil War.

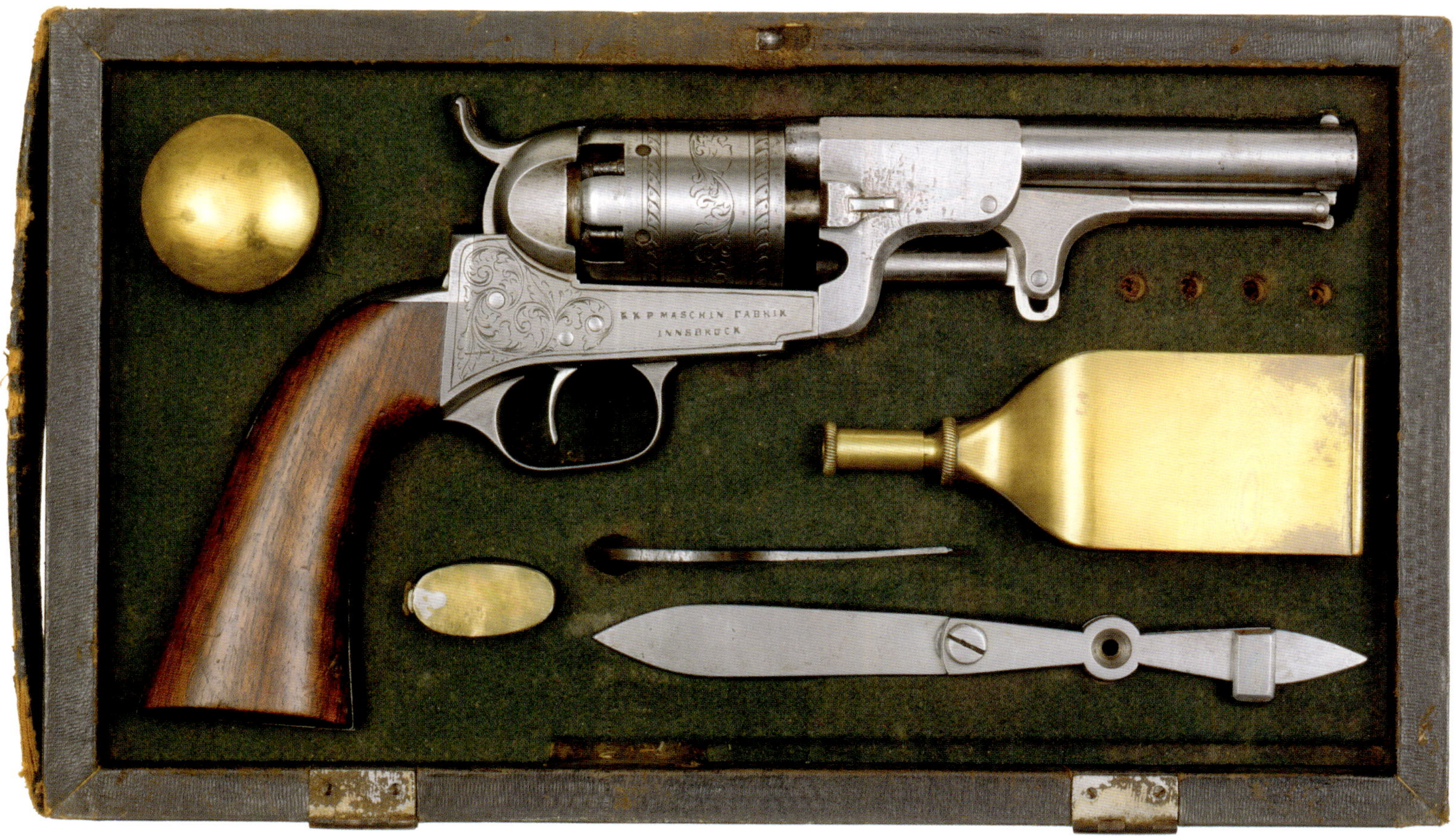

Muzzle-Loading Percussion Revolver

KKP Maschin Fabrik, Innsbruck, Austria

Colt Brevete "Third Model" Dragoon, circa 1855–1860, Caliber .36 Rifled Bore

Weitzenhoffer Collection, 2001.46.006

By the mid-1850s, Colt revolvers had gained sufficient notoriety and reputation in Europe that several Continental firearms manufacturers started replicating the principal design. Colt sought to preclude infringements by granting manufacturing licenses (brevetes) to specifically approved makers. This Austrian-made revolver presents a rather ungainly interpretation of the Dragoon configuration, departing from the original in its atypical, somewhat exaggerated profiles of the barrel lug, loading lever joint, hammer, grips, frame, and trigger guard. It is finished "bright," or "in-the-white," and is hand engraved in an open foliate-scroll pattern over a clear field on the cylinder and frame.

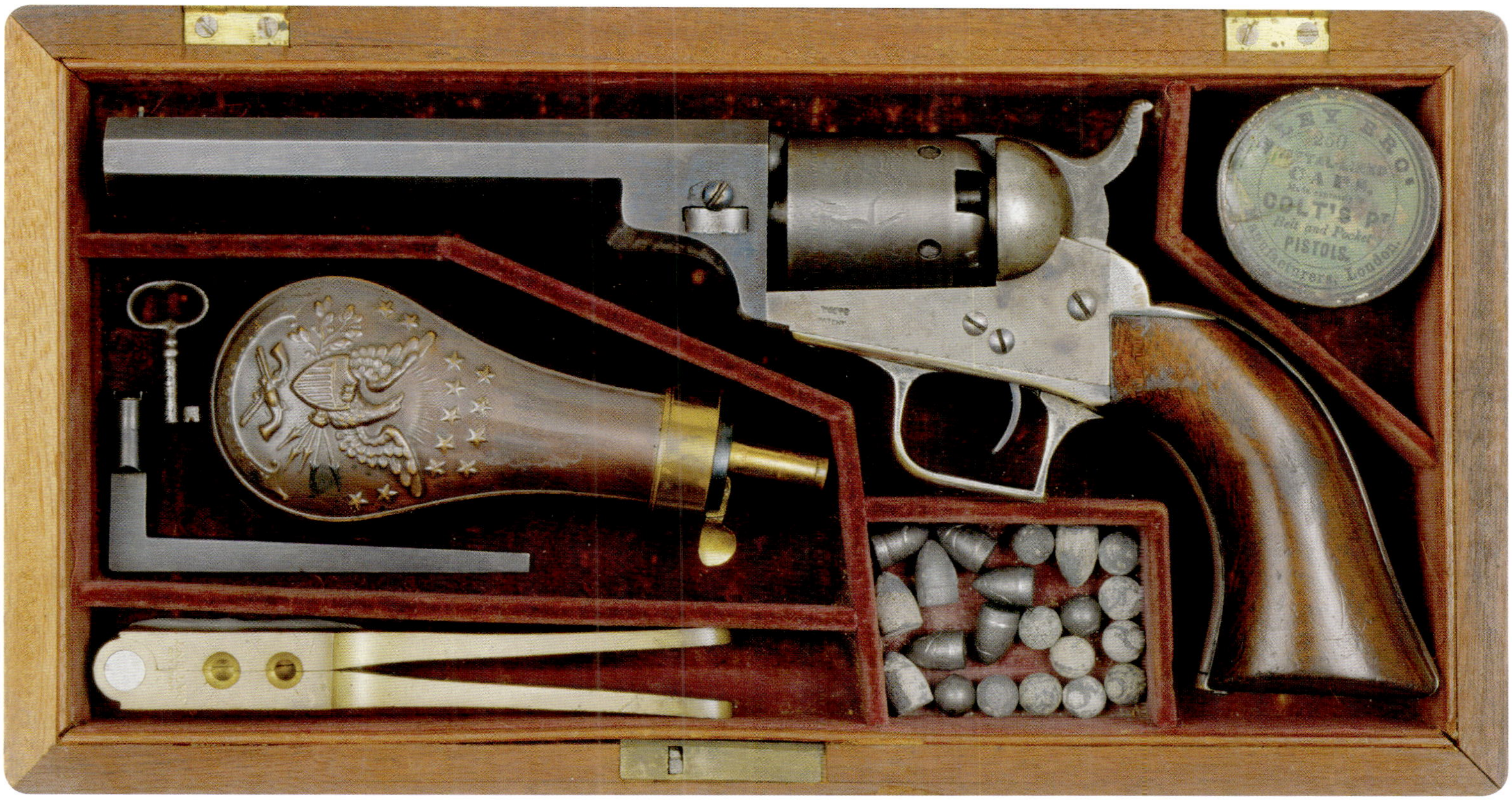

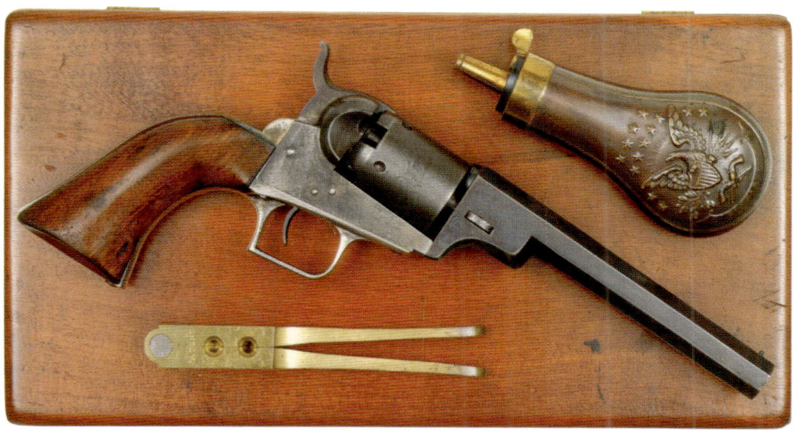

Muzzle-Loading Percussion Revolver

Colt's Patent Fire Arms Manufacturing Company, Hartford, Connecticut
Model 1848 Baby Dragoon, 1847–1850, Caliber .31 Rifled Bore

Weitzenhoffer Collection, 2001.46.007

The first of the pocket-model revolvers produced at Colt's Hartford facility, the Baby Dragoon won instant popularity in the civilian marketplace—some 15,000 were made before being superseded by the Model 1849 Pocket revolver. A diminutive interpretation of the massive Dragoon series, the type incorporated similar, round cylinder stop slots and a square-backed trigger guard, as well as the Texas Ranger–Indian fight cylinder scene. A special-order gun, this pistol is housed in a compartmented mahogany casing lined in purple velvet accompanied with complete loading accessories, including the key.

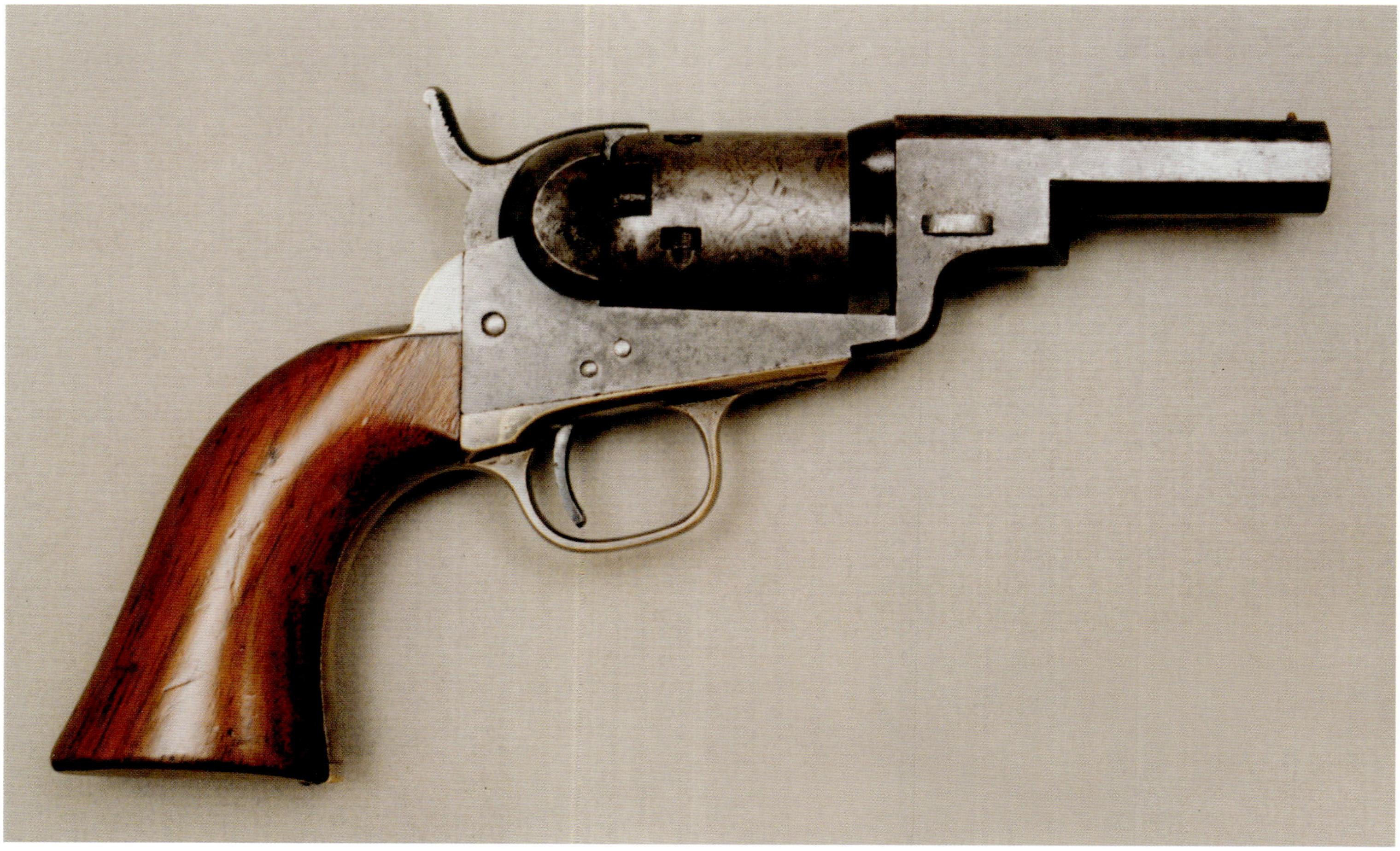

Muzzle-Loading Percussion Revolver

Colt's Patent Fire Arms Manufacturing Company, Hartford, Connecticut
Model 1849 "Wells Fargo" Pocket, circa 1851–1861, Caliber .31 Rifled Bore
Weitzenhoffer Collection, 2001.46.008

A quite scarce and very desirable variation of the standard Model 1849 Pocket revolver, only about 4,000 of the so-called Wells Fargo pattern were made in a total production of 325,000 revolvers. The five-shot type is distinguished by the absence of a loading lever under the barrel, suggesting a limited continuation of the Model 1848 Baby Dragoon pattern. Standard for these arms, this specimen carries Waterman L. Ormsby's stagecoach holdup scene roll-engraved over the cylinder (see page 171). Although the pistol would have made an ideal personal defense weapon for Wells Fargo messengers, there is no documented purchase of these guns by the famed California firm.

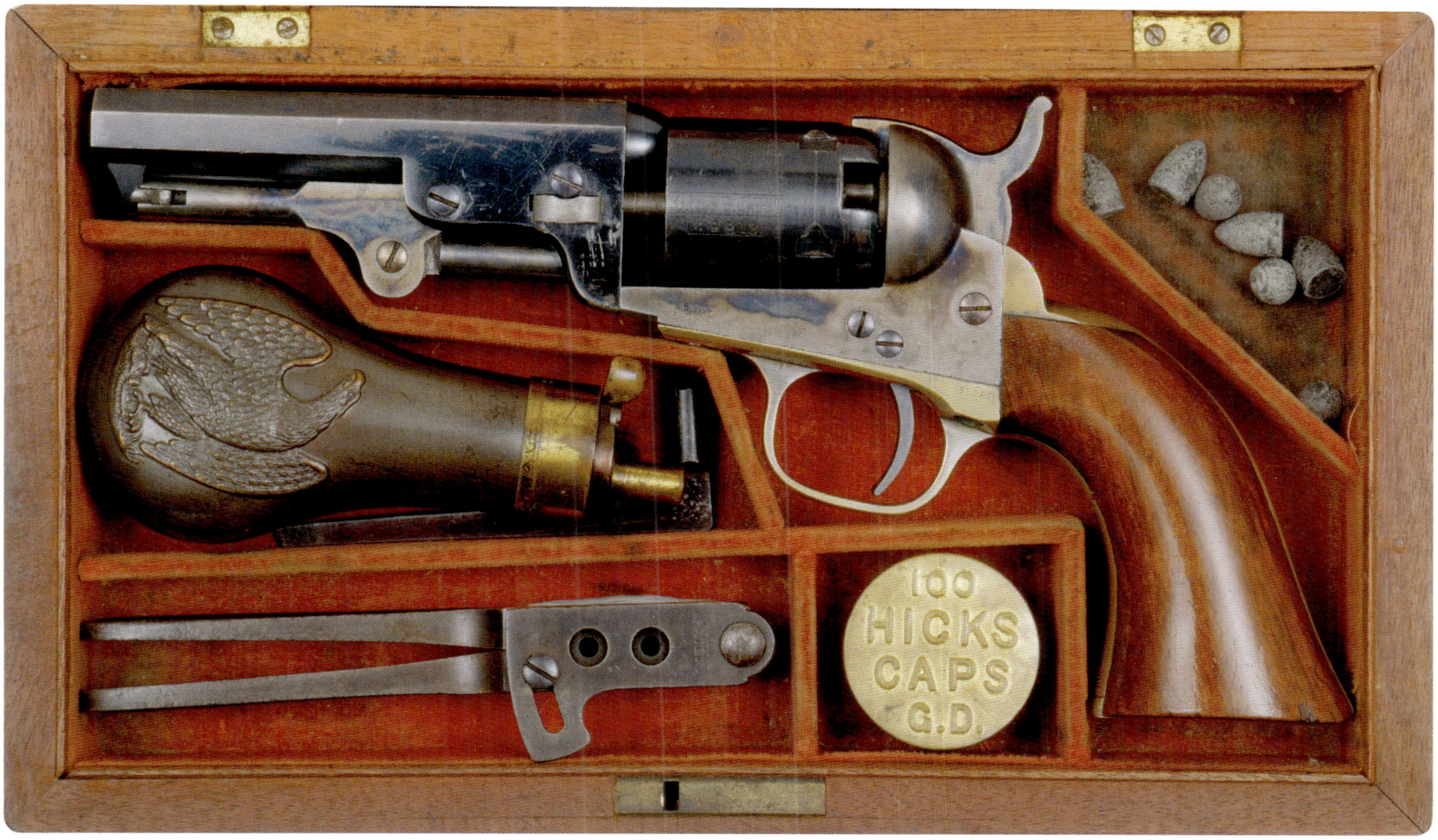

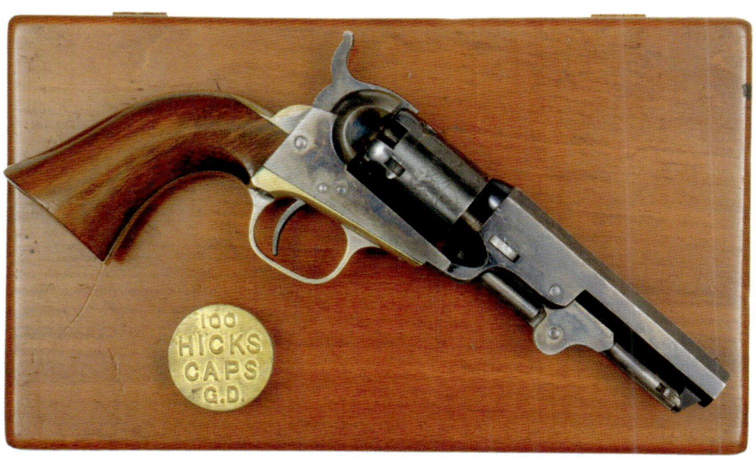

Muzzle-Loading Percussion Revolver

Colt's Patent Fire Arms Manufacturing Company, Hartford, Connecticut
Model 1849 Pocket, 1850–1873, Caliber .31 Rifled Bore

Weitzenhoffer Collection, 2001.46.010

The Model 1849 Pocket revolver enjoyed the largest production total among all the percussion-era Colts, reaching some 325,000 made at the Hartford facilities alone. This special-order set comprises a 4-inch-barreled revolver in a compartmented, velvet-lined, mahogany casing accompanied by a powder flask, steel bullet mold, combination tool, cap tin, and sampling of loose projectiles. The interior of the case lid carries a rare "S. Sutherland / Richmond, Virginia" trade label that, reckoned by the gun's serial number in the 279,000 range, suggests shipment to the South immediately following the close of the Civil War.

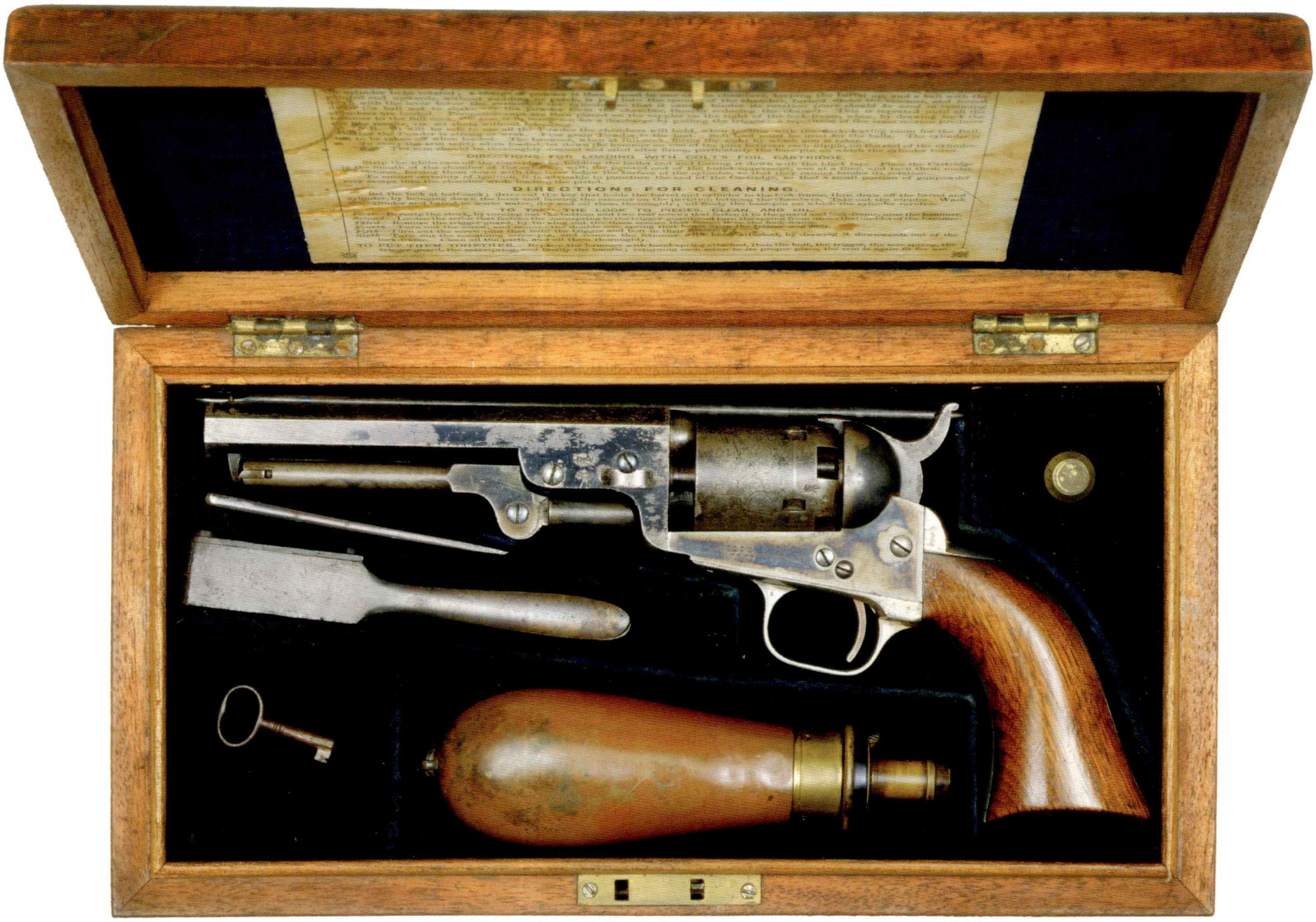

Muzzle-Loading Percussion Revolver

Colt's Patent Fire Arms Manufacturing Company, London, England
Model 1849 London Pocket, 1853–1856, Caliber .31 Rifled Bore
Weitzenhoffer Collection, 2001.46.011

London manufacture of the Model 1849 Pocket revolver totaled 11,000 units at the time Colt's English operation terminated in 1857. Ornamented with the roll-engraved, stagecoach holdup scene on the cylinder and carrying the barrel inscription "ADDRESS COL. COLT / LONDON," this 5-inch-barreled specimen is housed in an English casing of varnished mahogany having a compartmented interior lined in blue velvet and featuring a "Directions for Loading Colt's Pistols" trade label within the lid. Like much of Colt's English production, the quality, fitting, and finish of these London-made, Model 1849 Pocket revolvers typically equaled, and sometimes excelled, Hartford-made specimens.

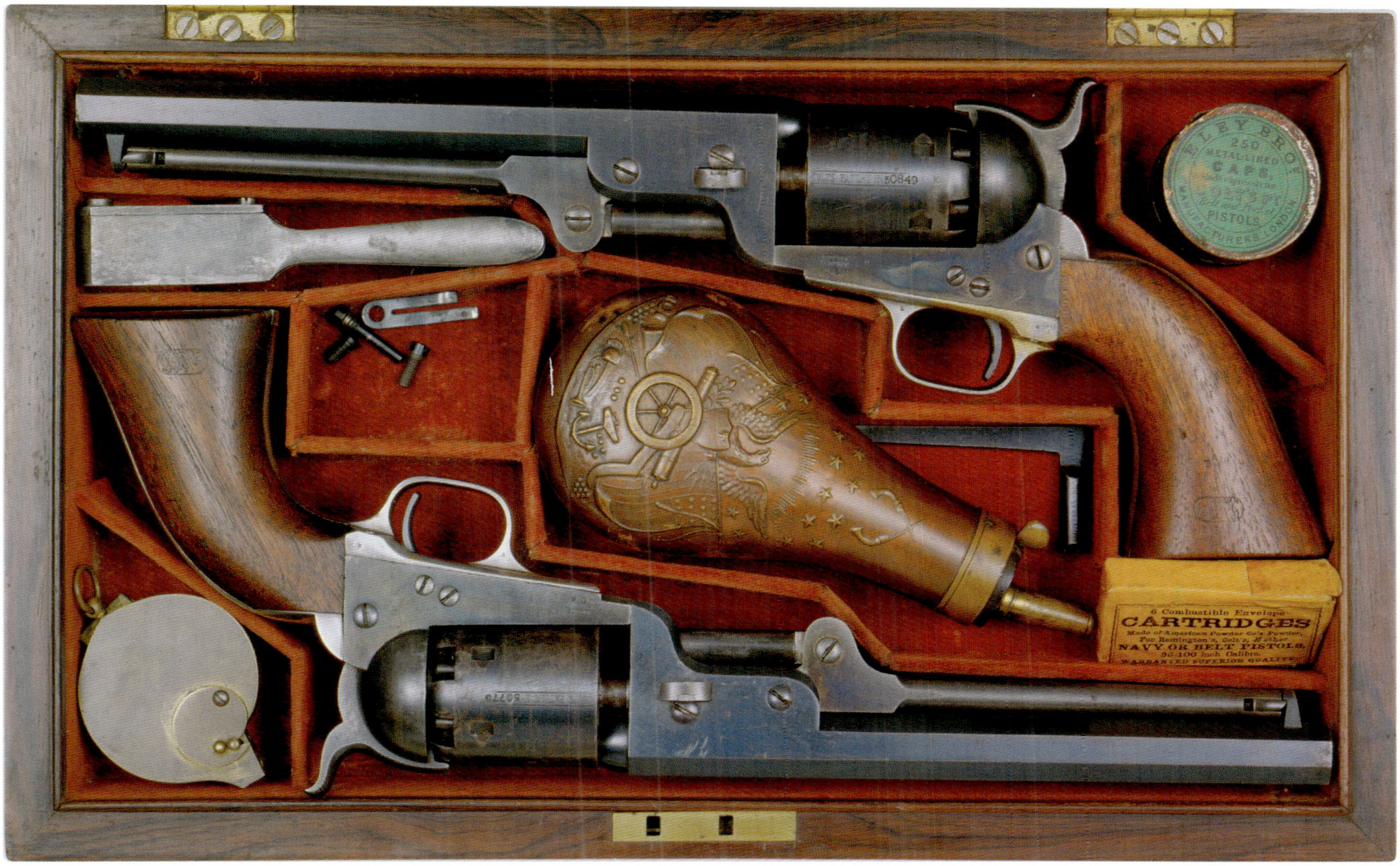

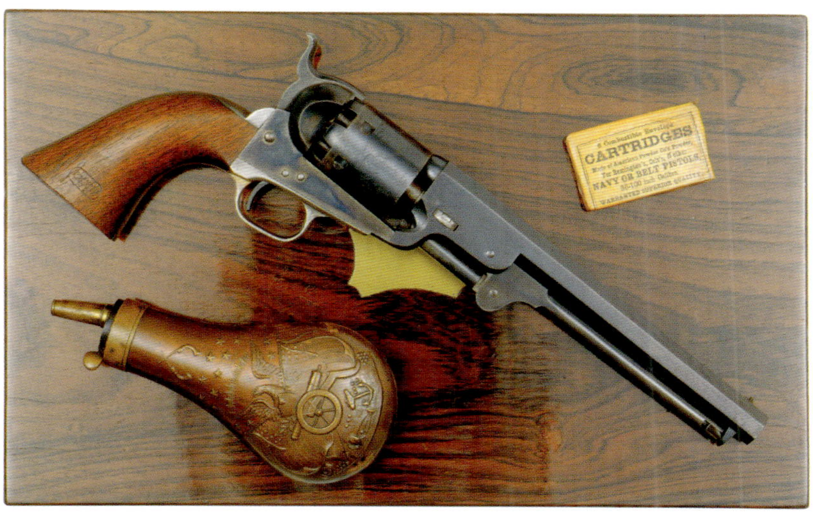

Muzzle-Loading Percussion Revolvers

Colt's Patent Fire Arms Manufacturing Company, Hartford, Connecticut
Model 1851 Navy, Third Variation, 1851–1858, Caliber .36 Rifled Bore
Weitzenhoffer Collection, 2001.46.012 A&B

Cased pairs of revolvers represent a significant rarity within Colt firearms production. This set exhibits crisp renditions of Waterman L. Ormsby's roll-engraved cylinder scene of a naval battle between Texas and Mexico in 1843 (see page 171). Martially marked, the revolvers carry ordnance inspector's initials on their one-piece grips and U.S. markings on their frames just below the "Colts / Patent" stampings. The revolvers are housed in a compartmented rosewood casing with burgundy velvet lining. Accessories include a handsomely embossed "stand-of-arms" powder flask, a steel bullet mold, a brass capping device, a cap tin, and a packet of combustible cartridges.

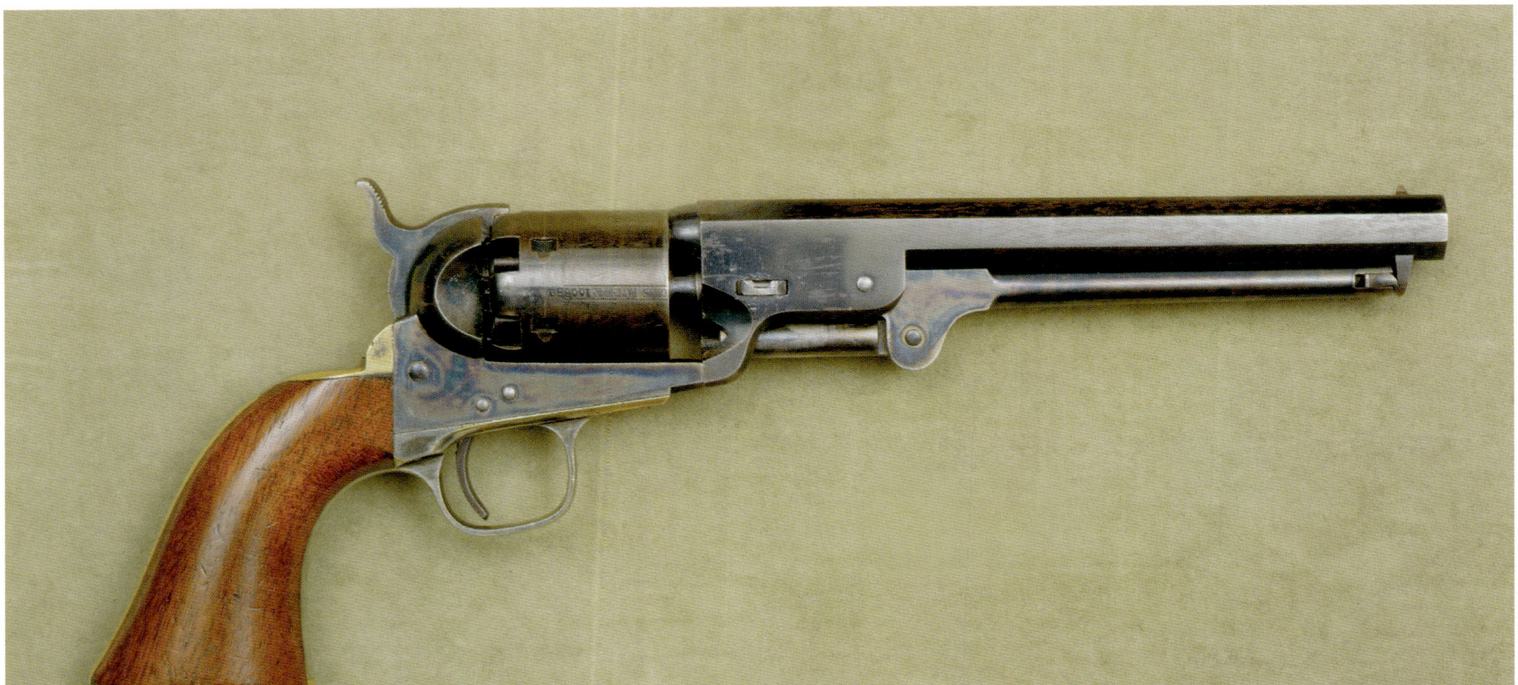

Muzzle-Loading Percussion Revolver
Colt's Patent Fire Arms Manufacturing Company, Hartford, Connecticut
Model 1851 Navy, Fourth Variation, 1858–1873, Caliber .36 Rifled Bore
Weitzenhoffer Collection, 2001.46.013

Among the most popular of all Colt percussion revolvers, the Model 1851 Navy had a production run of more than a quarter of a million during a period of twenty-three years. The design was fabricated in both Connecticut and London, copied in Europe, employed by the U.S. and British militaries, and embraced by frontiersmen across the West. This specimen, made in 1861 at the outbreak of the Civil War, retains a sharp rendering of Ormsby's naval-engagement scene roll-engraved on the six-shot cylinder (see page 171).

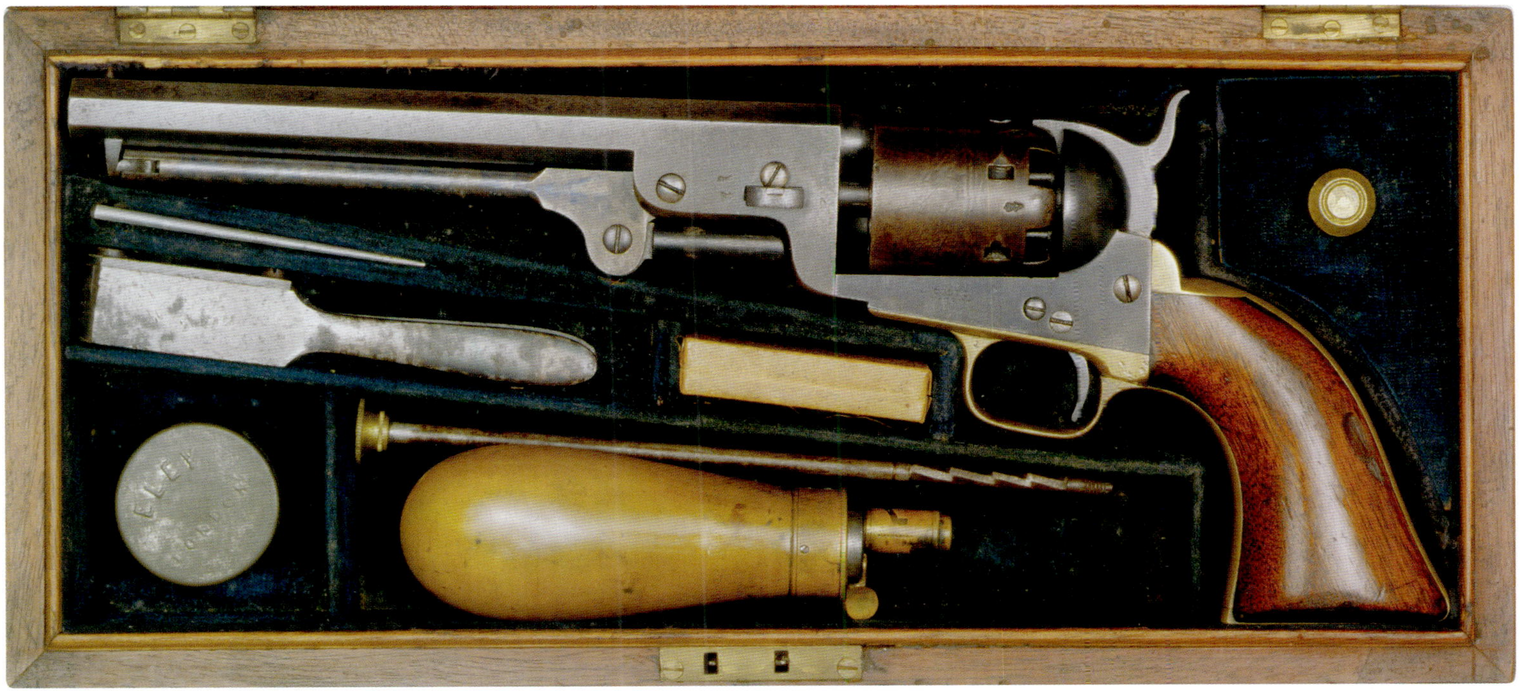

Muzzle-Loading Percussion Revolver

Colt's Patent Fire Arms Manufacturing Company, London, England

Model 1851 London Navy, 1853–1856, Caliber .36 Rifled Bore

Weitzenhoffer Collection, 2001.46.014

London manufacture of the Model 1851 Navy revolver reached 42,000 when the English factory closed in 1857. This casing holds a typical early production revolver with a small brass trigger guard and the "Colts / Patent" stamping centered on the left side of the frame. The barrel carries an "ADDRESS COL. COLT. LONDON" imprint, while British proof stamps appear on the left barrel lug and around the cylinder breech. Lined in blue velvet, the compartmented mahogany case also houses an English bag-type powder flask, a steel bullet mold, a combination screwdriver and nipple wrench, a cleaning rod, a percussion cap tin, and a packet of combustible cartridges.

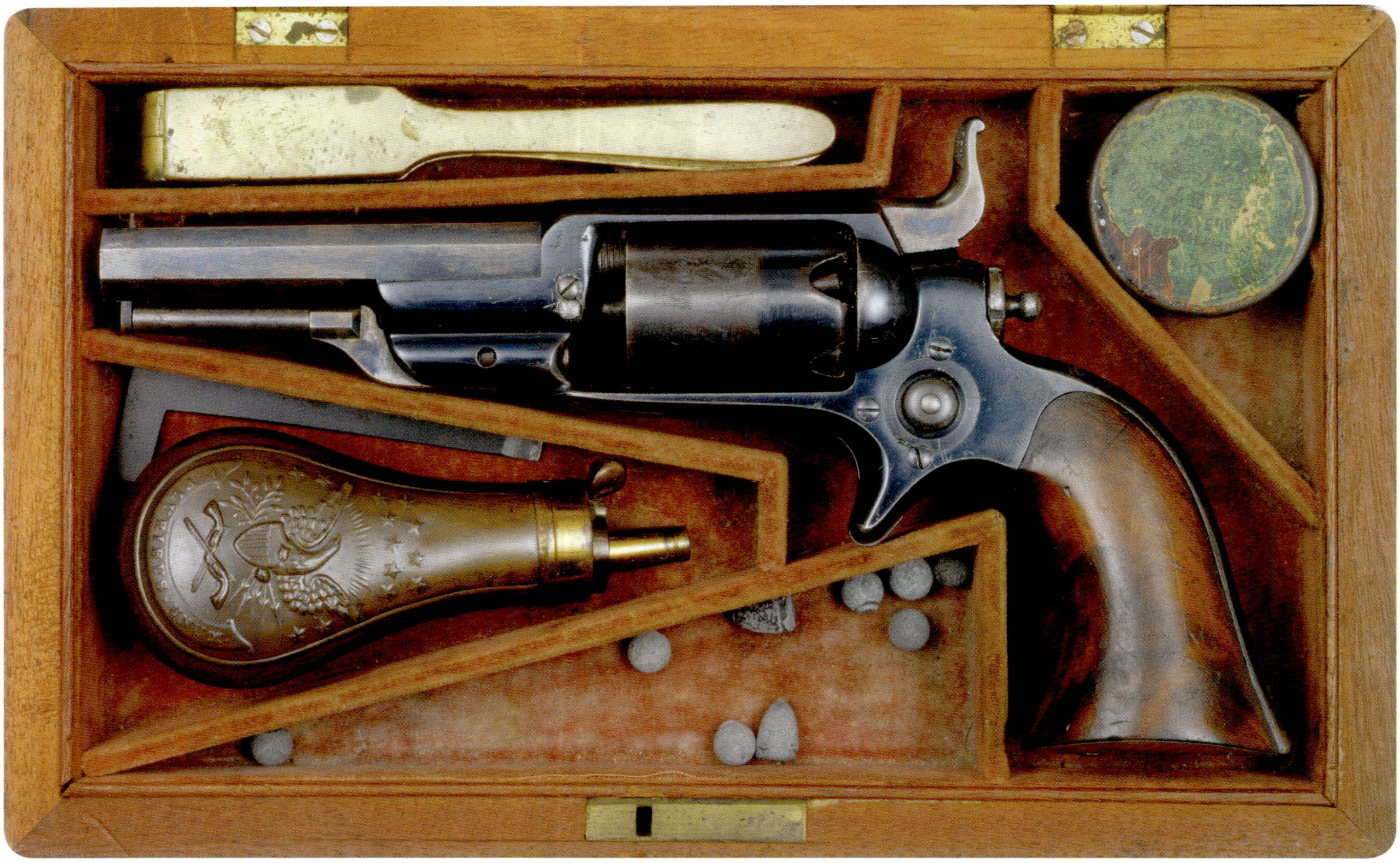

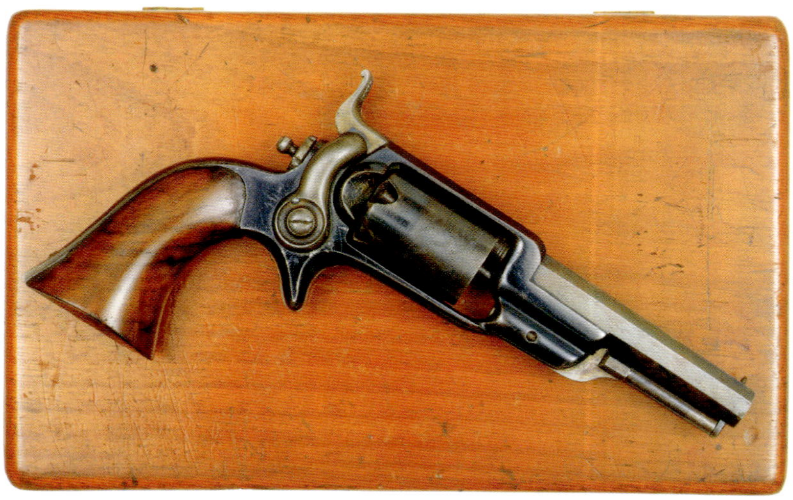

Muzzle-Loading Percussion Revolver
Colt's Patent Fire Arms Manufacturing Company, Hartford, Connecticut
Model 1855 "Root" Sidehammer Pocket, 1855–1870, Caliber .28 Rifled Bore
Weitzenhoffer Collection, 2001.46.018

Production of "Root" model pocket revolvers (in both .28 and .31 caliber) reached approximately 40,000 during fifteen years of manufacture. Introduced the year that Colt opened his state-of-the-art factory in Hartford, the model departed from earlier pocket-sized designs with its solid-frame, side-hammer, spur-trigger design. This mahogany-cased specimen presents the Model 2 variant, having an octagonal barrel with the Hartford address marking, and a round (unfluted) cylinder ornamented with Ormsby's roll-engraved log cabin and pioneer-Indian skirmish scene (see page 171). The casing also holds an assortment of loading accessories, including an embossed powder flask, percussion cap tin, bullet mold, and combination tool.

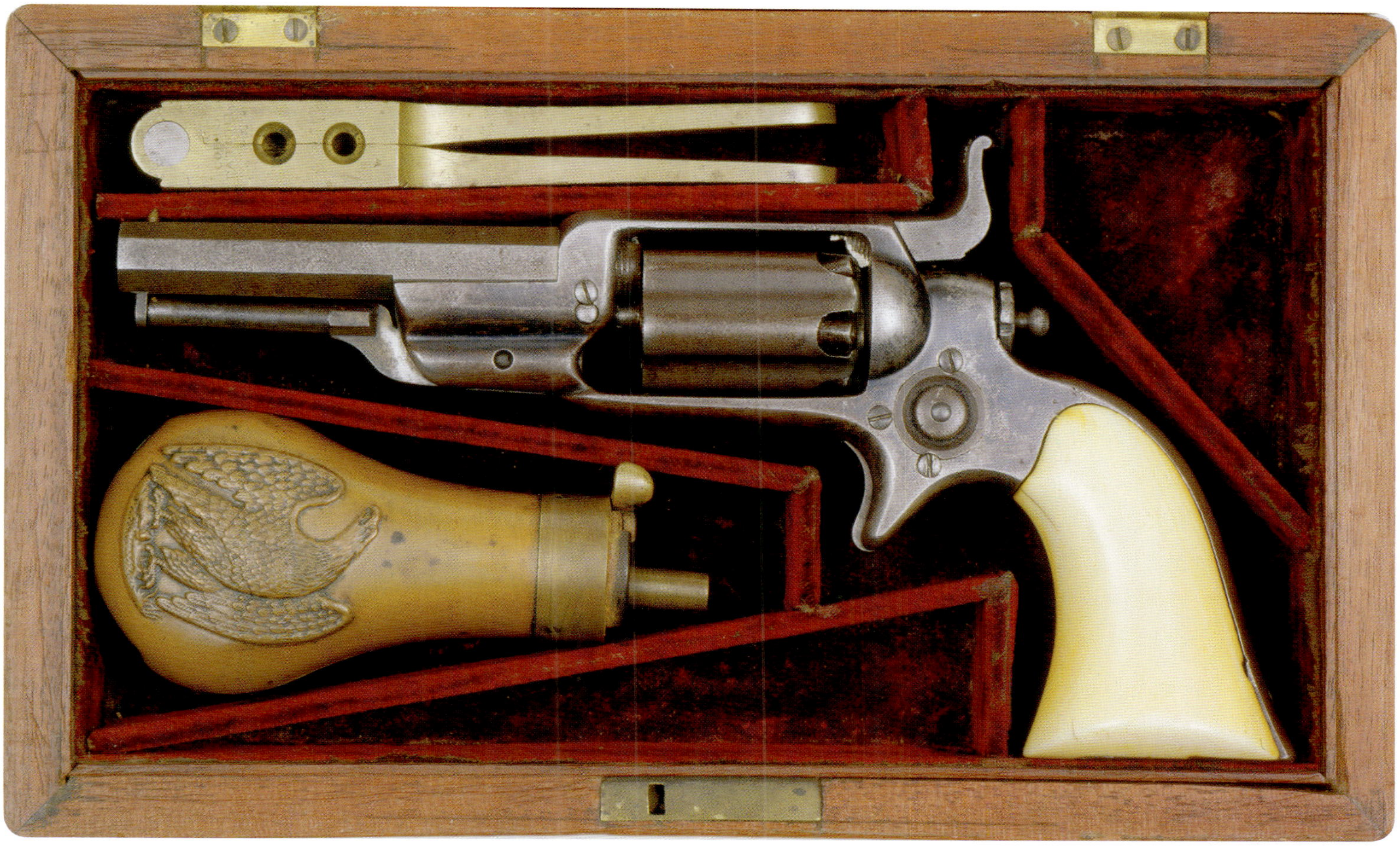

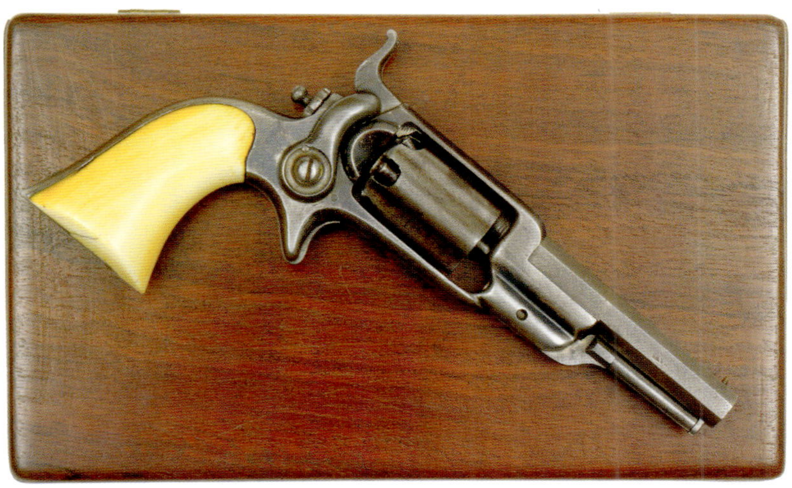

Muzzle-Loading Percussion Revolver

Colt's Patent Fire Arms Manufacturing Company, Hartford, Connecticut
Model 1855 "Root" Sidehammer Pocket, 1855–1870, Caliber .28 Rifled Bore
Weitzenhoffer Collection, 2001.46.019

This ivory-gripped, Model 3 variation of the Root revolver differs from the previous example in having a fully fluted cylinder without a roll-engraved scene. It resides in a compartmented casing that also holds an eagle-embossed powder flask and a brass, two-cavity bullet mold. Esteemed as a rare and desirable commodity, the elephant ivory used for stocking this revolver was considered a special-order embellishment at the Hartford factory. Throughout most of the 1850s and 1860s, for example, the factory price of plain ivory grips ($5) exceeded the cost of standard-coverage, scroll engraving ($4) on the Model 1849 and Model 1855 revolvers. Not a few of these handguns served as personal sidearms during the Civil War.

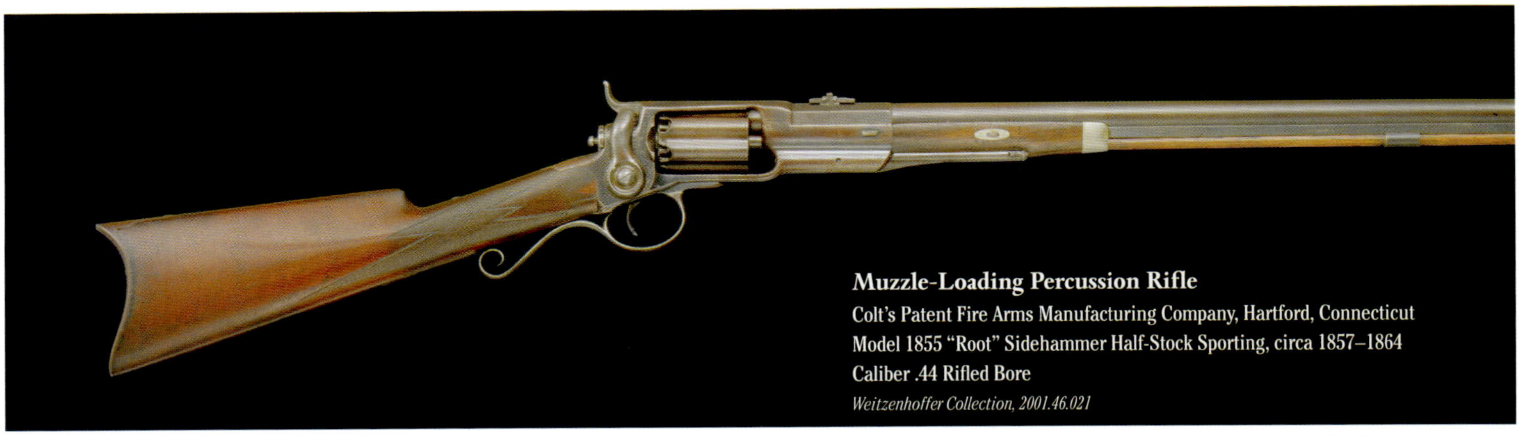

Muzzle-Loading Percussion Rifle
Colt's Patent Fire Arms Manufacturing Company, Hartford, Connecticut
Model 1855 "Root" Sidehammer Half-Stock Sporting, circa 1857–1864
Caliber .44 Rifled Bore
Weitzenhoffer Collection, 2001.46.021

The "Root" solid-frame, side-hammer design premiered Colt's production of revolving-cylinder longarms at Hartford. The most popular of the Model 1855 Sporting rifles, the half-stock was made with several variations in caliber, barrel length, sights, trigger, trigger guard pattern, and finish. This specimen is distinguished by an unusually long, 31¼-inch barrel and a forestock of varnished walnut finished with a German silver endcap. It also features a hickory cleaning rod, a scroll-pattern trigger guard, and checkering over the wrist of the stock. While the manufacture of all Model 1855 Sporting rifle patterns approached 3,000 arms over nine years, the production of similar military-pattern muskets and carbines for service in the Civil War amounted to nearly 14,000 weapons by 1864, when the Hartford plant burned, temporarily halting work. First model variations of this arm had round (unfluted) cylinders adorned with W. L. Ormsby's deer-hunting tableau (see page 171).

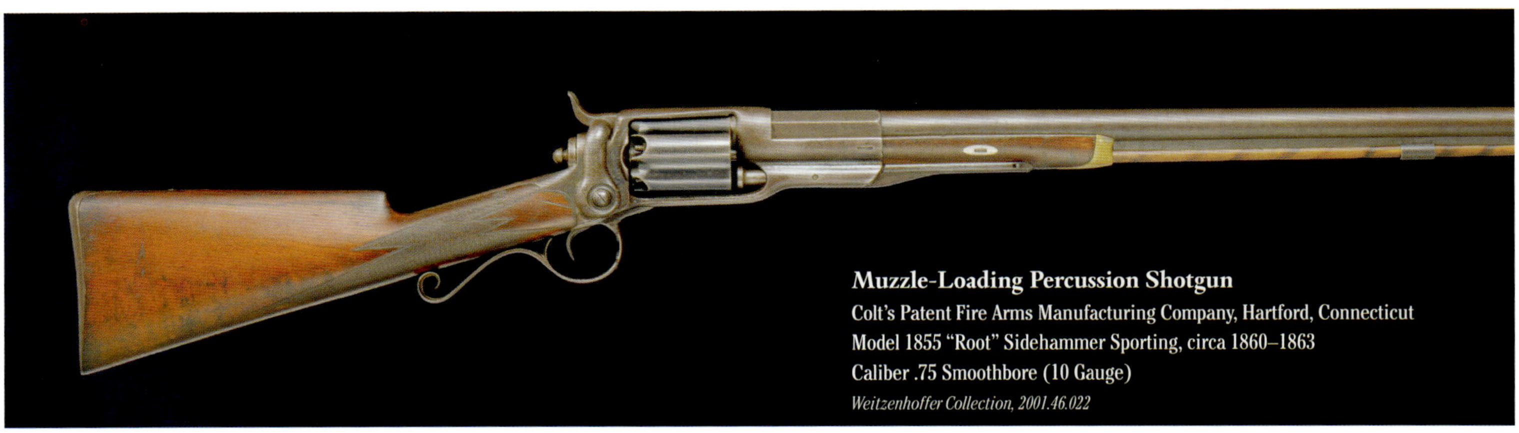

Muzzle-Loading Percussion Shotgun
Colt's Patent Fire Arms Manufacturing Company, Hartford, Connecticut
Model 1855 "Root" Sidehammer Sporting, circa 1860–1863
Caliber .75 Smoothbore (10 Gauge)
Weitzenhoffer Collection, 2001.46.022

Distinguished by its smoothbore barrel, shotgun-style buttplate, and absence of any rear sight, this firearm was otherwise virtually identical to the Model 1855 Half-Stock Sporting rifle in its overall form and function. In practical field use, these longarms received mixed reviews because burning powder gases and fragments of lead sometimes blew out between the cylinder and the breech of the barrel. A Colt sales representative observed after one demonstration, "[It took] me all evening to pick the powder and pieces of lead out of my face. No use telling a person to hold their head back, might as well tell them to hold it under their arm."

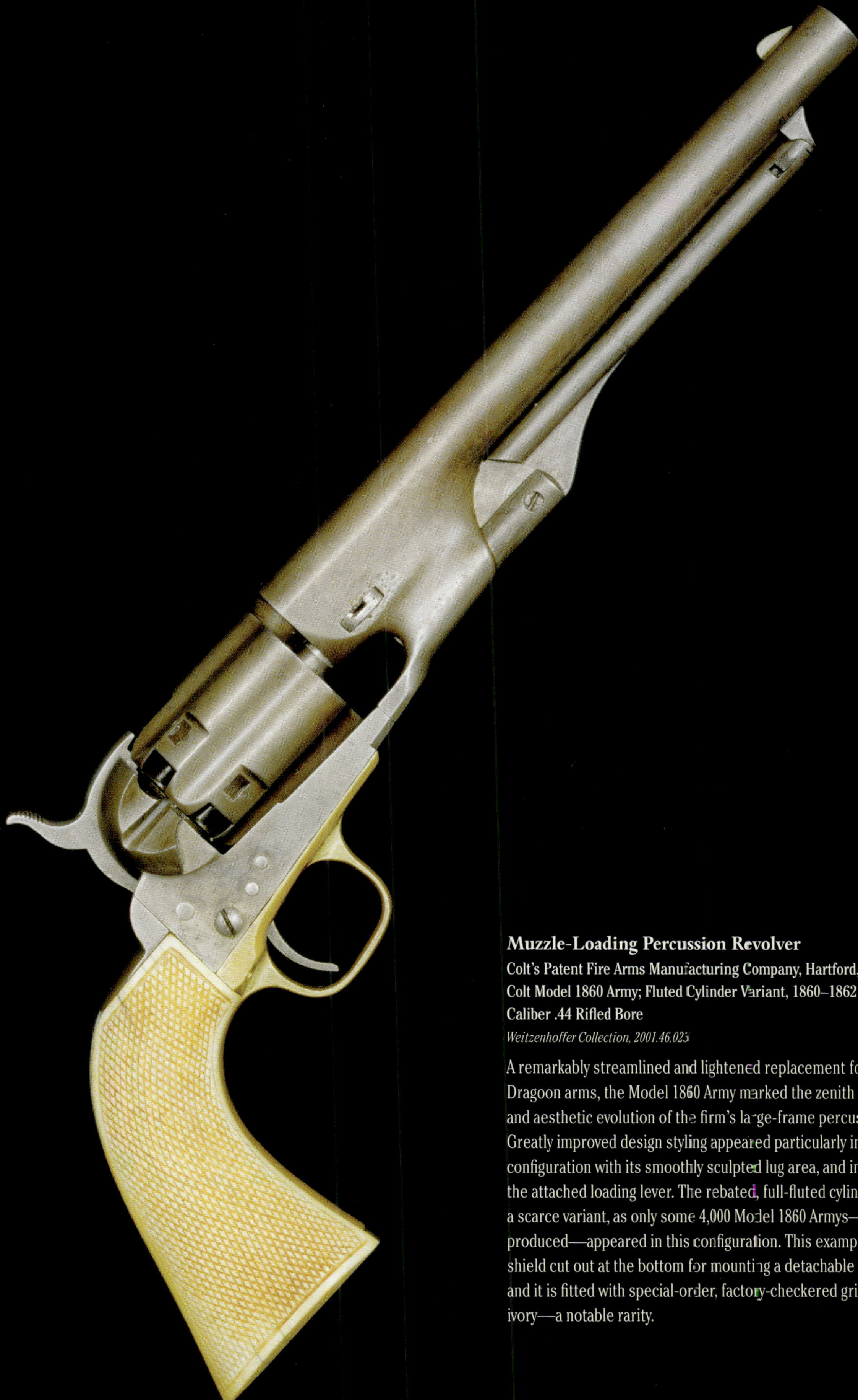

Muzzle-Loading Percussion Revolver

Colt's Patent Fire Arms Manufacturing Company, Hartford, Connecticut
Colt Model 1860 Army; Fluted Cylinder Variant, 1860–1862
Caliber .44 Rifled Bore

Weitzenhoffer Collection, 2001.46.025

A remarkably streamlined and lightened replacement for Colt's earlier Dragoon arms, the Model 1860 Army marked the zenith of the technical and aesthetic evolution of the firm's large-frame percussion revolvers. Greatly improved design styling appeared particularly in the round barrel configuration with its smoothly sculpted lug area, and in the sleek lines of the attached loading lever. The rebated, full-fluted cylinder is considered a scarce variant, as only some 4,000 Model 1860 Armys—of some 200,000 produced—appeared in this configuration. This example has the recoil shield cut out at the bottom for mounting a detachable shoulder stock, and it is fitted with special-order, factory-checkered grips of elephant ivory—a notable rarity.

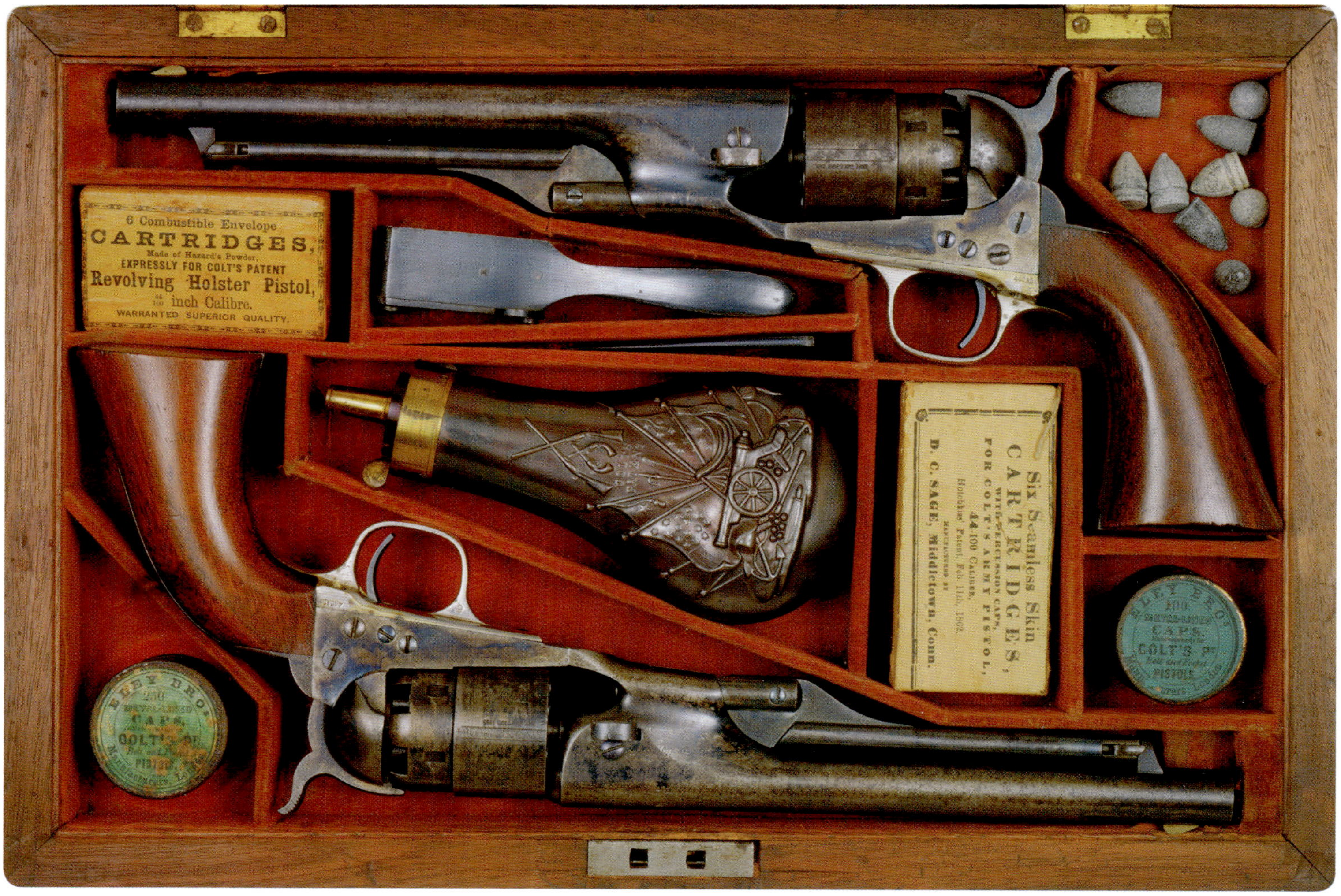

Muzzle-Loading Percussion Revolvers

Colt's Patent Fire Arms Manufacturing Company, Hartford, Connecticut
Model 1860 Army, 1860–1873, Caliber .44 Rifled Bore

Weitzenhoffer Collection, 2001.46.024 A&B

This matched and cased pair of inscribed, presentation revolvers represents something of a rarity in Colt production. Serially numbered 50,081 and 50,084, the handguns were fabricated in 1862, purchased by the men of the Hartford City Guard, and presented to an admired Civil War officer. Both revolvers carry factory-engraved backstrap inscriptions:

Lieut. John H. Burnham, Adjt. 16th Regt. C. V.
From Hartford City Guard

Housed in a compartmented mahogany casing lined in burgundy velvet, both guns are decorated with the standard, naval battle scenes roll-engraved on their cylinders. They are accompanied by a large "stand-of-arms" powder flask, a steel bullet mold, two percussion cap tins, and several packets of combustible cartridges.

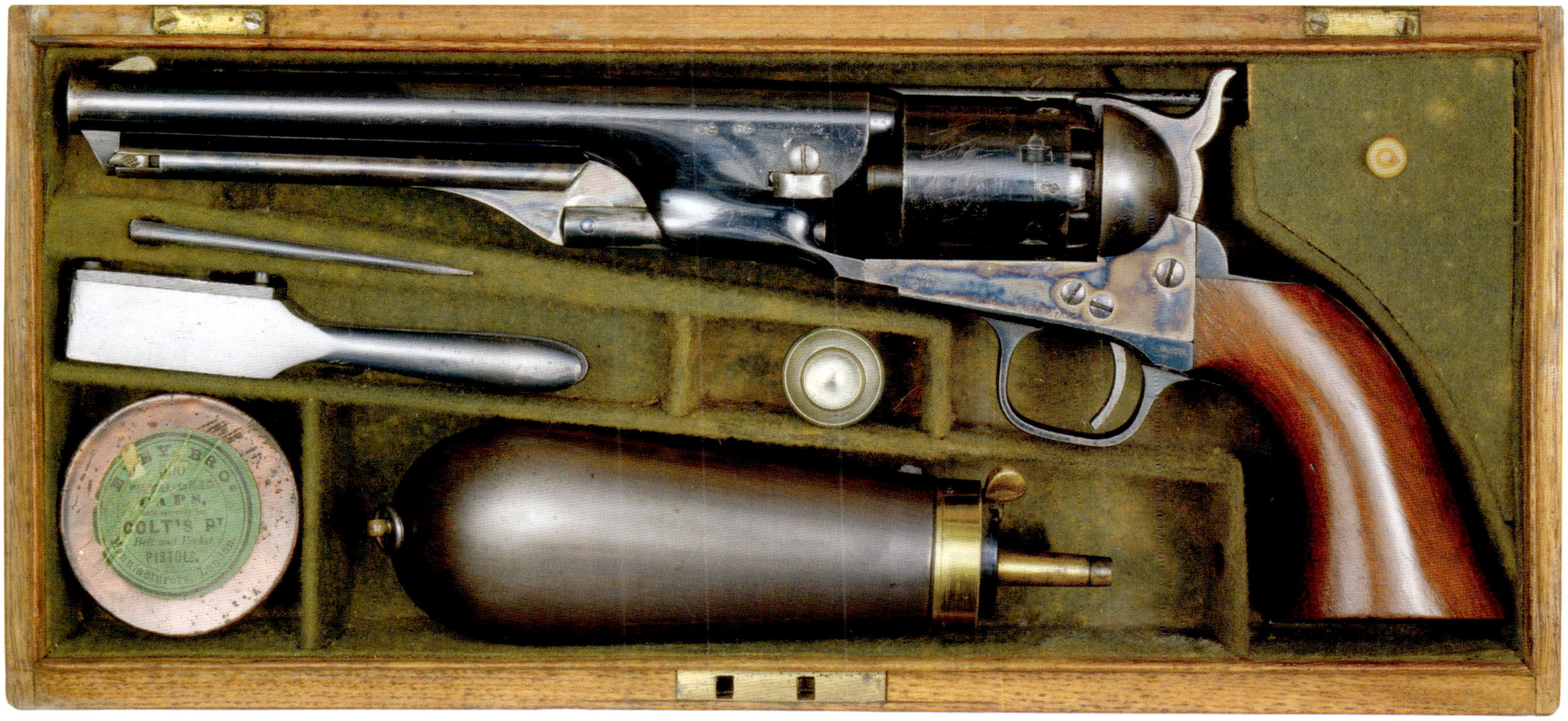

Muzzle-Loading Percussion Revolver

Colt's Patent Fire Arms Manufacturing Company, Hartford, Connecticut
Model 1861 Navy, 1861–1873, Caliber .36 Rifled Bore
Weitzenhoffer Collection, 2001.46.025

Combining sleek design and excellent balance, this successor to the Model 1851 Navy is perhaps the most gracefully styled of all Colt revolvers in overall scale and line. This specimen is among the finer examples known, retaining superb condition overall and an extremely sharp cylinder-roll scene of the Texas-Mexican naval battle originally engraved by W. L. Ormsby. Exported to Great Britain, the pistol carries English proof stamps and is housed in a compartmented, London oak casing with a lining of green felt and a "Colt's Patent Fire-Arms Mfg. Co., Pall Mall, London" trade label inside the lid. An English, bag-type powder flask, together with other loading and cleaning accessories, accompanies the revolver.

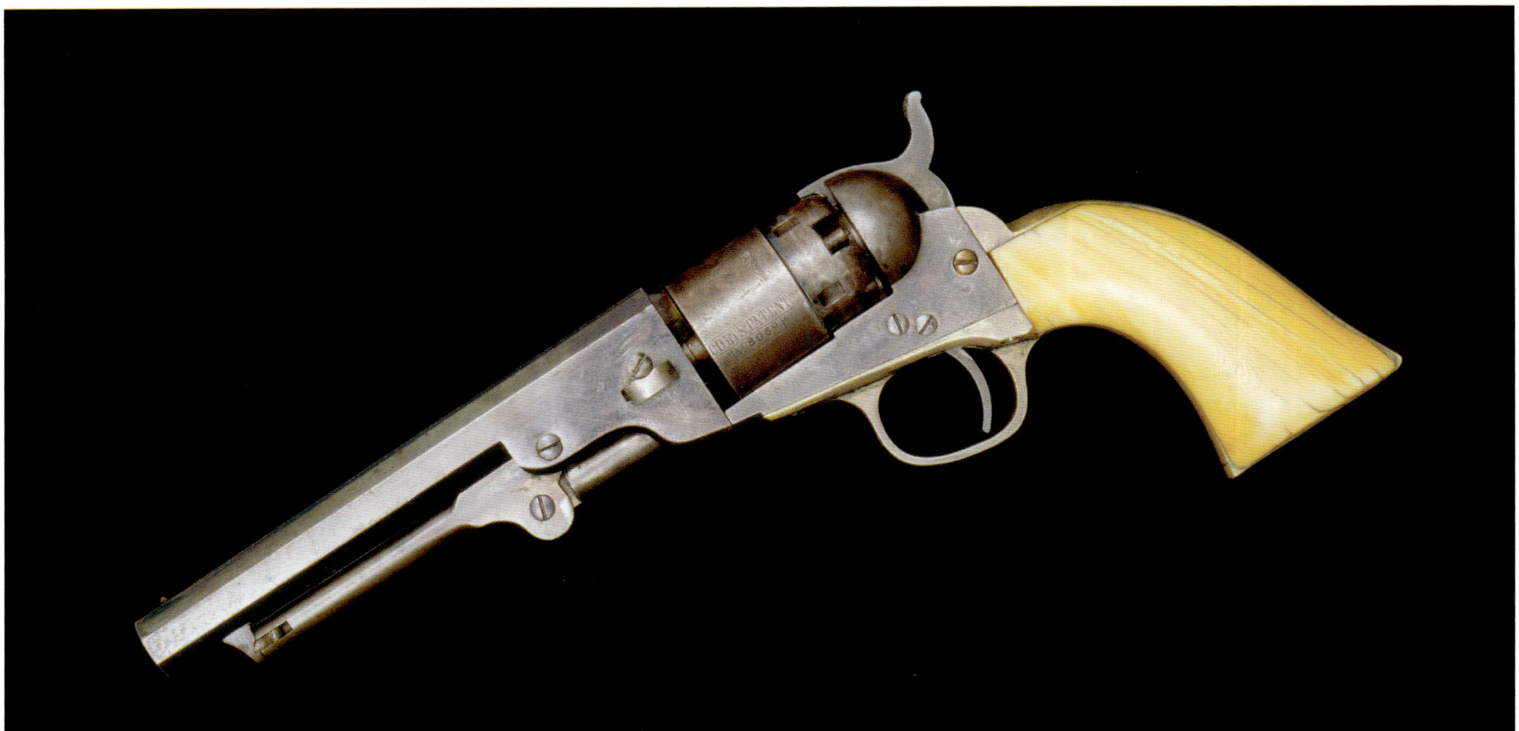

Muzzle-Loading Percussion Revolver
Colt's Patent Fire Arms Manufacturing Company, Hartford, Connecticut
Model 1862 [1865] Pocket Navy, 1865–1873, Caliber .36 Rifled Bore
Weitzenhoffer Collection, 2001.46.027

Something of a design regression, the Pocket Navy represented a scaled-up version of the earlier, Model 1849 Pocket pistol, now having a five-shot, rebated cylinder chambering the more potent, .36 caliber (Navy-size) projectile. Around the cylinder, a fairly distinct impression of Ormsby's classic, roll-engraved stagecoach holdup scene remains. Relatively few Model 1862 [1865] Pocket Navy revolvers appear with special-order enhancements. This specimen, mounted with one-piece grips of mellowed elephant ivory, presents a scarce instance of custom adornment in this model series. Large numbers of these revolvers were converted to (or made as) metallic-cartridge arms during the 1870s.

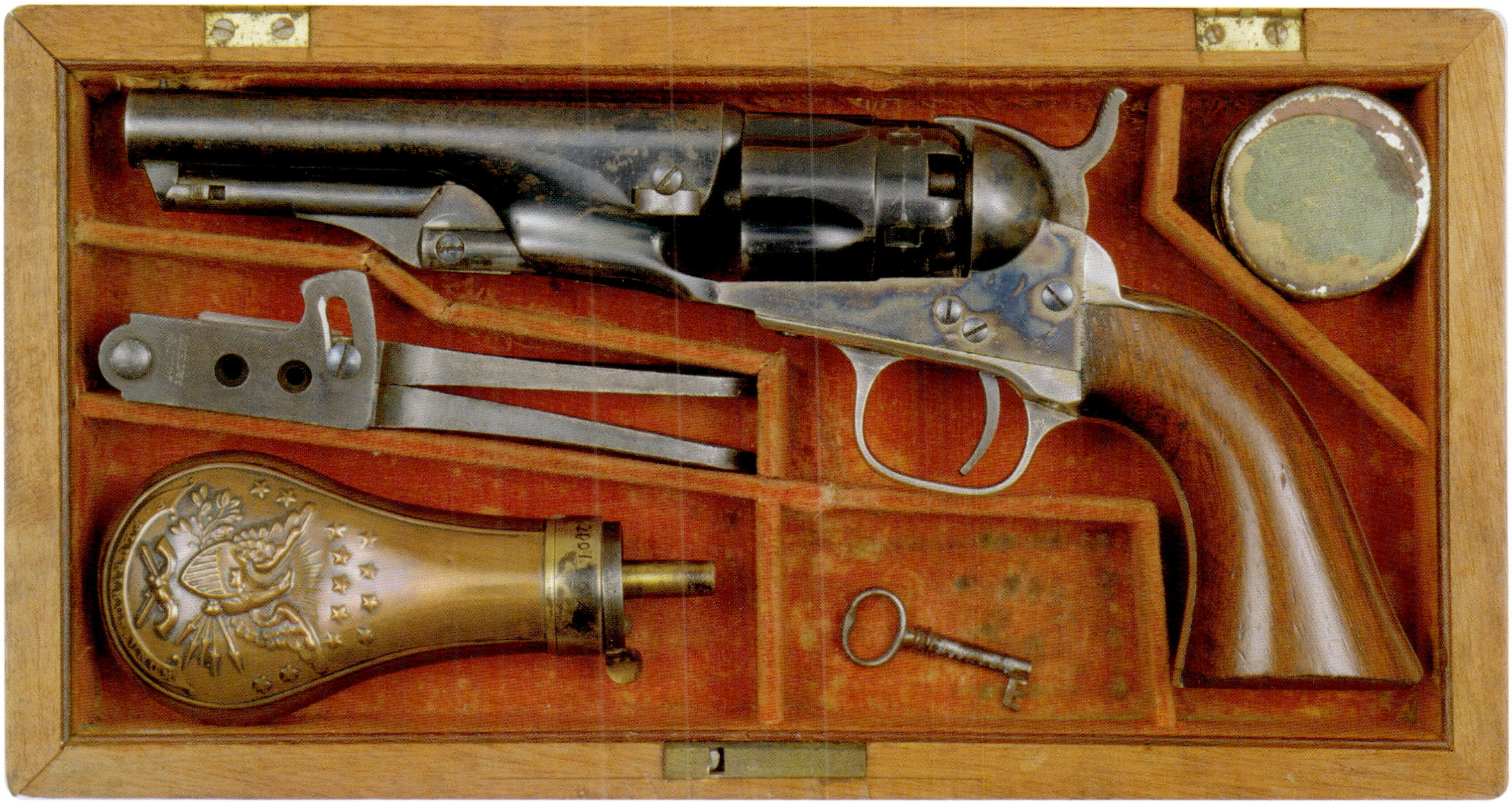

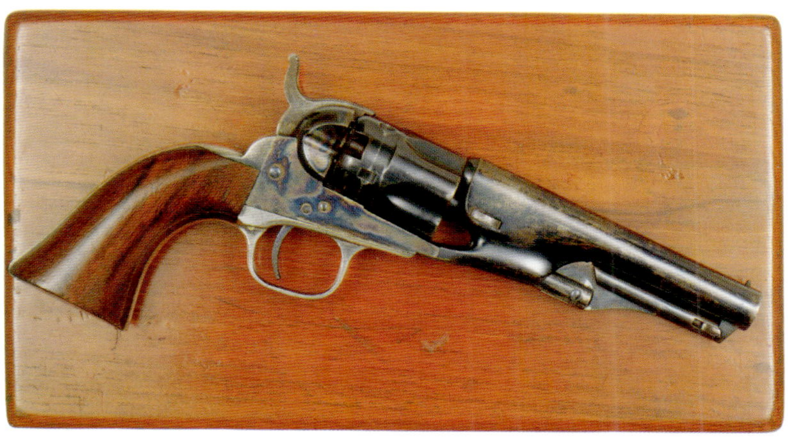

Muzzle-Loading Percussion Revolver

Colt's Patent Fire Arms Manufacturing Company, Hartford, Connecticut
Model 1862 Police, 1861–1873, Caliber .36 Rifled Bore

Weitzenhoffer Collection, 2001.46.028

The smallest of Colt's redesigned New Model series, the five-shot Model 1862 Police provided a handy and handsomely streamlined personal protection weapon. Approximately 28,000 of these revolvers saw manufacture over thirteen years. This cased, 4½-inch-barreled specimen retains fine overall condition with still-vibrant, case-hardened colors on the frame. The compartmented mahogany casing also includes the original key, powder flask, two-cavity steel bullet mold, and percussion-cap tin.

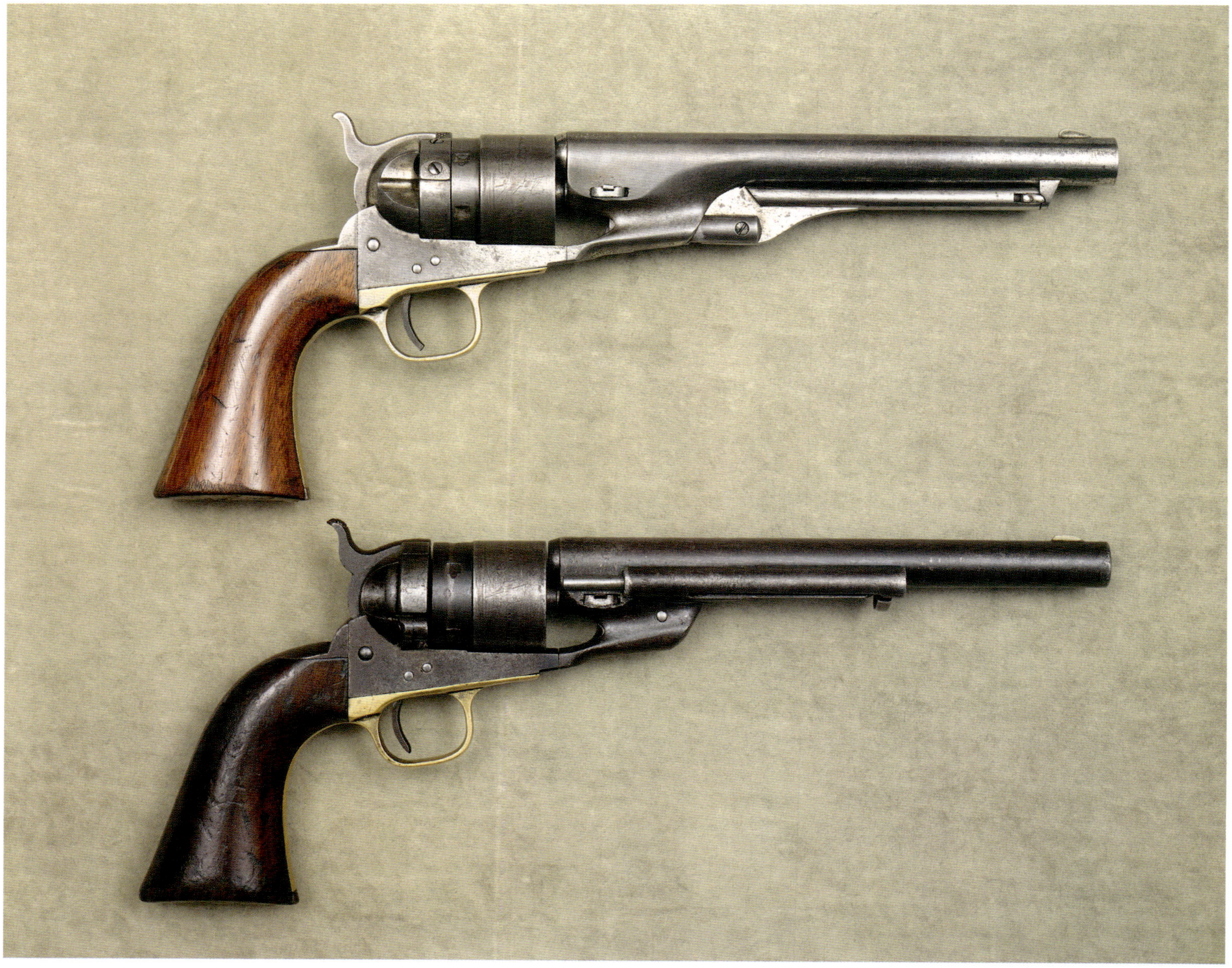

Muzzle-Loading Cartridge Revolver (top)
Colt's Patent Fire Arms Manufacturing Company, Hartford, Connecticut
Model 1860 Army Thuer Conversion, 1869–1872, Caliber .44 Thuer Centerfire
Weitzenhoffer Collection, 2001.46.029

The Colt-Thuer conversion utilized a tapered, rimless, centerfire metallic cartridge that loaded at the front of the pistol's cylinder. The change from percussion to cartridge ignition involved the removal of the cylinder's breech end and the insertion of a steel ring incorporating the ejector mechanism and a rebounding firing pin. Patented by F. Alexander Thuer in September 1868, only about 5,000 of these weapons were made from available stocks of 1849 Pocket, 1851 Navy, 1860 Army, 1861 Navy, and 1862 Police and [1865] Pocket Navy revolvers. Relative scarcity—coupled with the gun's status as the first metallic-cartridge revolver produced by Colt—contributes to the Thuer conversion's desirability among arms collectors and connoisseurs.

Breech-Loading Cartridge Revolver (bottom)
Colt's Patent Fire Arms Manufacturing Company, Hartford, Connecticut
Model 1860 Army Richards-Mason Conversion, 1873–1878
Caliber .44 Colt Centerfire
Weitzenhoffer Collection, 2001.46.033

A technical improvement over the Thuer-pattern conversion, this revolver incorporates the design alterations of C. B. Richards and William Mason. These changes included removing the breech of the cylinder and through-boring the chambers, adding a standing breechplate with a hinged loading gate on the frame, removing the loading lever assembly beneath the barrel, and adding an ejector rod assembly. This transitional specimen, joining both Richards and Richards-Mason alterations, has 1871 and 1872 patent dates on the left frame indicating its fabrication largely from modified and unassembled parts—rather than from the total reworking of a complete or nearly finished percussion revolver.

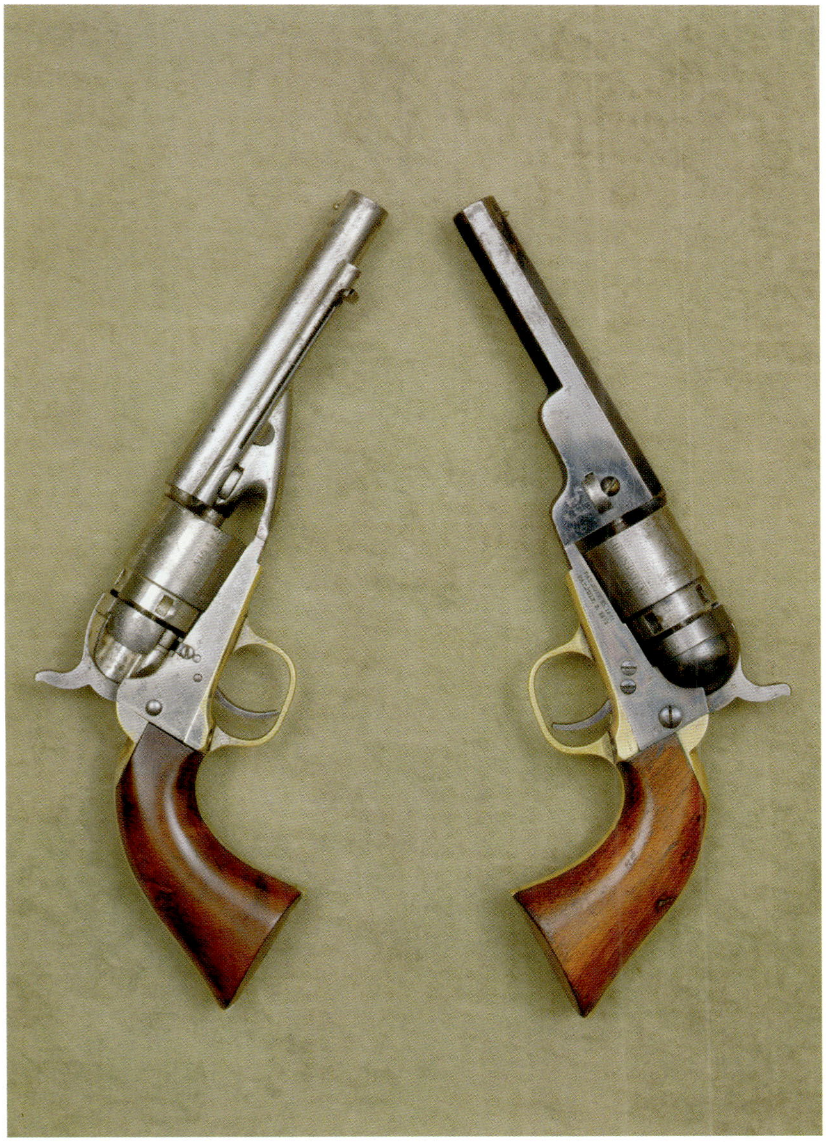

Breech-Loading Cartridge Revolver (left)

Colt's Patent Fire Arms Manufacturing Company, Hartford, Connecticut
Model 1862 Police/Pocket Navy Conversion, 1873–1875, Caliber .38 Colt Rimfire
Weitzenhoffer Collection, 2001.46.031

Fabricated from virtually obsolete parts, this revolver combines the standard Model 1862, Richards-Mason conversion frame with a round, Navy-pattern cylinder and a round, Police-pattern barrel fitted with an ejector rod assembly. The specimen, like so many Colt conversions exported to Mexico and Central and South America, received a full nickel-plated factory finish, perhaps in an effort to increase the visual appeal of an all-but-antiquated weapon. However, the high polish typically required by Colt's for its plated guns was not applied to an arm considered little better than surplus. Approximately 6,500 such arms were produced in varying barrel lengths.

Breech-Loading Cartridge Revolver (right)

Colt's Patent Fire Arms Manufacturing Company, Hartford, Connecticut
Model 1862 [1865] Pocket Navy Conversion, 1873–1875, Caliber .38 Colt Rimfire
Weitzenhoffer Collection, 2001.46.030

Stamped with 1871 and 1872 patent dates on the left side of the frame, this revolver incorporates the design alterations of C. B. Richards and William Mason in converting Colt percussion pistols to metallic-cartridge ignition. This specimen actually was made from unassembled and modified parts—rather than from the reworking of a complete or nearly finished percussion Pocket Navy. Changes included the removal of the cylinder's breech and through-boring of each chamber, the addition of a breechplate with a hinged loading gate on the frame, and the elimination of the loading lever assembly. About 4,000 revolvers of this particular pattern were produced with the 4½-inch octagonal barrel.

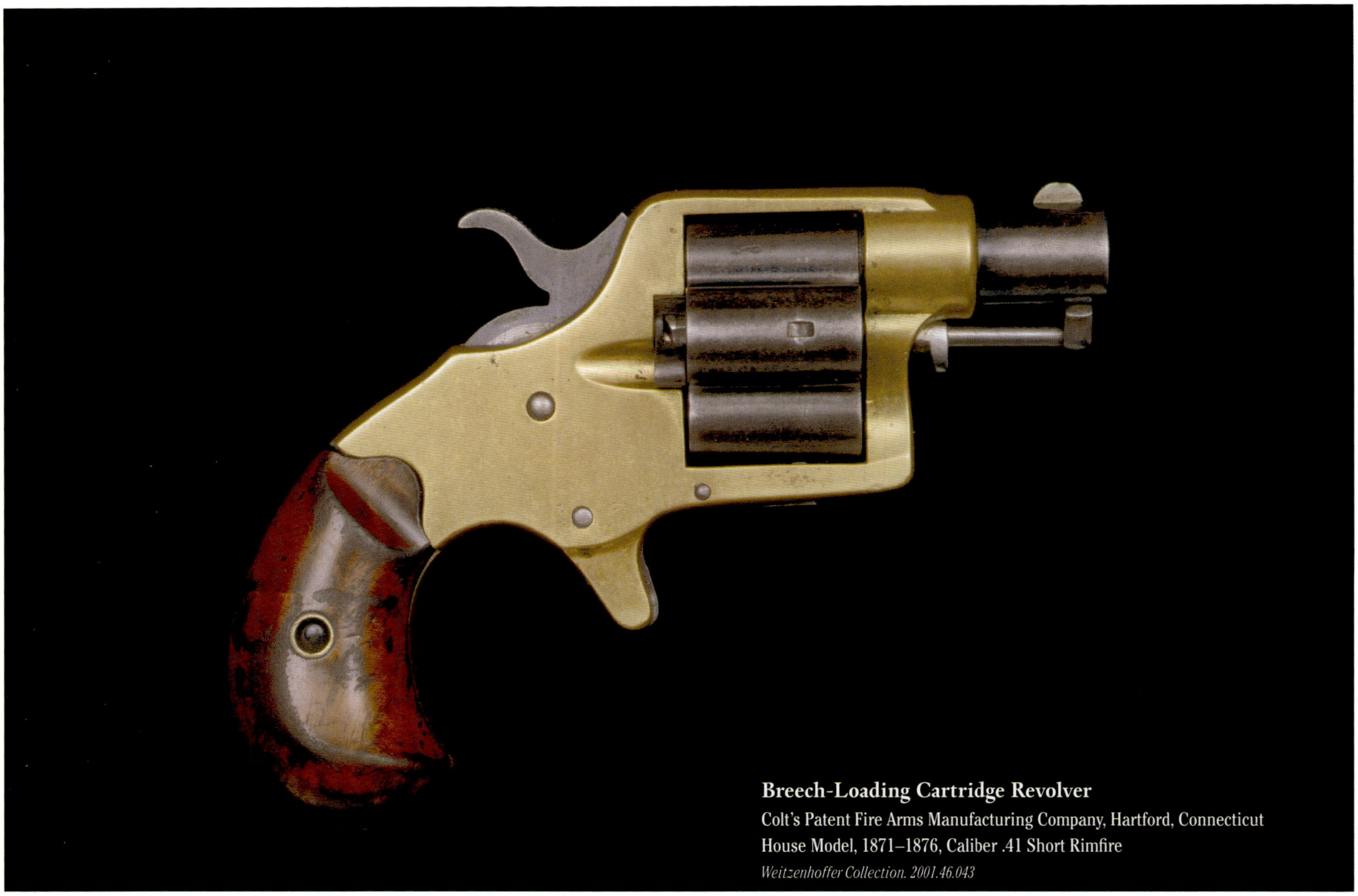

Breech-Loading Cartridge Revolver
Colt's Patent Fire Arms Manufacturing Company, Hartford, Connecticut
House Model, 1871–1876, Caliber .41 Short Rimfire
Weitzenhoffer Collection. 2001.46.043

A significant milestone in Colt firearms manufacture, the House Model was the first revolver pattern designed specifically for metallic cartridges. Often termed the Cloverleaf revolver for the shape of its four-chambered cylinder, the pistol incorporated a bird's-head-pattern grip and spur trigger on a milled and broached frame of brass. The gun presented one particularly original attribute—the rear face of the cylinder was countersunk around the chambers to receive the rims of the cartridges, a design feature that was years ahead of its time and not utilized again by Colt until the twentieth century. This example is considered scarce because of its snub-nosed, 1½-inch barrel, marked simply "COLT."

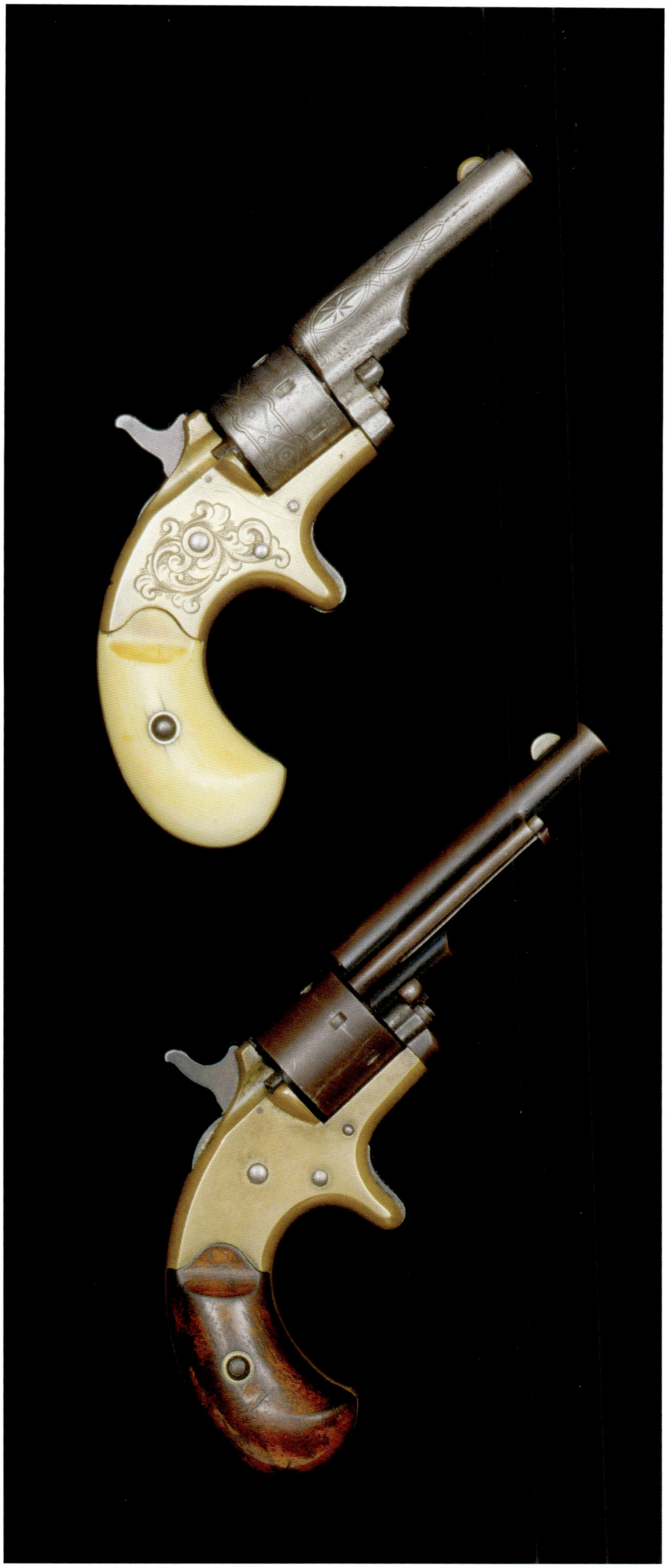

Breech-Loading Cartridge Revolvers

Colt's Patent Fire Arms Manufacturing Company, Hartford, Connecticut
Open Top Pocket Models, 1871–1877, Caliber .22 Short or Long Rimfire
Weitzenhoffer Collection, 2001.46.044 and 2001.46.045

Although this open-topped, three-piece pattern (having separate barrel, cylinder, and grip-frame elements secured with a pivoting wedge) represented an antiquated design from the company's percussion era, it nevertheless sold extremely well. More than 114,000 Open Top Pocket revolvers were marketed over a span of seven years, even though Colt's New Line Pocket Model rendered them obsolete after 1873. The design was notable as Colt's first seven-shot pistol. *Above*, a rare example engraved in either a Nimschke or a "New York" style on the frame, cylinder, and barrel; nickel-plated over cylinder and barrel; and fitted with grips of elephant ivory. *Below*, a scarce, early production piece fitted with a 3-inch barrel and ejector rod assembly.

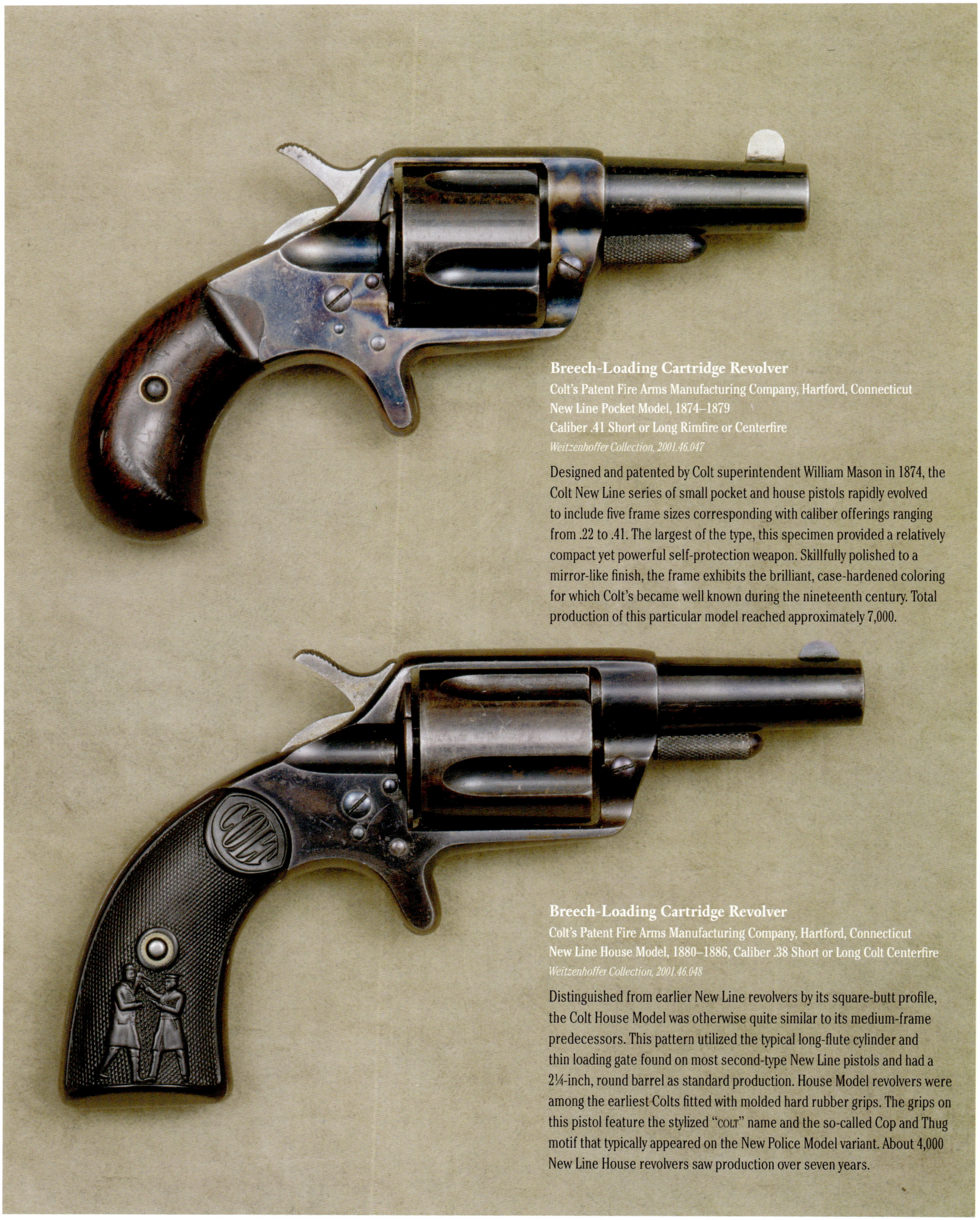

Breech-Loading Cartridge Revolver
Colt's Patent Fire Arms Manufacturing Company, Hartford, Connecticut
New Line Pocket Model, 1874–1879
Caliber .41 Short or Long Rimfire or Centerfire
Weitzenhoffer Collection, 2001.46.047

Designed and patented by Colt superintendent William Mason in 1874, the Colt New Line series of small pocket and house pistols rapidly evolved to include five frame sizes corresponding with caliber offerings ranging from .22 to .41. The largest of the type, this specimen provided a relatively compact yet powerful self-protection weapon. Skillfully polished to a mirror-like finish, the frame exhibits the brilliant, case-hardened coloring for which Colt's became well known during the nineteenth century. Total production of this particular model reached approximately 7,000.

Breech-Loading Cartridge Revolver
Colt's Patent Fire Arms Manufacturing Company, Hartford, Connecticut
New Line House Model, 1880–1886, Caliber .38 Short or Long Colt Centerfire
Weitzenhoffer Collection, 2001.46.048

Distinguished from earlier New Line revolvers by its square-butt profile, the Colt House Model was otherwise quite similar to its medium-frame predecessors. This pattern utilized the typical long-flute cylinder and thin loading gate found on most second-type New Line pistols and had a 2¼-inch, round barrel as standard production. House Model revolvers were among the earliest Colts fitted with molded hard rubber grips. The grips on this pistol feature the stylized "COLT" name and the so-called Cop and Thug motif that typically appeared on the New Police Model variant. About 4,000 New Line House revolvers saw production over seven years.

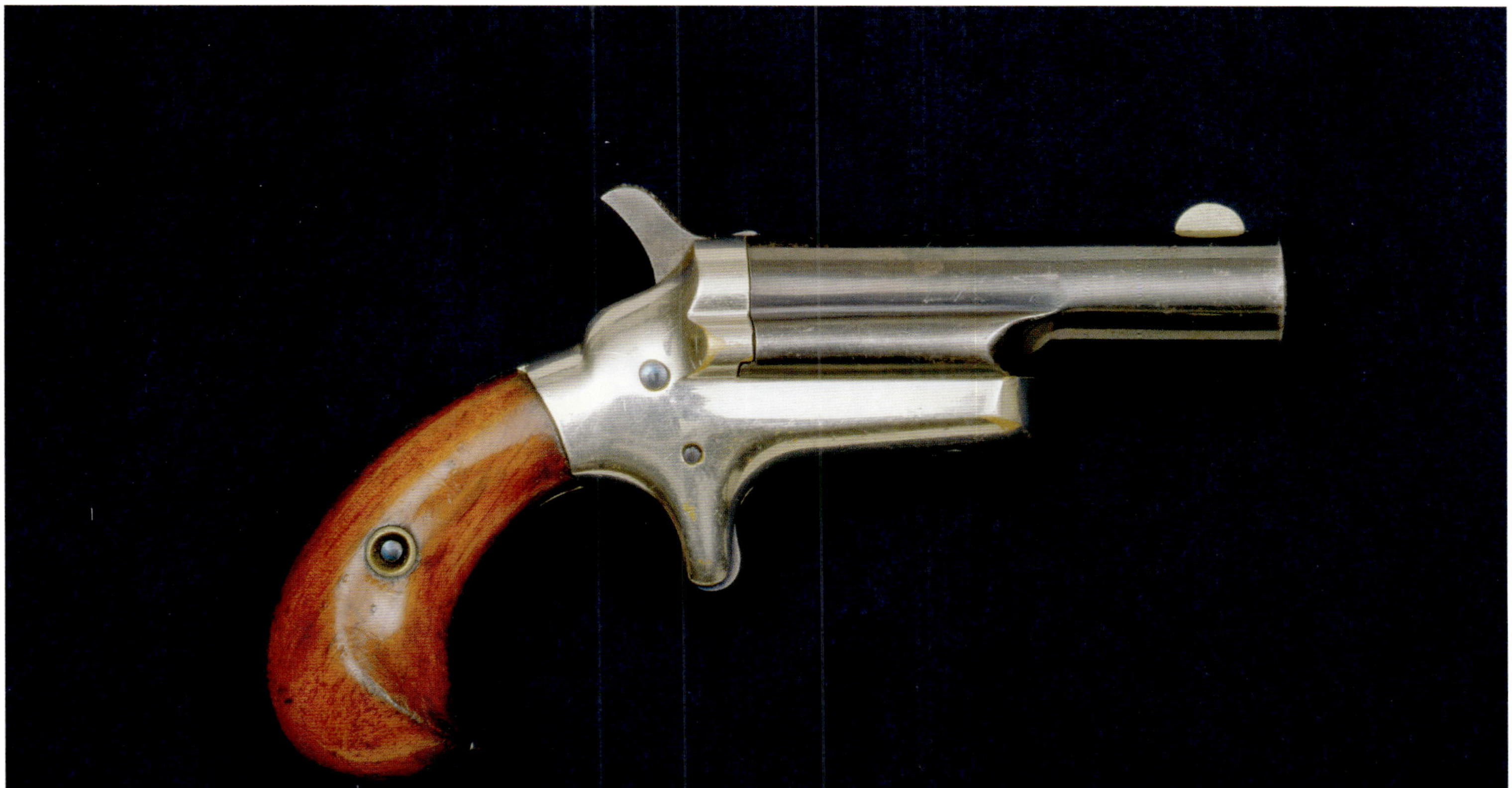

Breech-Loading Cartridge Pistol
Colt's Patent Fire Arms Manufacturing Company, Hartford, Connecticut
Third Model (Thuer) Deringer, 1875–1912, Caliber .41 Short Rimfire
Weitzenhoffer Collection, 2001.46.942

Designed by Colt engineer F. Alexander Thuer, the Third Model Deringer proved by far the most popular of the firm's diminutive, single-shot pistols. Approximately 48,000 were produced over thirty-seven years. The design featured a graceful, bird's-head grip profile, a spur trigger, and a pivoting-barreled breech-loading system. Stamped "-COLT-" on its blued barrel, this specimen is finished with a bright, nickel-plated frame, a color-case-hardened hammer, and varnished grips of walnut. The Third Model Deringer was the first firearm to carry the "Rampant Colt" trademark—if only in limited numbers. The firm's London agency requested the logo to distinguish its arms from the cut-priced deringers sold in England by various New York City exporters.

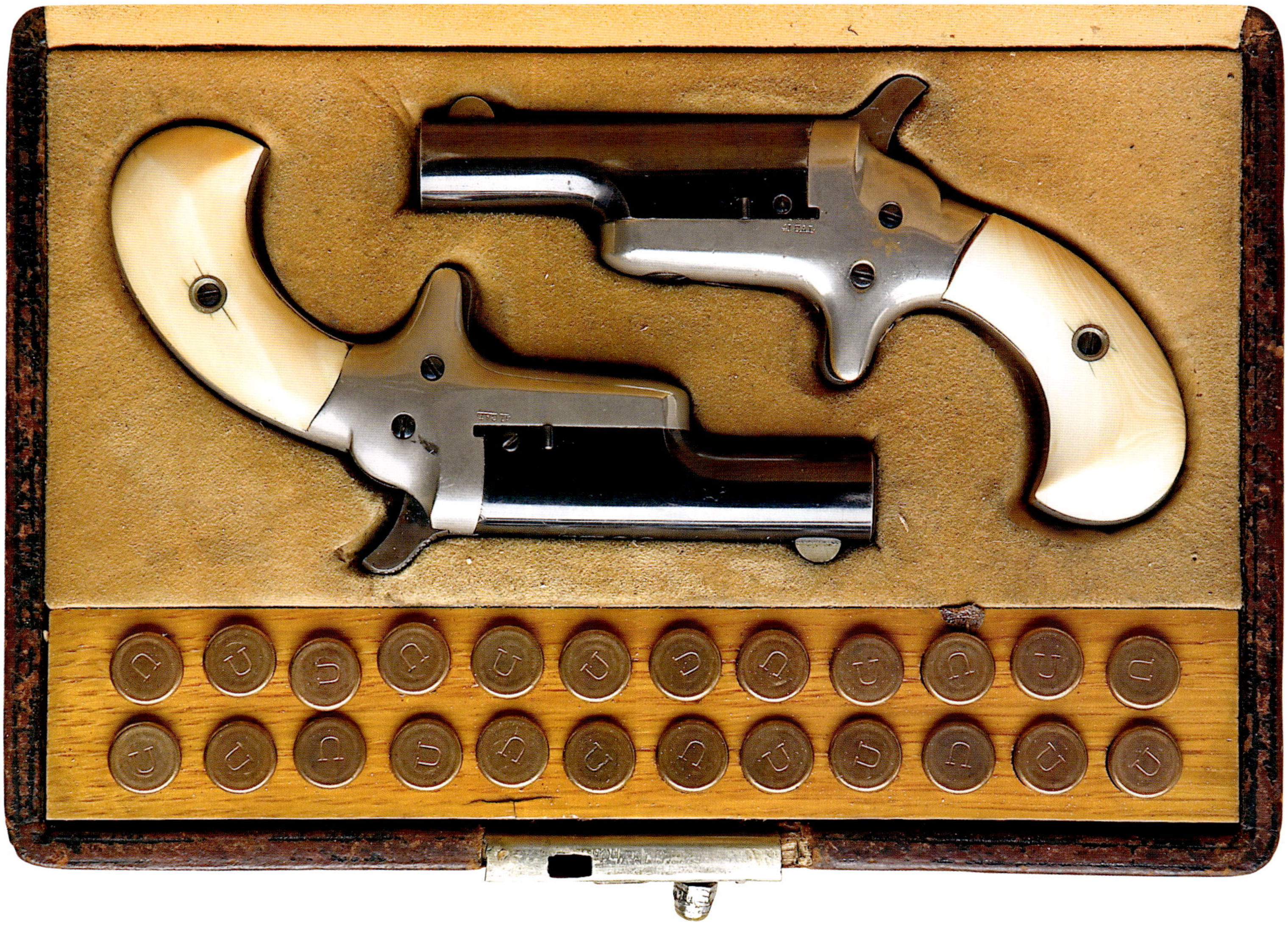

Breech-Loading Cartridge Pistols

Colt's Patent Fire Arms Manufacturing Company, Hartford, Connecticut
Third Model (Thuer) Deringers, 1875–1912, Caliber .41 Short Rimfire
Weitzenhoffer Collection, 2006.06.1 A–C

These special-order, Colt Third Model deringers reflect the discriminating taste of a Gilded Age gentleman. The pistols are housed in a leather-covered wooden case with French-style, contour-fitted interior finished in tan suede and with an oak cartridge block accommodating twenty-four rounds. Finished as a matched pair, the guns incorporate bright-blued barrels, nickel-plated frames, case-hardened hammers, and custom grips of elephant ivory. The backstrap of each weapon is engraved with the original owner's name: "MATHEW CORBETT." The ensemble probably was gotten up for Mr. Corbett by the dealer, whose business is identified in gold-embossed letters on the interior of the case lid: "J. B. CROOK & CO. / GUNS & PISTOLS / 1180 BROADWAY, N.Y."

Image courtesy Greg Martin Auctions, San Francisco, California

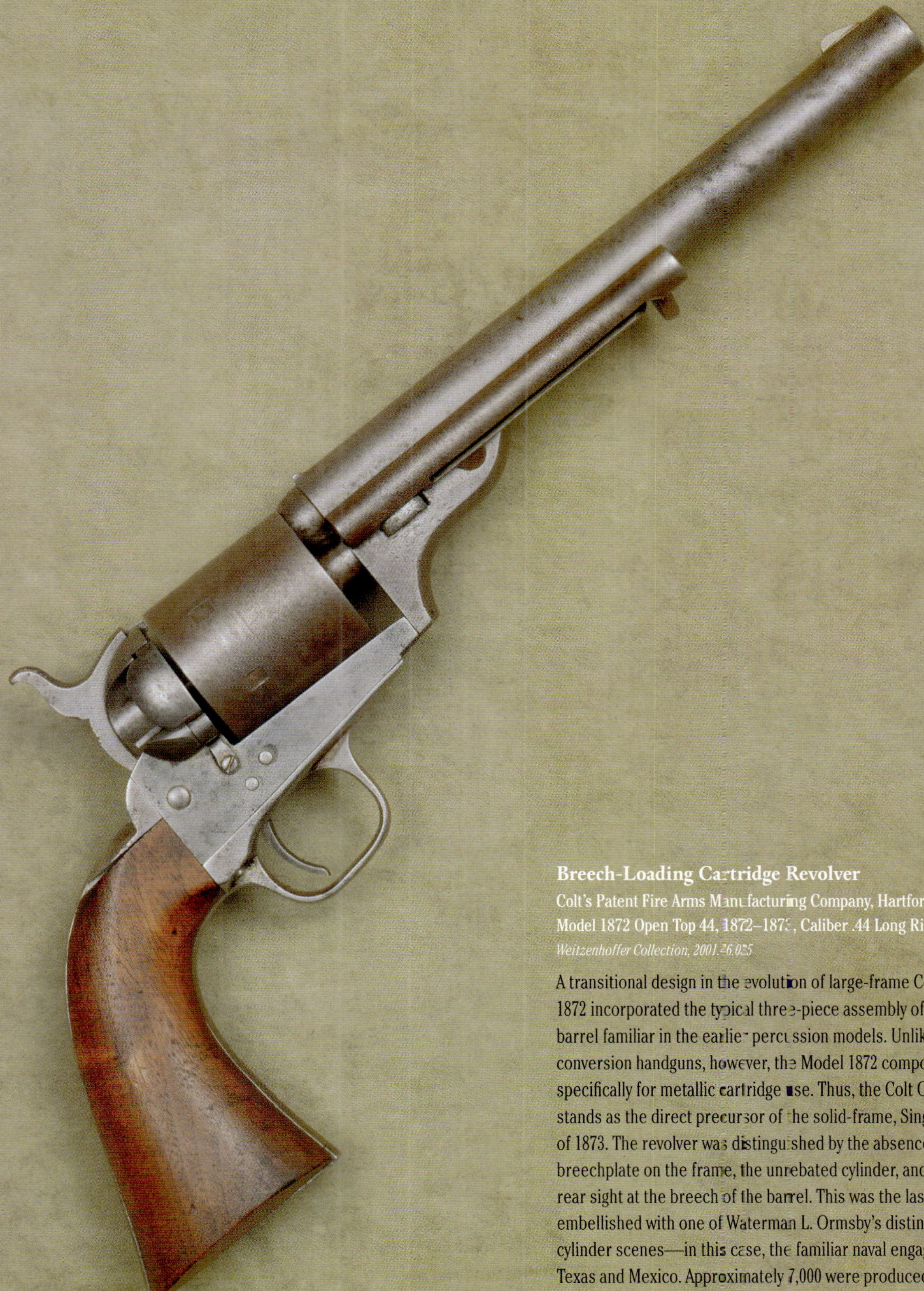

Breech-Loading Cartridge Revolver

Colt's Patent Fire Arms Manufacturing Company, Hartford, Connecticut
Model 1872 Open Top 44, 1872–1873, Caliber .44 Long Rimfire
Weitzenhoffer Collection, 2001.26.025

A transitional design in the evolution of large-frame Colt pistols, the Model 1872 incorporated the typical three-piece assembly of frame, cylinder, and barrel familiar in the earlier percussion models. Unlike the percussion and conversion handguns, however, the Model 1872 components were made specifically for metallic cartridge use. Thus, the Colt Open Top revolver stands as the direct precursor of the solid-frame, Single Action Army model of 1873. The revolver was distinguished by the absence of a conversion breechplate on the frame, the unrebated cylinder, and the integral rear sight at the breech of the barrel. This was the last Colt revolver embellished with one of Waterman L. Ormsby's distinctive, roll-engraved cylinder scenes—in this case, the familiar naval engagement between Texas and Mexico. Approximately 7,000 were produced.

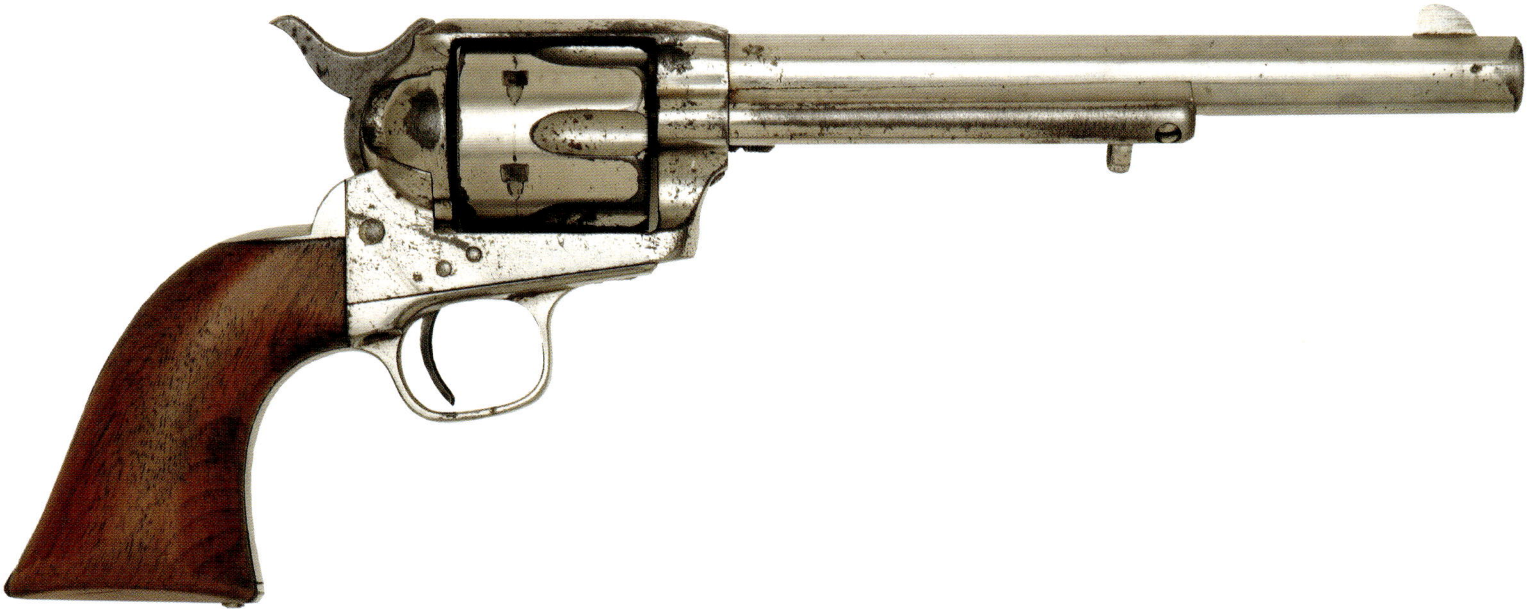

Breech-Loading Cartridge Revolver
Colt's Patent Fire Arms Manufacturing Company, Hartford, Connecticut
Model 1873 Single Action Army, 1872–1940, Caliber .45 Long Colt Centerfire
Gift of Sharon Hicks, 2009.25

Colt Single Action Army revolvers with 7½-inch-barrels provided their devotees with a well-balanced handgun having natural "pointability." This was the standard barrel length issued to U.S. mounted troops, and the pattern proved a favorite with many frontier civilians as well. (Other standard, commercial barrel lengths for the Model 1873 Colt were 5½ inches and 4¾ inches, though almost any length could be had on special order.) Manufactured in 1876, this specimen boasts a special-order, full nickel-plated factory finish, but it is otherwise a standard revolver with one-piece grips of walnut. This Colt model perhaps most exemplifies the classic handgun of the American West.

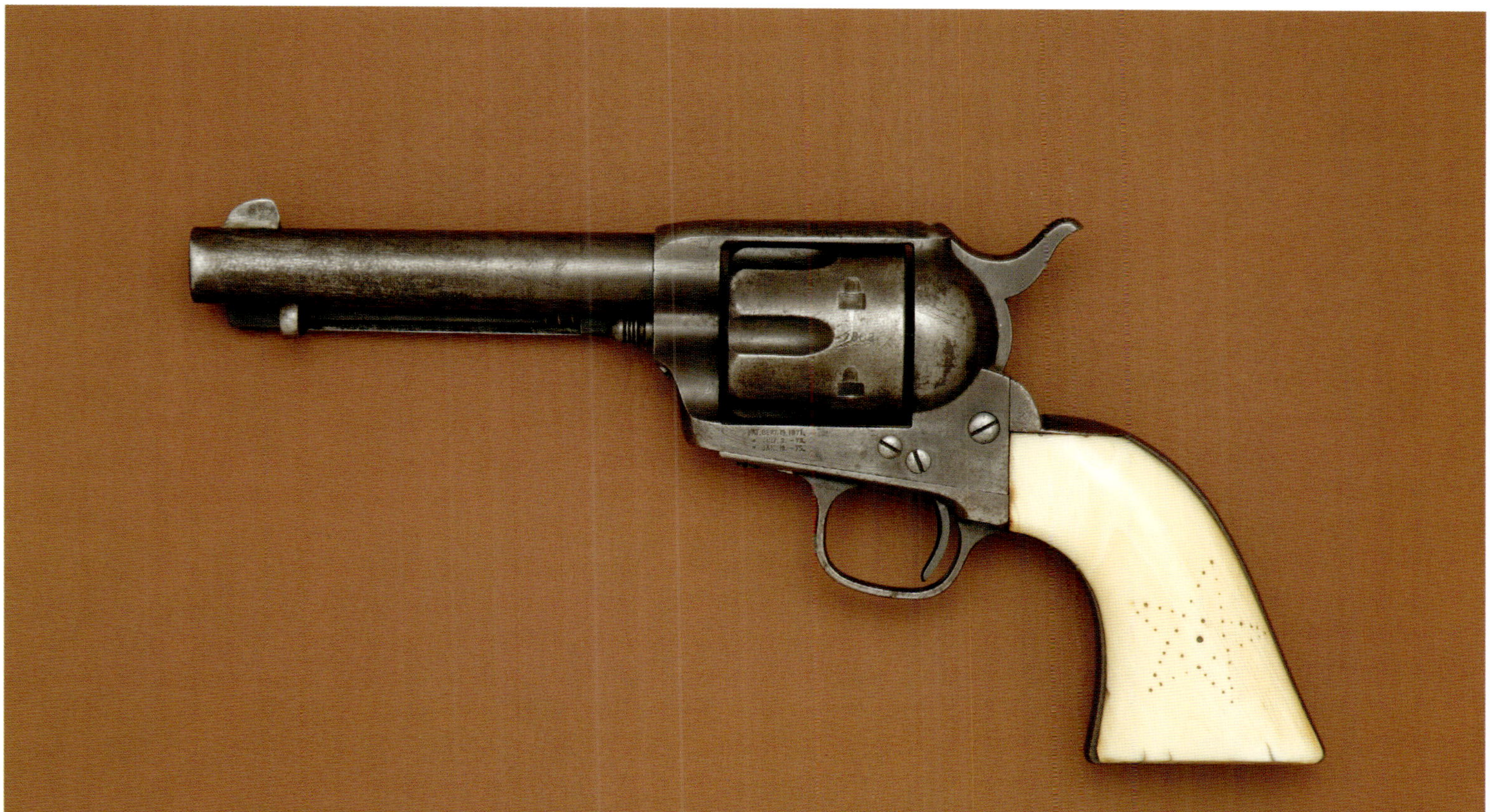

Breech-Loading Cartridge Revolver
Colt's Patent Fire Arms Manufacturing Company, Hartford, Connecticut
Model 1873 Single Action Army, 1872–1940, Caliber .45 Long Colt Centerfire
Permanent Collection, H.175

Manufactured in 1883, this Colt Single Action Army revolver was carried by N. T. Stubbs, U.S. deputy marshal in the Indian Territory and later a deputy sheriff of Atoka County, Oklahoma. The piece left the factory with a 4¾-inch barrel; standard, case-hardened and blued finish; and special-order, one-piece grips of elephant ivory. Stubbs was probably the one who neatly embellished the left grip panel with a five-pointed star—likely using a heated pin—to commemorate his career in law enforcement. Now with mellowed grips and well patinated from years of service, this piece retains fine western and historical provenance.

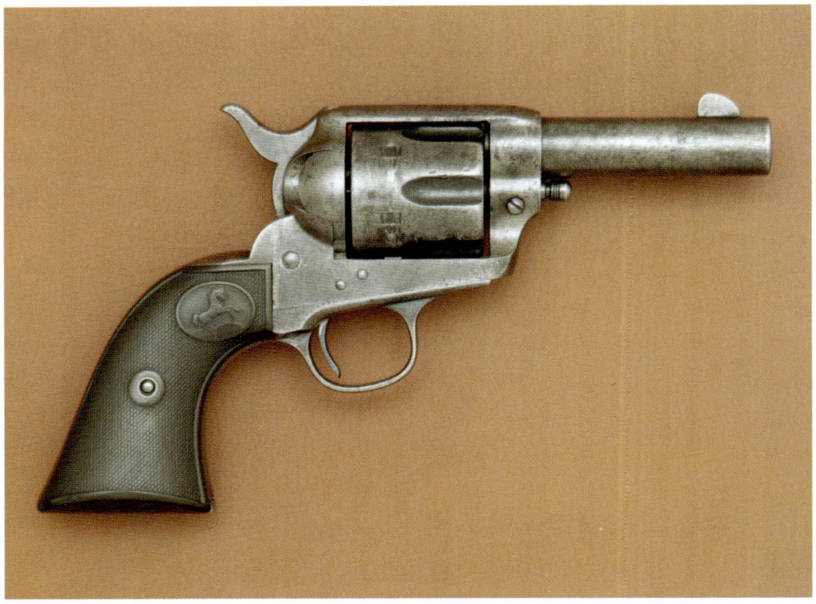

Breech-Loading Cartridge Revolver

Colt's Patent Fire Arms Manufacturing Company, Hartford, Connecticut
Model 1873 Sheriff's or Storekeeper's Single Action Army, 1882–1940
Caliber .45 Long Colt Centerfire
Weitzenhoffer Collection, 2001.46.038

Among the scarcer variations of the Colt Single Action Army revolver, the so-called Sheriff's or Storekeeper's model is distinguished by the absence of an ejector rod assembly on the right side of the barrel. First introduced in 1882, only about 1,200 of this pattern were manufactured—most with barrels ranging from only 2½ to 4¾ inches. The pistol was conceived and marketed as a powerful "hideaway" gun ideal for concealment in one's coat pocket or waistband. Made about 1904, this specimen was supplied with a 3½-inch barrel and standard hard rubber grips having the "Rampant Colt" trademark of the period.

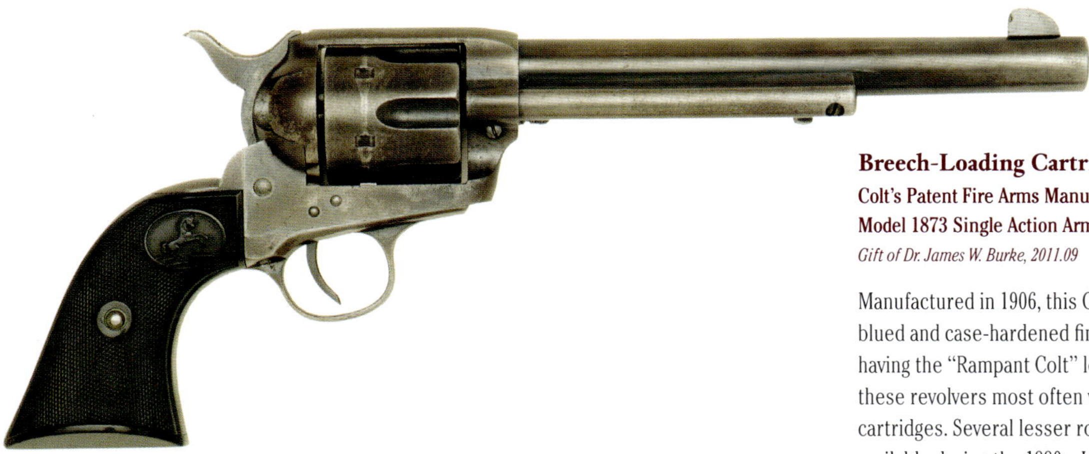

Breech-Loading Cartridge Revolver

Colt's Patent Fire Arms Manufacturing Company, Hartford, Connecticut
Model 1873 Single Action Army, 1872–1940, Caliber .38 Colt Centerfire
Gift of Dr. James W. Burke, 2011.09

Manufactured in 1906, this Colt Single Action Army revolver presents standard blued and case-hardened finish with two-piece, checkered hard rubber grips having the "Rampant Colt" logo at the top. During the late nineteenth century, these revolvers most often were chambered for the larger .45- and .44-caliber cartridges. Several lesser rounds, such as the .38 Colt employed here, became available during the 1880s. In all, Colt's chambered this long-serving revolver for some thirty different cartridges, from the diminutive .22 rimfire to the powerful .476 Eley centerfire. In total, some 357,000 Single Action Armys saw manufacture from 1872 to their first production cessation in 1940.

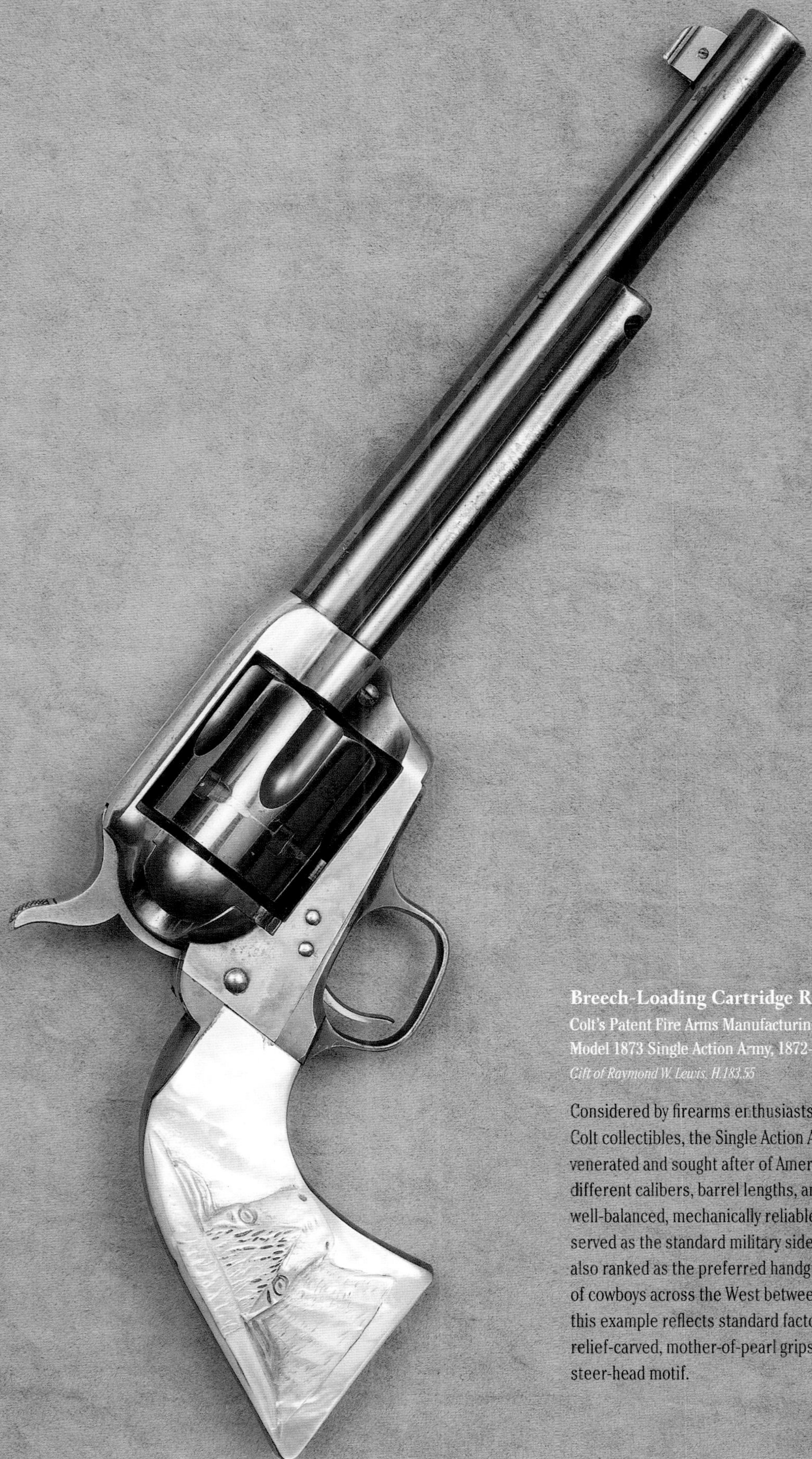

Breech-Loading Cartridge Revolver

Colt's Patent Fire Arms Manufacturing Company, Hartford, Connecticut
Model 1873 Single Action Army, 1872–1940, Caliber .45 Long Colt Centerfire
Gift of Raymond W. Lewis, H.183.55

Considered by firearms enthusiasts and connoisseurs as one of the classic Colt collectibles, the Single Action Army revolver remains among the most venerated and sought after of American weapons. The revolver (in its many different calibers, barrel lengths, and design variations) was exceptionally well-balanced, mechanically reliable, powerful, and accurate. The pistol served as the standard military sidearm during the late Indian Wars era and also ranked as the preferred handgun of outlaws, lawmen, and the majority of cowboys across the West between 1875 and 1900. Manufactured in 1907, this example reflects standard factory finish except for its special-order, relief-carved, mother-of-pearl grips, which are adorned with the popular steer-head motif.

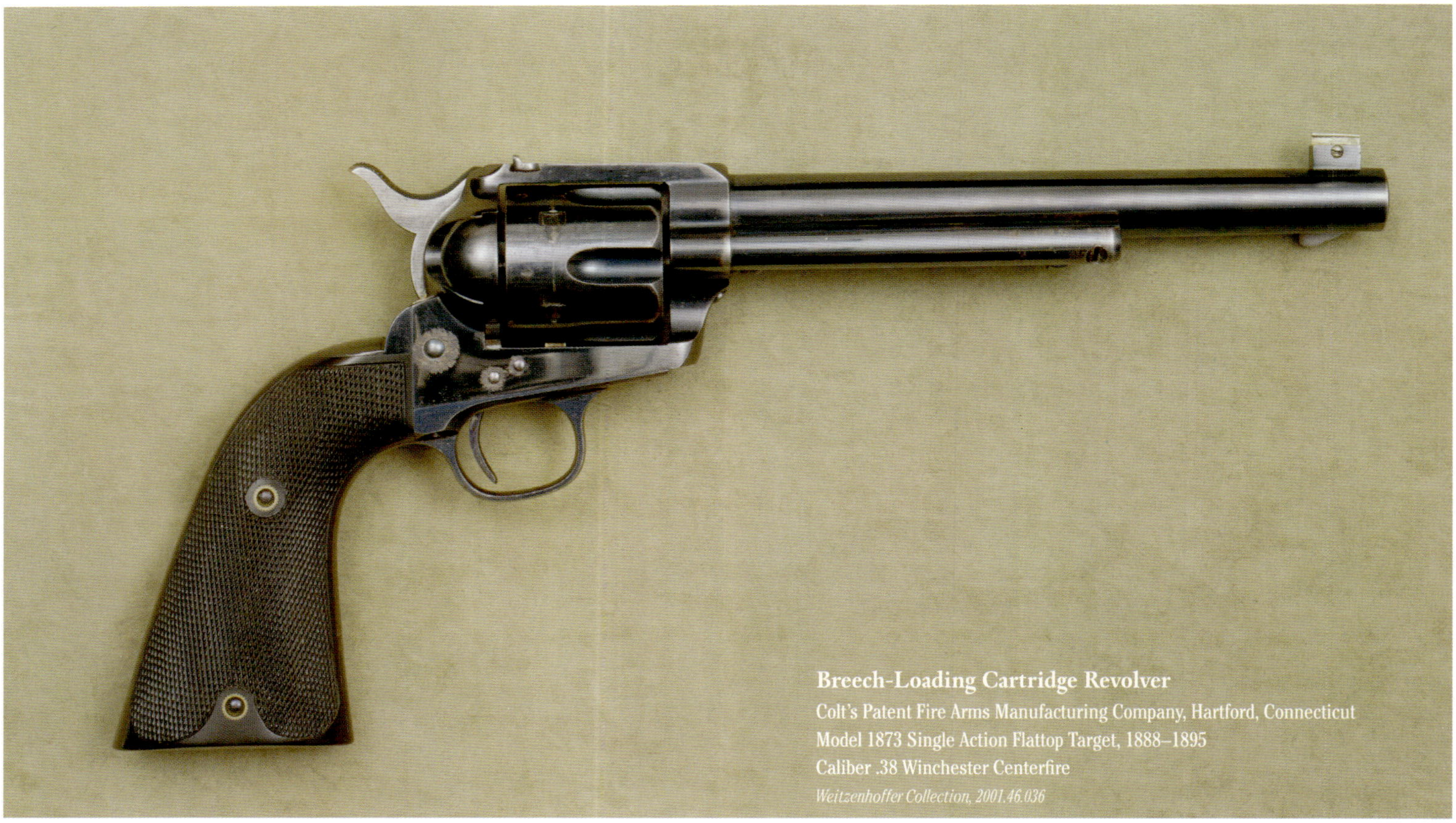

Breech-Loading Cartridge Revolver
Colt's Patent Fire Arms Manufacturing Company, Hartford, Connecticut
Model 1873 Single Action Flattop Target, 1888–1895
Caliber .38 Winchester Centerfire
Weitzenhoffer Collection, 2001.46.036

A distinctive and highly prized variant of the celebrated Colt Single Action Army, the Flattop Target revolver is distinguished by its flat, ungrooved topstrap with dovetailed rear sight and by a slotted, front sight base accommodating a variety of inserts (here a German silver bead type). This specimen also has an extended grip profile with finely checkered composition grips, and a hand-checkered trigger—both special-order features that increase its rarity. The barrel is marked "Colt's PT. F. A. MFG. Co. Hartford, CT. U.S.A. / Depot 14, Pall Mall London"—yet another detail contributing to the revolver's desirability. The type is highly sought after by Colt connoisseurs, as only about 925 saw manufacture and even fewer incorporated these features.

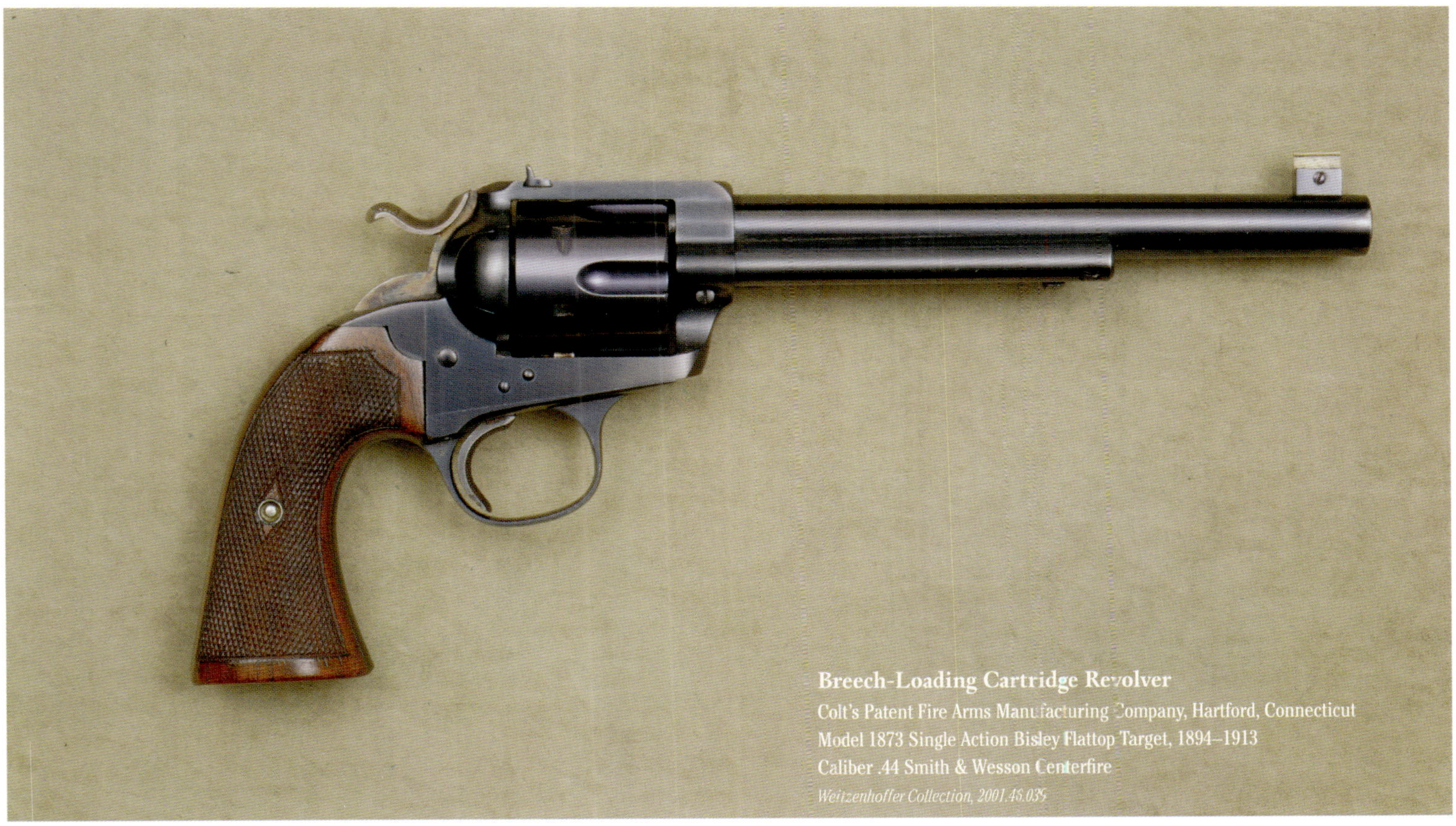

Breech-Loading Cartridge Revolver
Colt's Patent Fire Arms Manufacturing Company, Hartford, Connecticut
Model 1873 Single Action Bisley Flattop Target, 1894–1913
Caliber .44 Smith & Wesson Centerfire
Weitzenhoffer Collection, 2001.46.039

Although the standard Single Action Flattop Target pistol satisfied most competitive shooters, Colt's decided to further improve the arm by redesigning three particular elements. An extended, hunchback-pattern grip, positioned at nearly a right angle to the barrel, improved the revolver's "pointability." A redesigned low-profile hammer spur allowed cocking with less disruption of the shooter's "hold" on the grip. And a larger trigger guard and slightly longer trigger provided the shooter easier trigger access and finer shot control. Incorporating the characteristic flattop design with dovetailed rear and slotted front sights, only 976 of these "Special Target Model of 1894" revolvers saw production.

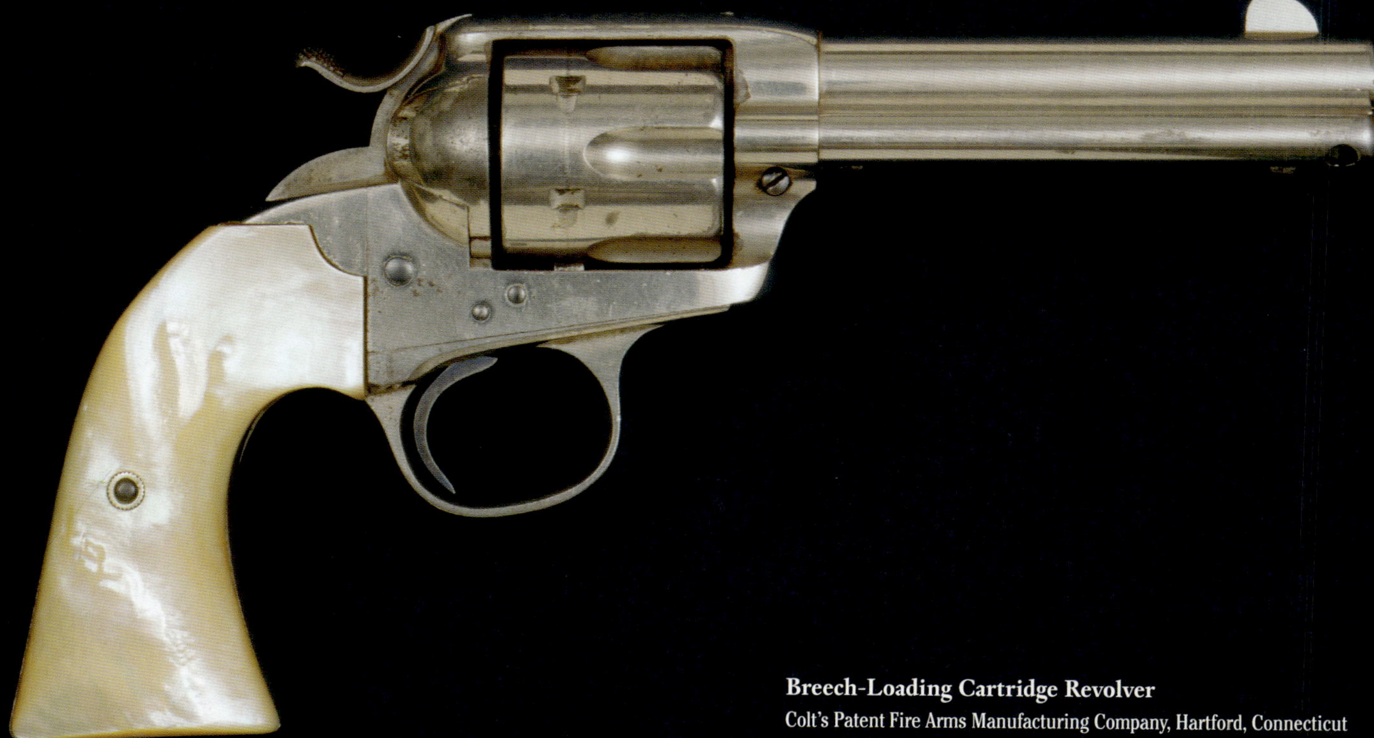

Breech-Loading Cartridge Revolver
Colt's Patent Fire Arms Manufacturing Company, Hartford, Connecticut
Model 1873 Single Action Bisley Target, 1896–1915,
Caliber .44 Winchester Centerfire
Gift of Bureau of Alcohol, Tobacco & Firearms, 1983.38.03

Increasing demand among American and British target shooters for an accurate and economical target rendition of the standard Single Action Army revolver induced Colt's to create the Bisley pattern, a modified design that replaced the earlier Flattop Target variation. Named for a famed English target range, the Bisley incorporated the extended grip, low hammer spur, longer trigger, and larger trigger guard of its flattop antecedent. This special-order, factory-finished gun boasts full nickel plating and iridescent mother-of-pearl grips. Designating both model and caliber, the barrel is roll-stamped "(BISLEY MODEL) / COLT FRONTIER SIX SHOOTER." Although this model appeared too late for much use in the frontier West, the Bisley nevertheless sold better in the United States than in Great Britain. In all, some 44,000 were manufactured.

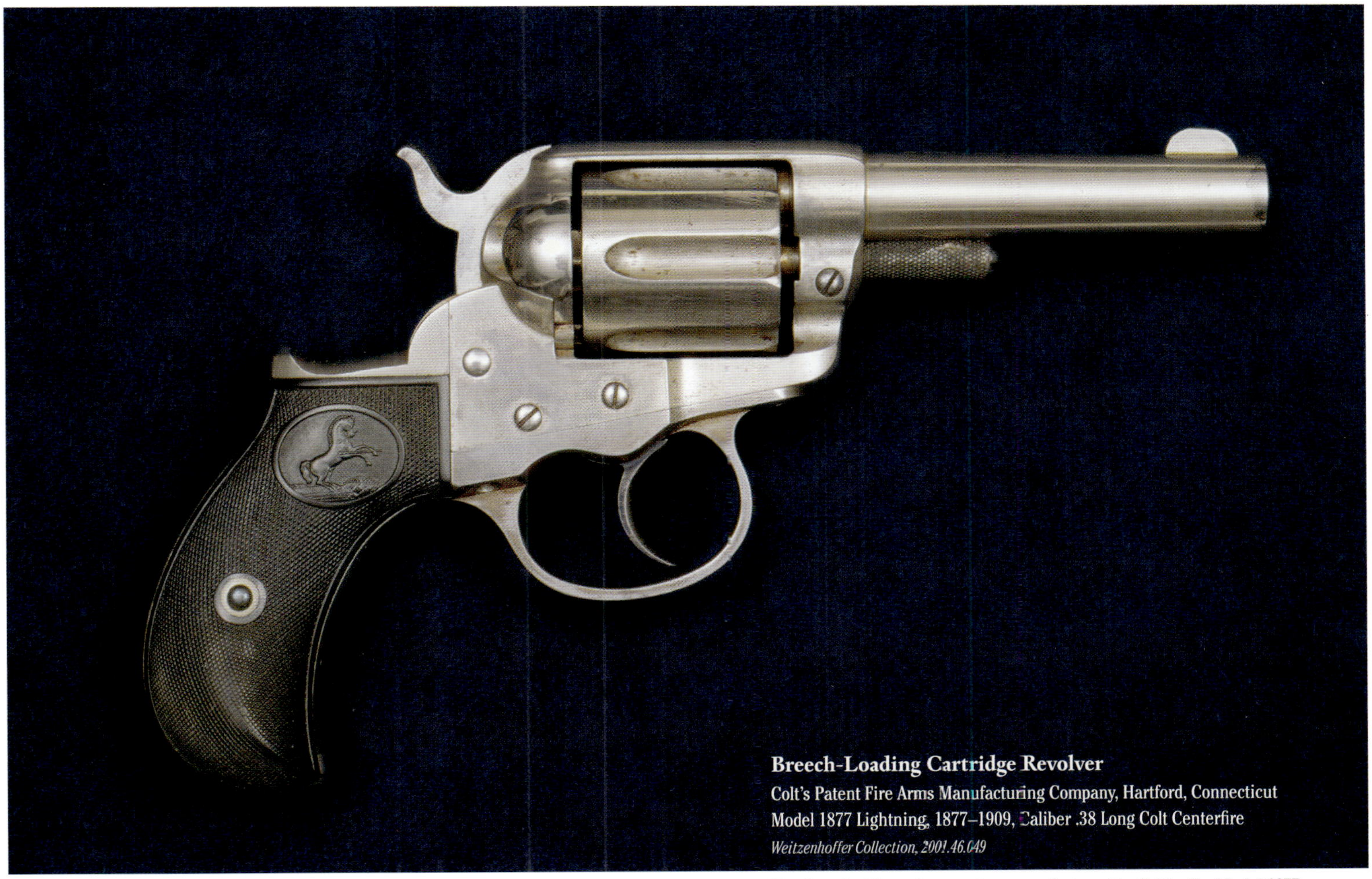

Breech-Loading Cartridge Revolver
Colt's Patent Fire Arms Manufacturing Company, Hartford, Connecticut
Model 1877 Lightning, 1877–1909, Caliber .38 Long Colt Centerfire
Weitzenhoffer Collection, 2001.46.049

The first double-action revolver manufactured by Colt's, the Model 1877 combined the general design of the Model 1873 Single Action Army with a tasteful bird's-head grip and a distinctively sculpted frame. This example, having a relatively short, 3½-inch barrel without an ejector rod assembly, is reminiscent of the scarce Sheriff's or Storekeeper's variation of the Single Action Army revolver. This specimen's full nickel-plated finish, though not uncommon in this model, remained a special-order embellishment that commanded a modest premium in retail cost. The hard rubber grips are finished with likenesses of the firm's famous "Rampant Colt" trademark. Approximately 166,000 Lightning revolvers were manufactured over thirty-two years.

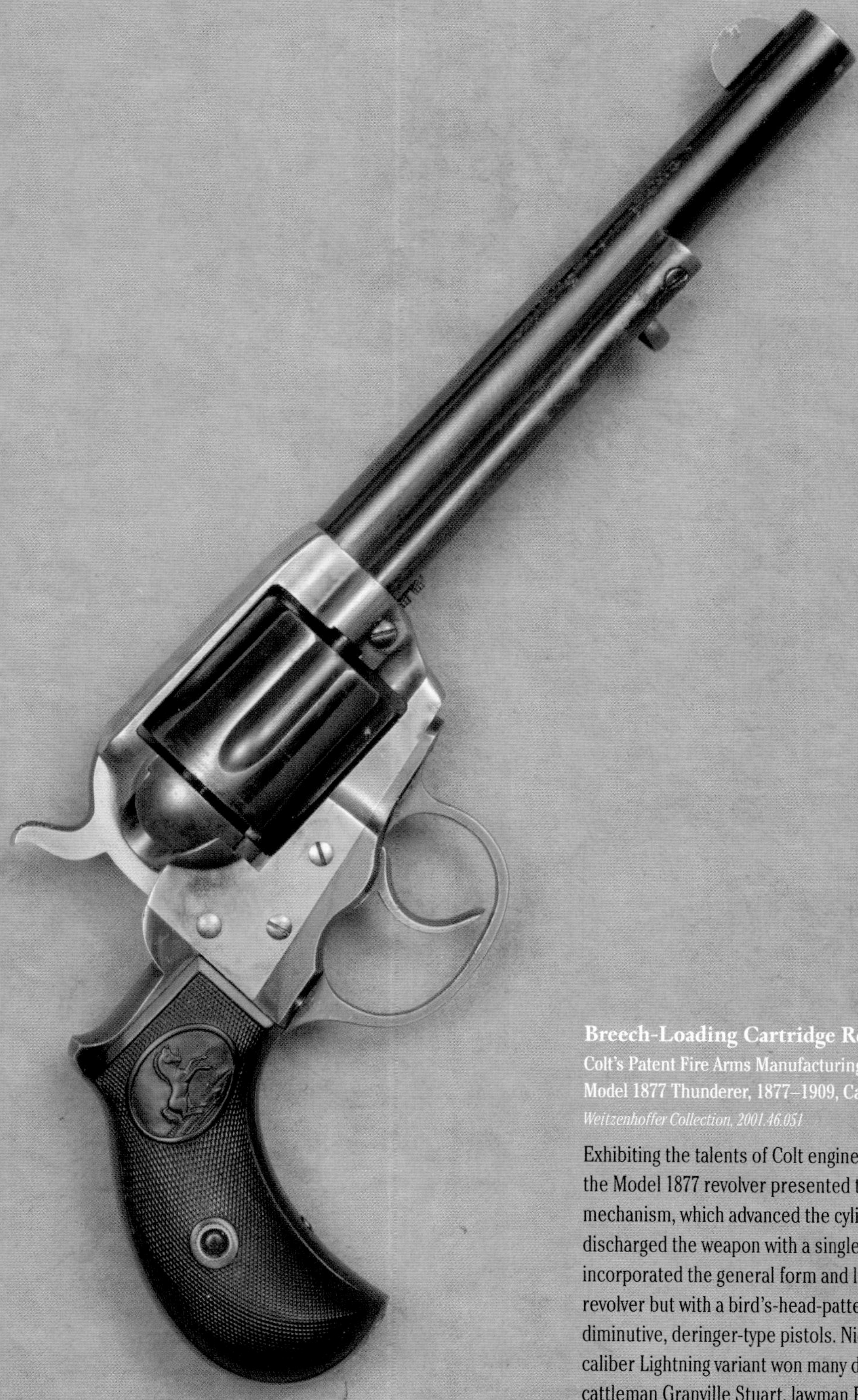

Breech-Loading Cartridge Revolver
Colt's Patent Fire Arms Manufacturing Company, Hartford, Connecticut
Model 1877 Thunderer, 1877–1909, Caliber .41 Long Colt Centerfire
Weitzenhoffer Collection, 2001.46.051

Exhibiting the talents of Colt engineers William Mason and C. B. Richards, the Model 1877 revolver presented the company's first double-action mechanism, which advanced the cylinder, cocked the hammer, and discharged the weapon with a single squeeze of the trigger. The design incorporated the general form and line of the larger Single Action Army revolver but with a bird's-head-pattern grip design reminiscent of diminutive, deringer-type pistols. Nicknamed the Thunderer, this larger-caliber Lightning variant won many devotees in the Old West—among them cattleman Granville Stuart, lawman Pat Garrett, and outlaws John Wesley Hardin and Billy the Kid.

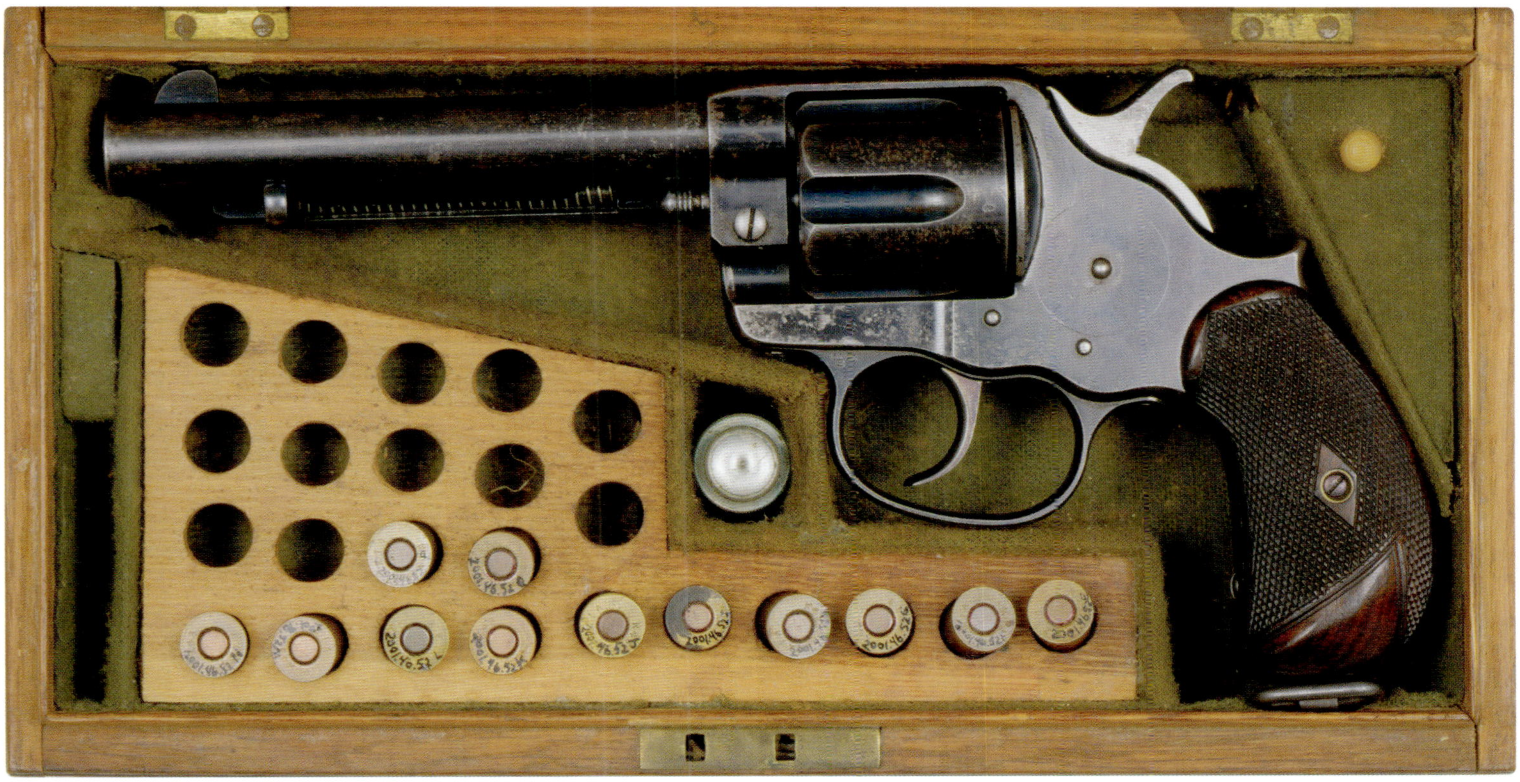

Breech-Loading Cartridge Revolver

Colt's Patent Fire Arms Manufacturing Company, Hartford, Connecticut
Model 1878 Frontier, 1878–1905, Caliber .450 Boxer Centerfire
Weitzenhoffer Collection, 2001.46.052

Among the largest-framed metallic-cartridge revolvers ever produced by Colt, the Model 1878 "Frontier" represented the company's double-action rendition of the very popular Single Action Army model. This specimen is considered scarce both for its early-production, checkered walnut grips (in contrast to the more common hard rubber variety) and for its chambering for a large, British military cartridge. The piece is housed in a compartmented London oak casing having a green felt lining. The interior of the case lid retains its Colt trade label with the well-known "14 Pall Mall, London" address, and the barrel also carries the London address inscription.

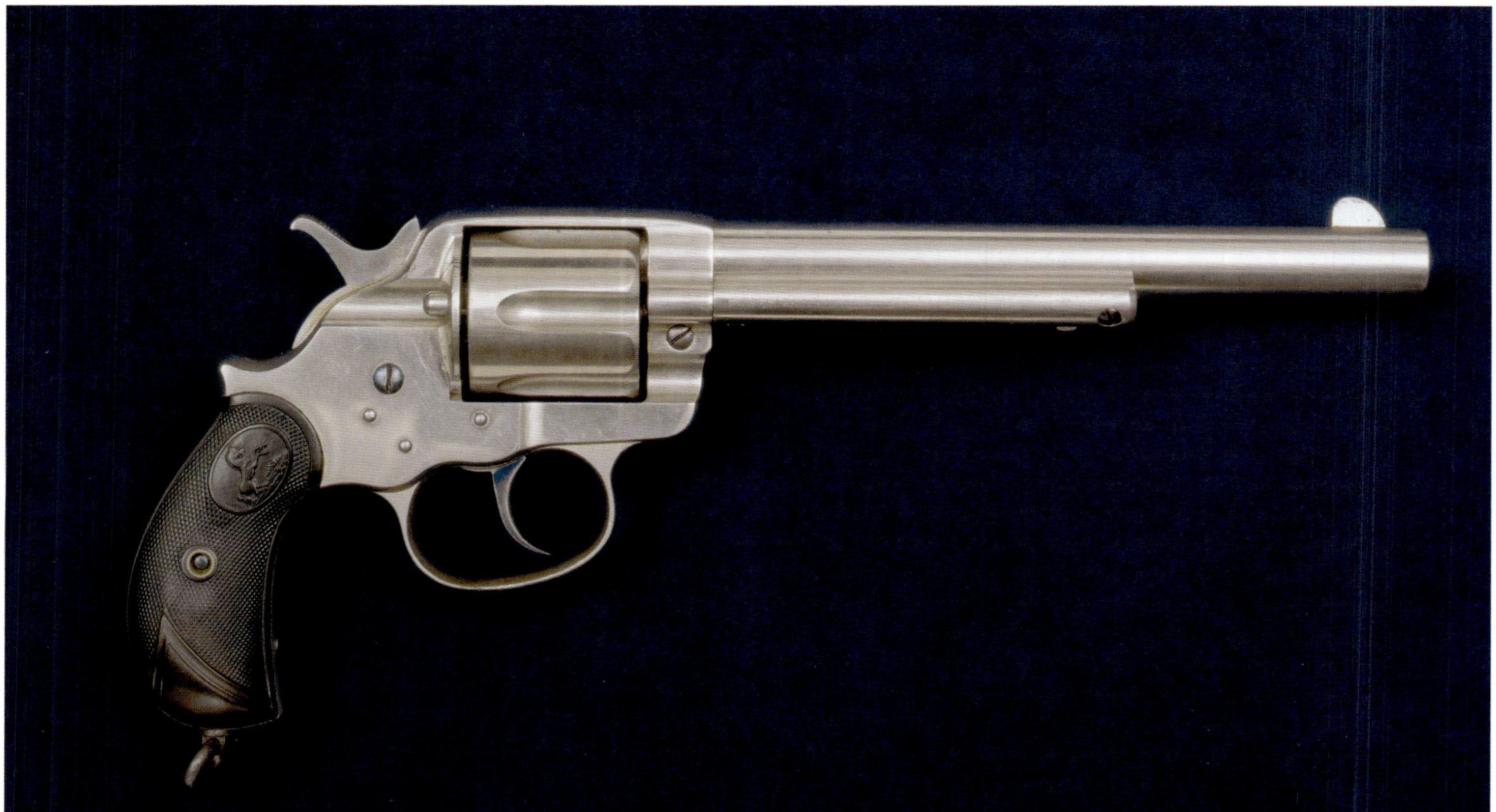

Breech-Loading Cartridge Revolver
Colt's Patent Fire Arms Manufacturing Company, Hartford, Connecticut
Model 1878 Frontier, 1878–1905, Caliber .38 Winchester Centerfire
Weitzenhoffer Collection, 2001.46.053

Manufactured in 1893, this large-frame, double-action revolver boasts a special-order, full nickel-plated finish direct from the Colt factory. Nickel plating demanded expert handling in the dry polishing shop to achieve the required mirror-smooth finish—particularly on the intricate surface contours of the frame. Although somewhat common in this model, nickel plating still exacted a nominal premium in price when compared with the standard blued finish. The model's chamberings in .32, .38, and .44 Winchester also were available in Colt's medium-frame Lightning model rifles, allowing Colt devotees to carry a single type of cartridge for both sidearms and longarms.

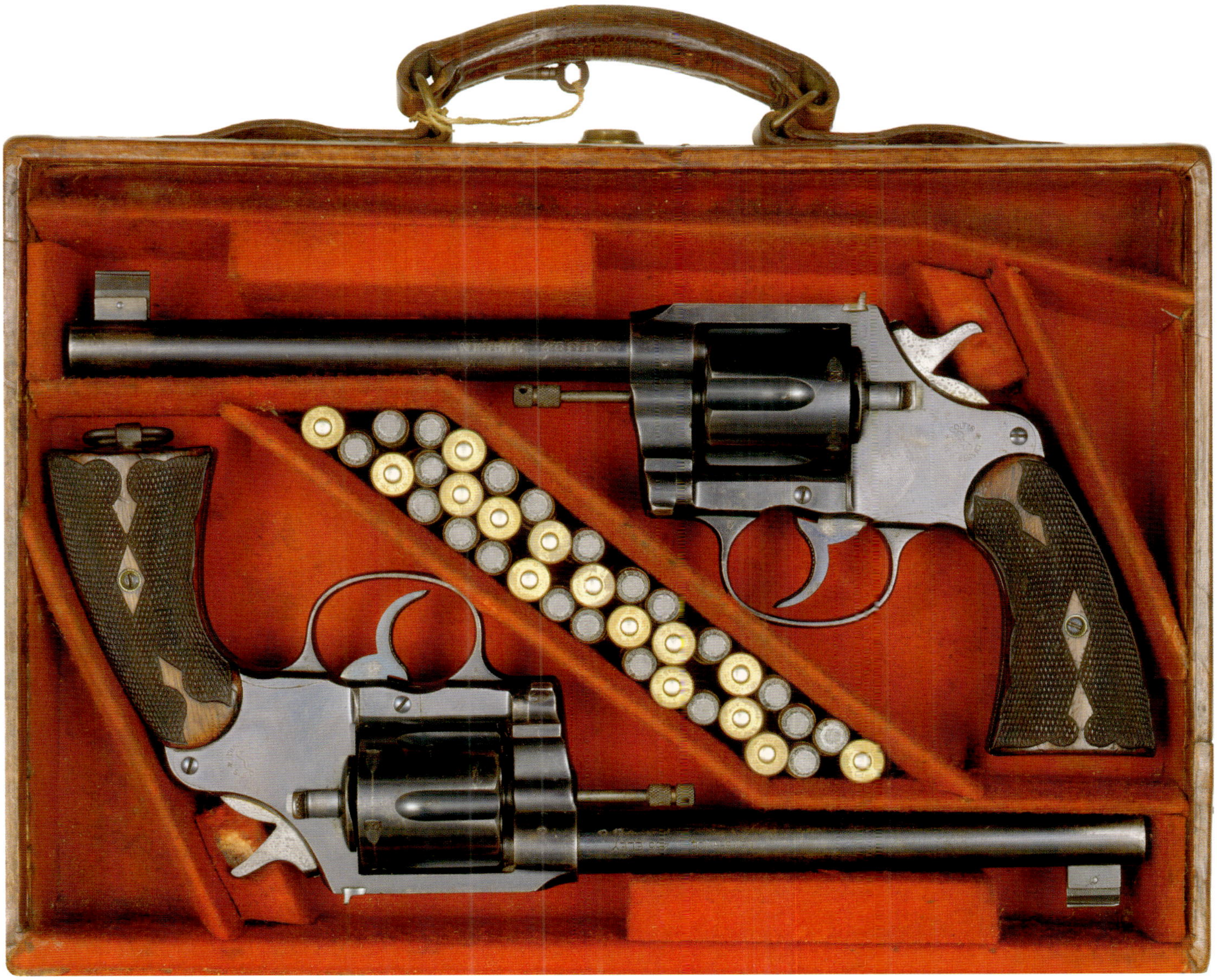

Breech-Loading Cartridge Revolvers

Colt's Patent Fire Arms Manufacturing Company, Hartford, Connecticut
New Service Flattop Target Model, 1898–1944, Caliber .455 Eley Centerfire
Weitzenhoffer Collection, 2001.060.6 A&B

Reflecting nineteenth-century etiquette in arms presentation, this pair of Colt target revolvers resides in a compartmented, leather-and-canvas casing lined in burgundy felt with a Colt trade label in the lid. A distinctive variant of the large-frame, swingout-cylinder, double-action New Service Model, these Flattop Target revolvers feature flat, ungrooved topstraps with dovetailed rear sights and slotted front sight bases with long-bead inserts. Probably made and presented in 1908, the topstraps of both guns are hand engraved: "CAPT. TOLLEY / QUEEN VICTORIA'S RIFLES." The pieces have hand-checkered gripstraps and triggers, as well as finely checkered grips of varnished walnut.

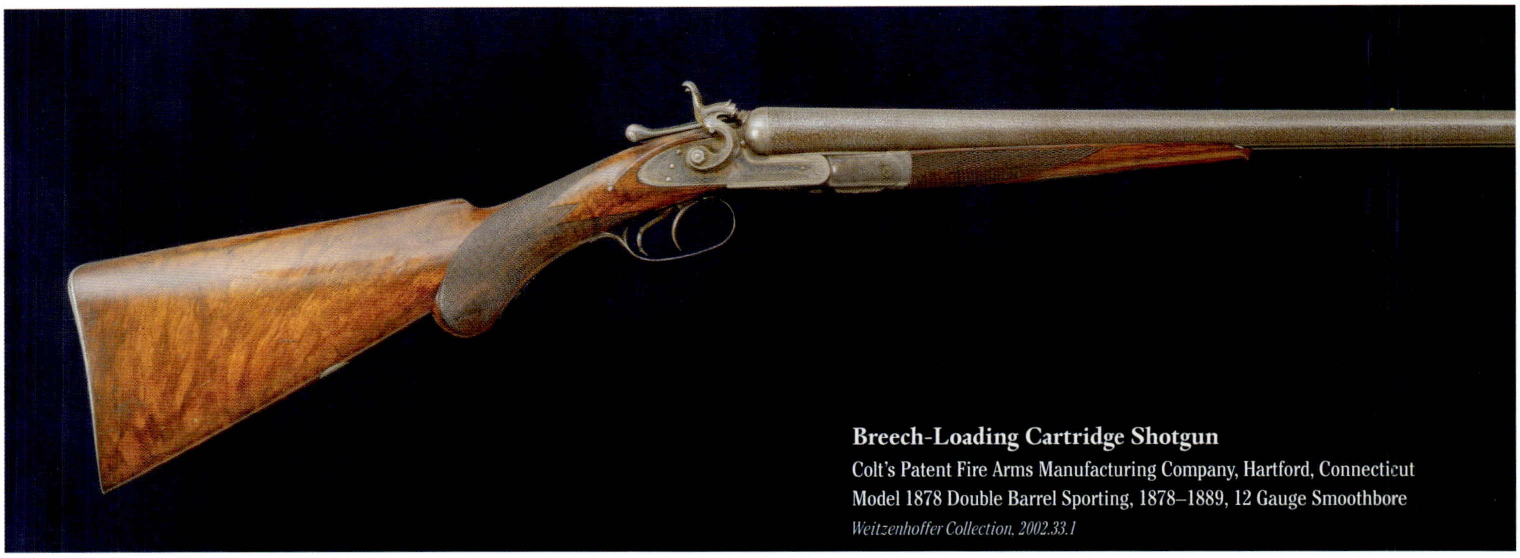

Breech-Loading Cartridge Shotgun
Colt's Patent Fire Arms Manufacturing Company, Hartford, Connecticut
Model 1878 Double Barrel Sporting, 1878–1889, 12 Gauge Smoothbore
Weitzenhoffer Collection, 2002.33.1

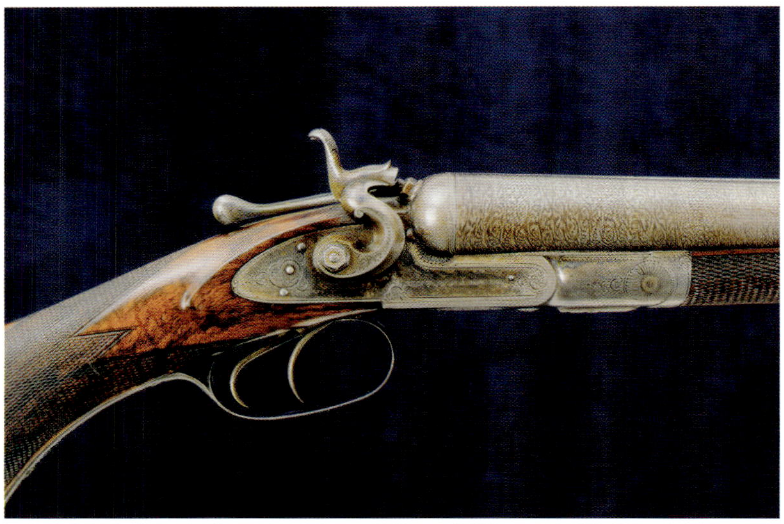

Colt's initial entry into the domestic double-barreled shotgun market, the Model 1878 side-by-side established a superior standard of quality rivaled only by the early Parker Brothers firm of nearby Meriden, Connecticut. Colt advertised the model as "made of the best materials, with the best workmanship on the interchangeable plan. . . . In beauty of finish, these guns are unexcelled." This example boasts vivid, color-case-hardened locks and nicely patterned Damascus barrels imported from Liege, Belgium. The frame, locks, and other hardware are engraved in a tight, English-style scroll of medium coverage. The select walnut stocking is of semi-pistol-grip pattern and exhibits fine checkering over the wrist and forepiece. In various gradations, approximately 22,700 Model 1878 double guns saw production over twelve years.

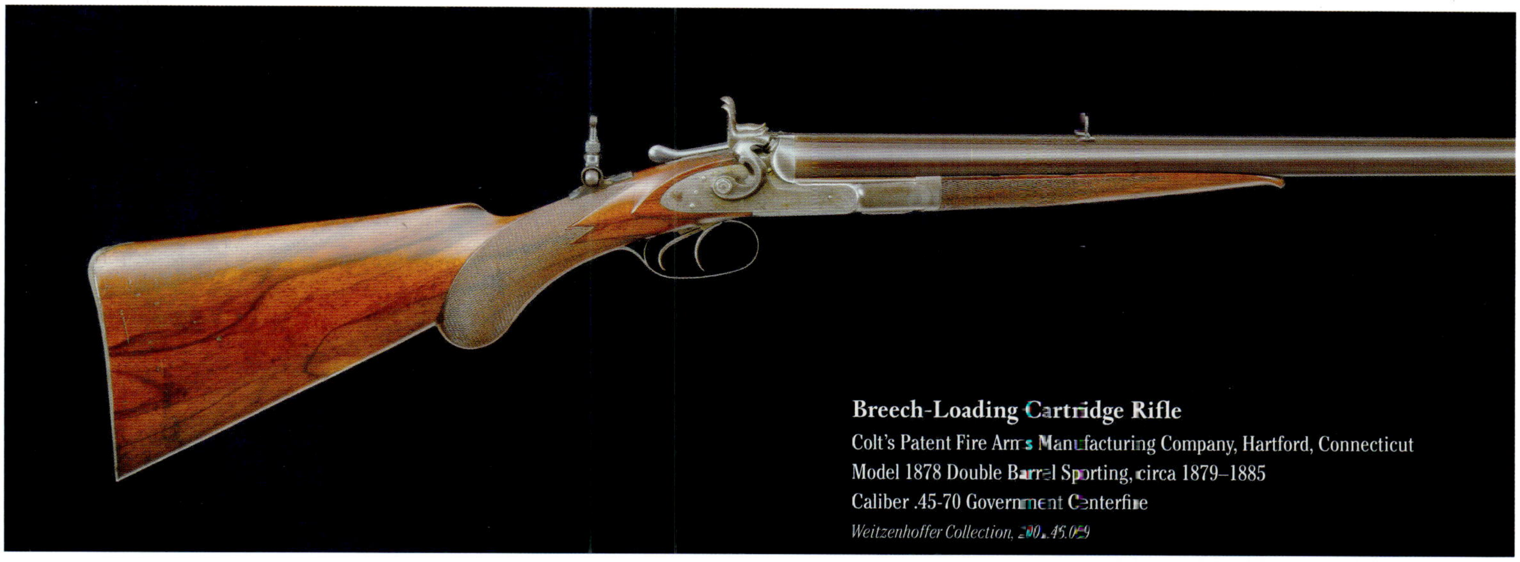

Breech-Loading Cartridge Rifle
Colt's Patent Fire Arms Manufacturing Company, Hartford, Connecticut
Model 1878 Double Barrel Sporting, circa 1879–1885
Caliber .45-70 Government Centerfire
Weitzenhoffer Collection, 200[illegible].45.059

The inspiration of Caldwell Colt (arms enthusiast, sportsman, and son of Samuel Colt), fewer than forty double rifles were produced. The rifle was built on the firm's Model 1878 double-barreled shotgun frame and was intended to rival the "best quality" sporting guns fabricated in England and Europe. This piece has standard, hand-cut border engraving on the thumb release, frame, locks, trigger guard, forestock mounts, and buttplate, while the walnut stocking is finely checkered over the forestock and pistol grip. The blued steel barrels, rifled in opposite directions for a close point of impact at 100 yards, are fitted with ivory bead and Lyman folding leaf sights on a striated rib. An adjustable, folding peep sight is mounted at the wrist. Because of their very limited production, such double rifles are among the true rarities in Colt collecting.

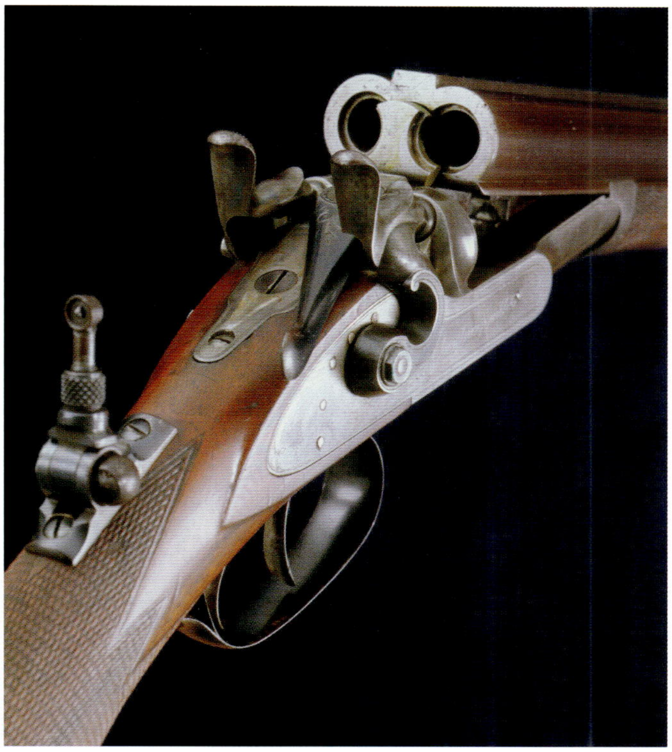

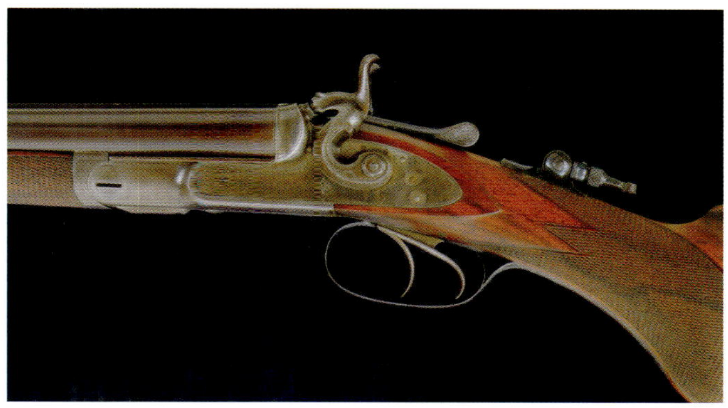

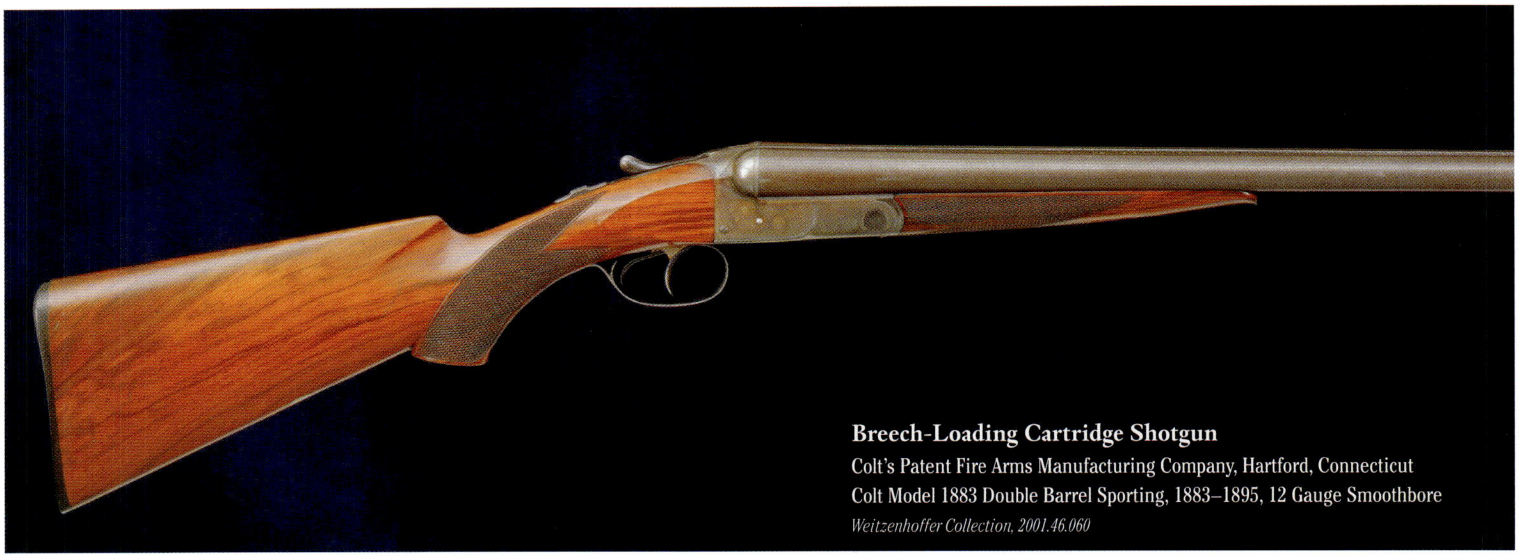

Breech-Loading Cartridge Shotgun
Colt's Patent Fire Arms Manufacturing Company, Hartford, Connecticut
Colt Model 1883 Double Barrel Sporting, 1883–1895, 12 Gauge Smoothbore
Weitzenhoffer Collection, 2001.46.060

Among all varieties of sporting longarms, the classic side-by-side, double-barreled hammerless shotgun typically manifests the most graceful lines and provides the finest handling and shooting characteristics. Many firearms connoisseurs consider the pattern the most aesthetically pleasing of all gun types and designs. The Colt Model 1883 double shotgun—a box-lock action mounted with Belgian-made Damascus barrels—certainly met such standards. Colt advertised the model as "made of the best materials and workmanship . . . in all respects equal to the most costly and best English sporting guns." Produced in 1894, this standard-grade gun features select walnut stocking with checkered forepiece and pistol grip, factory-engraved border work on the frame, and intricately patterned twist-steel barrels. Fabricated in several different grades, only some 7,300 Model 1883 shotguns were manufactured over thirteen years.

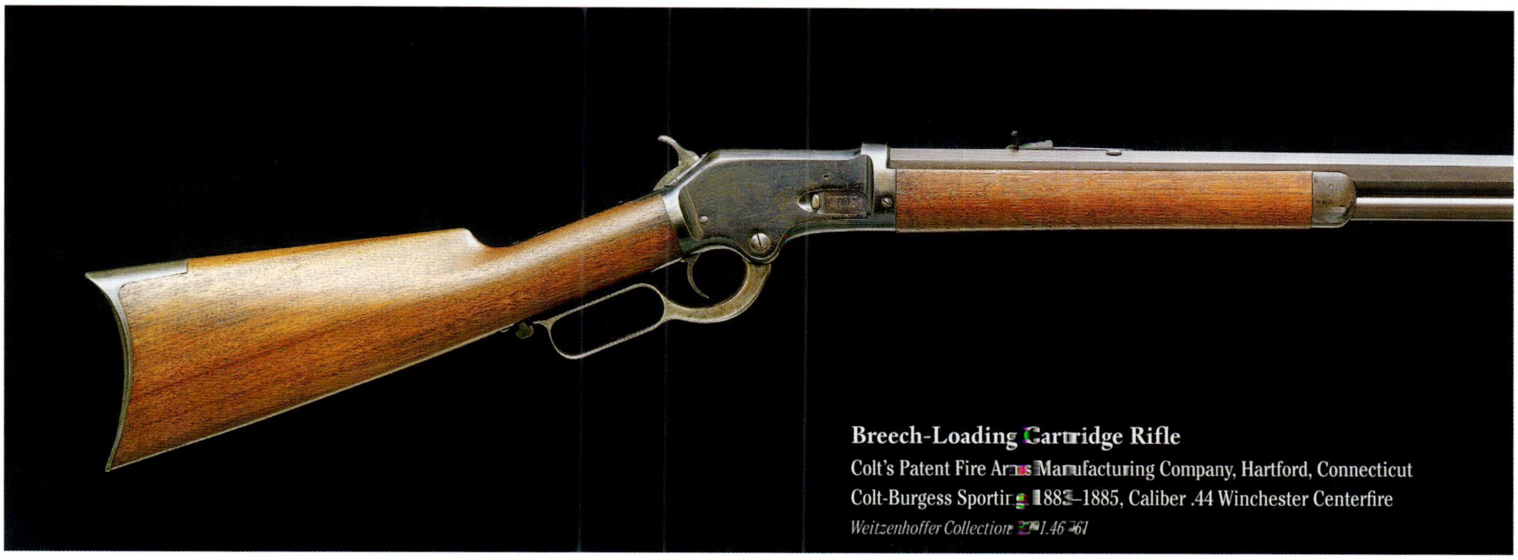

Breech-Loading Cartridge Rifle
Colt's Patent Fire Arms Manufacturing Company, Hartford, Connecticut
Colt-Burgess Sporting, 1883–1885, Caliber .44 Winchester Centerfire
Weitzenhoffer Collection [illegible]1.46.61

Devised by Andrew Burgess (perhaps second only to John Browning as America's most prolific nineteenth-century arms designer), this lever-action, repeating magazine rifle was produced for only two years—supposedly because of competitive business pressure brought by the Winchester Repeating Arms Company. A standard production model, the piece features a blued receiver with browned barrel and magazine tube, case-hardened hammer and lever, and oil-finished stocking of walnut. The full-octagon barrel is outfitted with a German silver, knife-blade front sight and an adjustable, semibuckhorn rear sight. Colt's answer to the popular Model 1873 Winchester, the Burgess rifle chambered the ubiquitous .44-40 cartridge—then also available in the Colt Single Action Army revolver. Only some 6,400 of these rifles were made; today they are quite sought after by Colt longarm collectors.

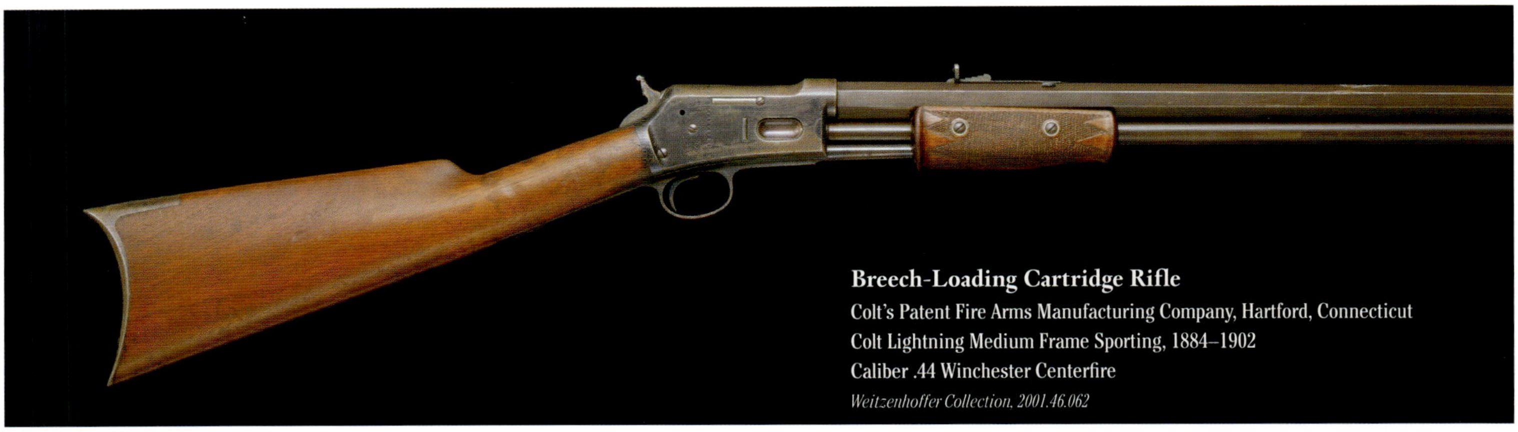

Breech-Loading Cartridge Rifle
Colt's Patent Fire Arms Manufacturing Company, Hartford, Connecticut
Colt Lightning Medium Frame Sporting, 1884–1902
Caliber .44 Winchester Centerfire
Weitzenhoffer Collection, 2001.46.062

Designed by Colt engineers William Elliot and Carl J. Ehbets, the Lightning magazine rifle was the first slide-action repeater produced in America. In company with the Model 1878 and Model 1883 shotguns and the Burgess lever-action rifle, the Lightning series represented Colt's attempt to broaden its position in the late nineteenth-century marketplace. This medium-frame variation shared common chamberings (.32, .38, and .44 WCF) with the firm's Model 1873 Single Action Army and 1878 Double Action Frontier revolvers, and such versatility doubtless contributed to a production run of nearly 90,000 rifles over eighteen years. Presenting standard, unadorned finish, this particular specimen nonetheless would command special collector enthusiasm because of its single-digit serial number—6.

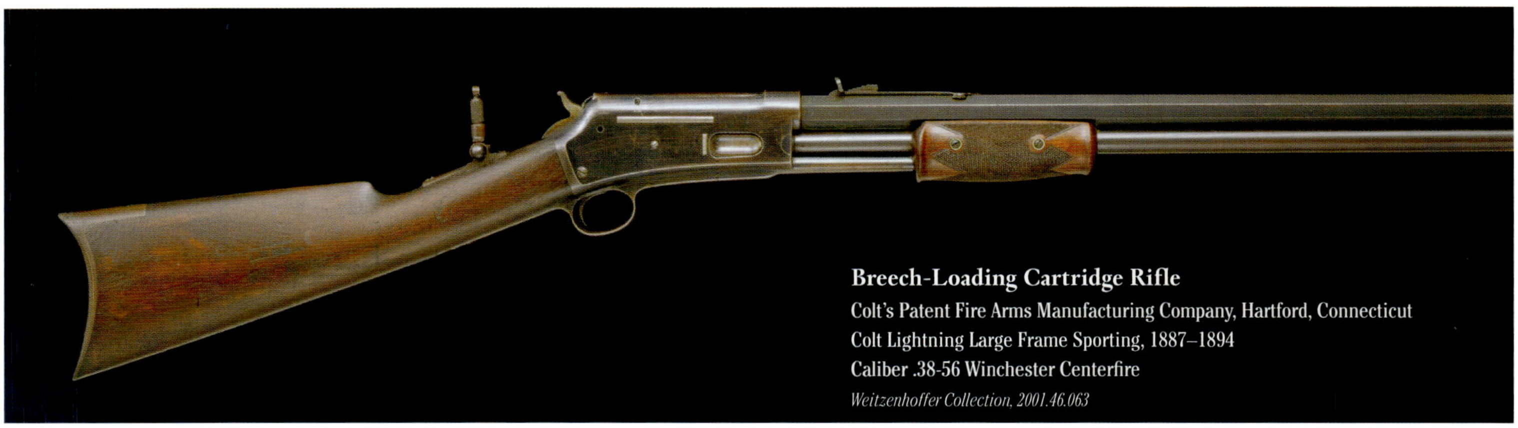

Breech-Loading Cartridge Rifle
Colt's Patent Fire Arms Manufacturing Company, Hartford, Connecticut
Colt Lightning Large Frame Sporting, 1887–1894
Caliber .38-56 Winchester Centerfire
Weitzenhoffer Collection, 2001.46.063

Often called the "Express Model" for its accommodation of high-powered big-game cartridges (from .38-56 up to .50-95 Express), this large-frame variation of the Colt Lightning slide-action magazine rifle is sought after by Colt collectors because of its scale, its use by sportsmen in the American West, and its comparative scarcity (only 6,500 produced). Like its medium- and small-frame cousins, this rifle had an extremely fast and smooth action. Company literature observed that "by holding the trigger back, and using a reciprocating motion with the left hand, the Rifle can be fired with great rapidity and without further use of the trigger." Although the design provided strong technical competition to the lever-action repeaters marketed by Winchester and Marlin, it never gained extensive popularity in the marketplace before Colt discontinued production.

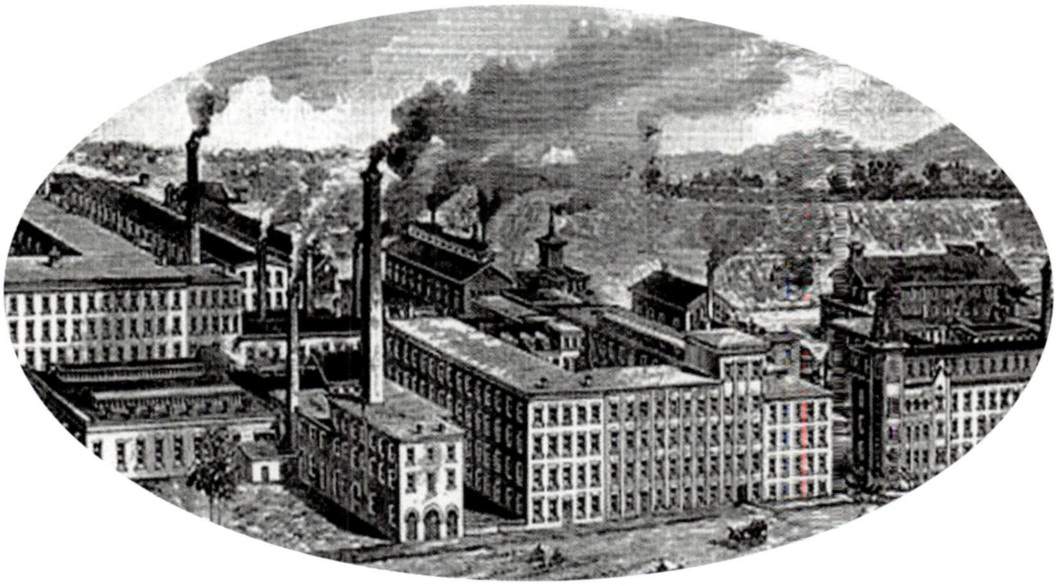

Chapter 4

REMINGTON FIREARMS

FOUNDED AS A BARREL-MAKING enterprise in 1816 by Eliphalet Remington, Jr. (1793–1861), E. Remington & Sons and the later Remington Arms Company of Ilion, New York, commenced the mass production of entire firearms around 1850 under military contract. With the expiration of Colt's master patent in 1857, the firm started making .31- and .36-caliber percussion pocket, belt, and police model revolvers for commercial sale. Made under the patents of Fordyce Beals and Joseph Rider for single-action and double-action arms, respectively, some 58,000 of these civilian-market handguns saw manufacture between 1857 and 1873; like their rival, Colt's, many were converted from percussion to metallic-cartridge ignition in the latter years.[1]

With the overriding demands of the Civil War, the Remington company turned largely to military production for the Union, producing the contract "Zouave" percussion rifle and a series of large-frame percussion revolvers. Of solid-frame, topstrap design, these .36- and .44-caliber Remington-Beals, Model 1861, and New Model Navy and Army revolvers proved structurally superior to Colt's traditional, open-top-frame design, and they provided the Hartford firm with its greatest competition during the war years. Between 1861 and 1865, more than 133,000 of these handguns saw production, and manufacture of the New Model pattern continued into the late 1870s. Many of these firearms, too, were converted to metallic-cartridge ignition in the postwar years, and many ultimately served frontier civilians in the West.[2]

In the early 1860s, Remington introduced its first metallic-cartridge arms in a variety of diminutive single-shot, multishot, and repeating deringer-type pistols utilizing several William Elliot and Joseph Rider patents. Remington-Elliot .22 and .32 rimfire, ring-trigger pepperbox pistols made from 1861 into the late 1880s, saw production of about 25,000 guns. These were far surpassed by the classic,

The Remington factory, woodcut print, circa 1885
Image courtesy private collection

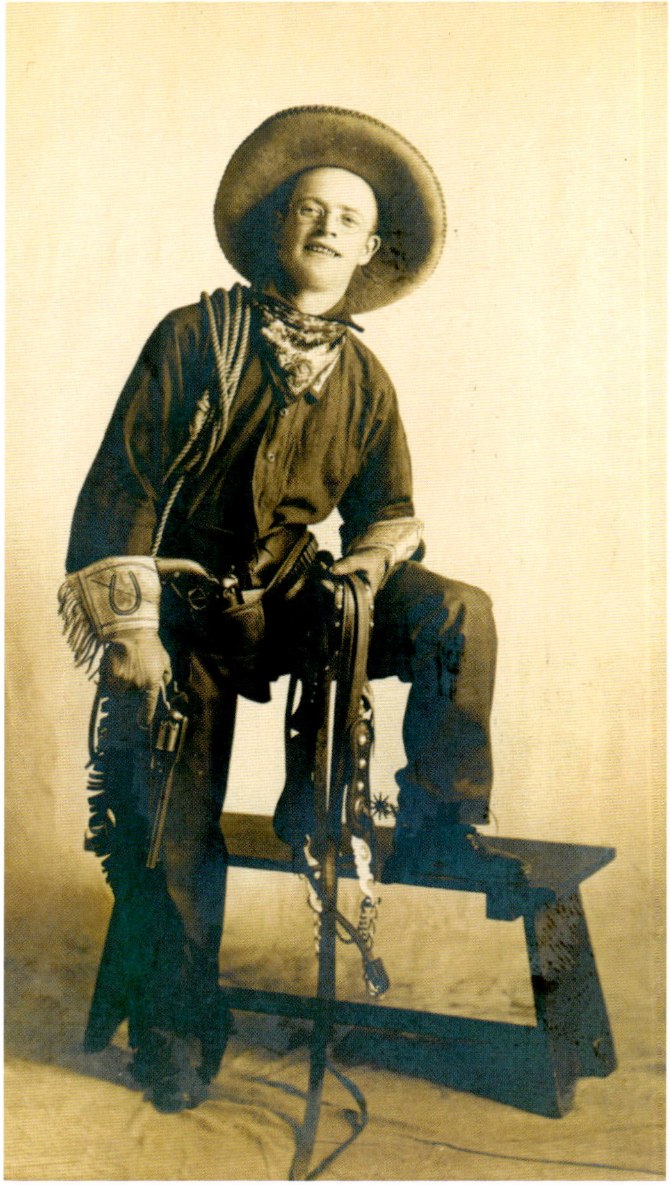

Photographer unknown. [Untitled.] Photographic postcard, America, circa 1905–1910. Dressed and accoutered in cowboy fashion, this bespectacled young man flaunts a brace of Remington New Model Army percussion revolvers. The outdated means of firearm ignition suggests that the revolvers—and perhaps all the clothing and gear—are studio props for cowpokes of the drugstore variety.

Dickinson Research Center, National Cowboy & Western Heritage Museum, 2005.019

superposed-barreled Remington Double Deringer, a .41 short rimfire Elliot design produced from 1866 to about 1935 in a quantity estimated at 150,000. (If deringer-type weapons appear in Western films—for example, in the opening scene of *The Shootist*—they are almost invariably of this pattern.) Certainly the most intriguing in its design and function, the Remington-Rider Magazine Pistol, manufactured circa 1871–1888 in a quantity of about 15,000, presented one of the only (and last) true metallic-cartridge repeating pistols employing a tubular magazine (holding five .32 rimfire extra-short rounds) beneath the barrel.[3]

With the 1869 expiration of Rollin White's patent for entirely bored-through chambers in revolving-cylinder arms (controlled by Smith & Wesson), Remington commenced production of metallic-cartridge revolvers. Designed by William Smoot, these medium-frame handguns ranged from .30 to .38 caliber and featured spur triggers and bird's-head or saw-handle-pattern grips. In all, some 33,000 Remington-Smoot revolvers saw production in three models between 1873 and 1888.[4] Two large-caliber, single-action revolvers followed—the Model 1875 (Improved Army or Frontier Army) and the Model 1890 (New Model Army). Intended to compete with the popular Colt Model 1873 Single Action Army revolver, neither attained government contracts, though the former enjoyed some popularity in the West. One who favored the Model 1875 over rivals Colt and Smith & Wesson was

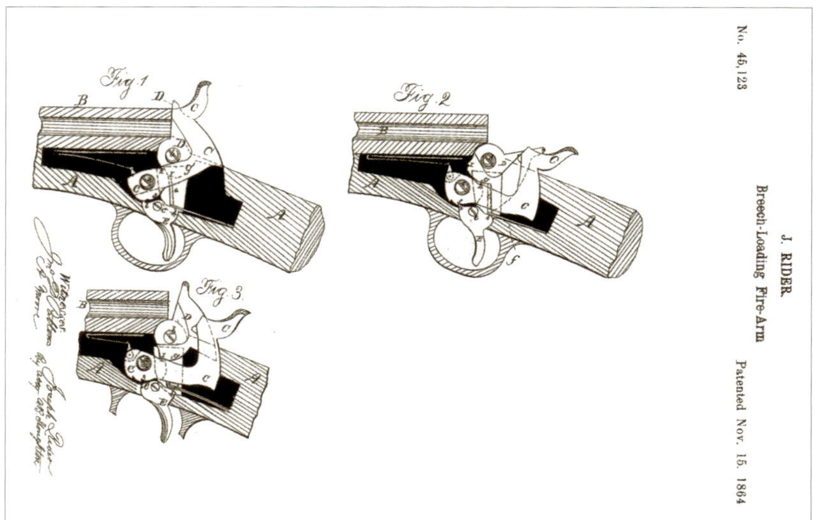

Patent drawing illustrating the internal workings of the Remington rolling-block action as refined by company designer Joseph Rider.

U.S. Patent and Trademark Office, uspto.gov

outlaw Jesse James, who praised the weapon "[b]ecause the Remington is the hardest and the surest shooting pistol made, and because it carries exactly the same cartridge that a Winchester rifle does."[5] Some 25,000 to 30,000 Model 1875 revolvers saw production in .44 and .45 caliber between 1875 and 1888. Only about 2,000 Model 1890s were manufactured from 1891 to 1894, largely because they provided little competition to the more advanced, Colt and Smith & Wesson double-action and swingout-cylinder revolvers then beginning to dominate the commercial and military markets.[6]

Though the Sharps Rifle Manufacturing Company won an early and appreciative market for its single-shot, falling-block rifle action, E. Remington & Sons captured equal fame—and far greater sales—with its single-shot, "rolling-block" action. Patented in 1863 by Leonard M. Geiger and improved at Remington's by Joseph Rider in 1864, this remarkably simple camming design combined an interlocking, center-hung hammer with a semicircular, rotating breechblock fully capable of chambering an expansive range of rimfire and centerfire cartridges. Over 1.5 million rolling-block firearms (rifles, carbines, shotguns, and pistols) saw production from 1868 to circa 1933, the majority for domestic and foreign military service—including the Papal Guard. Among commercial models, the No. 1 Sporting rifle won acclaim in the American West with sport and market hunters alike. Available in several hunting and target configurations (and in a plethora of calibers, barrel patterns and weights, stock patterns, and sight and trigger types), the No. 1 Sporting rifle—and its several successor models—reached a total production in excess of 70,000 arms.[7]

In 1880 Remington introduced yet another breech-loading single-shot rifle using Lewis L. Hepburn's patented falling-block design, which was actuated by a thumb lever on the right side of the frame. Remington-Hepburn No. 3 sporting and target rifles, available in calibers ranging from .22 centerfire to .50-90 centerfire and with myriad special order features, proved popular with both competitive target shooters in the East and latter-day big-game hunters in the West. Some 10,000 of these rifles left the factory before their discontinuance around 1907. And from the mid-1870s until about 1910, the firm also mass-produced and marketed a succession

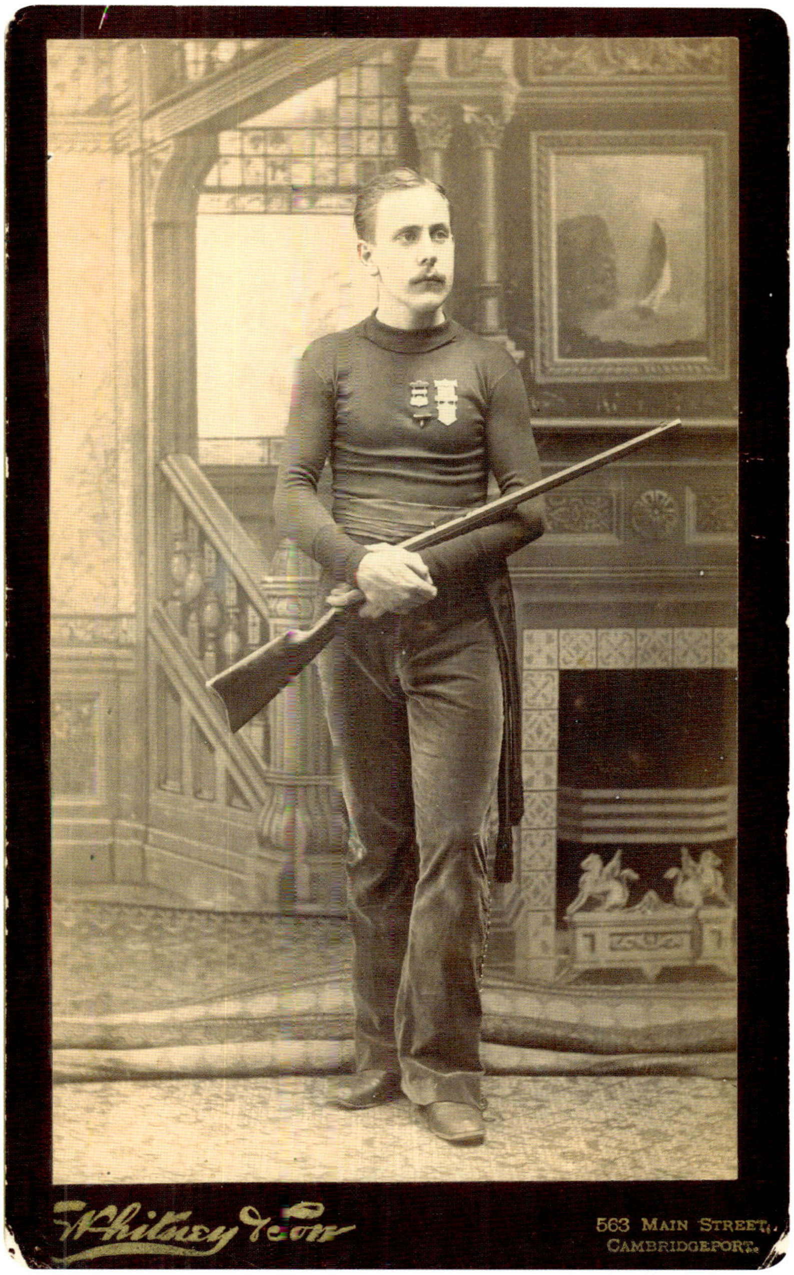

Whitney & Son, photographer. [Untitled.] Cabinet card, Cambridgeport, Massachusetts, circa 1895–1900. Striking a dignified attitude with his shooting medals on his chest, this performing marksman cradles what appears to be a Remington Model 1½ rolling-block sporting rifle—the weapon with which he very likely captured his various metallic accolades.

Dickinson Research Center, National Cowboy & Western Heritage Museum, 2003.035

of double-barrel hammer shotguns in eight models (and numerous gradations) amounting in total to nearly 40,000 arms.[8]

Poor business decisions led E. Remington & Sons into receivership in 1886. After languishing for two years, the company was revived in 1888 as the Remington Arms Company under joint ownership of Hartley & Graham (the Union Metallic Cartridge Company) and the Winchester Repeating Arms Company. (Winchester sold its interest to Hartley in 1896, but it had ensured in the intervening years that no competitive lever-action rifles were made in Ilion.) By 1910, the Remington enterprise stood on the threshold of a revitalized prosperity, girded by the new semiautomatic and slide-action designs of John M. Browning and John D. Pedersen, respectively. A century later, the venerable Remington Arms Company continues the manufacture of fine rifles and shotguns at Ilion, New York, and is rightly recognized as "America's Oldest Gunmaker."[9]

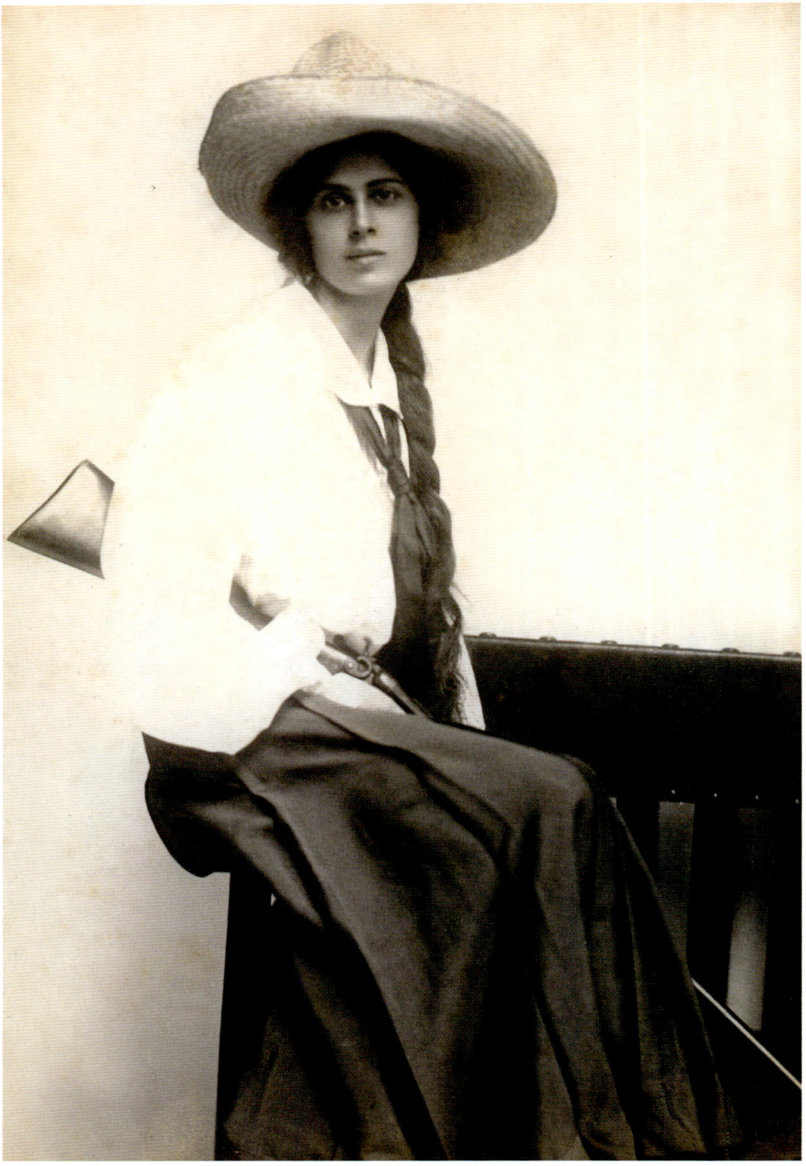

Photographer unknown. [Untitled.] Cabinet card, America, circa 1900–1905. Seated on the arm of a Mission-style settle, this rather dreamy young lady cradles a Remington Model 1889 double-barreled, hammer shotgun across her lap. Though little of the firearm is visible, the model can be discerned by the distinctive pattern of the visible hammer (see page 98).
Image courtesy private collection

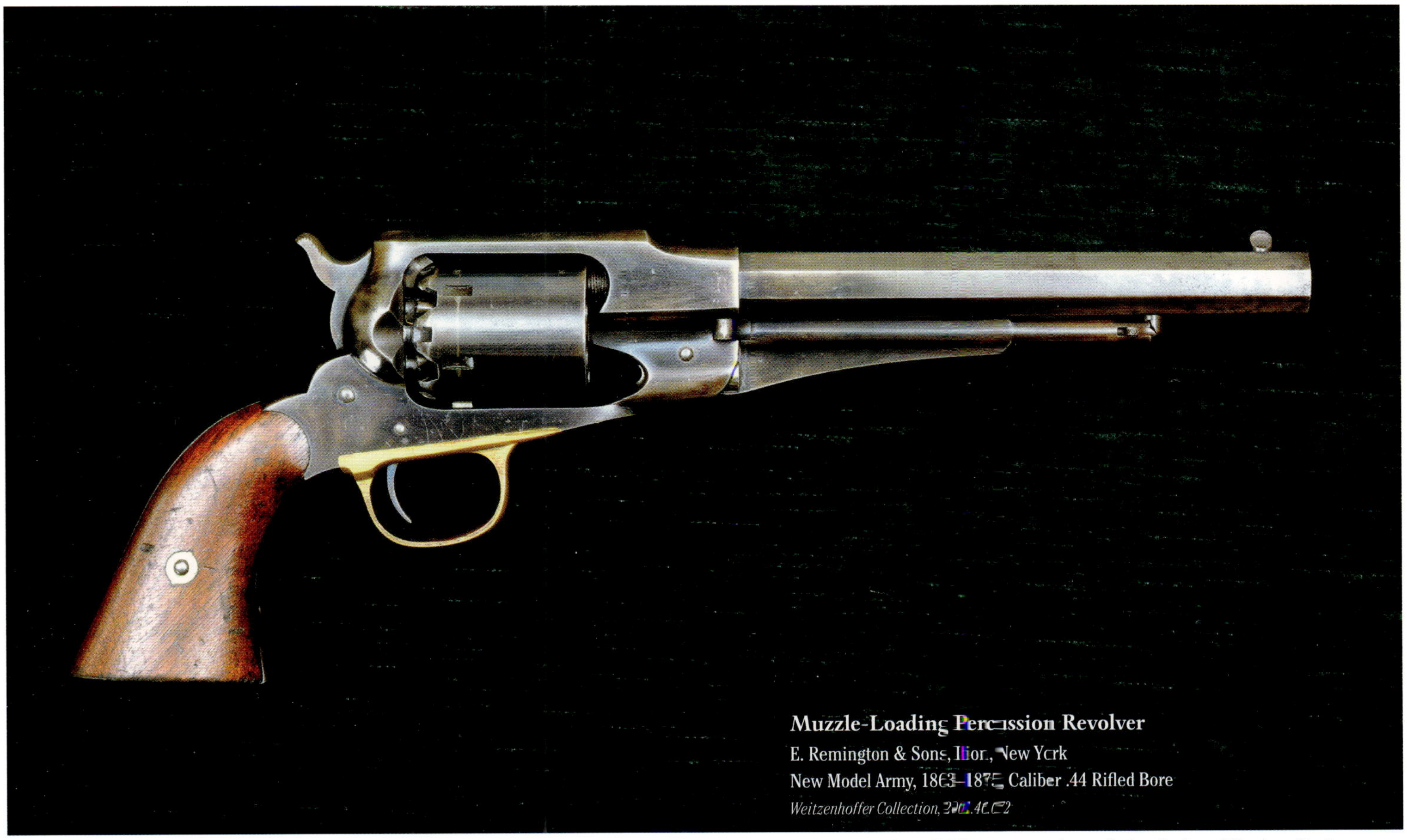

Muzzle-Loading Percussion Revolver
E. Remington & Sons, Ilion, New York
New Model Army, 1863–1875, Caliber .44 Rifled Bore
Weitzenhoffer Collection, 2002.40.02

Founded on the solid-frame, revolving-cylinder design of Fordyce Beals, Remington's fabrication of handguns commenced in 1857, coincident with the expiration of Samuel Colt's patent rights. The demands of the Civil War initiated the mass production of large-frame Remington revolvers, such as this .44-caliber specimen. The pattern was second only to Colt's Model 1860 Army in contract purchases, and its solid-frame design proved more rugged and dependable than the Hartford competition. Approximately 122,000 New Model Army revolvers were produced over thirteen years. (Unlike Colt's, standard-production Remington revolvers had no adornment in the form of a roll-engraved cylinder scene.)

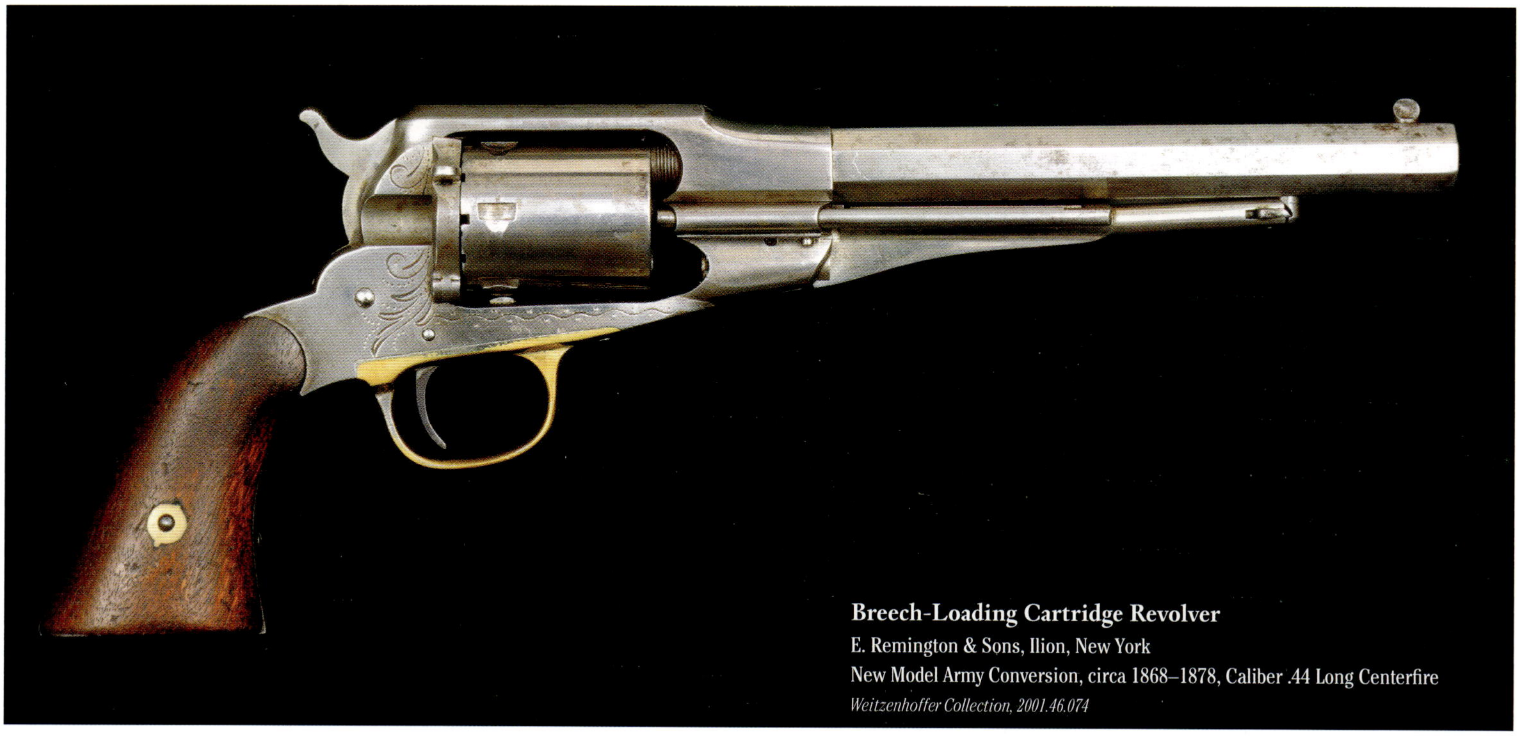

Breech-Loading Cartridge Revolver
E. Remington & Sons, Ilion, New York
New Model Army Conversion, circa 1868–1878, Caliber .44 Long Centerfire
Weitzenhoffer Collection, 2001.46.074

Like Colt's, the Remington firm adapted to changes in firearms technology—particularly to the adoption of breech-loading systems utilizing metallic-cartridge ammunition. This New Model Army revolver presents the typical factory alteration—milling of the right side of the frame with a cartridge loading groove; addition of a loading gate on the frame; fitting of a new, bored-through cylinder; and addition of an ejector rod assembly. Unlike Colt's, Remington conversions left the now-superfluous percussion loading lever in place beneath the barrel. During conversion at the factory, this piece was sparsely engraved and nickel-plated to increase its market appeal.

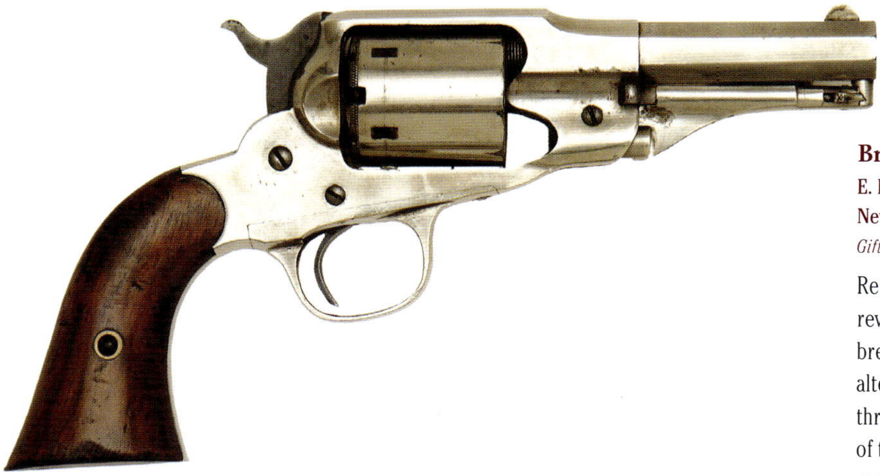

Breech-Loading Cartridge Revolver
E. Remington & Sons, Ilion, New York
New Model Police Conversion, circa 1870–1875, Caliber .38 Short or Long Rimfire
Gift of Raymond W. Lewis, H.183.13

Referred to as the Improved New Model Police in company literature, this revolver was, in later production, manufactured strictly as a metallic-cartridge breechloader and generally did not include a separate percussion cylinder. The alteration to cartridge ignition here involved the use of a two-piece, bored-through cylinder with a special backplate. Like many other Remington handguns of the period, this revolver has a full nickel-plated factory finish, which probably represented little if any premium cost over the standard finish. Approximately 18,000 Police model pistols, made first in percussion and later in metallic-cartridge ignition, saw production between their initial issuance in 1865 and discontinuance around 1875.

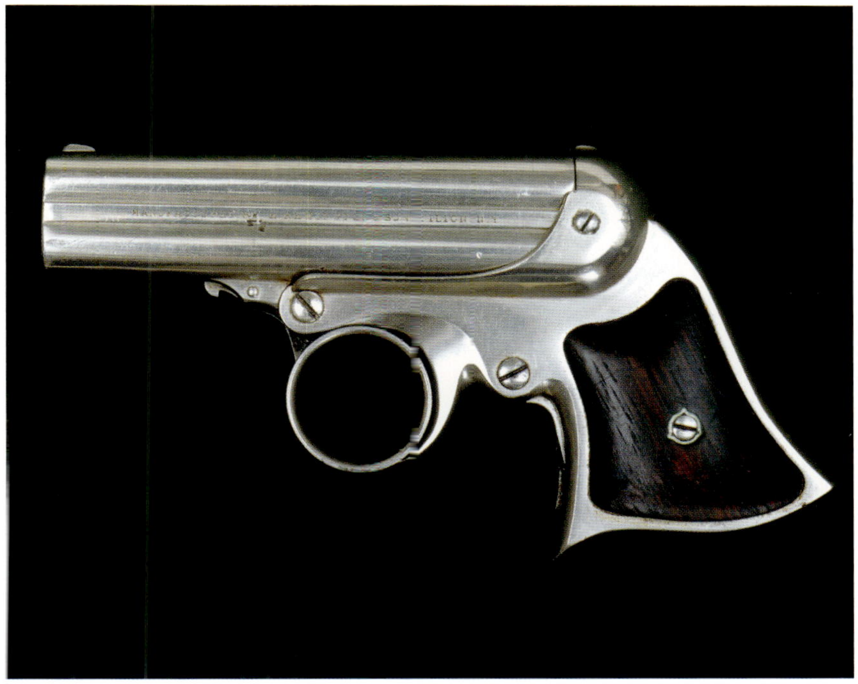

Breech-Loading Cartridge Pistol
E. Remington & Sons, Ilion, New York
Remington-Elliot Pepperbox Deringer, 1863–1888 Caliber .32 Short Rimfire
Weitzenhoffer Collection, 2001.46.080

Patented in 1861 by one of Remington's most prolific inventors, William Elliot, this four-barreled "pepperbox" deringer was similar in function to the contemporary Sharps pepperbox pistol. Both incorporated a stationary cluster of barrels discharged by a rotating firing pin. The Remington-Elliot featured a barrel cluster that tipped down for loading and used a hammerless, double-action mechanism operated by a ring trigger. Limited strictly to close-range shooting, the small repeater made an ideal personal-protection weapon. Like the small self-defense weapons manufactured by Colt's, this pistol was driven from the market by inexpensive competitors.

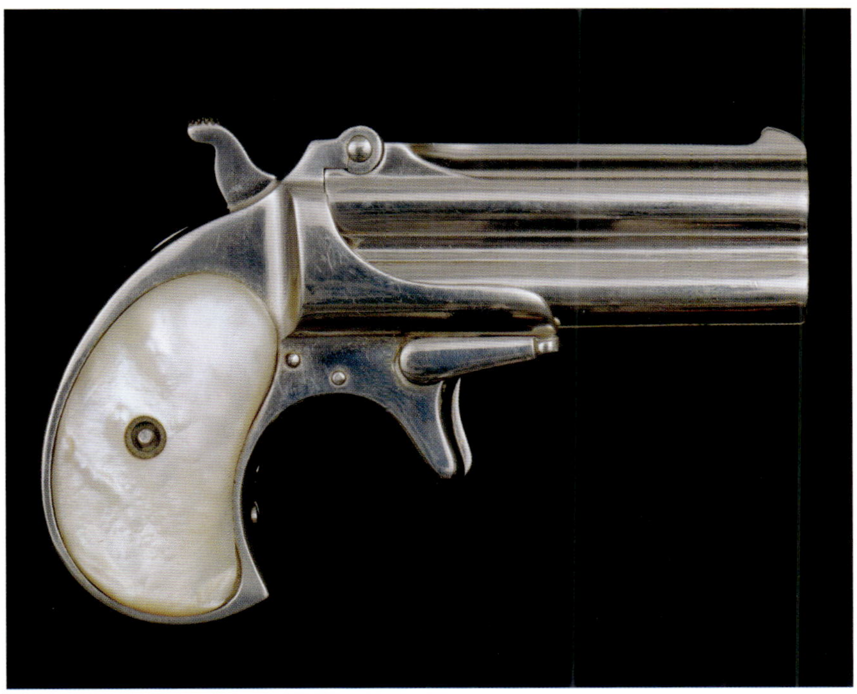

Breech-Loading Cartridge Pistol
Remington Arms–Union Metallic Cartridge Company, Ilion, New York
Remington-Elliot Double Deringer, Type III, 1912–1935
Caliber .41 Short Rimfire
Weitzenhoffer Collection, 2001.46.073

Designed and patented by Remington engineer William Elliot in 1865, this diminutive two-shot pistol certainly ranks as the best-known and most widely produced deringer-pattern pistol of the late nineteenth and early twentieth centuries. The pattern enjoyed great popularity among many denizens of the frontier West, and its robust practicality ultimately resulted in a production run of almost seventy years and some 150,000 units. The weapon consists of a forged and milled frame (with a bird's-head grip profile and a spur trigger) fitted with hinged, superposed barrels that rotate up at the muzzle for loading and extraction. (The firing pin in the hammer switches from one barrel to the other when the hammer is cocked.) This specimen has a full nickel-plated finish (probably a later renickel by the factory) and original special-order grips of iridescent mother-of-pearl.

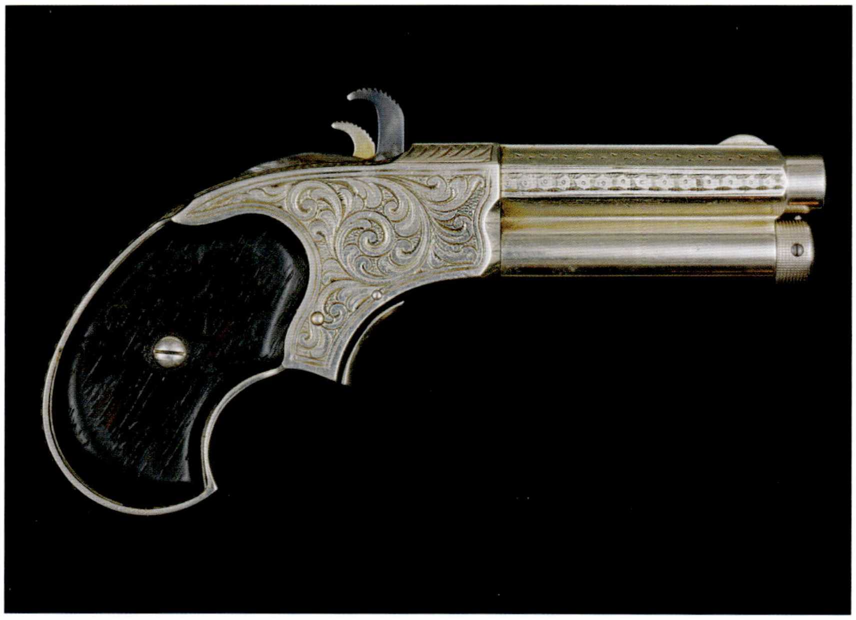

Breech-Loading Cartridge Pistol
E. Remington & Sons, Ilion, New York
Remington-Rider Magazine Deringer, 1871–1888
Caliber .32 Extra Short Rimfire
Weitzenhoffer Collection, 2001.46.079

This unusual deringer-type design by Joseph Rider is somewhat reminiscent of the earlier Volcanic magazine pistol, incorporating a similar five-shot, tubular magazine beneath the octagonal barrel with a thumb-actuated loading lever adjacent to the hammer. This specimen features bright nickel plating with contrasting grip plates of dark rosewood. The pistol's frame, with a modified bird's-head grip profile and a spur trigger, is engraved with a bold and flowing scroll pattern over a punched-dot ground in the style of freelance artisan Louis D. Nimschke. The Remington-Rider magazine pistol presents one of the very rare instances of nineteenth-century American arms manufacture in which finely engraved pieces are encountered with far greater frequency than standard—unembellished—examples.

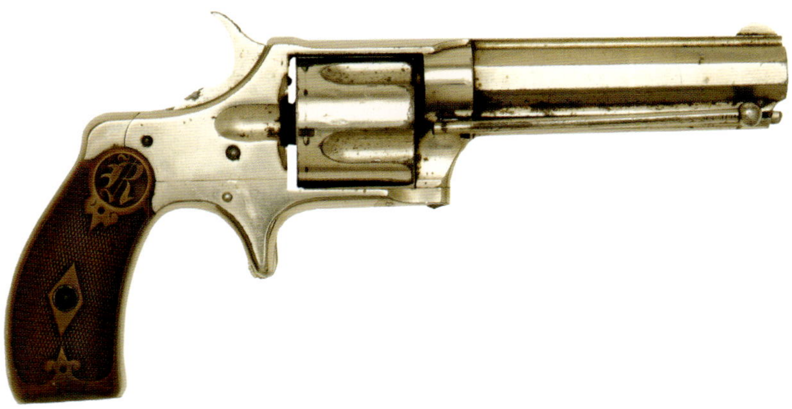

Breech-Loading Cartridge Revolver
E. Remington & Sons, Ilion, New York
Smoot New Model No. 3, 1878–1888, Caliber .38 Short Rimfire
Gift of Raymond W. Lewis, H.183.43

Designed by William Smoot, the Remington-Smoot series of medium-frame, metallic-cartridge revolvers was marketed as a direct competitor of Colt's contemporaneous New Line series. This five-shot New Model No. 3 variant is finished in full, factory nickel plating, as was common for the entire series. This piece features the rather scarce saw-handle-pattern grip (as opposed to the common bird's-head pattern), with checkered hard rubber grips featuring scales surmounted with the company's initial. About 28,000 of this model saw production within the entire Smoot series, which totaled some 33,000 over fifteen years.

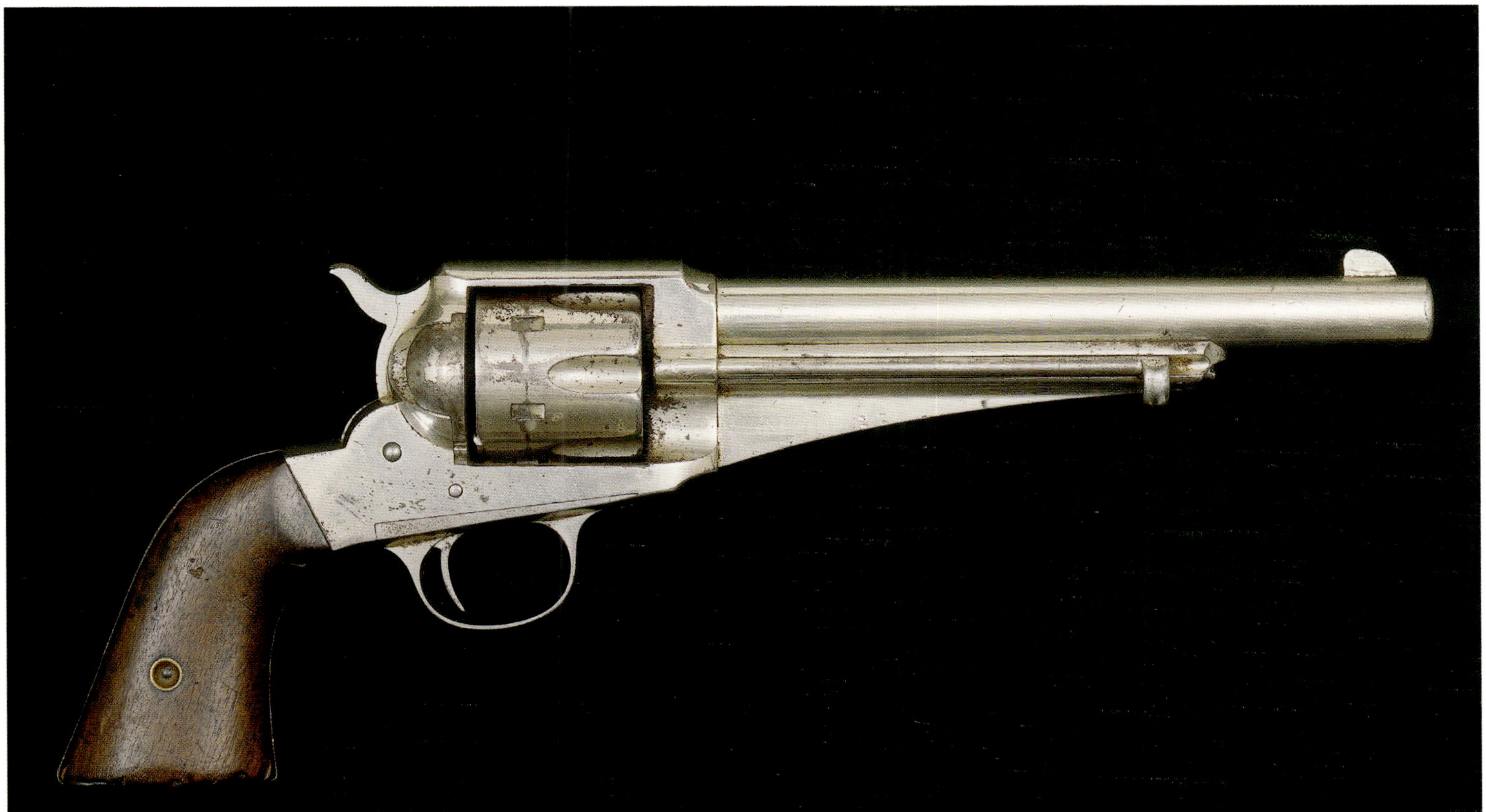

Breech-Loading Cartridge Revolver
E. Remington & Sons, Ilion, New York
Model 1875 Single Action Army, 1875–1889, Caliber .44 Winchester Centerfire
Grandee Collection, 19[illegible].01.[illegible]

Often referred to as the "Improved Army" or "Frontier Army," the Model 1875 Remington revolver was marketed as a competitor to the widely popular Colt Single Action Army. This effort, however, proved rather unsuccessful—while Remington fabricated and sold some 25,000 to 30,000 Model 1875 revolvers between 1875 and 1889, Colt's produced and sold approximately 115,000 of its famed Model 1873 Single Action Peacemakers during the same period. Colt's enjoyed the advantage of ongoing government purchases, while Remington contracted for only some 600 of these revolvers for U.S. Indian Police service in 1883. The weapon, nonetheless, won many devotees, such as outlaw Jesse James, in the waning frontier West.

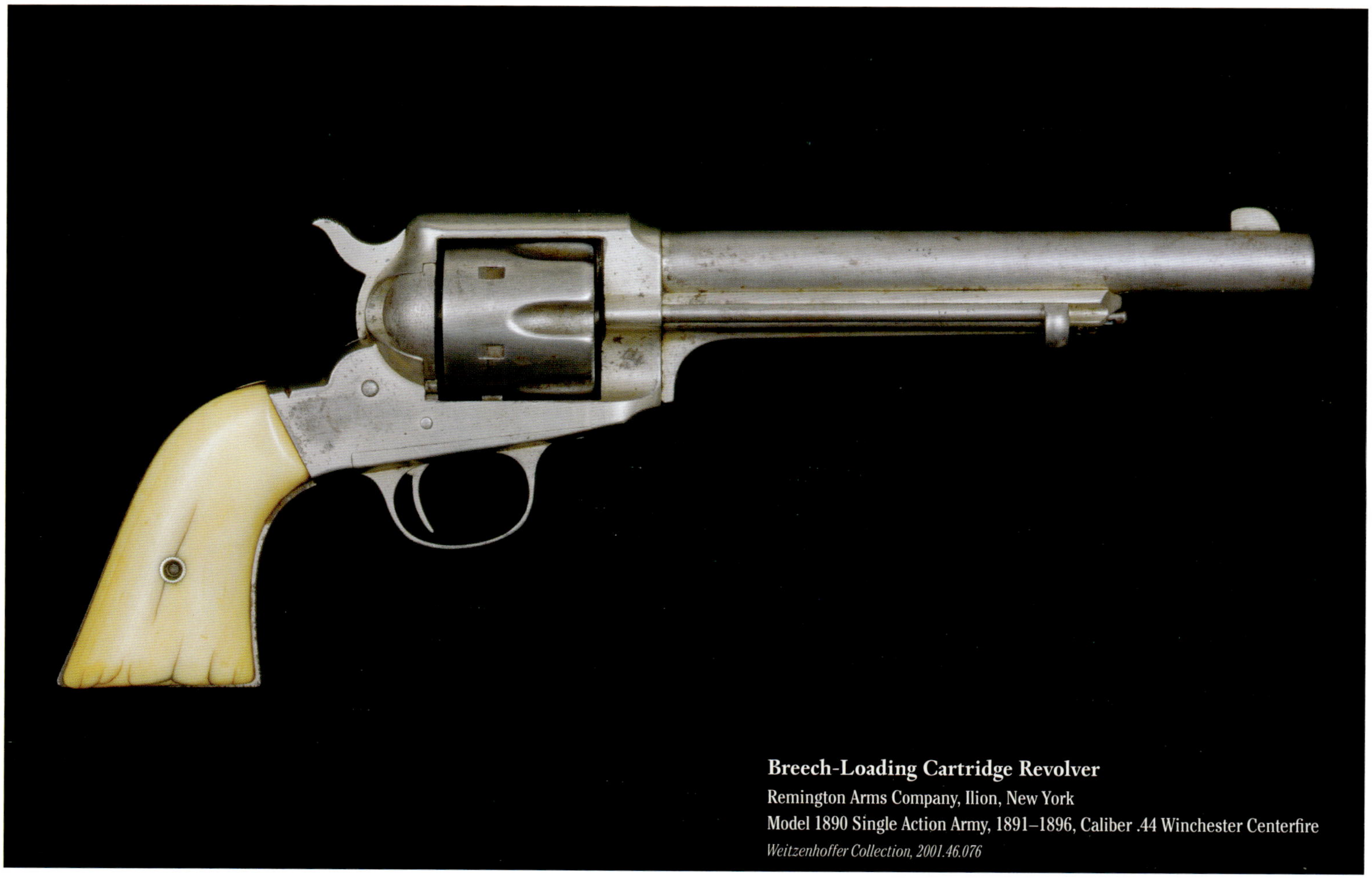

Breech-Loading Cartridge Revolver
Remington Arms Company, Ilion, New York
Model 1890 Single Action Army, 1891–1896, Caliber .44 Winchester Centerfire
Weitzenhoffer Collection, 2001.46.076

Remington's last attempt to compete with Colt's in the large-frame revolver market, the Model 1890 appeared remarkably similar in its overall design configuration to the venerable Colt Single Action Army Peacemaker. The pattern, in fact, represented a revamped successor of the Model 1875 Remington Army, and it met with even less success in the market, with only about 2,000 pistols produced over five years. (Quite a few Model 1875 Remington revolvers actually have been spuriously reworked to resemble the Model 1890 because of its far greater rarity and worth.) This special-order example left the factory with a full nickel-plated finish and custom ivory grips.

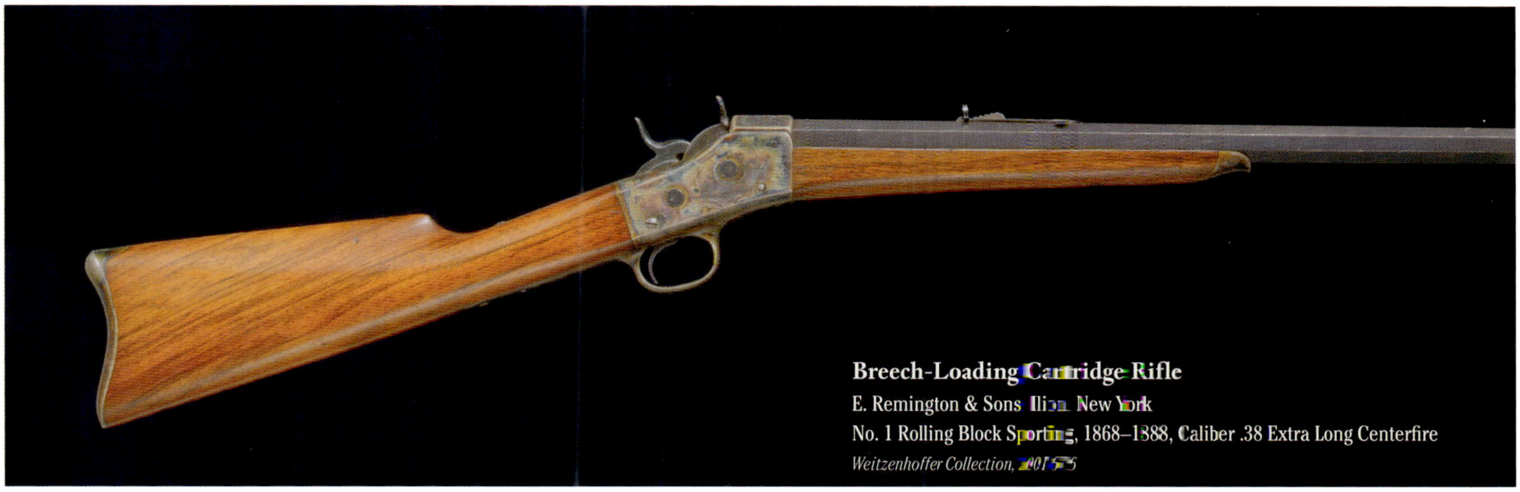

Breech-Loading Cartridge Rifle
E. Remington & Sons, Ilion, New York
No. 1 Rolling Block Sporting, 1868–1888, Caliber .38 Extra Long Centerfire
Weitzenhoffer Collection, 2001.5.[illegible]

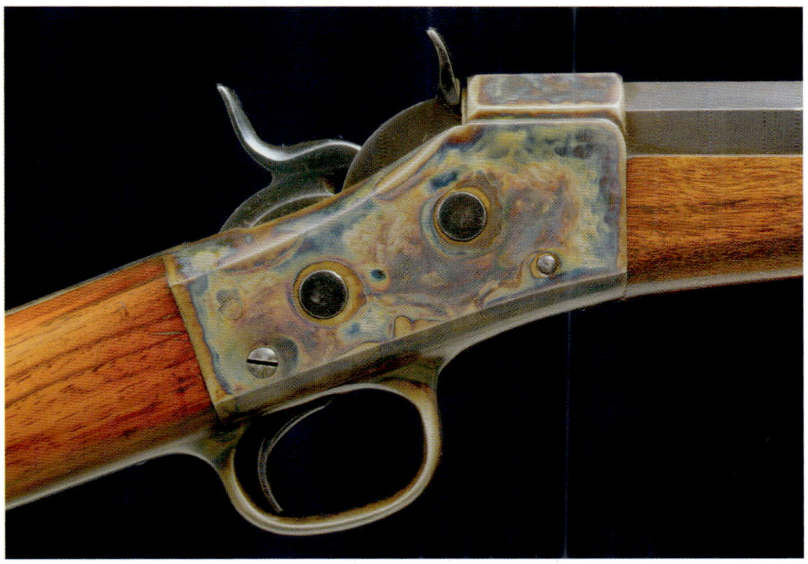

Utilizing the patents of Joseph Rider for a semicircular, camming breechblock interlocked with the hammer, the famous Remington rolling-block system proved to be one of the simplest and strongest single-shot actions ever devised. It won widespread military acclaim, and it became a favorite of American sportsmen. An 1872 Remington advertisement extolled it as "the preferred arm for hunting purposes on the Plains; its simplicity and durability especially commencing it for frontier use." The rifle pattern was made in several hunting and target configurations with a wide variety of optional calibers, barrel styles and weights, stock patterns, sights, and triggers. This example, a short-range variant produced circa 1878–1888, is a standard grade having plain walnut stocks; a blued, full-octagon barrel with knife-blade front and adjustable, semibuckhorn rear sights; and a color-case-hardened frame, forestock tip, and buttplate.

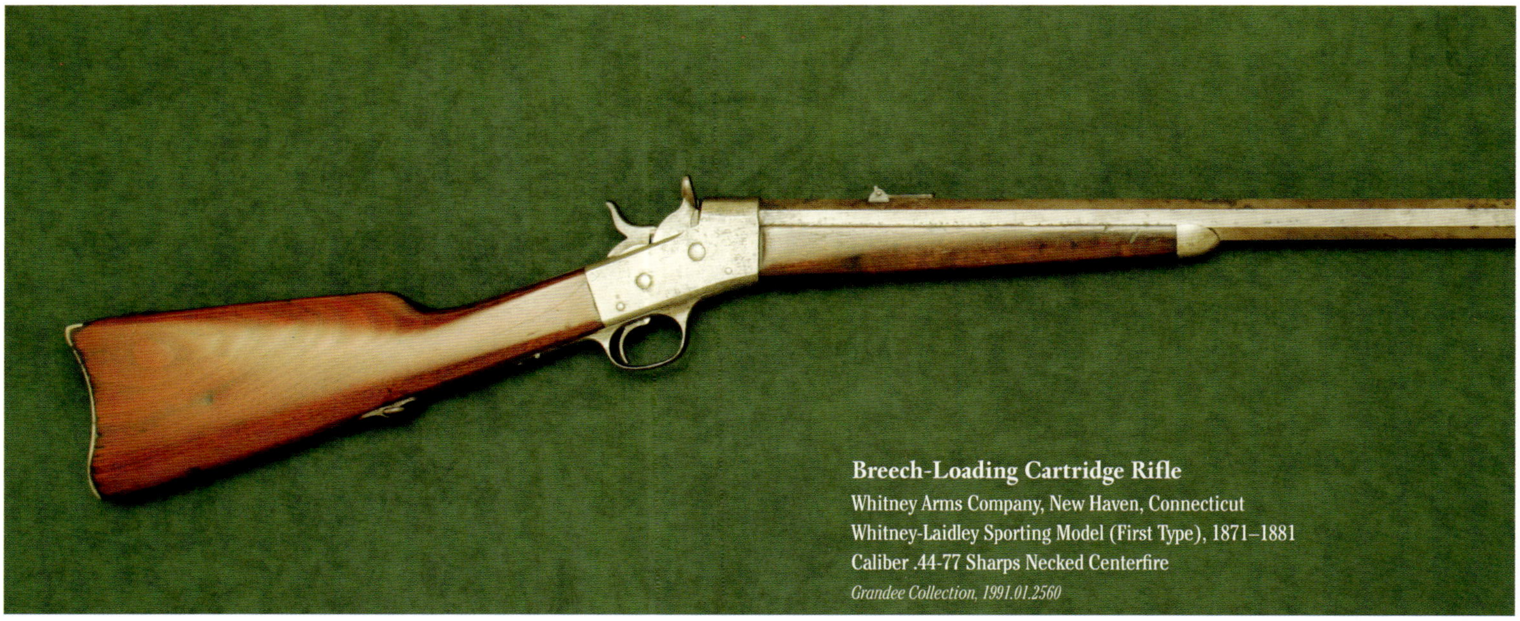

Breech-Loading Cartridge Rifle
Whitney Arms Company, New Haven, Connecticut
Whitney-Laidley Sporting Model (First Type), 1871–1881
Caliber .44-77 Sharps Necked Centerfire
Grandee Collection, 1991.01.2560

Manufactured to compete with Remington's popular rolling-block action, this "split-breech" variation presented a functional, but more complicated, four-piece breech system consisting of the hammer, locking cam, thumb-piece plate, and breechblock. Although far less common than the Remington rolling block, the Whitney-Laidley single-shot rifle nonetheless won devotees among sport and market hunters in the West who appreciated its accommodation of a variety of the high-powered cartridges. This seasoned example features a midweight, full-octagon barrel, open sights, and a plain trigger.

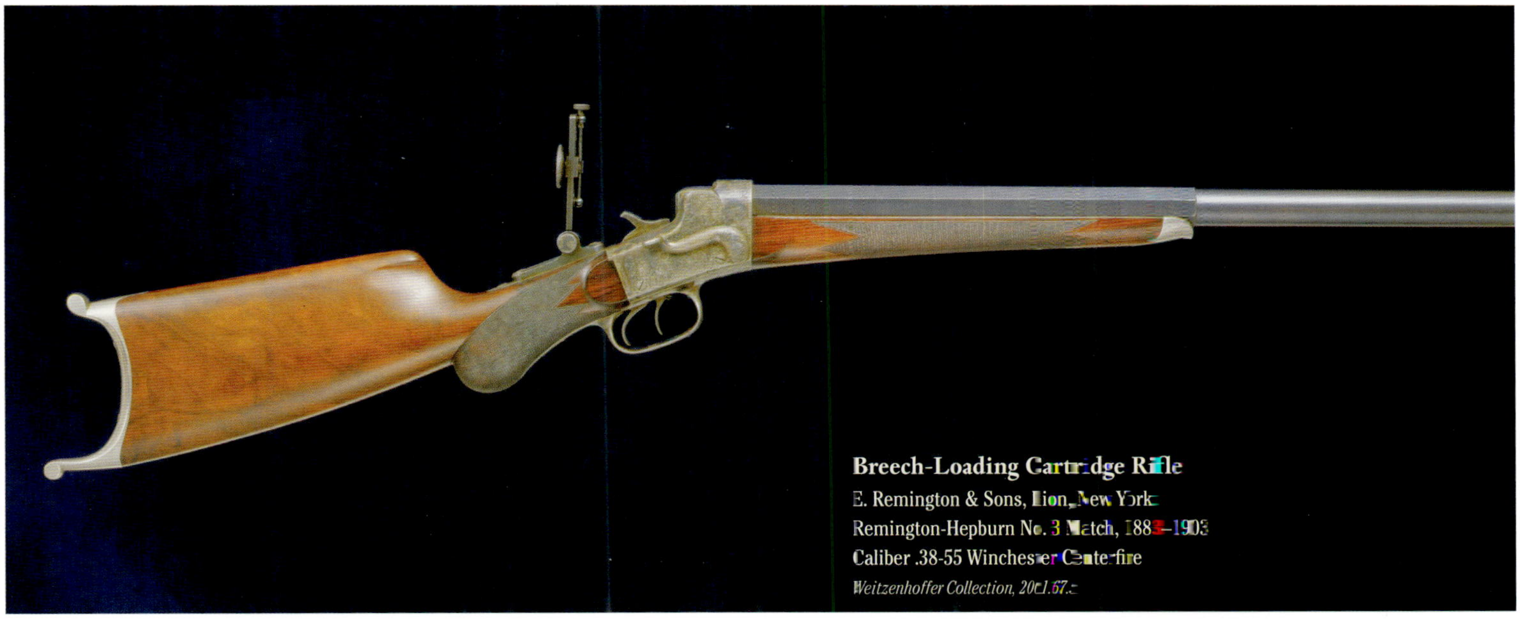

Breech-Loading Cartridge Rifle
E. Remington & Sons, Ilion, New York
Remington-Hepburn No. 3 Match, 1883–1903
Caliber .38-55 Winchester Centerfire
Weitzenhoffer Collection, 2011.67.1

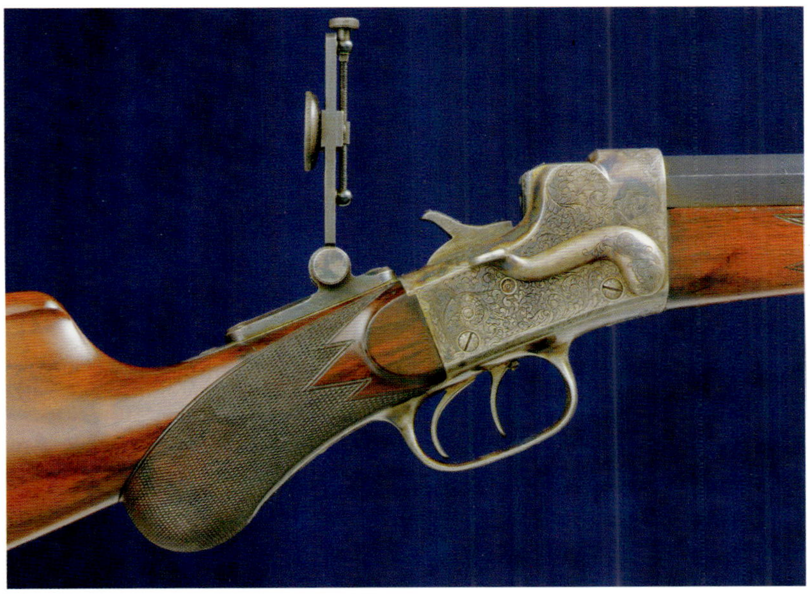

Patented by Lewis L. Hepburn, the No. 3 Remington rifle employed a side-lever-actuated, falling-block design similar to those of British gun makers Daniel Fraser and John Rigby. A very strong action, the No. 3 could handle centerfire cartridges ranging from .22 Winchester to .50-90 Sharps Straight and including numerous smokeless powder rounds. Like other domestic, single-shot rifles of the period, the No. 3 Hepburn was available in many styles or variations, including sporting, target, match, midrange and long-range Creedmoor, and military musket. Fewer than 1,000 of this Match variation were made for medium- and short-range target competition. (Further detail on this piece appears in chapter 10, page 184.)

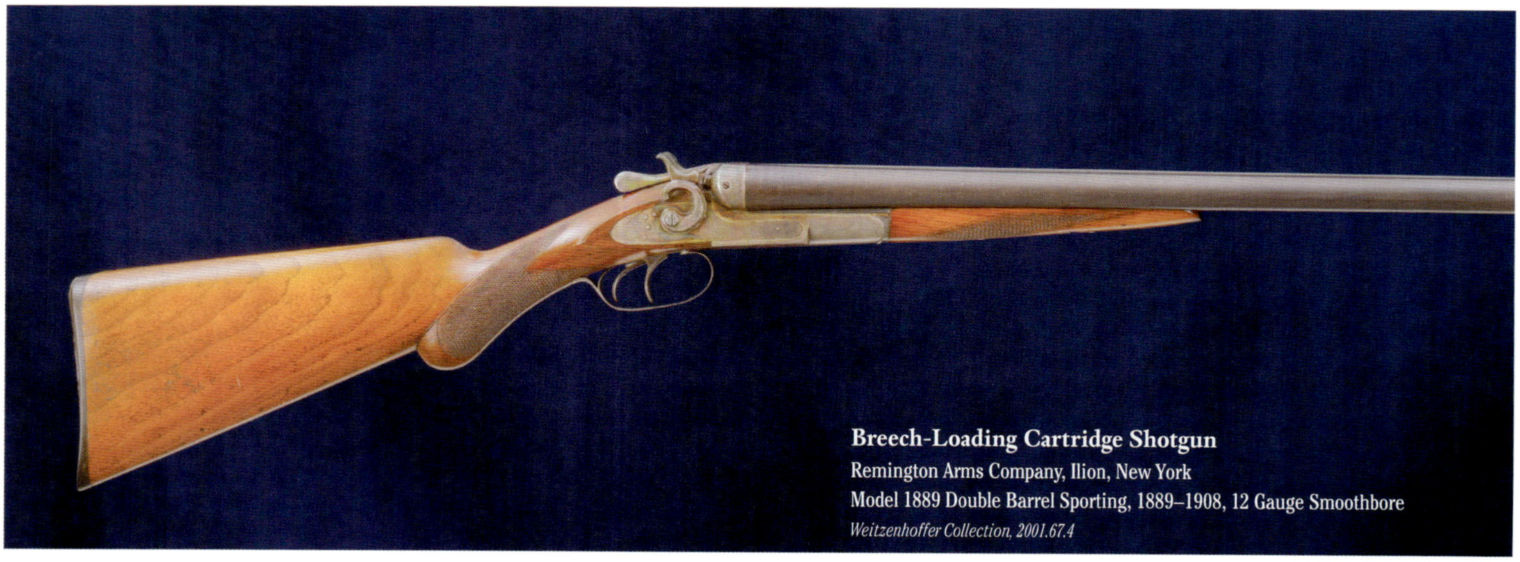

Breech-Loading Cartridge Shotgun
Remington Arms Company, Ilion, New York
Model 1889 Double Barrel Sporting, 1889–1908, 12 Gauge Smoothbore
Weitzenhoffer Collection, 2001.67.4

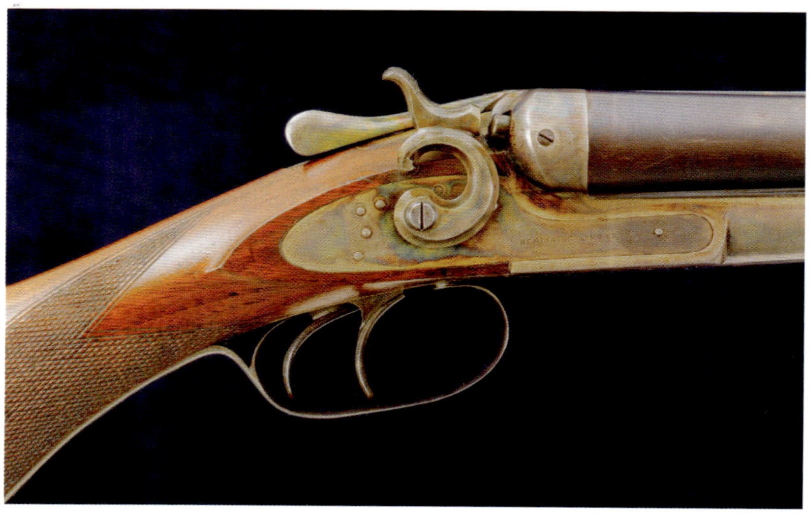

Commencing about 1874, the Remington firm was among the first substantial, domestic firearms manufacturers to market mass-produced, reasonably priced, breech-loading double guns. These utilized a conventional top-lever action and appeared in several grades depending on the quality of the barrels, stocking, checkering, and engraving. Pricing during the early 1890s ranged from $40 to $100. This specimen reflects the standard, or lowest, gradation—a simply checkered, half-pistol-grip stock of plain walnut; color-case-hardened frame and locks without engraving; and blued fluid steel barrels with matted rib and a single, bead-front sight.

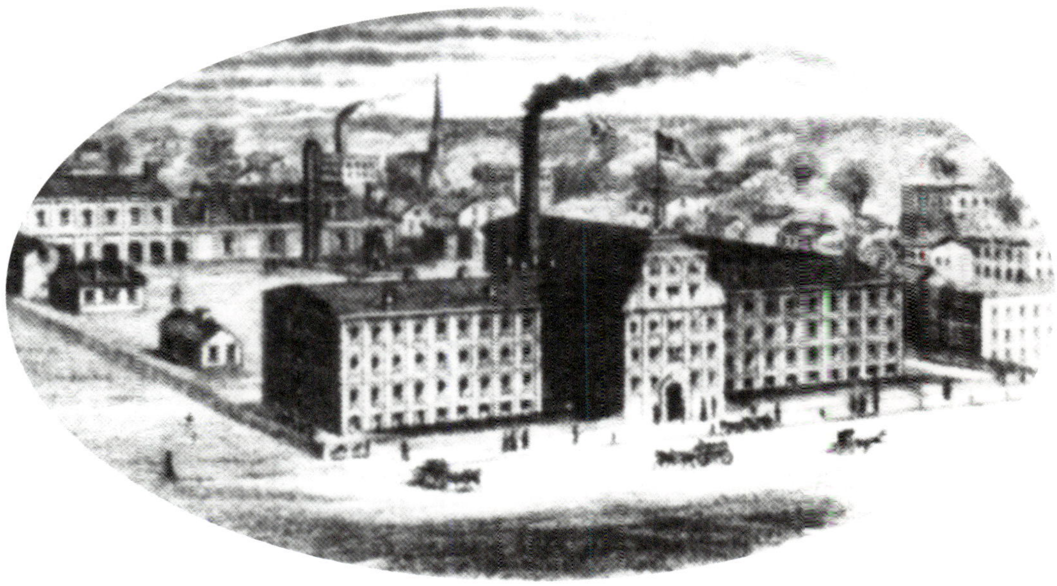

Chapter 5

SMITH & WESSON FIREARMS

EARLIER ASSOCIATED IN MANUFACTURING the iron-frame, lever-action Volcanic repeating pistol at Norwich, Connecticut, Horace Smith (1808–1893) and Daniel B. Wesson (1825–1906) formed a second partnership in 1856 at Springfield, Massachusetts. There, having secured the use of the Rollin White patent for revolvers with completely bored-through cylinder chambers loading at the breech, they began production of their "Seven Shooter," which fired self-contained metallic cartridges of their own design and manufacture (the original .22 Short rimfire). This benchmark innovation in revolver design and function began the famed line of Smith & Wesson handguns that continues to this day; it ultimately also came to dominate all of American revolver production after 1869, when White's patent protection expired.[1]

The early Smith & Wesson revolvers incorporated a "tip-up" barrel design for loading and chambered underpowered .22 and .32 rimfire cartridges in single-action, pocket and belt-sized models

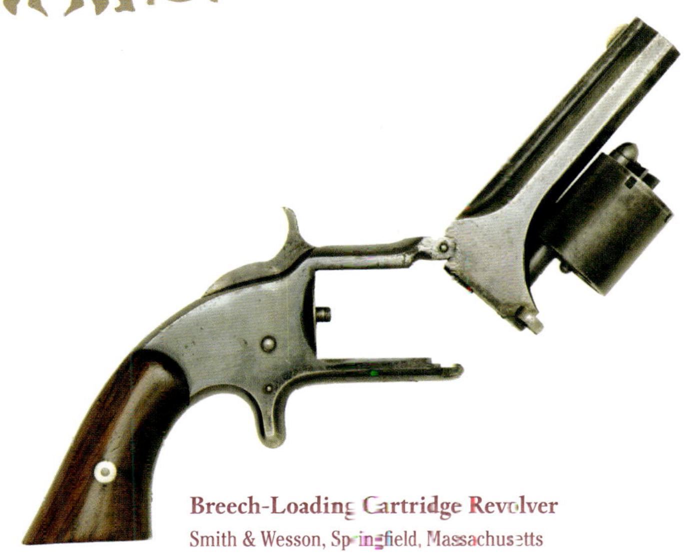

Breech-Loading Cartridge Revolver
Smith & Wesson, Springfield, Massachusetts
Model No. 1½ Pocket First Issue, 1865–1868, Caliber .32 Short Rimfire
Gift of Raymond W. Lewis H. 83.46

Hinged at the front of the topstrap, the "tip-up" design of the earliest Smith & Wesson revolvers allowed easy removal of the cylinder for cartridge loading at the breech. As illustrated here, the extraction of spent cartridge cases was achieved by passing the stationary ejector rod beneath the barrel through each of the cylinder chambers—muzzle first—in turn.

The Smith & Wesson factory, lithograph print, circa 1895.
Image courtesy private collection

with spur-pattern triggers. Traveling west in 1861, Samuel Clemens went heeled with the diminutive .22-caliber version, of which he recalled, "I was armed to the teeth with a pitiful little Smith & Wesson's seven-shooter, which carried a ball like a homeopathic pill, and it took the whole seven to make a dose for an adult. But I thought it was grand. It appeared to me to be a dangerous weapon."[2] Though certainly lacking in knockdown power, these early Smith & Wesson handguns nonetheless found an appreciative market as personal defense weapons among hundreds of overlanders like Clemens and thousands of Union soldiers serving in the Civil War. Production of all spur-trigger, tip-up models (Nos. 1, 1½, and 2) and their improved variations amounted to more than 460,000 revolvers from their inception in 1857 to their discontinuance in 1881.[3]

More significant in the frontier West (and worldwide) were Smith & Wesson's large-frame, "top-break," Model 3 American, Russian, and New Model single-action revolvers. These powerful .44-caliber weapons allowed for the simultaneous ejection of all spent cartridge cases when the topstrap over the cylinder was unlatched at the rear of the frame and the barrel was rotated downward. This provided a decided advantage over their Colt and Remington counterparts, in which the expended cartridge cases had to be ejected from the cylinder one at a time. (The only marginally competitive multiejection revolver was marketed by Merwin Hulbert & Company of New York City and in comparatively limited numbers.) Though the Model 3 Russian revolver was commercially viable in America, foreign contract sales far outstripped domestic sales. Between 1871 and 1878, some 131,000 of the three Russian model variants (or about 76 percent) went to the Czarist government.[4]

Testimonial and sometimes-cajoling letters from westerners to the Smith & Wesson firm regarding their big revolvers proved generally laudatory. From Ellsworth, Kansas, in May 1874, for instance, a gent signing off as "Cimmarron" wrote that he appreciated their weapon's "apparatus for throwing out the hulls [spent cases]" and observed, "I have been shooting your make and find it a perfect revolver. I can kill a man at 100 yds with my revolver every time[!] I like all frontiersmen like the Smith & Wesson better than Colts." In a later epistle, Cimmarron offered the company "as fine a wild horse" as he could catch in exchange for "one of your Russian Model Smith & Wesson Revolvers [in] 44 calibre."

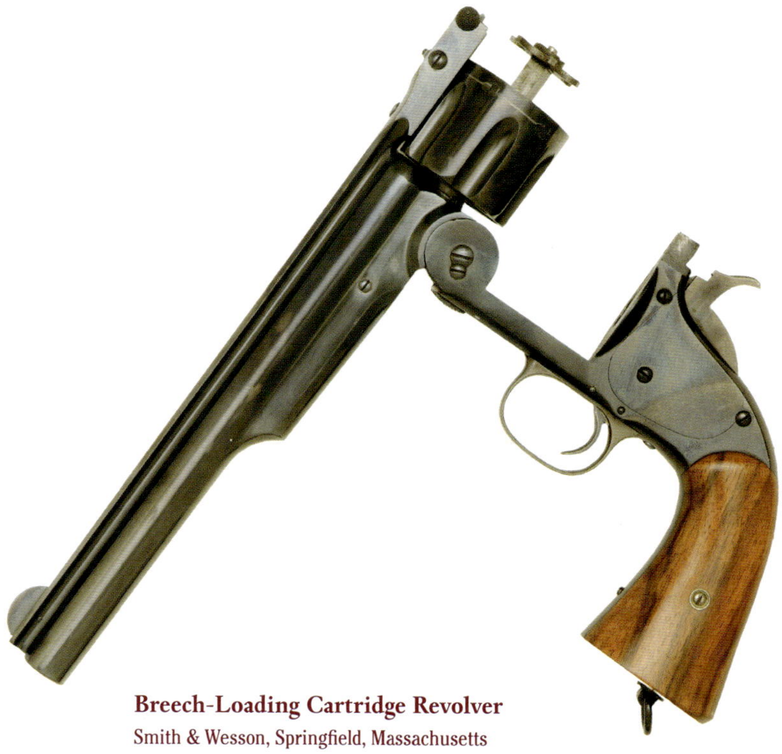

Breech-Loading Cartridge Revolver
Smith & Wesson, Springfield, Massachusetts
Model No. 3 Russian First Model, 1871–1874, Caliber .44 S&W Russian Centerfire
Gift of Raymond W. Lewis, H.183.12

Smith & Wesson's innovative Model 3 revolver design incorporated a "top-break" system that allowed the simultaneous ejection of all spent cartridge cases. As illustrated here, the topstrap has been unlatched from the frame and the barrel-and-cylinder assembly rotated downward, forcing the ejector upward with the (imagined) cartridge cases. With the barrel fully rotated, the ejector snapped back into the cylinder face, extracting the cartridge hulls in unison.

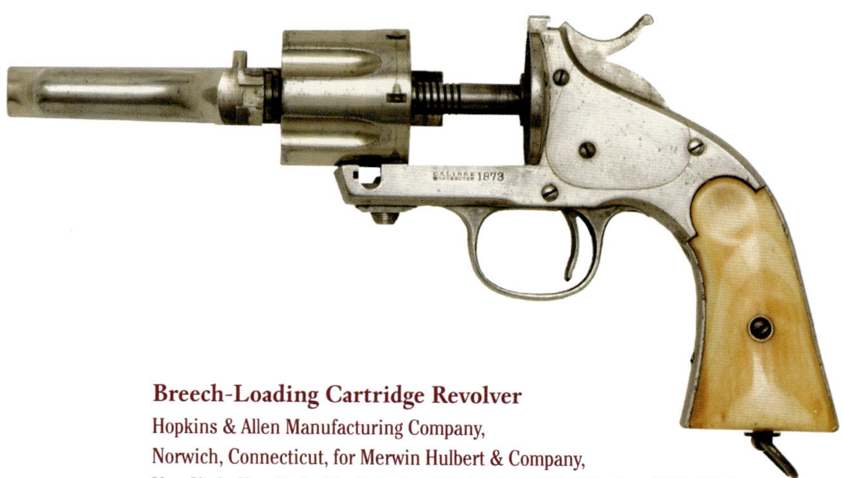

Breech-Loading Cartridge Revolver
Hopkins & Allen Manufacturing Company,
Norwich, Connecticut, for Merwin Hulbert & Company,
New York, New York, Single Action "Pocket Army" Model, circa 1880–1890
Caliber .44 Winchester Centerfire
Gift of Raymond W. Lewis, H.183.15

Although it never offered truly significant competition to the Smith & Wesson "top-break" cartridge-ejection mechanism, the large-frame Merwin Hulbert revolver also employed a simultaneous cartridge-extraction feature. Here the barrel and topstrap are unlocked, rotated clockwise, and then drawn forward—with the cylinder—to extract spent cartridges. This "Pocket Army" variant features a fluted cylinder, a square-butt profile, and grips of elephant ivory.

D. B. Wesson sent the revolver, but he never saw the horse, and he thereafter responded to such hopeful deal-makers with notable asperity.[5] Other, more-prominent frontier devotees of the formidable, fast-loading arms included railroad-builder General William J. Palmer, noted frontiersmen and showmen Buffalo Bill Cody and Texas Jack Omohundro, and notorious badmen Jesse James and John Wesley Hardin.[6]

Smith & Wesson's Schofield model revolver, a potent .45-caliber Model 3 variant modified for U.S. military service by Colonel George W. Schofield, proved its facility for fast reloading among cavalrymen in the field, but it nonetheless failed to supplant the Colt Model 1873 Single Action Army revolver. Many First Model Schofields, sold as surplus by the government, were purchased and refurbished by dealers Schuyler, Hartley & Graham, usually with shortened barrels and often with nickel plating. Many of these saw service in the West with the agents and guards of Wells Fargo & Company Express, and they were so marked and numbered. Production of all the various Model 3 revolvers, which commenced in 1870 and ceased around 1910, approached nearly 250,000, including the Schofield and Target variations as well as those manufactured in the fulfillment of several foreign military contracts. And although they met with little success in the U.S. military, an appreciable number of the nearly 9,000 First and Second Model Schofield revolvers did see some action on the late frontier with practitioners as varied as outlaw Cole Younger and lawman Bill Tilghman.[7]

By the early 1880s, Smith and Wesson employed about 500 workers and turned out more than 120,000 revolvers per annum. Contending with the rival Colt firm, in 1881 the company introduced a large-frame, "top-break," double-action revolver (a "self-cocker," in D. B. Wesson's parlance) chambered for either the .44-Russian or .44-Winchester (Frontier model) centerfire cartridge. Nearly 70,000 of these revolvers saw production before their discontinuance in 1913. By then the Smith & Wesson enterprise was well under way in the manufacture of its enduring series of swingout-cylinder, or "hand-ejector," double-action model revolvers—the basic design that still commands a substantial share of domestic handgun production today.[8] In the twenty-first century, high-quality revolvers and semiautomatic pistols continue to issue in the hundreds of thousands from the Springfield, Massachusetts, facilities of the Smith & Wesson Holding Corporation.

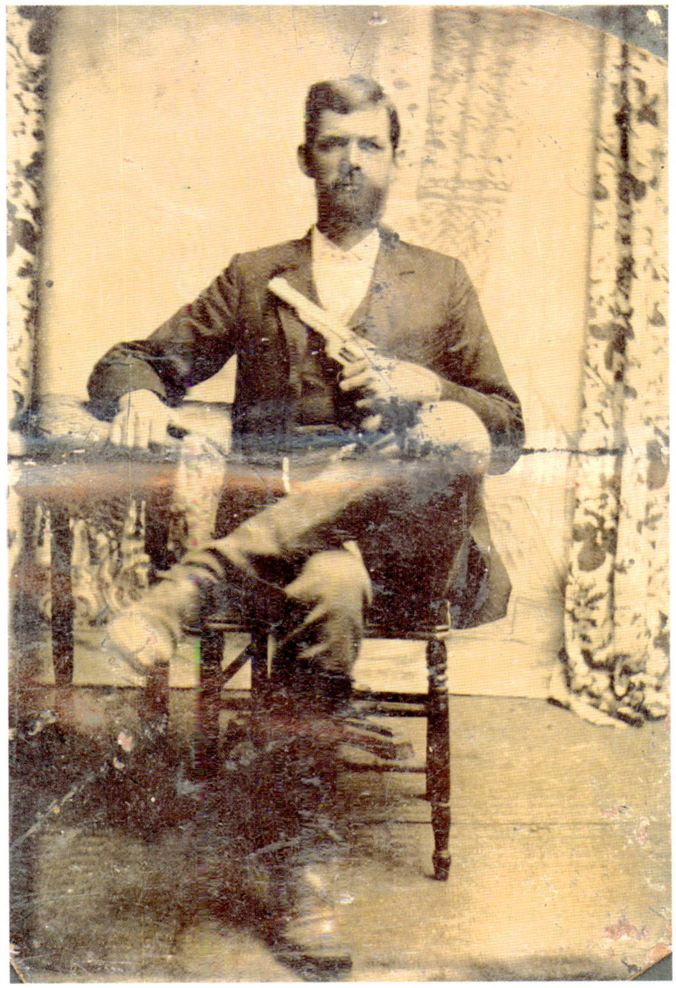

Photographer unknown. [Untitled.] Sixth-plate tintype, America, circa 1875–1880. Appearing every bit the frontier westerner in his high-topped boots, this bearded gent displays his shooting iron front-and-center. The revolver appears to be a Smith & Wesson Model 3 Second Model with full nickel-plated finish.

Dickinson Research Center, National Cowboy & Western Heritage Museum, 2003.052

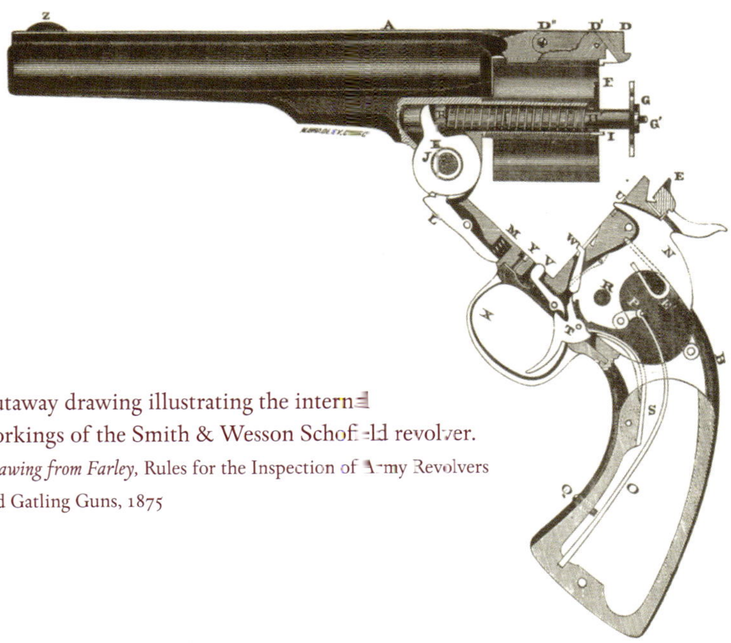

Cutaway drawing illustrating the internal workings of the Smith & Wesson Schofield revolver.

Drawing from Farley, Rules for the Inspection of Army Revolvers and Gatling Guns, 1875

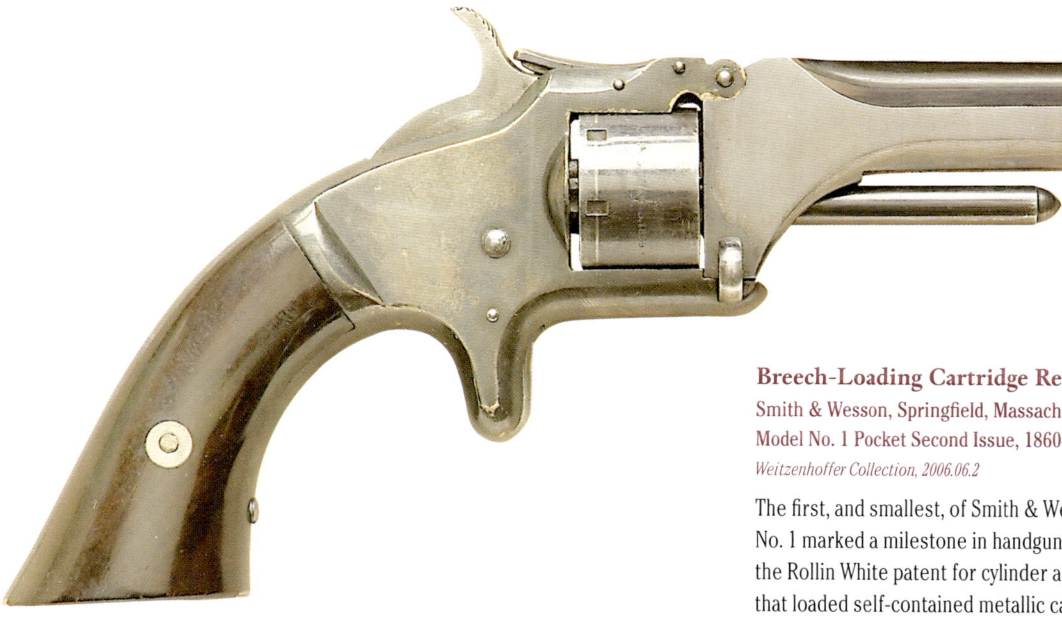

Breech-Loading Cartridge Revolver
Smith & Wesson, Springfield, Massachusetts
Model No. 1 Pocket Second Issue, 1860–1868, Caliber .22 Short Rimfire
Weitzenhoffer Collection, 2006.06.2

The first, and smallest, of Smith & Wesson's "tip-up" revolvers, the Model No. 1 marked a milestone in handgun design and technology with its use of the Rollin White patent for cylinder arms having bored-through chambers that loaded self-contained metallic cartridges at the breech. Though markedly underpowered in comparison to contemporary percussion revolvers, this tiny pocket pistol won an enthusiastic market, attaining a production total of about 117,000 units in a mere eight years. Presenting the model's standard two-tone finish, this specimen has a silver-plated frame with a blued barrel and cylinder. The grips are varnished rosewood.

Image courtesy Greg Martin Auctions, San Francisco, California

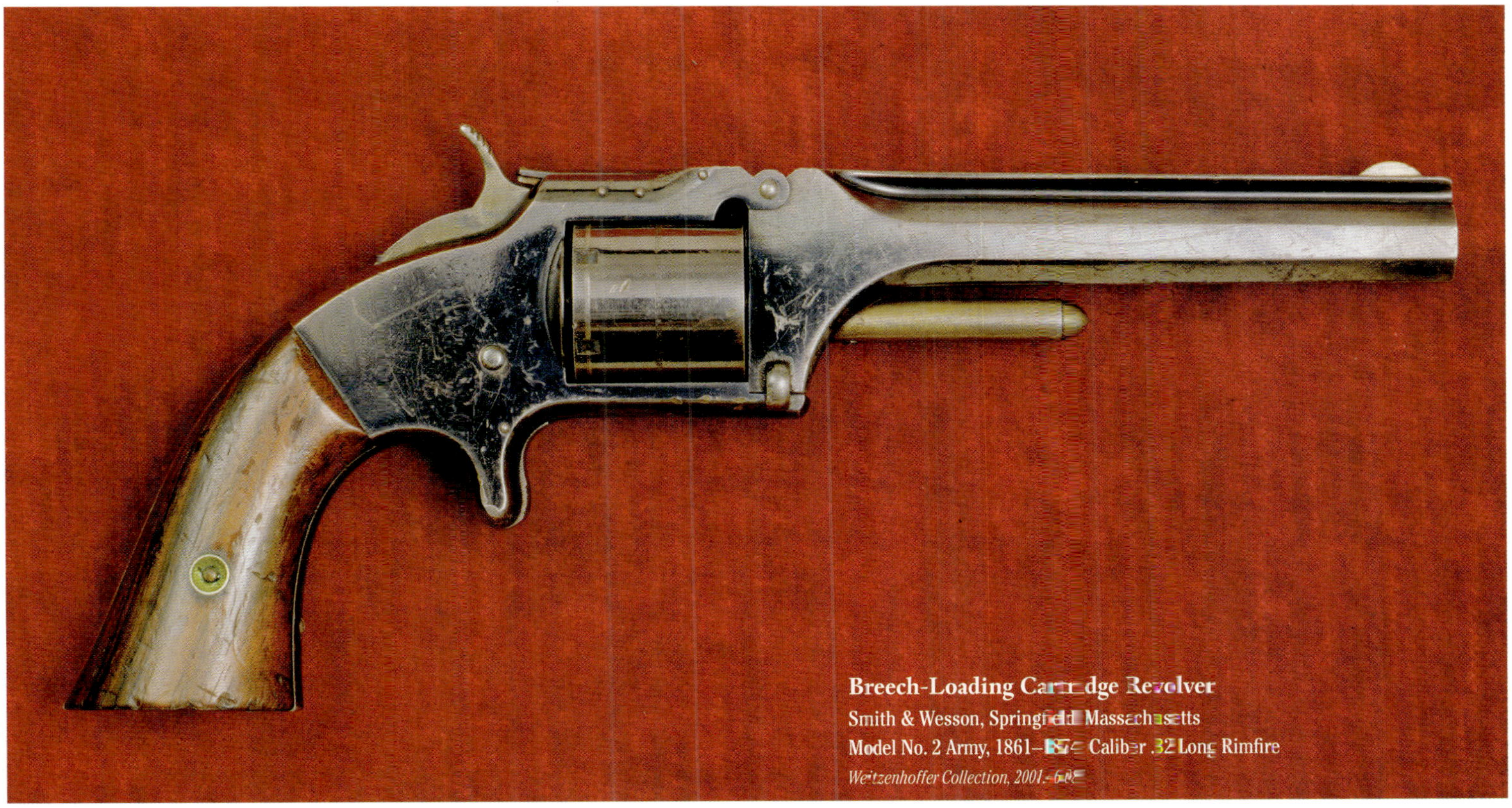

Breech-Loading Cartridge Revolver
Smith & Wesson, Springfield, Massachusetts
Model No. 2 Army, 1861–[illegible], Caliber .32 Long Rimfire
Weitzenhoffer Collection, 2001.[illegible]

Realizing that a .22-caliber revolver—even one utilizing metallic cartridges—could not compete with larger-caliber percussion handguns, Smith & Wesson introduced this scaled-up and slightly more potent .32-caliber weapon in 1861. The largest of Smith & Wesson's "tip-up" revolvers, the Model No. 2 Army had a frame of forged and milled iron rather than cast brass. With the Civil War just getting under way, the pistol won widespread favor as a personal sidearm among Union officers and enlisted men, and the firm enjoyed such unprecedented demand that by mid-1862, orders for the Model 2 filled the factory's production capacity through 1865. This 6-inch-barreled specimen exhibits the standard, full-blued finish with grips of varnished rosewood. (For embellished specimens, see chapter [illegible].)

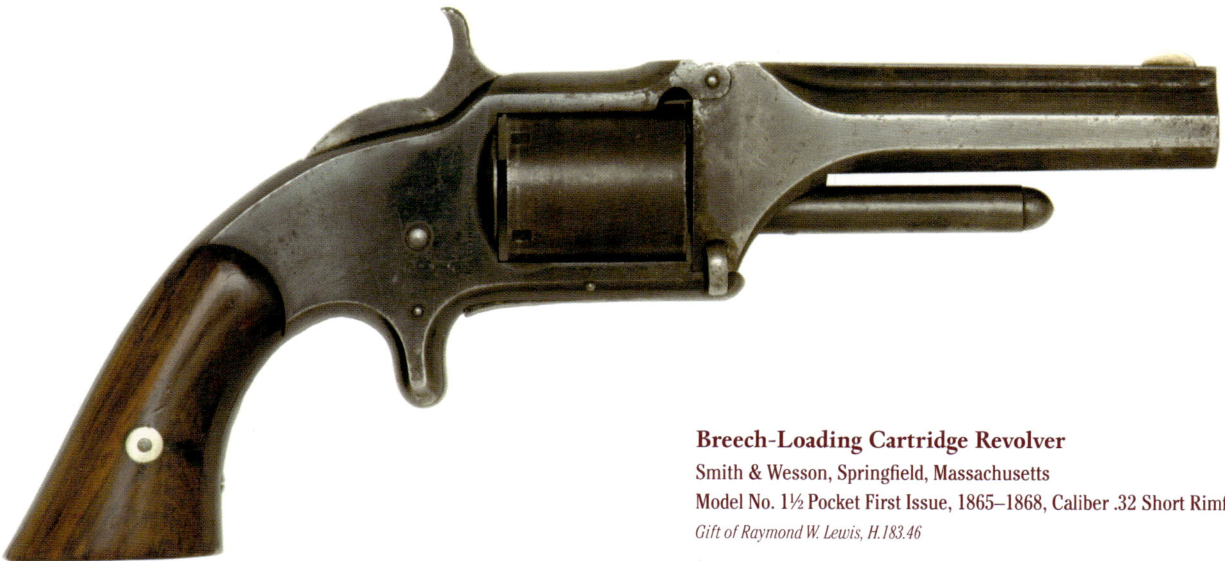

Breech-Loading Cartridge Revolver
Smith & Wesson, Springfield, Massachusetts
Model No. 1½ Pocket First Issue, 1865–1868, Caliber .32 Short Rimfire
Gift of Raymond W. Lewis, H.183.46

A slightly more robust version of the Model 1 "tip-up" design, this single-action, five-shot piece accommodated fewer—but more powerful—cartridges than its antecedent. Interestingly, when the pattern was introduced, Smith & Wesson lacked sufficient shop space for its production. Most of the parts for this revolver actually were fabricated under contract by gun and tool makers King & Smith of Middletown, Connecticut, with the final fitting and assembly performed by Smith & Wesson in Springfield. The pattern evidently was well received, as some 26,000 renditions saw production in less than four years.

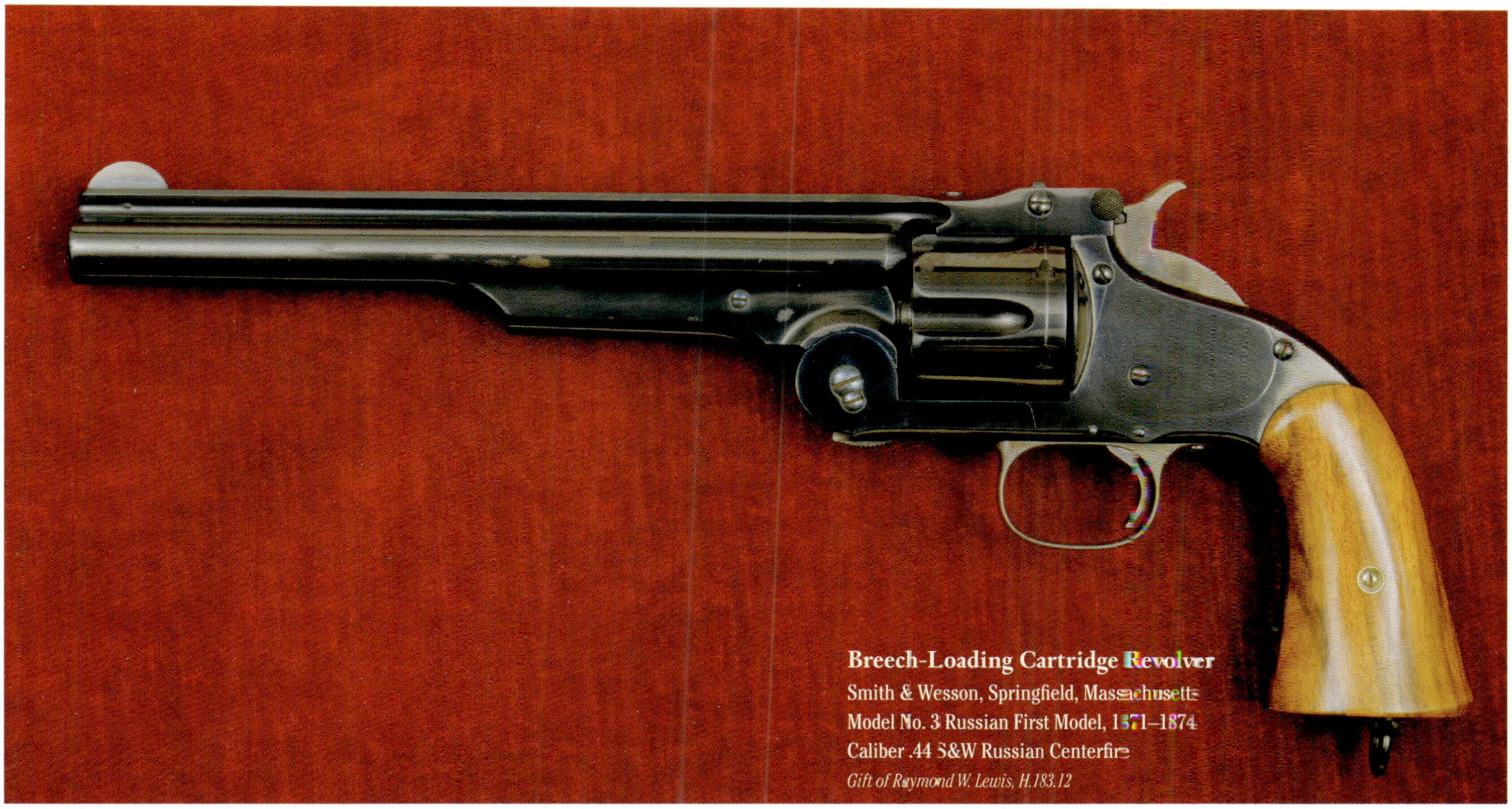

Breech-Loading Cartridge Revolver
Smith & Wesson, Springfield, Massachusetts
Model No. 3 Russian First Model, 1871–1874
Caliber .44 S&W Russian Centerfire
Gift of Raymond W. Lewis, H.183.12

Smith & Wesson's introduction in the early 1870s of the Model 3, American and Russian-pattern revolvers solidified the company's position as a leading contender with Colt in domestic handgun manufacture. These large-framed, .44-caliber pistols won many devotees in the frontier West, and they captured many domestic and foreign military contracts. The innovative new design incorporated a "top-break" system that allowed the simultaneous ejection of all spent cartridge cases, greatly facilitating the reloading process (see above discussion). In the far West, the commercial type found particular favor with outlaw John Wesley Hardin and western showman Texas Jack Omohundro.

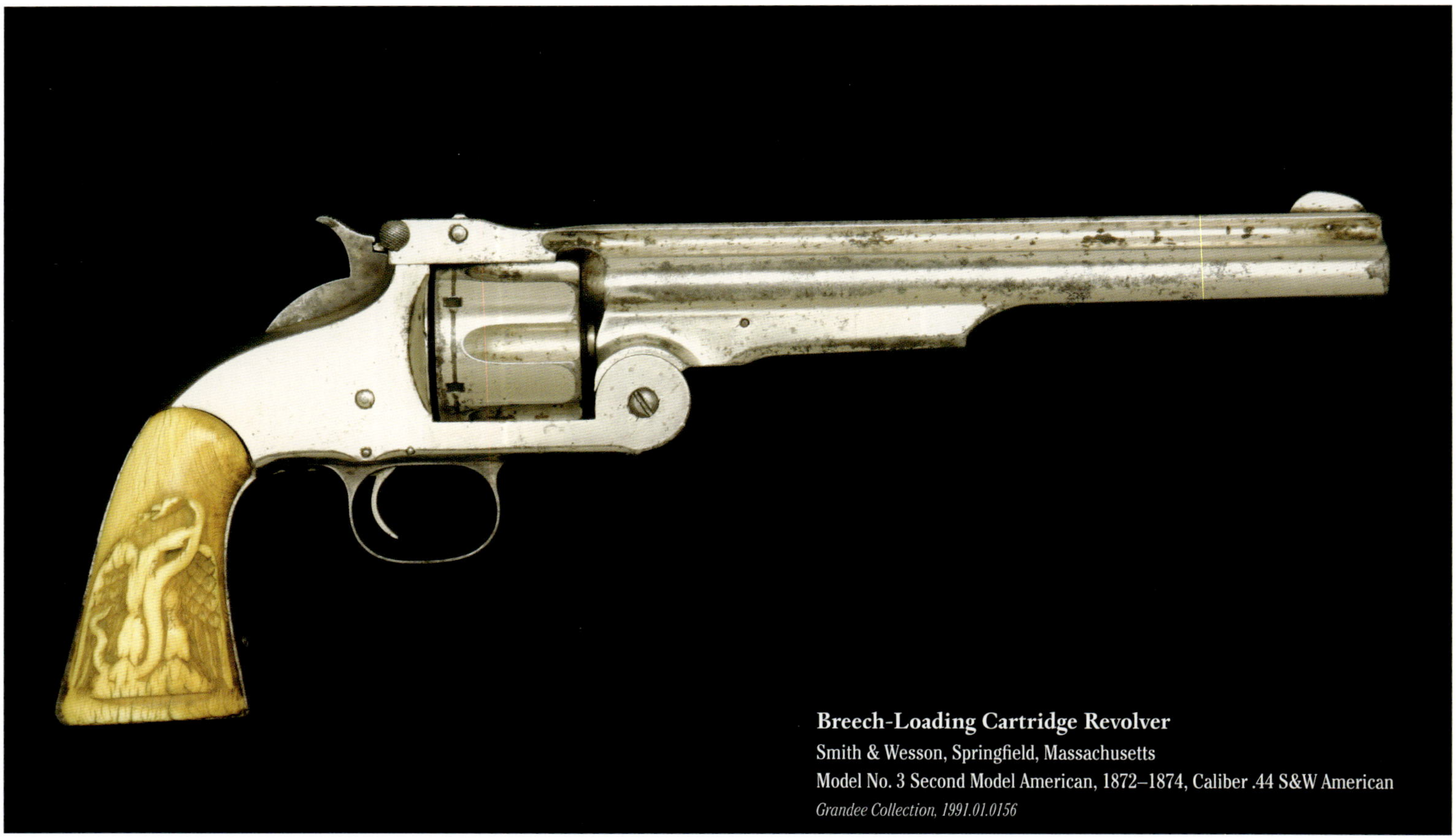

Breech-Loading Cartridge Revolver
Smith & Wesson, Springfield, Massachusetts
Model No. 3 Second Model American, 1872–1874, Caliber .44 S&W American
Grandee Collection, 1991.01.0156

With the creation of its Model 3 single-action revolver series, Smith & Wesson made a dramatic and highly competitive entry into the commercial and military markets for large-frame repeating handguns. This special-order, Second Model American revolver boasts full, factory nickel plating and high-relief-carved grips of elephant ivory bearing the Mexican national motif. These custom embellishments suggest that the piece very likely was first marketed in the Mexican Republic or in Texas. Virtually identical to the Model No. 3 Russian First Model, approximately 25,900 revolvers of both commercial models saw manufacture over four years.

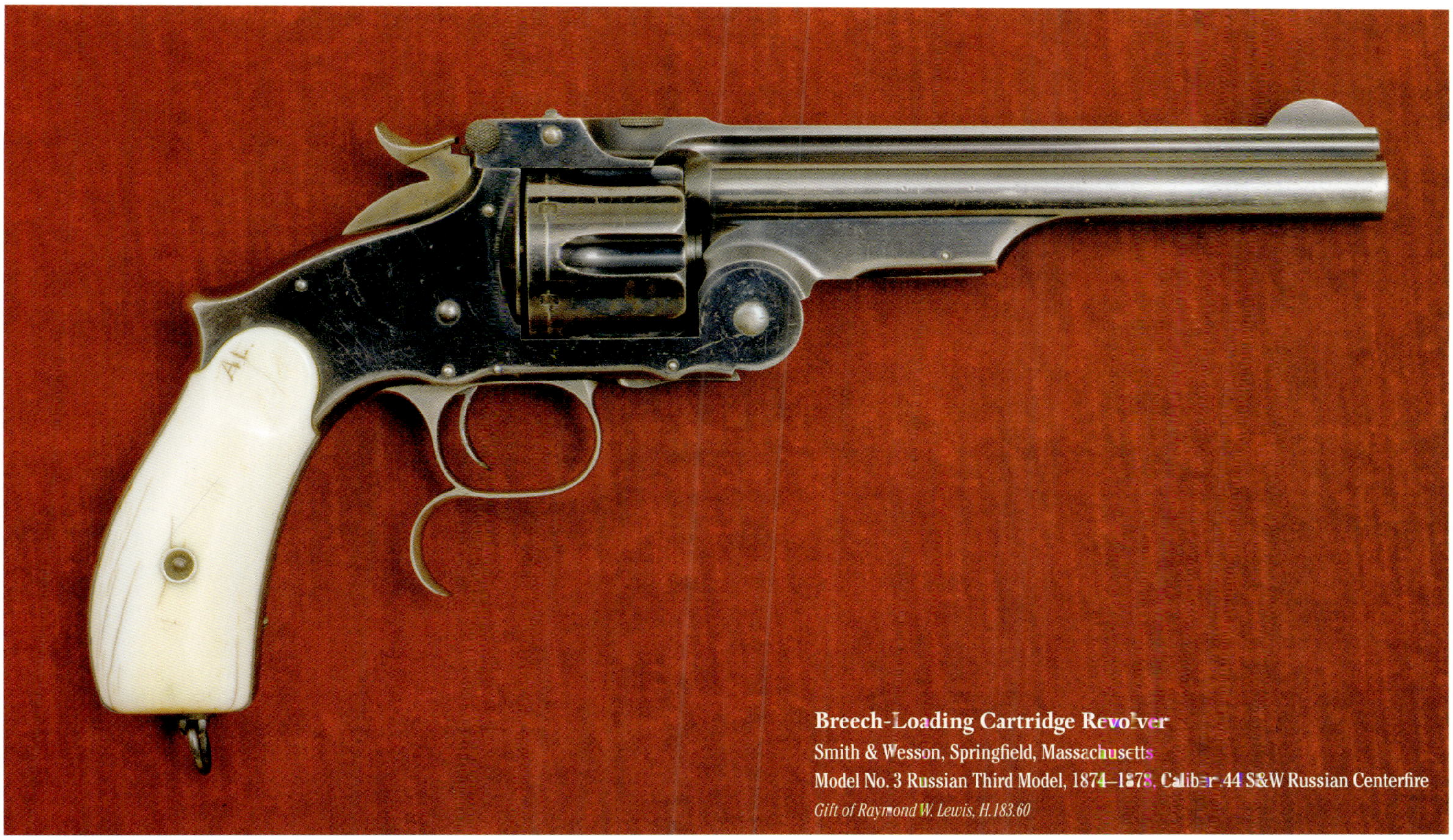

Breech-Loading Cartridge Revolver
Smith & Wesson, Springfield, Massachusetts
Model No. 3 Russian Third Model, 1874–1878, Caliber .44 S&W Russian Centerfire
Gift of Raymond W. Lewis, H.183.60

The culmination of Smith & Wesson's Russian-pattern revolvers, the third variation incorporated several design modifications from the previous rendition, as well as some improvements of its own. This model, like the second variation, featured a saw-handled grip profile with rounded butt and a spur-type finger rest (or sash or belt hook) on the rear of the trigger guard. Introduced were an improved (shorter) ejector system beneath the barrel and a standard 6½-inch barrel having an integral, steel-blade front sight. In some factory literature, this model was called the "Model 1874," or the "Cavalry Model," although it was never adopted by the American military. Fitted with special-order ivory grips at the factory, this commercial variant was one of about 13,500 produced.

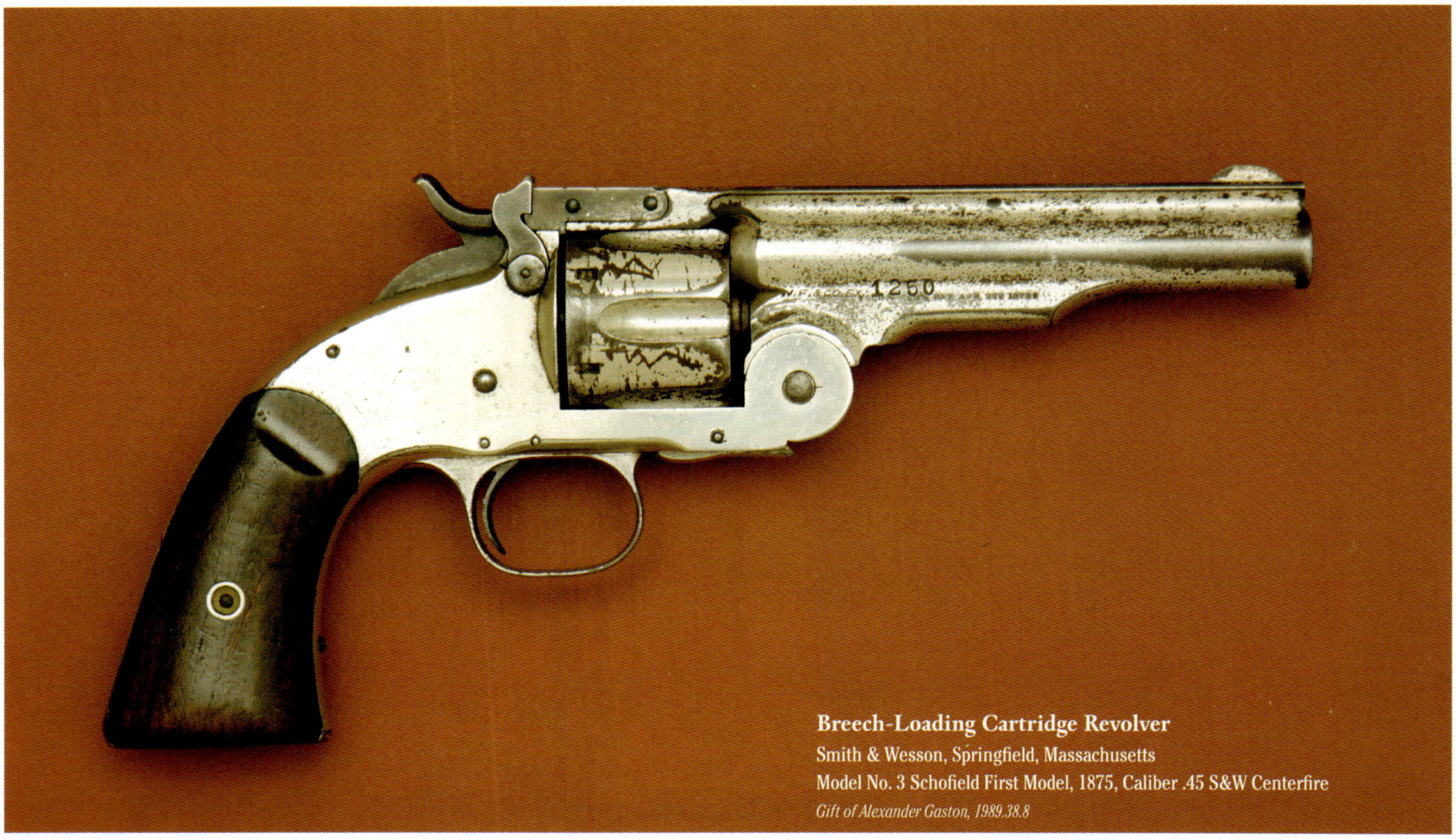

Breech-Loading Cartridge Revolver
Smith & Wesson, Springfield, Massachusetts
Model No. 3 Schofield First Model, 1875, Caliber .45 S&W Centerfire
Gift of Alexander Gaston, 1989.38.8

Incorporating design improvements patented by Major George W. Schofield, this modification of Smith & Wesson's Model 3 revolver introduced an improved latching system (fixed on the frame instead of the topstrap) that better secured the break-top action, an improved grip profile, and a slightly simplified ejector system. The army contracted for 3,000 of these revolvers in 1875, most of which served with the Fourth, Ninth, and Tenth Cavalry regiments. Later sold as surplus to gun dealers Schuyler, Hartley & Graham of New York City, many—like this piece—were refurbished with shortened barrels and nickel plating and were subsequently bought by Wells, Fargo & Company for their overland and railroad security guards. The ejector rod housing of this piece retains the over-stamped marking "W. F. & CO. EX. / 1260" in declaration of their ownership.

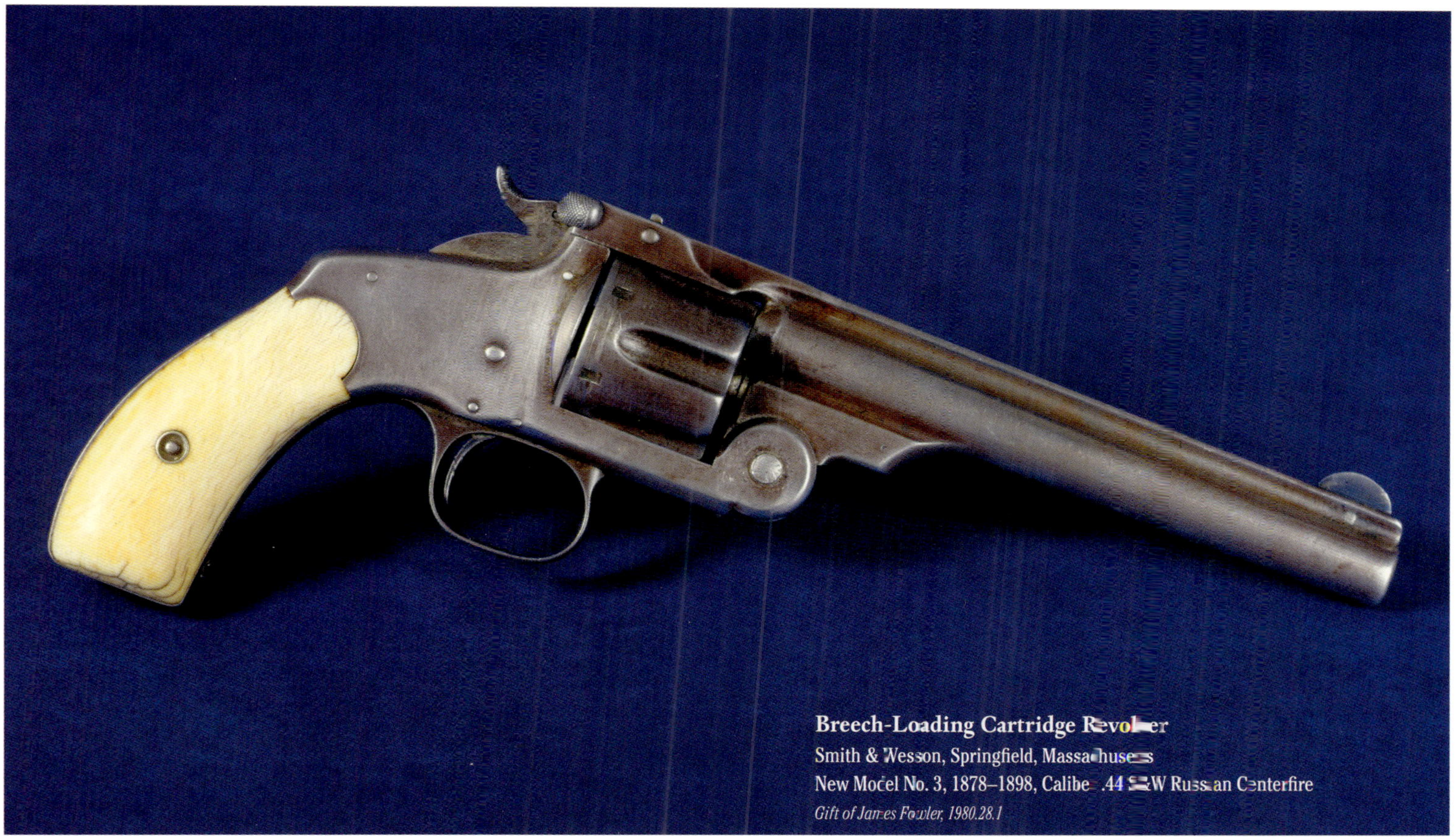

Breech-Loading Cartridge Revolver
Smith & Wesson, Springfield, Massachusetts
New Model No. 3, 1878–1898, Caliber .44 S&W Russian Centerfire
Gift of James Fowler, 1980.28.1

Successor to the Model 3 American and Russian patterns, the New Model No. 3 saw the longest production run of any Smith & Wesson large-frame, "top-break," single-action revolver. The model was also the first to be offered in a special target variation and as a revolving rifle with detachable shoulder stock. A favorite among some latter-day westerners, this particular example with its two-piece grips of elephant ivory is attributed—though undocumented—to Virgil Earp, one-time U.S. deputy marshal in Arizona Territory and a participant in the OK Corral gunfight at Tombstone in 1881.

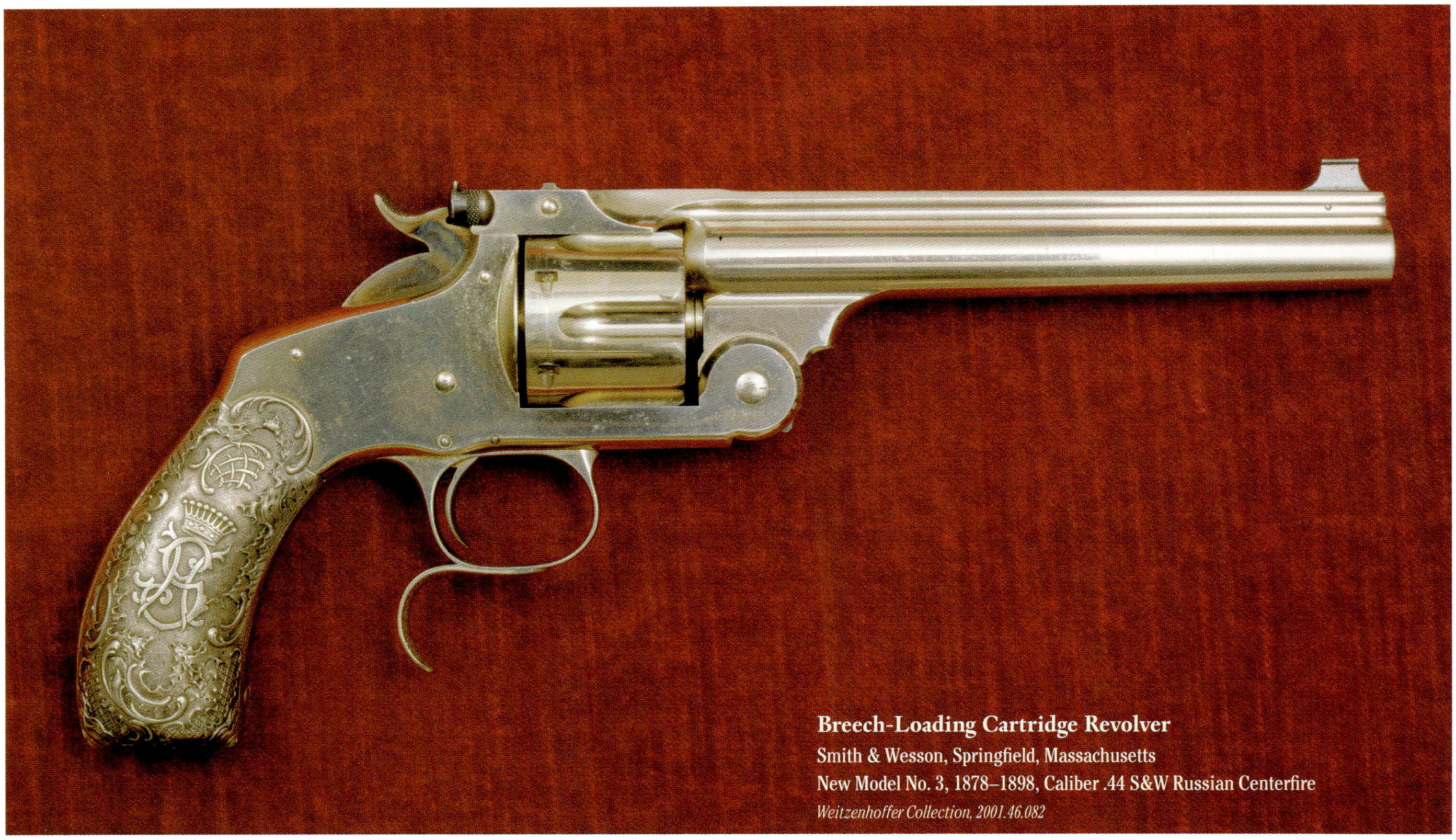

Breech-Loading Cartridge Revolver
Smith & Wesson, Springfield, Massachusetts
New Model No. 3, 1878–1898, Caliber .44 S&W Russian Centerfire
Weitzenhoffer Collection, 2001.46.082

In the late 1880s, Smith & Wesson began offering their No. 3 revolvers outfitted for competitive target shooting—both to prove their superior accuracy and to promote a growing niche market. In addition to specialized .32- and .38-caliber New Model No. 3 Target revolvers made especially for competition, quite a few standard .44-caliber revolvers—like this specimen—were supplied by the factory with target accessories. This revolver features a rear sight adjustable for windage, a Paine-type bead-front sight, and a special-order hand-checkered finger spur fitted on the trigger guard in the style of the Russian Model. The piece is factory nickel-plated and fitted with monogrammed, rococo-style grips of cast silver in the "Tiffany" genre.

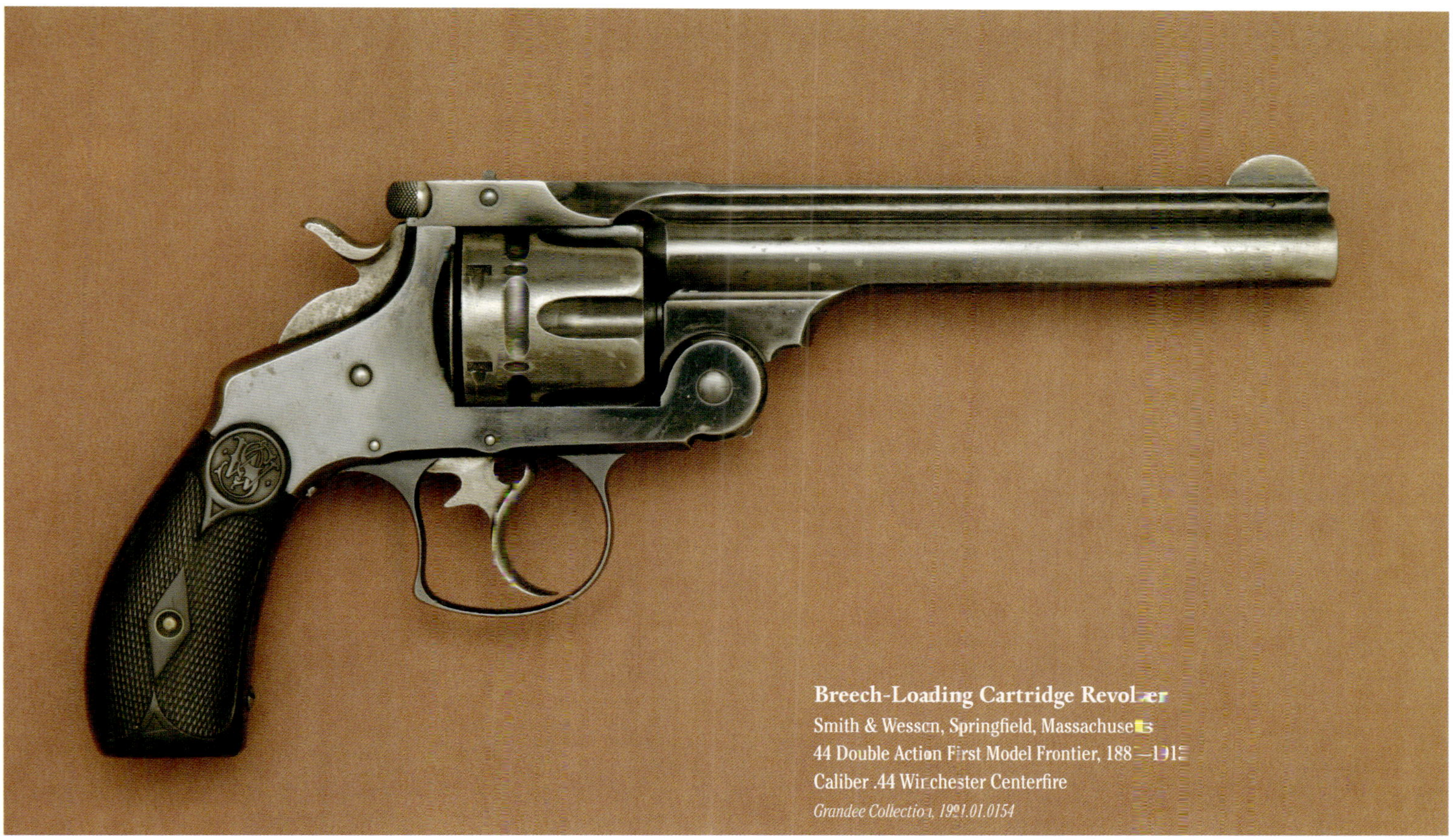

Breech-Loading Cartridge Revolver
Smith & Wesson, Springfield, Massachusetts
44 Double Action First Model Frontier, 188[illegible]–1913
Caliber .44 Winchester Centerfire
Grandee Collection, 1991.01.0154

The Smith & Wesson firm's first double-action revolver, this weapon continued the "top-break" design of the Model 3 series. Here, the fluted cylinder is distinguished by double stop slots with long "free grooves" around the middle to accommodate the movement of the double rocker-type locking mechanism in the bottom of the frame. This 6-inch-barreled specimen has a blued finish with case-hardened hammer and trigger—the latter of a rather ornamental, recurved and scalloped pattern unique among Smith & Wesson arms. It is fitted with checkered hard rubber grip plates with the S&W monogram impressed at the tops. Among notorious westerners, Indian Territory outlaw Belle Starr was known to favor this handgun model.

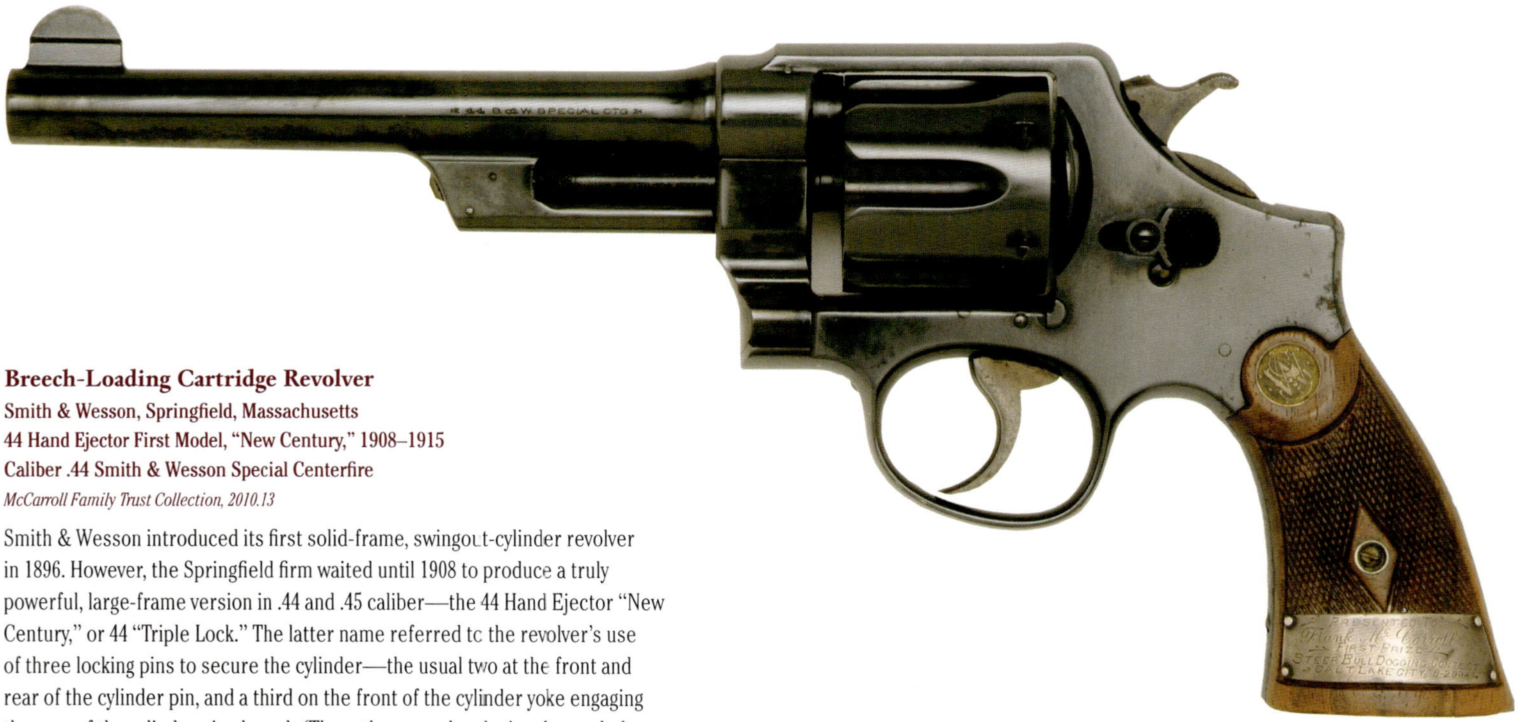

Breech-Loading Cartridge Revolver
Smith & Wesson, Springfield, Massachusetts
44 Hand Ejector First Model, "New Century," 1908–1915
Caliber .44 Smith & Wesson Special Centerfire
McCarroll Family Trust Collection, 2010.13

Smith & Wesson introduced its first solid-frame, swingout-cylinder revolver in 1896. However, the Springfield firm waited until 1908 to produce a truly powerful, large-frame version in .44 and .45 caliber—the 44 Hand Ejector "New Century," or 44 "Triple Lock." The latter name referred to the revolver's use of three locking pins to secure the cylinder—the usual two at the front and rear of the cylinder pin, and a third on the front of the cylinder yoke engaging the rear of the cylinder pin shroud. (The rather complex design demanded especially close tolerances in machining and assembly.) This example was awarded as a trophy to rodeo contestant Frank McCarroll for "First Prize / Steer Bull Dogging Contest" at Salt Lake City, Utah, in 1914.

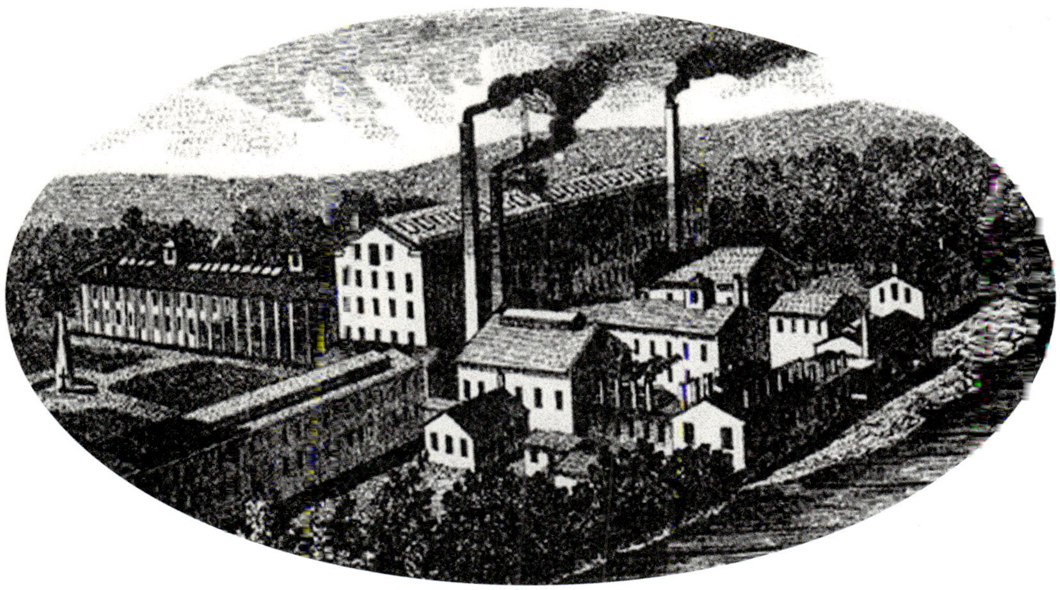

Chapter 6

SHARPS FIREARMS

AS RELATED IN CHAPTER 2, Christian Sharps (1810–1874) designed the first commercially successful breech-loading action in 1848. Contrary to several written accounts, he apparently never worked for John H. Hall at the Rifle Works in Harpers Ferry, Virginia, but he nonetheless recognized and understood the advantages and deficiencies of Hall's breech-loading system. As well, Sharps certainly appreciated the principles—and the decided advantages—of machine-based, interchangeable-parts manufacture in the fabrication of firearms demanding close tolerances in their action systems. Under Sharps's direction, contract gunsmith Albert S. Nippes of Mill Creek, Pennsylvania, began the halting production of Sharps's basic lever-actuated, falling-block design in 1849.[1]

Even the earliest Nippes-made slant-breech versions of the Sharps rifle (Models 1849 and 1850) won the attention of the civilian market. In March 1850, for example, a *Scientific American* article lauded the new firearm, noting that "in one instance it was fired nine times in one minute and all the balls [bullets] were placed within a circle of six inches diameter, at forty yards distance." The following month arms dealers Burtis & Brother advertised in the Saint Louis *Missouri Republican*, "Just Received By Express. Sharp's [*sic*] Patent Rifle, a splendid article for California emigrants. This rifle is of an entirely new construction, differing from all other fire arms; is loaded at the breech, caps [primes] itself, and is warranted to shoot ten times in a minute."[2] Considering that fewer than 250 rifles of both models saw manufacture between 1849 and 1851, their apparent notoriety doubtless stemmed from their technological and functional superiority when compared to the common muzzle-loading rifles of the day.[3]

Substantial production, however, did not commence until the Sharps Rifle Manufacturing Company incorporated at Hartford, Connecticut, in 1851, and contract manufacture commenced with the Robbins & Lawrence firm at Windsor, Vermont, the following year. The familiar slant-breech

The Sharps factory, woodcut print, circa 1865.
Image courtesy private collection

design underwent several modifications and improvements in succeeding models, having various gas-check systems fitted at the breech to retain explosive power and reduce fouling, and employing either Dr. Maynard's patented tape-priming system or Sharps's own patented pellet-priming system for percussion ignition. The Model 1851 carbines and rifles presented a singular design featuring a rounded frame with the hammer mounted inside the lockplate. Thereafter, Models 1852, 1853, and 1855 had outside hammers and flat frames with a prominent contour over the slanted breechblock mortise. Full production at the Hartford factory began in 1856, and total output of these four models amounted to about 18,150 carbines and rifles between 1852 and 1858. In a western prelude to the Civil War, about 900 Model 1853 carbines, today known as "Beecher's Bibles," were shipped to "Bleeding Kansas" in 1855–1856 to protect the antislavery "Free-Soilers" against the proslavery "Border Ruffians."[4]

In 1859 the Sharps Rifle Manufacturing Company introduced a much-improved design in which the breechblock moved vertically within the frame mortise (that is, directly perpendicular to the barrel). This alteration markedly reduced the escape of powder gases at the breech, and it made the weapon far more marketable. Between 1859 and 1866, production of these percussion, straight-breech arms (New Models 1859, 1863, and 1865) totaled some 114,000 units—virtually all were military-pattern carbines and rifles that served the Union cause during the Civil War. By 1862, the Sharps factory nearly rivaled its Hartford neighbor, Colt, employing about 450 laborers and turning out about 28,000 guns per annum. Immediately after the war, Sharps, like several of the other commercial arms makers, converted some 32,000 of their weapons from percussion to metallic-cartridge ignition. The majority were for military reissue, chambering the potent .50-70 Government centerfire cartridge (a .50-caliber bullet propelled by seventy grains of powder), but an appreciable number also were rebarreled as commercial hunting rifles.[5]

Among breech-loading, single-shot, sporting rifles, the already proven Sharps falling-block system adapted readily to metallic-cartridge use. In 1871 the Hartford firm introduced a rifle expressly for new long-cased, high-powered rounds—a weapon that garnered immediate and lasting notoriety on both the buffalo ranges of the frontier

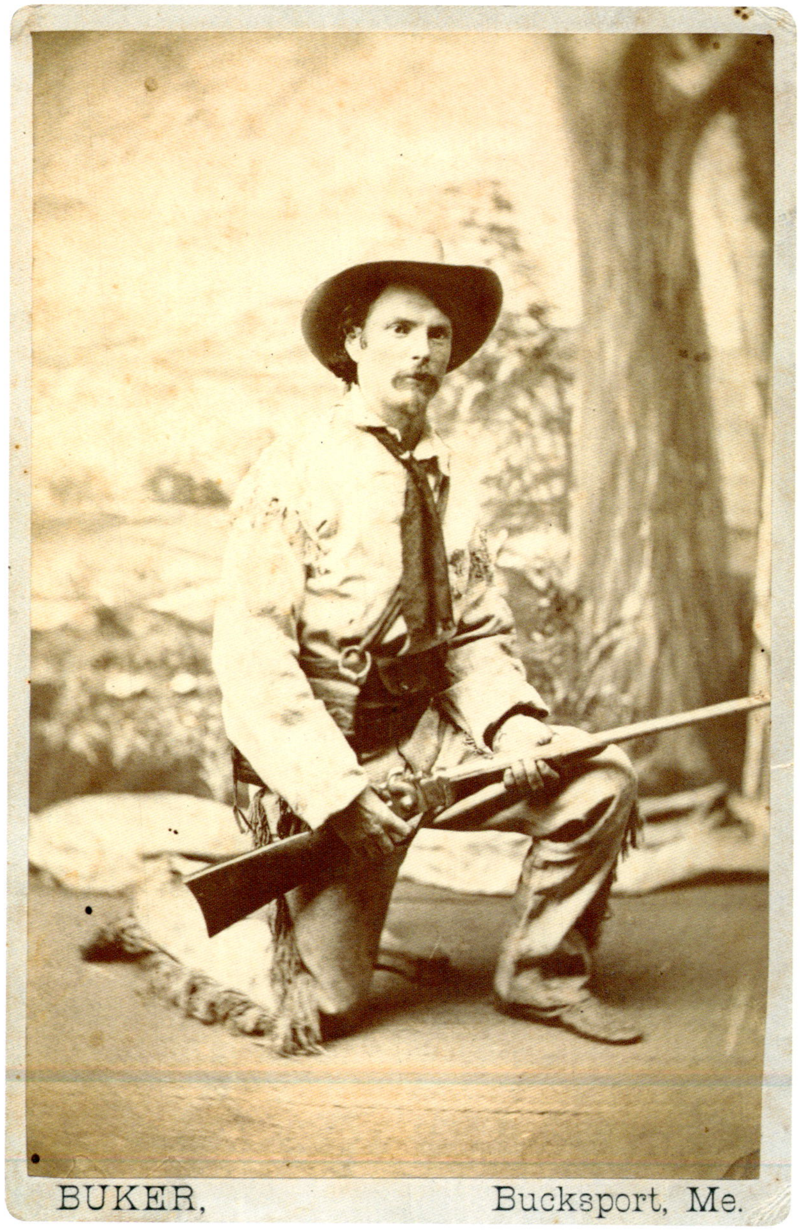

Buker, photographer. [Untitled.] Cabinet card, Bucksport, Maine, circa 1885–1895. Garbed in a fringed buckskin costume, this "western" poser grasps what appears to be a percussion, Sharps New Model 1863 straight-breech carbine without a patchbox. The ignition system, outmoded at the time the photograph was taken, suggests that the gun most likely served as a studio prop.

Dickinson Research Center, National Cowboy & Western Heritage Museum, 2003.111

West and the formal target ranges of America, Great Britain, and Ireland. Later designated the Model 1874, this remarkably robust breechloader ultimately accommodated such a broad range of powerful cartridges (from .40-70 and .44-90 to .45-100 and .50-120) that all manner of big-game hunters adopted it. One of these was P. C. Bicknell, who wrote to an eastern friend in 1876 from the buffalo range north of Fort Concho, Texas:

> [My companions] are men who have hunted and trapped all over the west from the Black Hills south to Texas. The gun they swear by is Sharps .44 cal. Just like yours they shoot 90 grains powder—some use the .50 cal. & 120 grains. The best gun in this part of the country is a Sharps .40 cal. with 90 grains of powder. . . . The hunter who has it says he can hit a bull as far almost as he can see it. It holds up wonderfully. [Still], the .44 cal. day before yesterday, . . . shot through several bulls in succession at 500 yds.[6]

Whether carried by professional hunters or by sport hunters, the Model 1874 Sharps sporting rifle (and its potent ammunition) did indeed "hold up wonderfully" over long distances, and the weapon more than lived up to its well-earned trade name, "Old Reliable."[7]

In addition to an evolving selection of cartridge calibers and loadings, Sharps offered the Model 1874 in an array of grades and styles based on barrel lengths, weights, and configurations; sight and trigger types; and stock patterns and finishes. Western subsistence and market hunters typically preferred a rather utilitarian rifle weighing fourteen to sixteen pounds with a thirty- or thirty-two-inch octagonal barrel, regular open sights, double set triggers, and a plain walnut, straight-grip stock with either a rifle- or a shotgun-pattern butt. In contrast, eastern gentlemen sport hunters and competitive target shooters usually selected a nine- to ten-pound rifle having a twenty-six to thirty-inch barrel. And they more often indulged in custom-order features such as half-octagon/half-round barrels, checkered pistol-grip stocking with shotgun-style butt, and various special-order sights including Globe, Beach, and wind-gauge front patterns in tandem with wrist-mounted, vernier-peep rear types adjustable for elevation and windage. Available in so many variations, the Model 1874 proved to be more

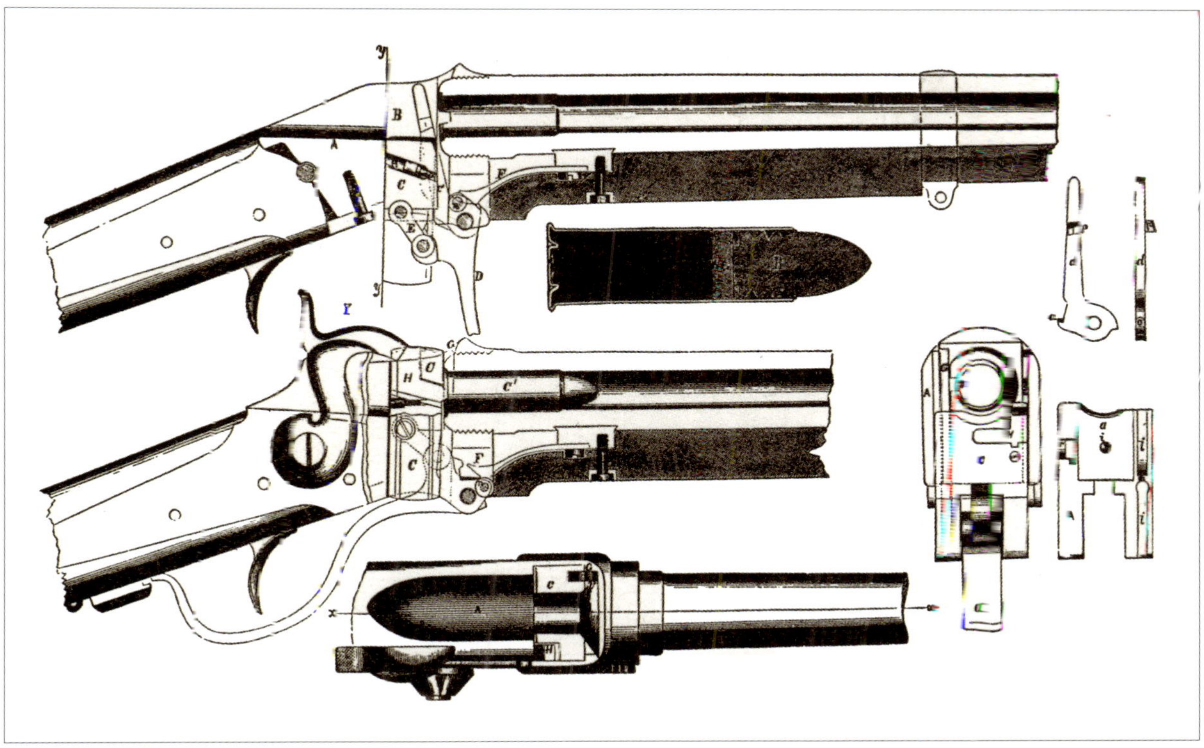

Cutaway drawings illustrating the internal workings of the Sharps Model 1874 single-shot rifle and its long-serving lever-actuated falling-block action—little changed from the 1850s except for its adaptation to metallic-cartridge ammunition.
Drawings from the Sharps Rifle Company's 1875 sales catalog

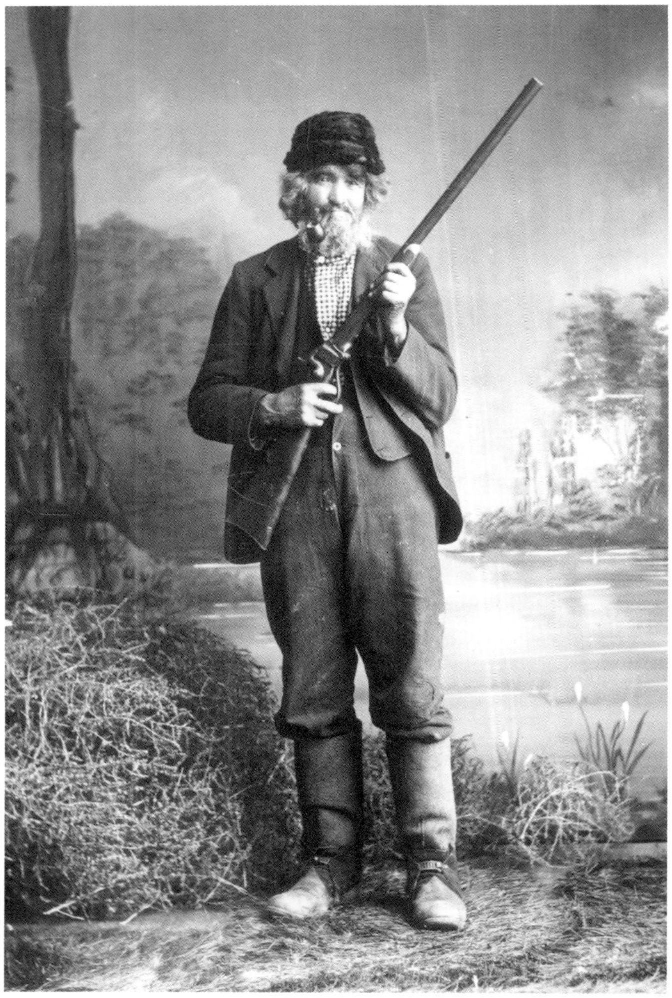

Photographer unknown. [*Henson Wiseman.*] Photographic print from a tintype, America, circa 1875–1880. A grizzled, pipe-smoking Henson Wiseman poses in a rustic studio setting with his Hartford-made, Sharps Model 1874 sporting rifle. Hartford manufacture is distinguished by the pewter forestock tip, which is typically absent in Bridgeport-made rifles.
Santangelo Collection, Dickinson Research Center, National Cowboy & Western Heritage Museum, 1994.10.1742

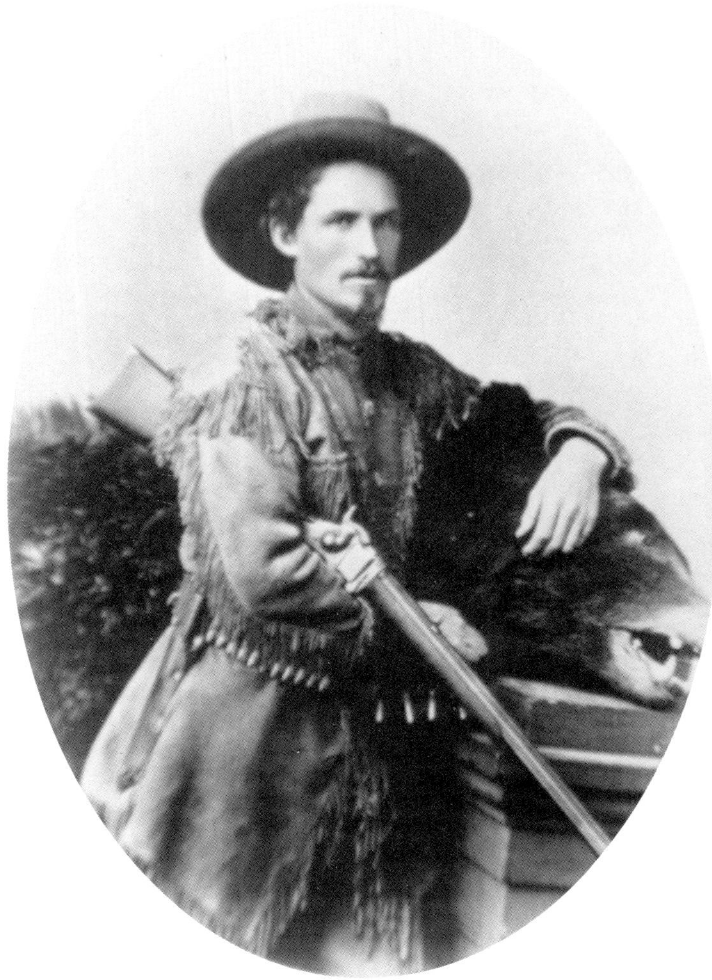

Photographer unknown. [Untitled.] Photographic print from a tintype, America, circa 1880–1885. This successful, buckskin-clad sport hunter proudly poses with his ursine trophy. Balanced over the subject's arm is the weapon that no doubt brought the bear to bag—a Bridgeport-made, Sharps Model 1874 sporting rifle. Long-cased metallic cartridges for the rifle are arrayed in a belt below the subject's elbow.
Santangelo Collection, Dickinson Research Center, National Cowboy & Western Heritage Museum, 1994.10.1744

a custom-built arm than an economically mass-produced weapon. Fewer than 9,500 sporting and target rifles of this model reached the market over a decade's output.[7]

Sharps's position as a manufacturer of but one firearm type (a single-shot breech-loader) resulted in a necessarily limited market and dwindling sales throughout the 1870s. In 1874 the firm reorganized as the Sharps Rifle Company and two years later moved manufacturing from Hartford to Bridgeport, Connecticut. There, in 1878, the company introduced the most advanced single-shot breech-loading rifle of the nineteenth century—the hammerless, falling-block action designed by German emigrant Hugo Borchardt. One who found the then-novel design of merit was sportsman George Oliver Shields, who observed in the early 1880s, "[A] large majority of the frontiersmen I met with—in fact, nearly all of them—used Sharp's rifles. . . . [Some] now in use are of the new hammerless model, forty-five caliber."[8] Few of these rifles, however, found market acceptance before the Sharps Rifle Company met financial collapse in 1881. At that time only some 2,460 Sharps-Borchardt sporting and target rifles had been produced. In all, the firm had manufactured some 172,650 rifles, carbines, and muskets over its thirty-year life span.[9]

Sharps Firearms

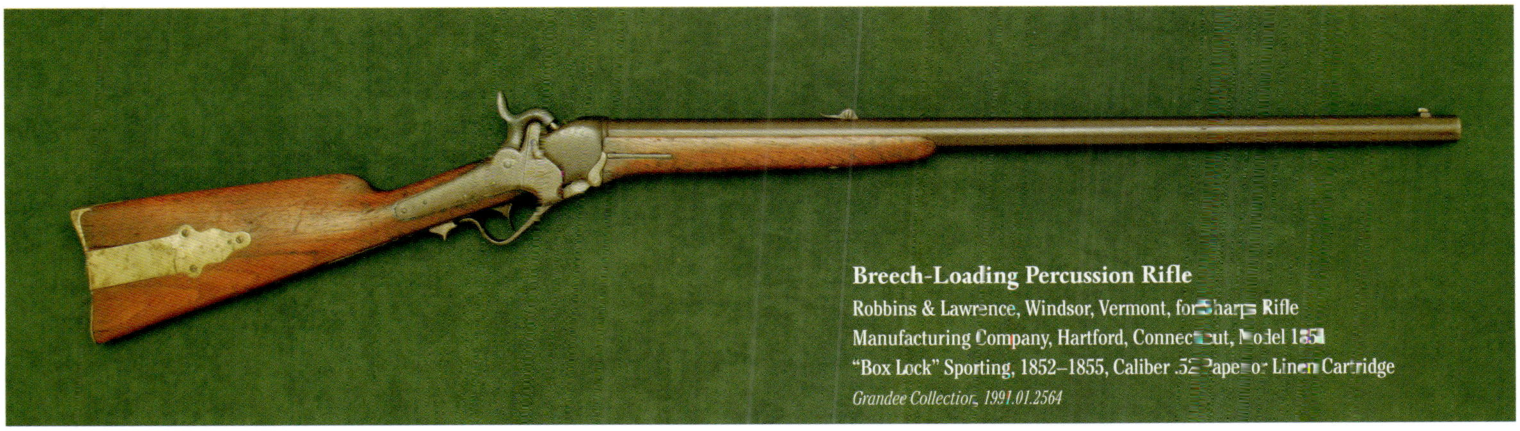

Breech-Loading Percussion Rifle
Robbins & Lawrence, Windsor, Vermont, for Sharps Rifle Manufacturing Company, Hartford, Connecticut, Model 1851 "Box Lock" Sporting, 1852–1855, Caliber .52 Paper or Linen Cartridge
Grandee Collection, 1991.01.2564

The first truly practical percussion breech-loading rifle available even in limited quantity, the Sharps Model 1851 falling-block design utilized Dr. Edward Maynard's patented tape-priming system for ignition. This device was inlet at the front of the lockplate, while the arm's hammer was mounted inside the lockplate—thus, the "box lock" appellation. This example has a 28-inch round barrel, and the stock is mounted with a brass patchbox and brass buttplate. Only about 180 sporting rifles saw production out of some 2,050 Model 1851s manufactured. Few such rifles likely saw use in the frontier West, though 60 military carbines of the pattern were issued to troops for trial at Fort Union, New Mexico.

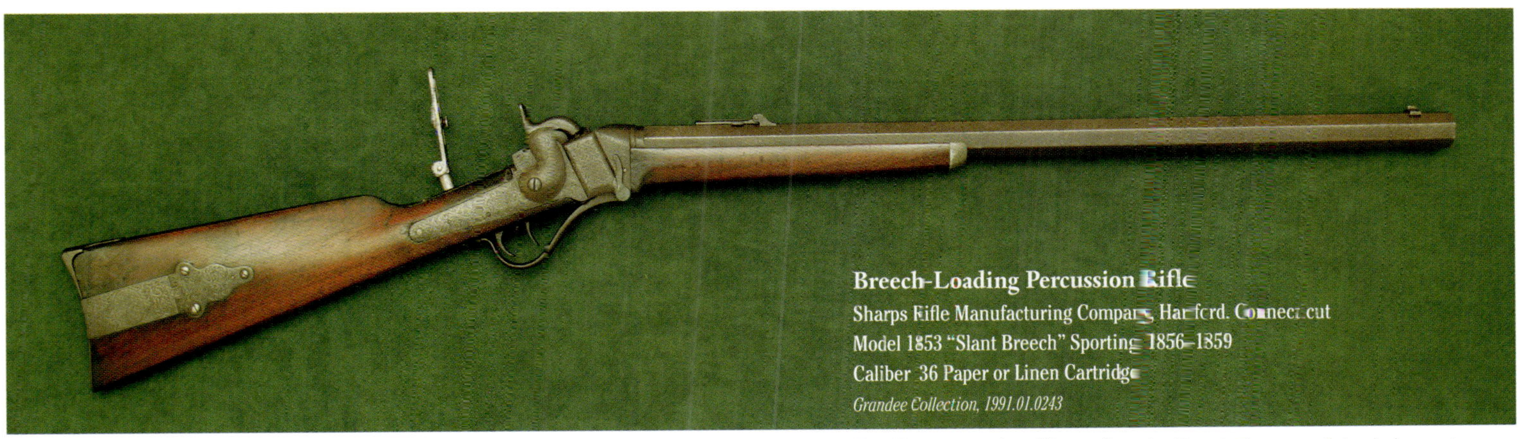

Breech-Loading Percussion Rifle
Sharps Rifle Manufacturing Company, Hartford, Connecticut
Model 1853 "Slant Breech" Sporting, 1856–1859
Caliber .36 Paper or Linen Cartridge
Grandee Collection, 1991.01.0243

This Sharps sporting rifle employs the inventor's own pellet priming system for ignition and incorporates a 27-inch full-octagon barrel, double set triggers, and a special-order folding and adjustable rear sight mounted at the wrist. First fabricated for Sharps by Robbins & Lawrence at Windsor, Vermont, this later example—fitted with iron rather than brass furniture—more than likely was produced in Hartford. This is probably the Sharps model of which gun dealer Horace Dimick proclaimed, "It is self evident that they are the best guns for persons exposed to the dangers of a frontier life." (The embellishment of this specimen is further illustrated and discussed in chapter 10.)

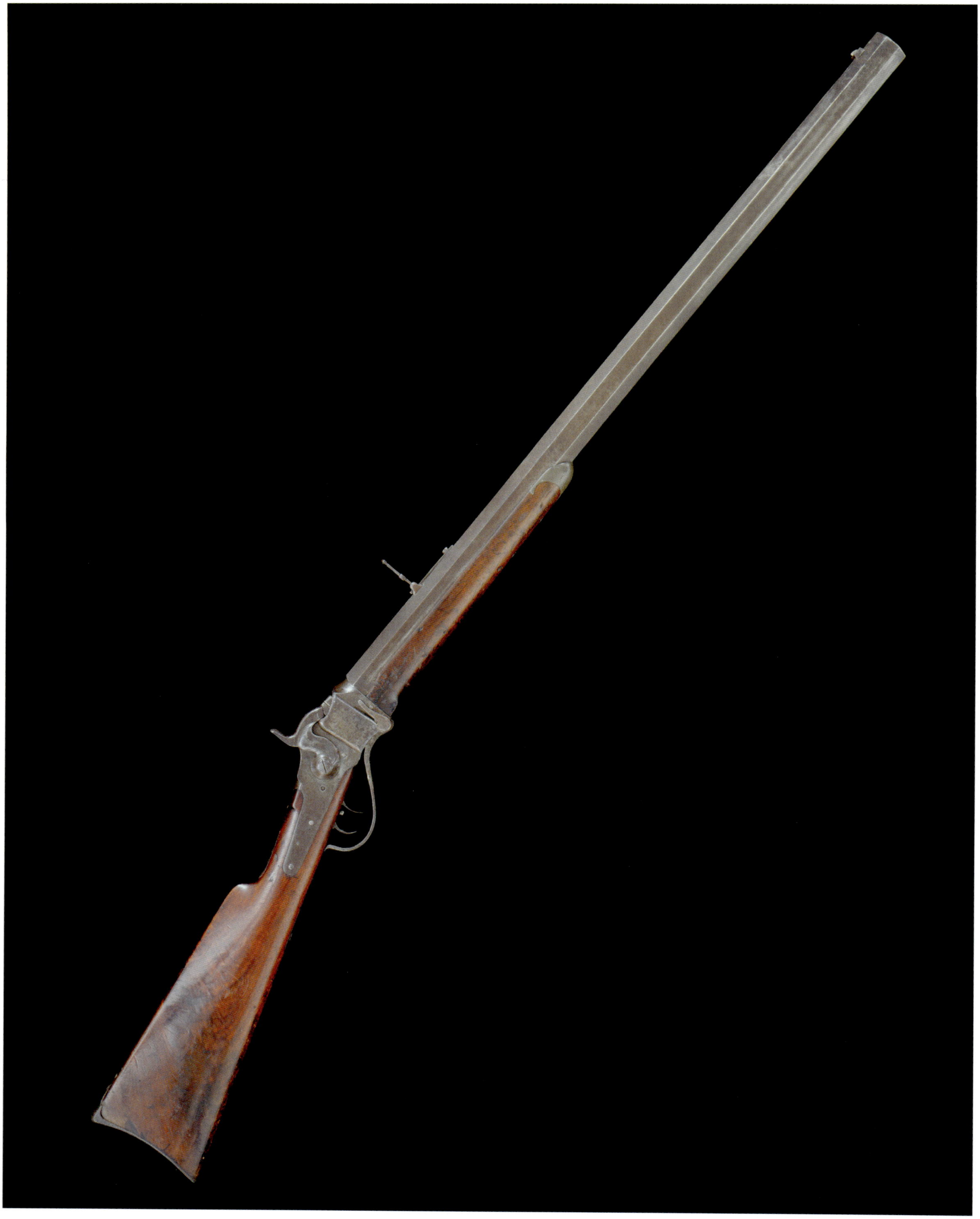

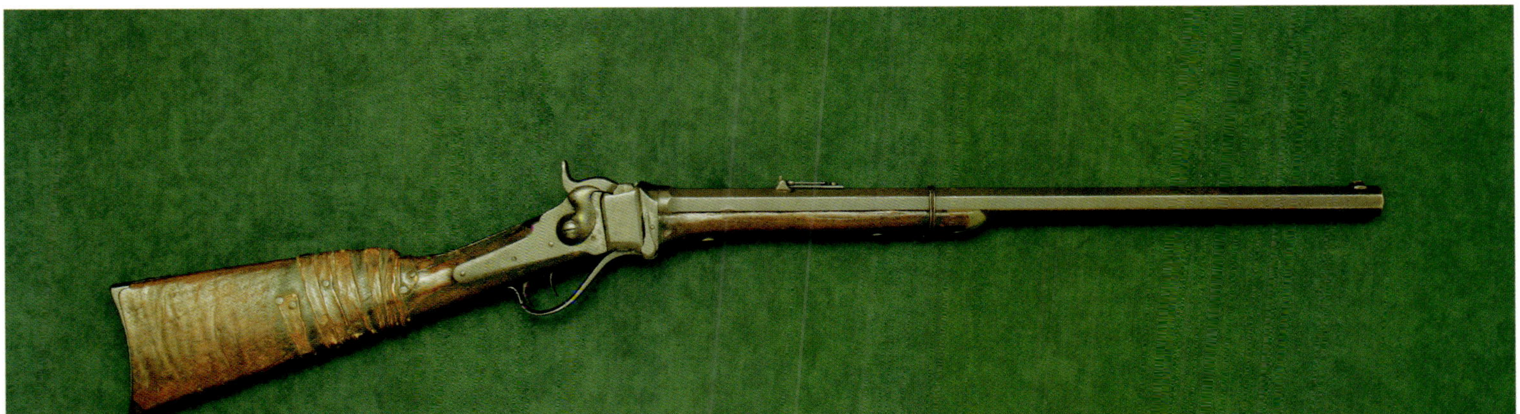

Breech-Loading Cartridge Rifle
Sharps Rifle Manufacturing Company, Hartford, Connecticut
Model 1874 Sporting, 1871–1876, Caliber .4[illegible]-70 Sharps Straight Centerfire
Loan courtesy Tench Tilghman, L85.6

Far more significant for its historical association than for its condition, this repaired and altered Sharps rifle was invoiced in June 1874 to gun dealer F. C. Zimmerman at Dodge City, Kansas. Bought by professional buffalo hunter and later-famed peace officer William "Bill" Tilghman, the weapon killed several thousand buffalo between 1874 and 187[illegible]. (Note the heavy wear on the bottom of the forestock from prolonged shooting over rest sticks.) Broken in a horse-fall out on the range, the buttstock was repaired by Tilghman using a strip of raw buffalo hide. Sometime after leaving the range, Tilghman had the octagonal barrel cut down from 32 inches to 24 inches to serve as a convenient saddle gun.

Breech-Loading Cartridge Rifle
Sharps Rifle Manufacturing Company, Hartford, Connecticut
Model 1874 Sporting, 1871–1876, Caliber .50-90 Sharps Straight Centerfire
Grandee Collection, 1991.01.2315

The choice of most professional hide hunters, Sharps falling-block, sporting rifles undoubtedly killed more buffalo than did any other make. This 16-pound rifle has an extra-heavy, octagonal barrel ideal for rapid, sustained shooting, as it allowed a steadier aim and slowed the heating and expansion that diminished accuracy. Shipped to noted Sharps dealer F. C. Zimmerman of Dodge City, Kansas, in December 1872, this classic buffalo rifle bears [illegible] provenance to the great slaughter on the prairies of Kansas and Texas. Often termed a "Poison Slinger" or "Big Fifty" by the hunters, the piece has the typically favored open sights and double set triggers.

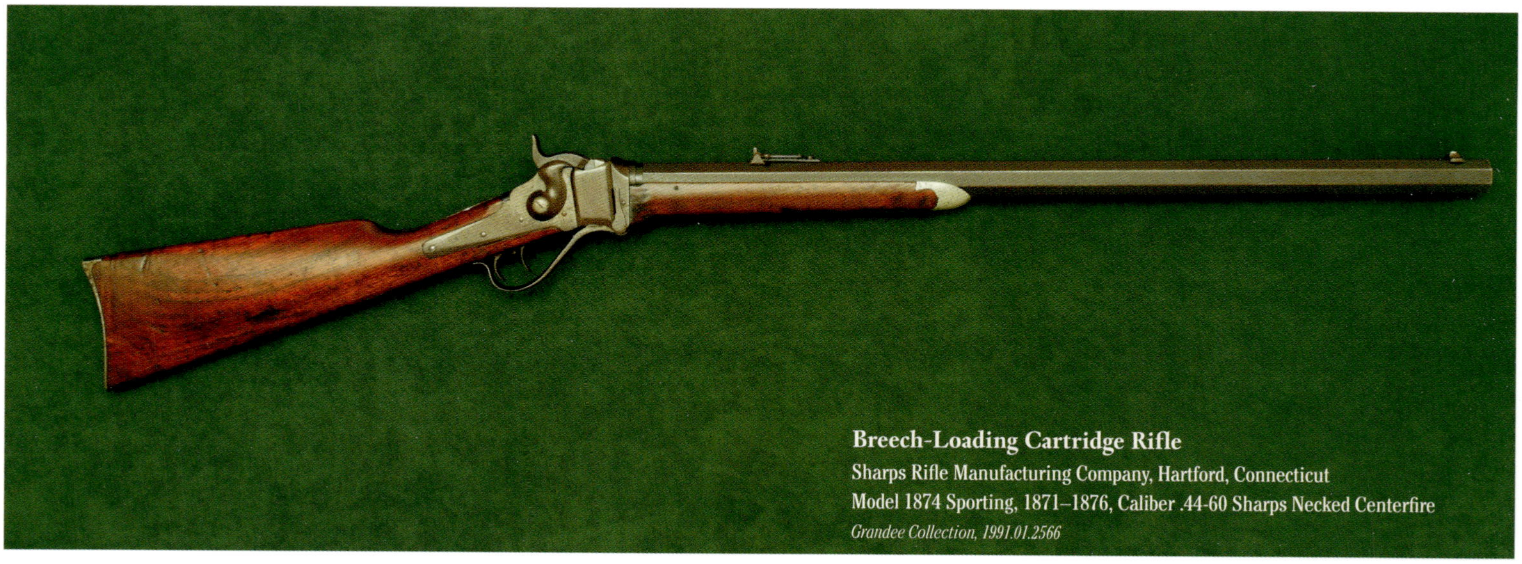

Breech-Loading Cartridge Rifle
Sharps Rifle Manufacturing Company, Hartford, Connecticut
Model 1874 Sporting, 1871–1876, Caliber .44-60 Sharps Necked Centerfire
Grandee Collection, 1991.01.2566

A continuing favorite of big-game hunters all over the West, the Sharps falling-block action proved readily adaptable to the powerful, long-cased metallic cartridges formulated during the early to mid-1870s. This Hartford-made specimen, with its pewter forestock tip, was shipped to gun dealer J. C. Grubb in Philadelphia in 1875 and was noted in the factory records as having a full 28-inch octagonal barrel, double set triggers, open sights, and oil-finished stocks. Available in various barrel lengths, weights, and calibers, these famed single-shot rifles delivered long-range accuracy and knockdown power superior to contemporary repeating rifles.

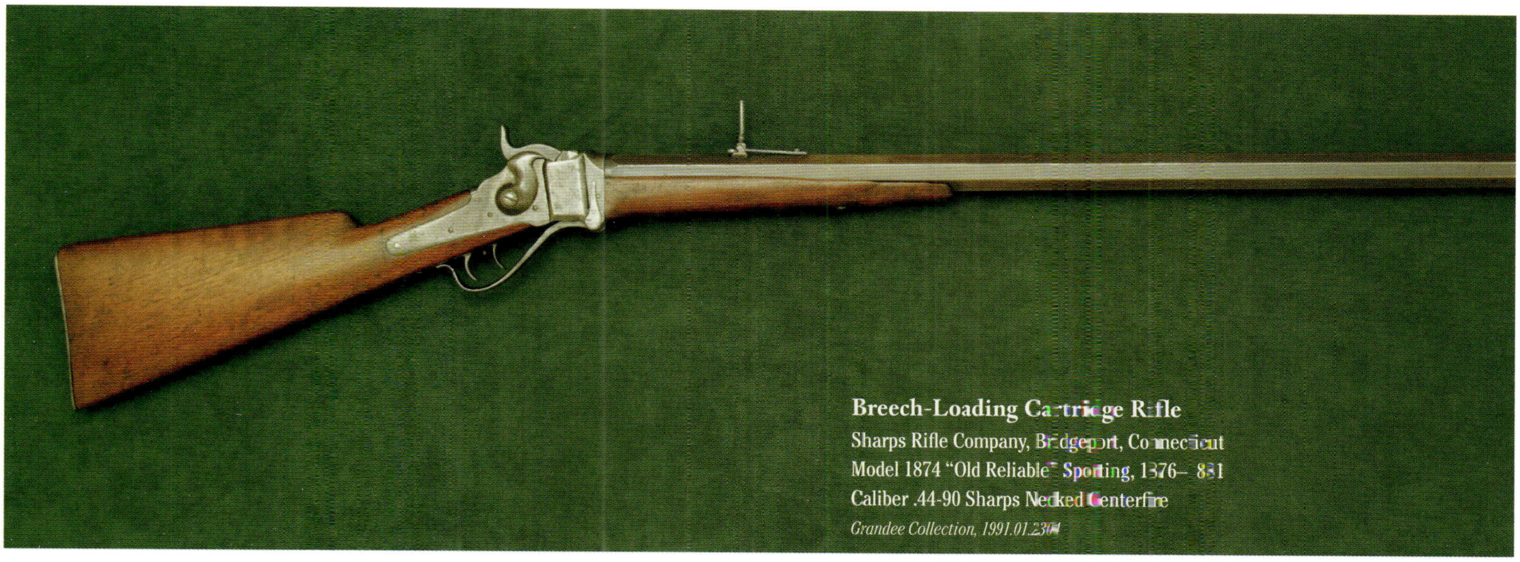

Breech-Loading Cartridge Rifle
Sharps Rifle Company, Bridgeport, Connecticut
Model 1874 "Old Reliable" Sporting, 1876–1881
Caliber .44-90 Sharps Necked Centerfire
Grandee Collection, 1991.01.230[illegible]

Powerful single-shot rifles stayed in use among professional market hunters even after acceptable repeaters became available. Sufficient for any big game in the West, this specimen—using the .44-90 cartridge—could attain a muzzle velocity of 1,300 feet per second and deliver as much as 1,850 foot-pounds of energy at the muzzle. Invoiced to the company's principal western jobber, B. Kittridge & Co. of Cincinnati, Ohio, in August 1876, this Model 1874 Sharps "Old Reliable" rifle is listed in the company records as having a full, 30-inch octagonal barrel; double set triggers; open sights; and a weight of 11 pounds, 10 ounces.

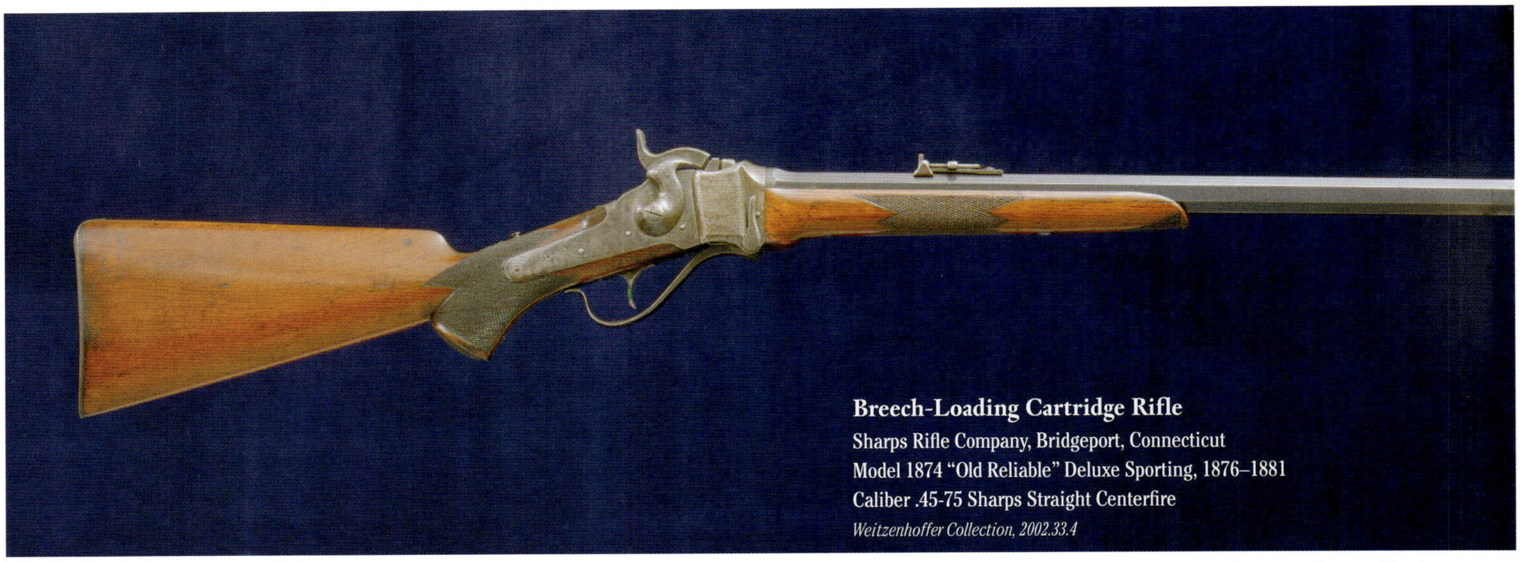

Breech-Loading Cartridge Rifle
Sharps Rifle Company, Bridgeport, Connecticut
Model 1874 "Old Reliable" Deluxe Sporting, 1876–1881
Caliber .45-75 Sharps Straight Centerfire
Weitzenhoffer Collection, 2002.33.4

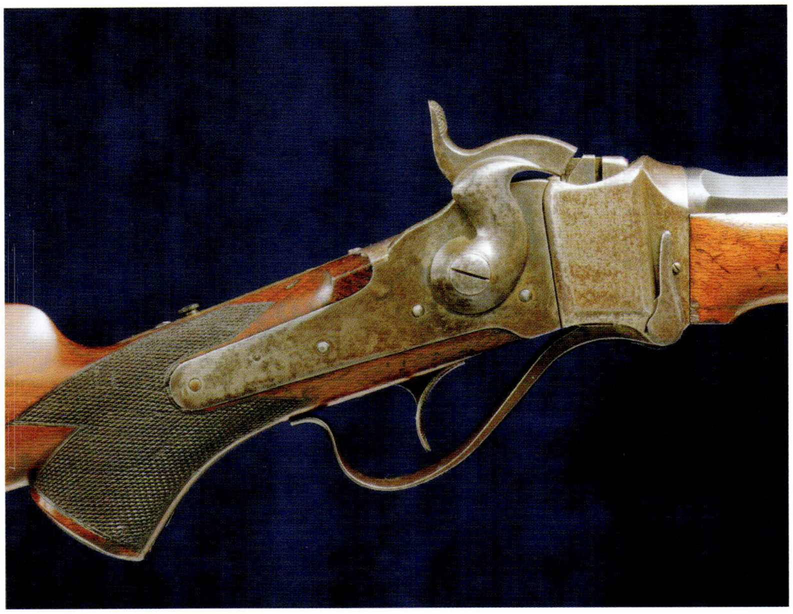

Among the best single-shot, breech-loading rifles manufactured in nineteenth-century America, the Model 1874 Sharps was available in several variations (Military, Sporting, Hunter's, Mid Range, Long Range, and Creedmore) and a multitude of calibers, barrel lengths, weights, stock patterns, finishes, and sights. This deluxe Sporting rifle incorporates a full-octagon barrel with small-blade, "Open" front and Lawrence high-notch, adjustable rear sights. Its special-order stocking features a shotgun-pattern butt with a checkered steel plate, a pistol grip, and hand-checkering at the wrist and forestock with detailing on the bottom of the grip and the tip of the semischnabel-pattern forestock. The piece also includes a special, finely checkered single trigger.

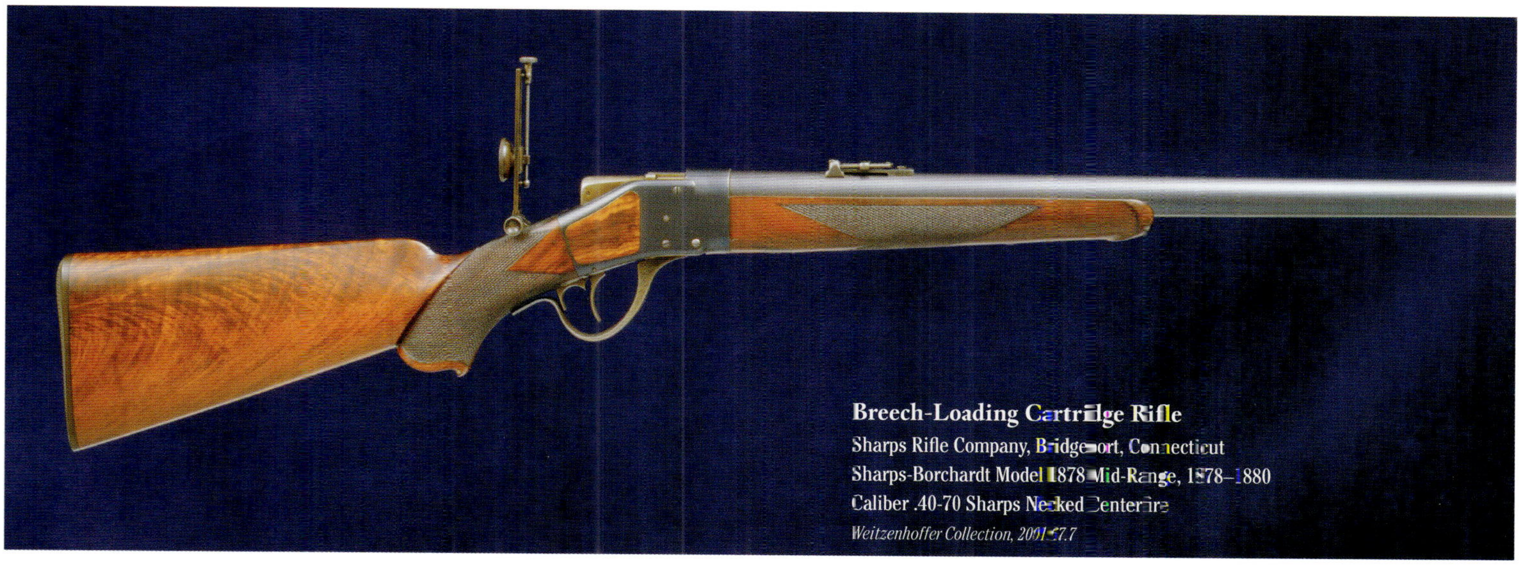

Breech-Loading Cartridge Rifle
Sharps Rifle Company, Bridgeport, Connecticut
Sharps-Borchardt Model 1878 Mid-Range, 1878–1880
Caliber .40-70 Sharps Necked Centerfire
Weitzenhoffer Collection, 2001.27.7

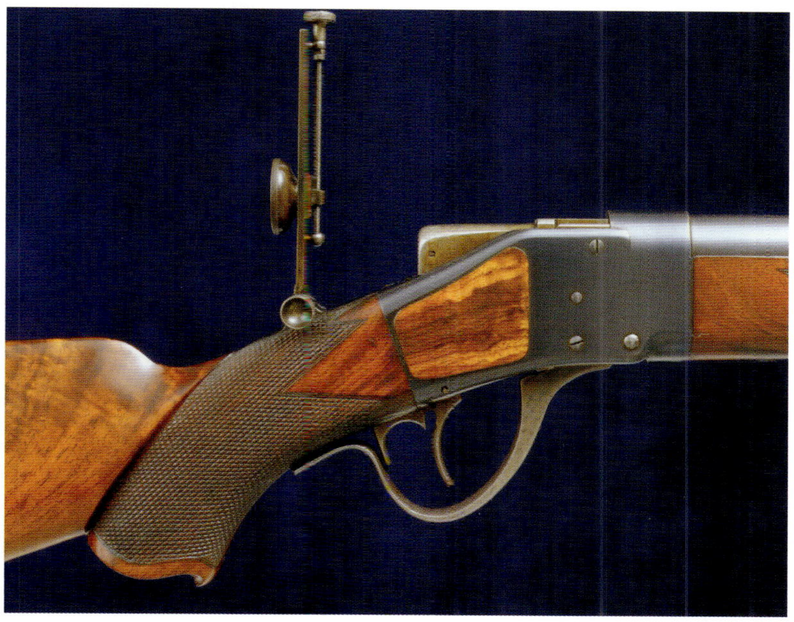

Designed by German emigrant Hugo Borchardt, the hammerless, falling-block system of the Model 1878 Sharps rifle presented the definitive development of the single-shot action in late nineteenth-century America. Far ahead of its time technologically, it remains one of the strongest rifle actions ever conceived. It became a favorite of precision target shooters, and some 450 were finished as custom long-range and midrange rifles for match competition. This midrange rifle has a heavyweight, round barrel with a wind-gauge, spirit-level front sight with sighting tube; an adjustable, ladder-type rear sight; and a patented Sharps Mid-Range vernier peep sight mounted at the wrist. The piece is finished with finely checkered, select walnut stocking featuring a shotgun-pattern butt, a pistol grip, and a schnabel-pattern forestock.

Breech-Loading Cartridge Rifle
United States Armory, Springfield, Massachusetts
Model 1873 "Trapdoor" Sporting Alteration by Browning Brothers,
Ogden, Utah Territory, 1880–1890, Caliber .40-65 Winchester Centerfire
Grandee Collection, 1991.01.2313

A sometime rival of Sharps and Remington in the commercial single-shot rifle market, Springfield "Trapdoor" military rifles and carbines found some appeal among westerners who appreciated their strong action, widely available ammunition, and comparatively low cost. Indeed, a few frontier gunsmiths, such as the later-famed Browning Brothers, found some profit in repurposed "Trapdoor" rifles that they altered to sporting pattern. This rifle's action has been refitted with a 28-inch, midweight, full-octagon barrel mounted with blade front and adjustable-leaf, buckhorn rear sights. The original full stock has been reduced to a half-stock pattern with hand-checkering over the wrist and forestock and with a decorative ebony inlay at the schnabel-pattern forestock tip.

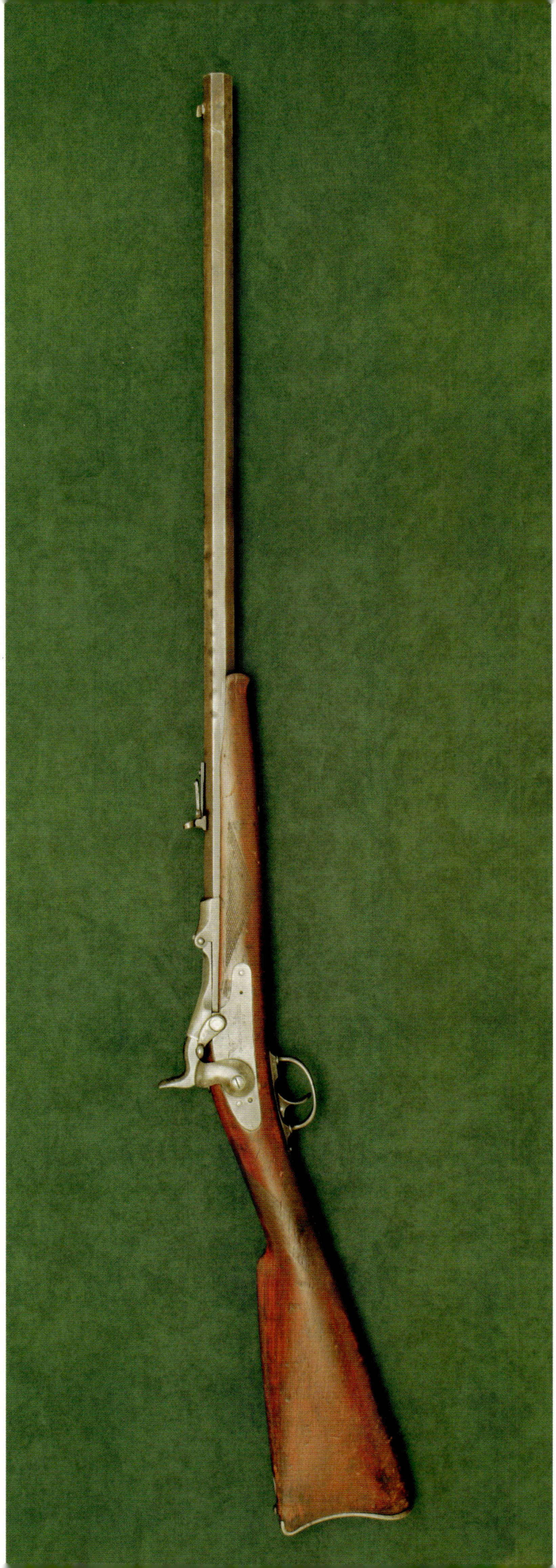

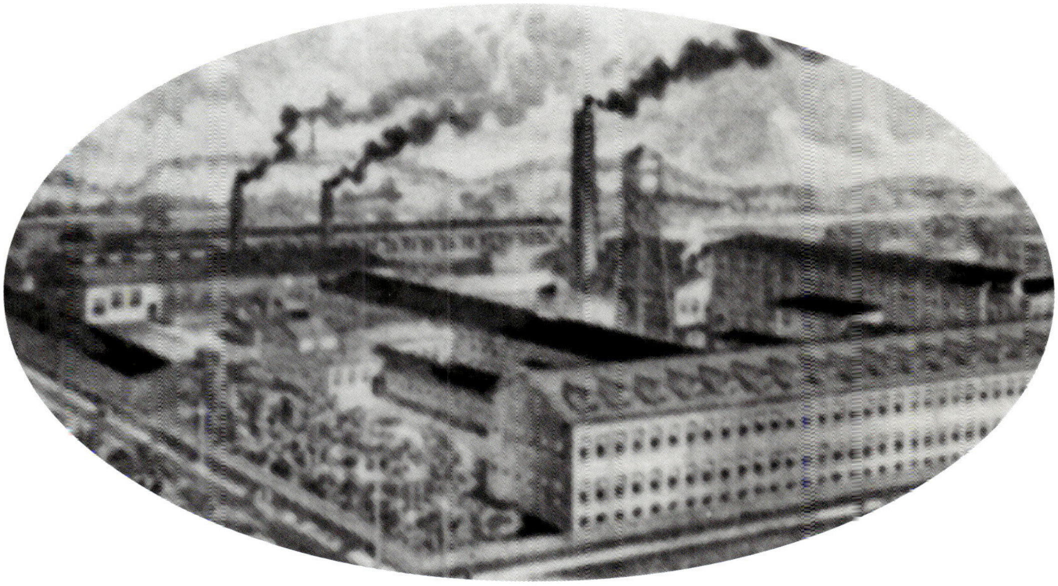

The Marlin factory, woodcut print, circa 1900.
Image courtesy private collection

Chapter 7

MARLIN FIREARMS

AN EMPLOYEE OF COLT'S at the beginning of the Civil War, machinist and toolmaker John M. Marlin (1836–1901) established his own arms-making business at New Haven, Connecticut, in 1863. Initially operating under the name J. M. Marlin, the firm first manufactured a series of single-shot, metallic-cartridge, deringer-type pistols carrying marketing monikers such as OK, Victor, Nevermiss, and Stonewall. Nearly 18,000 of these small protection weapons, ranging in caliber from .22 to .41 rimfire, were produced from 1863 to 1881. Much like Colt's, Marlin also brought out a series of pocket revolvers upon the expiration of the Rollin White patent for bored-through cylinder chambers (controlled by Smith & Wesson until 1869). Reminiscent of the early Smith & Wesson product line, these handguns employed "tip-up" and "top-break" designs for loading and ranged in caliber from .22 rimfire to .38 centerfire. Between 1872 and 1899, about 82,000 of these pistols were manufactured.[1]

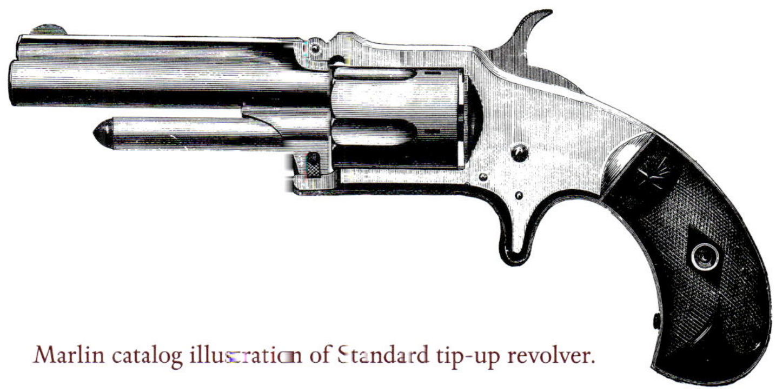

Marlin catalog illustration of Standard tip-up revolver.

In 1875, the Marlin company undertook manufacture of the Ballard-patent single-shot breech-loading rifle. Designed by Charles Ballard in 1861, this lever-actuated, receding-block design had been produced by five different contract manufacturers between 1862 and 1873, in all amounting to some 2[illegible],000 rifles. Over the following sixteen years, Marlin improved and greatly diversified the Ballard rifle line, surpassing even the offerings of rivals Remington and Sharps with some thirty model variations and subvariations including

Hunter's, Sporting, Mid and Long Range Target, Gallery, Creedmoor, and Schuetzen patterns. These rifles ranged from .22 rimfire to .50-70 Government centerfire. Among far-western frontier types, the Ballard No. 5 Pacific and No. 5½ Montana rifles found particular favor with their heavy barrels, double set triggers, and Rocky Mountain–pattern sights. In 1877 Marlin advertised its No. 5 Pacific sporting rifle in the *Colorado Chieftain* newspaper, boldly proclaiming it "Superior to all others. Greater accuracy, greater safety, less recoil. Especially adapted to Far West trade." In all, J. M. Marlin and the later Marlin Firearms Company manufactured an estimated 40,000 Ballard-patent rifles of all patterns between their inception in 1875 and their discontinuance in 1891.[2]

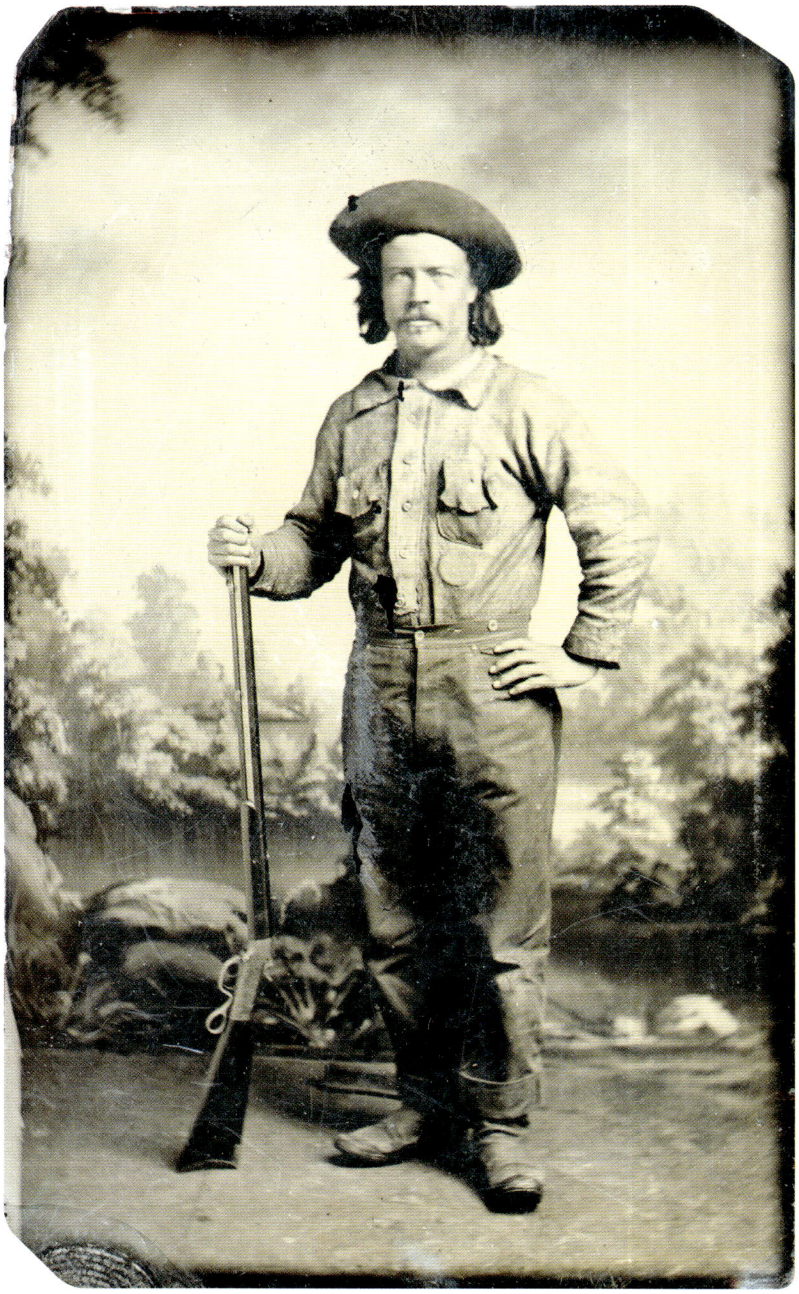

Photographer unknown. [Untitled.] Sixth-plate tintype, America, circa 1880–1885. Garbed in his well-worn field outfit, this anonymous commercial hunter strikes a dauntless pose with his trusty Marlin-Ballard No. 5 Pacific sporting rifle. The single-shot, breech-loading arm is distinguished by its wiping rod, double set trigger, and ring-shaped lever terminus.
Dickinson Research Center, National Cowboy & Western Heritage Museum, 2003.080

In 1881 the New Haven firm reorganized as the Marlin Fire Arms Company and turned its focus to the manufacture of a line of lever-action repeating rifles and carbines for which it would become—and remain—justly famed. Utilizing the patents of Andrew Burgess, Lewis L. Hepburn, himself, and others, Marlin marketed a series of arms that vied with—and sometimes excelled—Winchester's already established and popular line of lever-action repeaters. Like the majority of Winchesters, Marlin repeaters employed a spring-loaded tubular magazine beneath the barrel to house and feed cartridges to the breech. And much like its larger New Haven rival, the Marlin firm offered a complete array of custom-order and deluxe extras in trigger and sight types; barrel weights, lengths, and configurations; and takedown options, stocking patterns, finishes, and special embellishment. By the close of the nineteenth century, Marlin's repeating rifles held a well-deserved reputation for manufacturing excellence and dependable performance.[3]

The company's first offering, the Model 1881, incorporated a smooth-functioning, lever-bolt-locking system devised by Andrew Burgess that actually was superior to Winchester's then-traditional toggle-link system. Designed for long-cased cartridges, the rifle could handle a range of powerful, centerfire rounds from .32-40, .38-55, and .40-60 to .45-70 and .45-85. The latter chamberings endeared the Model 1881 to big-game hunters in the West, some of whom thought it a better repeating rifle than the rival Model 1876 Winchester. In March 1882, outfitters Broadwater, Hubbell & Company of Miles City, Montana Territory, put out an advertisement for the new Marlin repeaters, declaring (with some little exaggeration), "These are THE Buffalo Gun." Manufactured from 1881 to 1892, slightly over 20,500 of these large-frame rifles saw production.[4]

After an interval of seven years, the Marlin firm commenced the design and manufacture of a seemingly constant stream of new and improved lever-action repeaters (Models 1888, 1889, 1891, 1892, 1893, 1894, 1895, and 1897) to vie with Winchester. Intended to compete with the Model 1873 Winchester rifle, the first of these (Models 1888 and 1889) were medium-frame, short-action rifles designed for pistol-sized centerfire cartridges (.32-20, .38-40, and the very popular .44-40). The Model 1889 actually replaced its short-lived predecessor, introducing

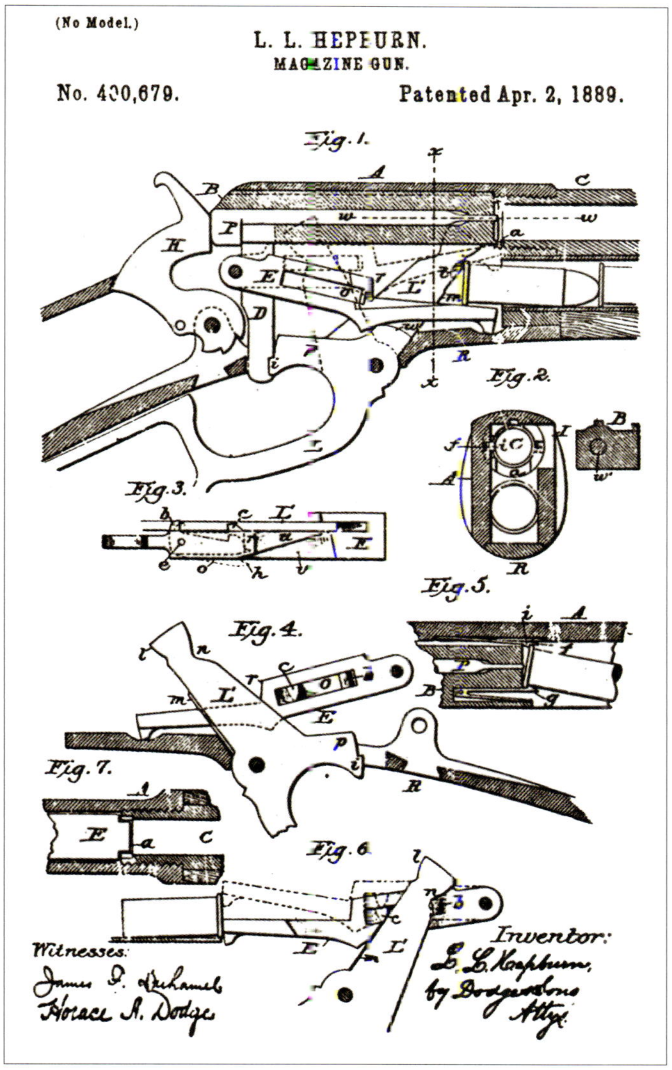

Patent drawing illustrating the internal workings of the Hepburn-designed Marlin Model 1889 lever-action rifle, which introduced the solid-top receiver with side ejection.

U.S. Patent and Trademark Office, uspto.gov

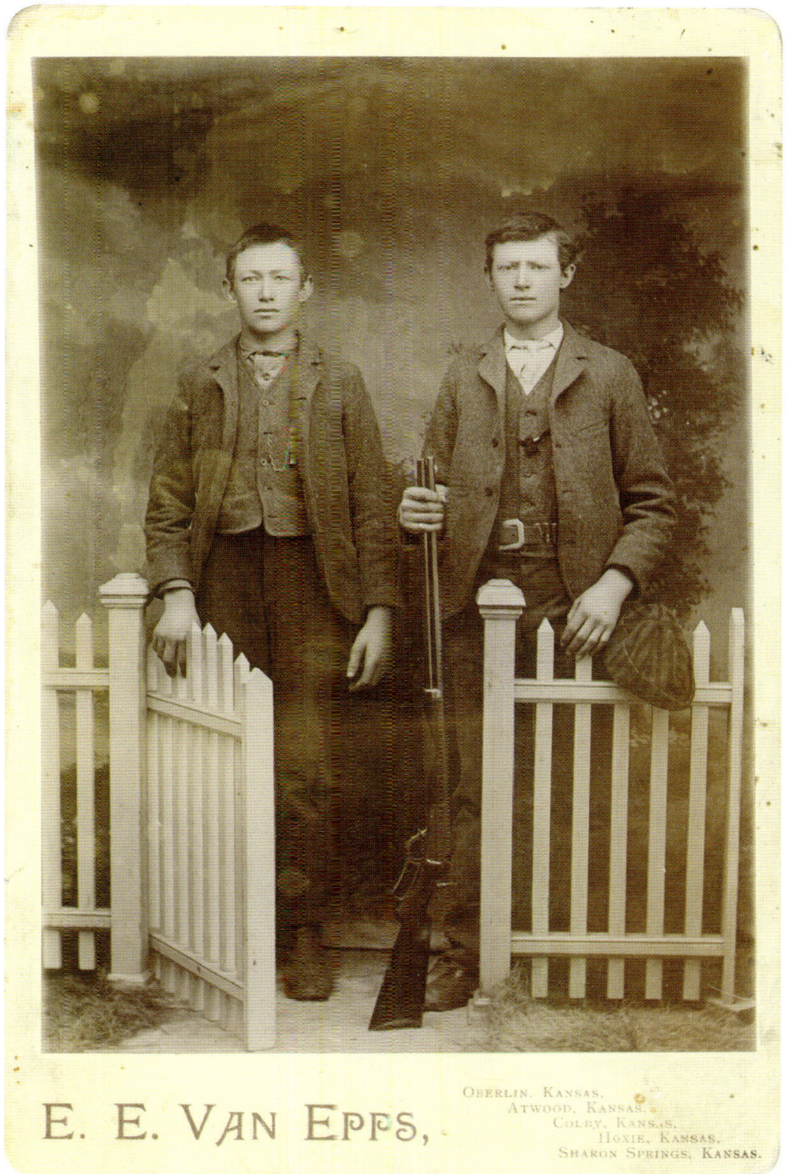

E. E. Van Epps, photographer. [Untitled.] Cabinet card, Oberlin et al., Kansas, circa 1890–1895. Posed at the studio's garden gate, these young Kansans demonstrate their shooting interests with a deluxe Marlin Model 1889 lever-action, repeating sporting rifle. The piece features custom-order checkered stocking with a pistol grip and a wrist-mounted peep sight.

Dickinson Research Center, National Cowboy & Western Heritage Museum, 2004.335

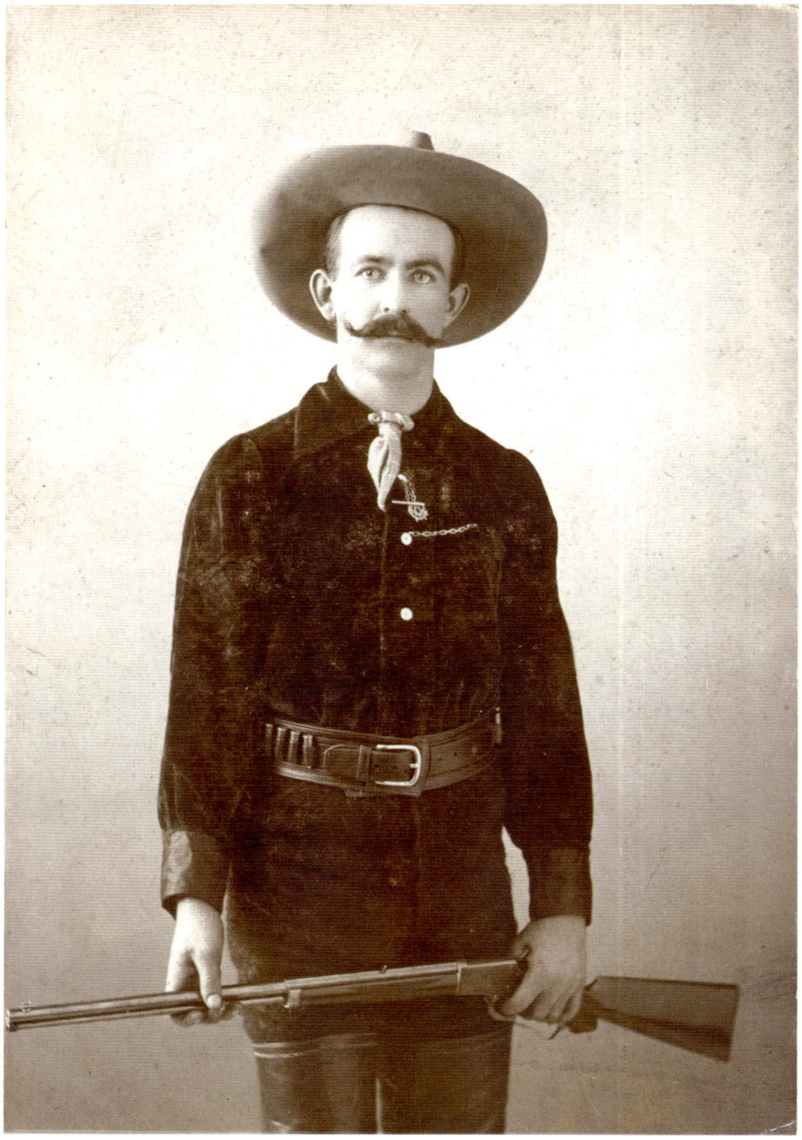

Shaeffer, photographer. [Untitled.] Cabinet card, Bellefonte, Pennsylvania, circa 1895–1905. Very likely a Wild West show or vaudeville marksman, this rather theatrical gent evokes the latter-day frontier with his western headgear and thigh-high boots à la Buffalo Bill Cody. He models with a .22-caliber Model 1891 Marlin lever-action repeating rifle having a special-order, half-octagon barrel and special-order checkered stocking with pistol grip.
Image courtesy private collection

a signature Marlin improvement designed and patented by Lewis L. Hepburn, who formerly had worked with Remington. This was the solid-top receiver design that featured the side ejection of spent cartridge cases. All of the later Marlin lever-actions incorporated the side-ejection design, which virtually eliminated a nuisance common to top-ejectors such as those in Winchester's product line—that of spent brass cartridge cases flying through one's line of sight and collecting in one's collar or hat brim. This fundamental design improvement also made the Model 1889 and later Marlin lever-action rifles far more adaptable to peep sight and scope mounting on the receiver top. In total, nearly 60,000 Model 1888 and 1889 rifles and carbines—along with a few muskets—were produced between 1888 and 1899.[5]

The Marlin Models 1891 and 1892 featured diminutive receivers for .22 Short, Long, or Long Rifle rimfire cartridges or .32 Long rimfire (or centerfire) cartridges. Early versions of the former model loaded into the receiver, while all later iterations loaded more conveniently into an aperture in the magazine tube. Some 63,000 of these small-caliber lever-actions saw production before the latter model's discontinuance in 1916. With the Model 1893, the company offered an improved, Hepburn-designed, medium-frame action for medium-game centerfire ammunition in calibers .25-36, .30-30, .32 Special, .32-40, and .38-55 (all ultimately accommodating the new smokeless powder loadings). Company literature noted improvements in the design of the firing pin, lever catch, locking bolt, and trigger mechanism—all of which would be carried forward into succeeding lever-action Marlins. Standard rifle barrel length was twenty-six inches, but barrels could be special-ordered as short as twenty inches or as long as thirty-two inches with full or half magazines. By 1905, when the Model 1893 designation was changed to the Model 93, nearly 70,000 such rifles and carbines had been manufactured. Today, the basic Model 1893 design continues in production, now as the Marlin Model 336 with total output numbering in the several millions.[6]

Marlin's next offering, the Model 1894, represented an improved rendition of the earlier short-action Model 1889, again chambering pistol-sized centerfire ammunition ranging from the .25-20 to the still-popular .44-40. This rifle represented the company's response to

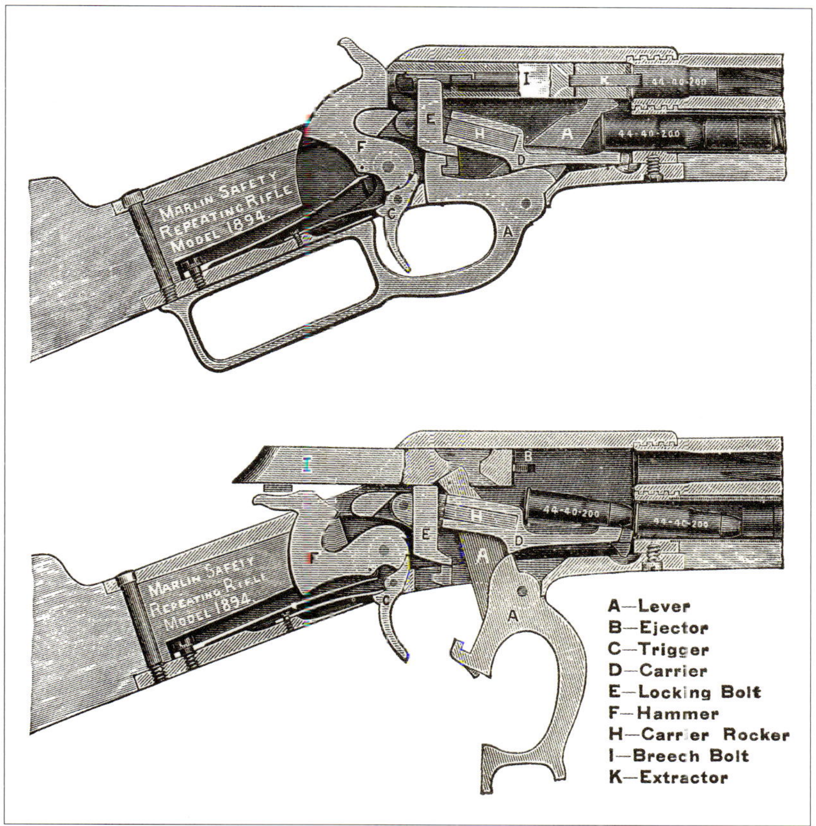

Cutaway drawings illustrating the internal workings of the Model 1894 Marlin lever-action repeating rifle, which featured several Hepburn improvements to the action system first introduced in the Model 1893.

Drawings from the Marlin Fire Arms Company's 1905 sales catalog

Winchester's very popular Model 1892. By the time of their discontinuance in 1935, total production of Model 1894 rifles and carbines would reach nearly a quarter-million units. Having discontinued its large-caliber Model 1881 in 1892, Marlin then introduced the Model 1895 in an effort to keep a big-game rifle in the marketplace and compete with Winchester's Model 1886. The long-action of the improved Model 1895 was adapted to smokeless-powder cartridges ranging from .38-56 to .45-90. Like its companion Models 1893 and 1894, this rifle, too, could be bought with a myriad of special-order features in stock and barrel configurations, sights, takedown, and finishes. In all, approximately 18,000 Model 1895 rifles and carbines saw manufacture by the time of their discontinuance in 1917.[7]

Marlin's last lever-action rifle offering of the nineteenth century was the Model 1897, an improved, takedown version of the .22-caliber Model 1892. Designated as the Model 39 in 1922, it reigns today as the oldest domestic lever-action rifle design in continuous production (now the Model 39A). In 2008, under the umbrella Freedom Arms holding company, the Remington Arms Company absorbed Marlin, and it since has moved much of the latter firm's current model production to Ilion, New York.[8]

Marlin Firearms

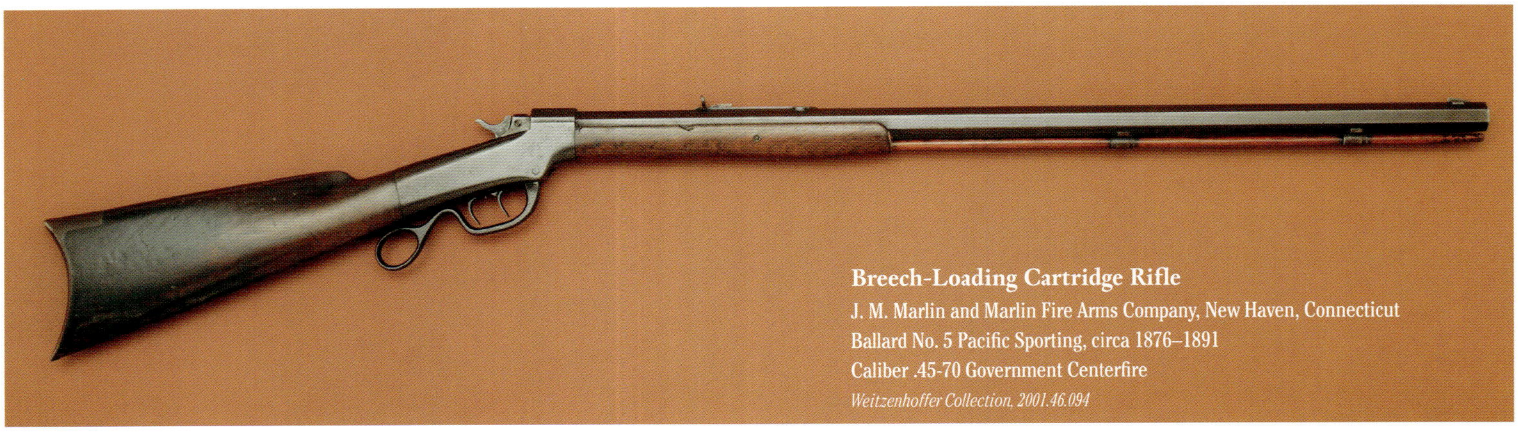

Breech-Loading Cartridge Rifle
J. M. Marlin and Marlin Fire Arms Company, New Haven, Connecticut
Ballard No. 5 Pacific Sporting, circa 1876–1891
Caliber .45-70 Government Centerfire
Weitzenhoffer Collection, 2001.46.094

Although ranked in popularity after the Sharps and Remington single-shots, the Ballard-Pacific rifle enjoyed considerable favor among sport and market hunters in the postbellum West. Like its short-lived twin, the No. 5½ Montana rifle, the No. 5 model was distinguished by its ring-pattern lever terminus, its wiping rod beneath the barrel, and the standard inclusion of double set triggers. Available in calibers ranging from .38-50 to .50-70 centerfire, the No. 5 Ballard rifle could be had in standard configuration with 30- or 32-inch, medium- or heavyweight octagonal barrels fitted with Rocky Mountain–pattern sights.

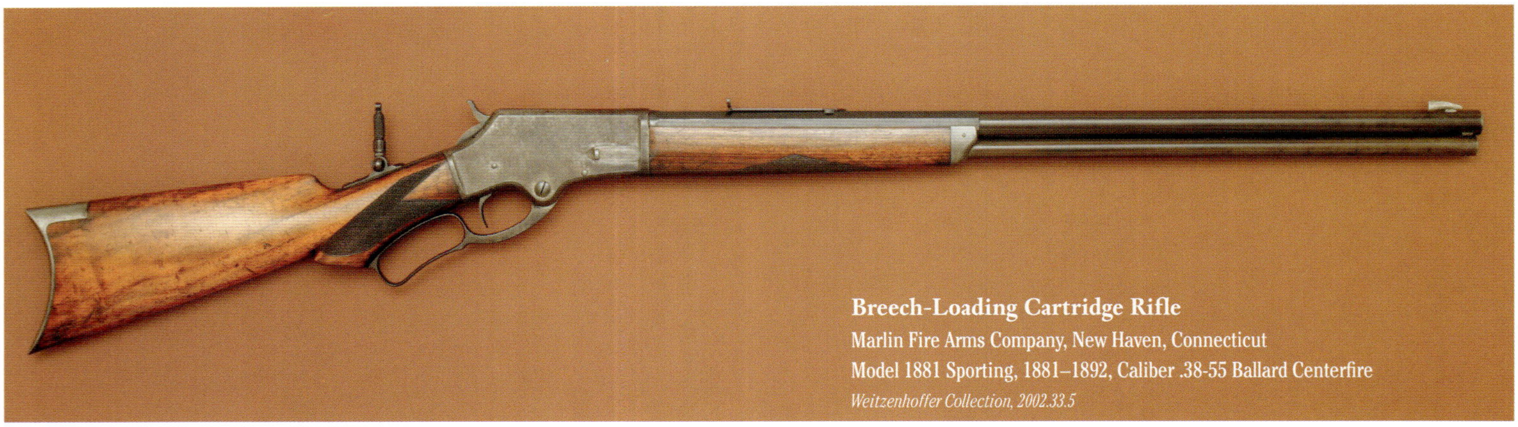

Breech-Loading Cartridge Rifle
Marlin Fire Arms Company, New Haven, Connecticut
Model 1881 Sporting, 1881–1892, Caliber .38-55 Ballard Centerfire
Weitzenhoffer Collection, 2002.33.5

Combining the patents of prolific arms designer Andrew Burgess with improvements by John M. Marlin himself, the Model 1881 was the company's first lever-action repeater. The large-frame, large-caliber design commanded an appreciative market among big-game hunters until it was overshadowed by the Model 1886 Winchester. This deluxe piece includes several special-order features, among them a 26-inch, half-octagon/half-round barrel; a Lyman adjustable peep sight mounted at the wrist; select, figured walnut stocking with a gracefully recurved pistol grip; and hand-checkering in Style A over the wrist and forestock—all adding about $14.00 to the retail cost of $22.50 in 1887.

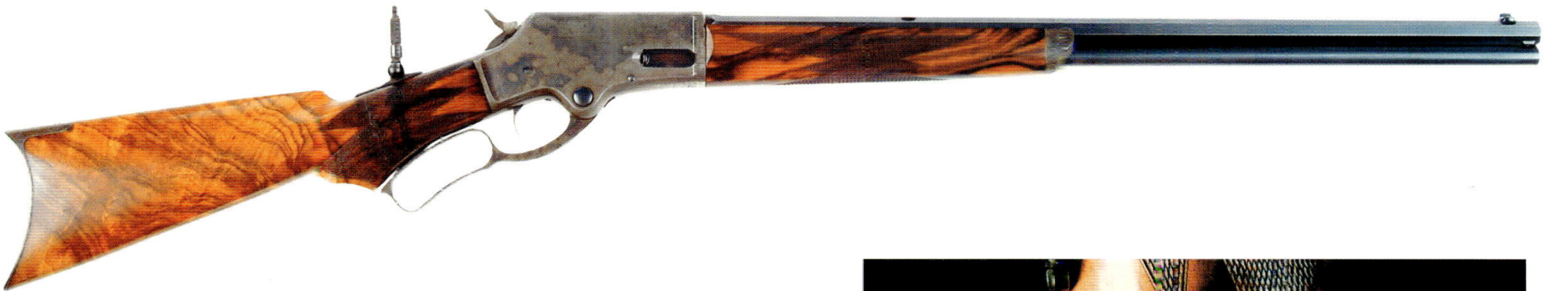

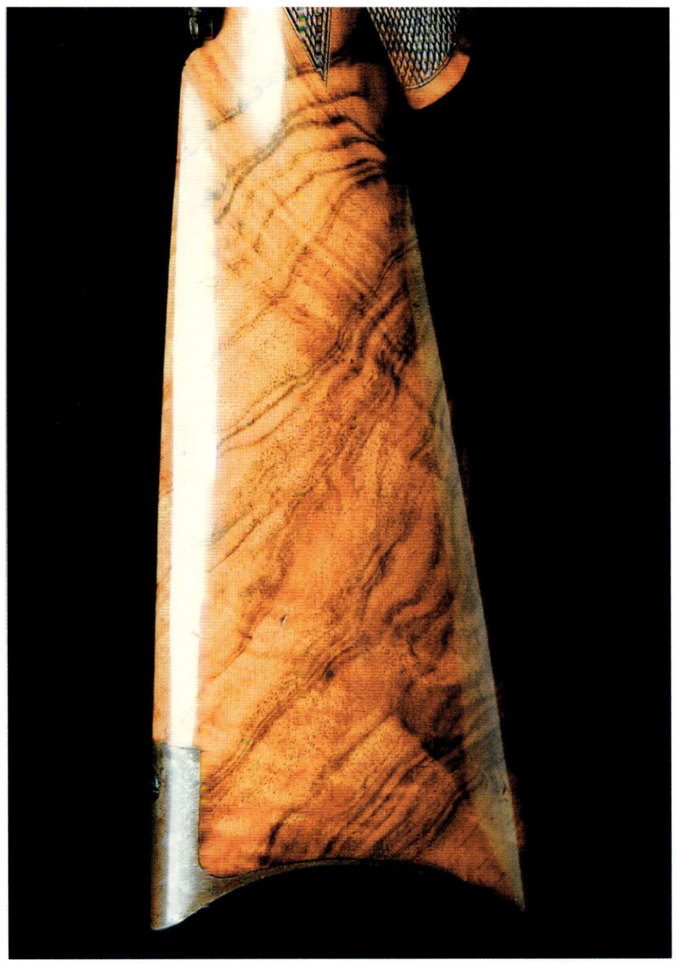

Breech-Loading Cartridge Rifle
Marlin Fire Arms Company, New Haven, Connecticut
Model 1881 Sporting, 1881–1892, Caliber .38-55 Ballard Centerfire
Weitzenhoffer Collection, 2004.282

Available in long-cased chamberings ranging from .32-40, .38-55, and .40-60 to .45-70 and .45-85, the Model 1881 Marlin lever-action repeater was a viable contender with Winchester Models 1873 and 1876 alike. Like the previous example, this special lightweight, deluxe rifle features several special-order enhancements, including an ivory bead-front sight; an early Lyman peep sight on the top tang; highly figured stocking of select French walnut with a pistol grip; and fine hand-checkering in the Style 3, double-diamond pattern at the wrist and forestock—in all adding about $12.50 to the $22.50 retail cost in 1887.

Images courtesy James D. Julia Auctions, Fairfield, Maine

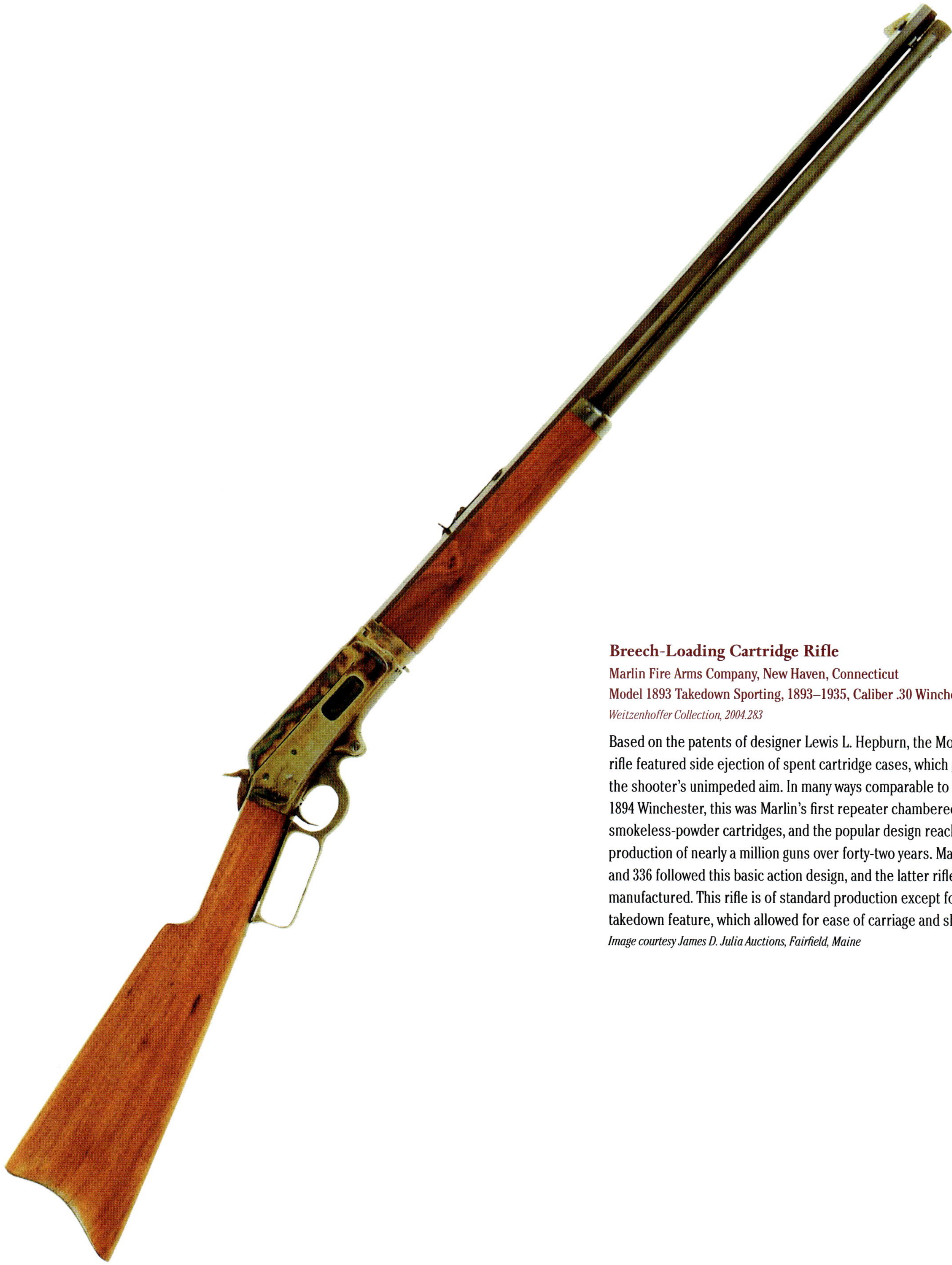

Breech-Loading Cartridge Rifle

Marlin Fire Arms Company, New Haven, Connecticut

Model 1893 Takedown Sporting, 1893–1935, Caliber .30 Winchester Centerfire

Weitzenhoffer Collection, 2004.283

Based on the patents of designer Lewis L. Hepburn, the Model 1893 Marlin rifle featured side ejection of spent cartridge cases, which greatly facilitated the shooter's unimpeded aim. In many ways comparable to the Model 1894 Winchester, this was Marlin's first repeater chambered for midsized smokeless-powder cartridges, and the popular design reached a total production of nearly a million guns over forty-two years. Marlin Models 1936 and 336 followed this basic action design, and the latter rifle is still being manufactured. This rifle is of standard production except for its special-order takedown feature, which allowed for ease of carriage and shipment.

Image courtesy James D. Julia Auctions, Fairfield, Maine

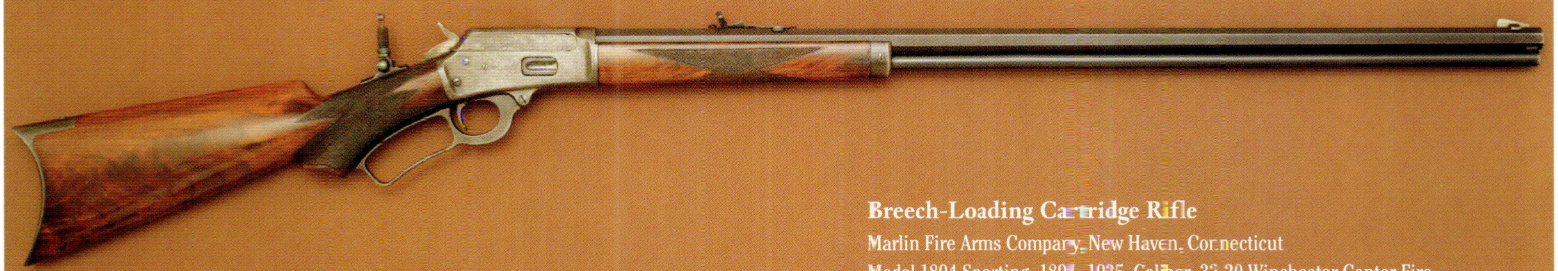

Breech-Loading Cartridge Rifle
Marlin Fire Arms Company, New Haven, Connecticut
Model 1894 Sporting, 1894–1935, Caliber .32-20 Winchester Center Fire
Weitzenhoffer Collection, 2002.[illegible]3[illegible]

Utilizing the patents of Lewis L. Hepburn, the Model 1894 Marlin incorporated a short-action receiver intended to accommodate several of the pistol-sized cartridges popular at the close of the nineteenth century. The design competed well against the comparable Winchester Model 1892, reaching a total production of about a quarter million units over forty years. This special-order specimen incorporates a longer-than-standard, 30-inch full-octagon barrel; King's patent Rocky Mountain–pattern rear sight and adjustable Lyman peep sight; and select, figured walnut stocking with the recurved pistol grip and forestock hand-checkered in Style B—amounting to about $18.00 over the standard retail cost of $[illegible]2[illegible]50 in 190[illegible].

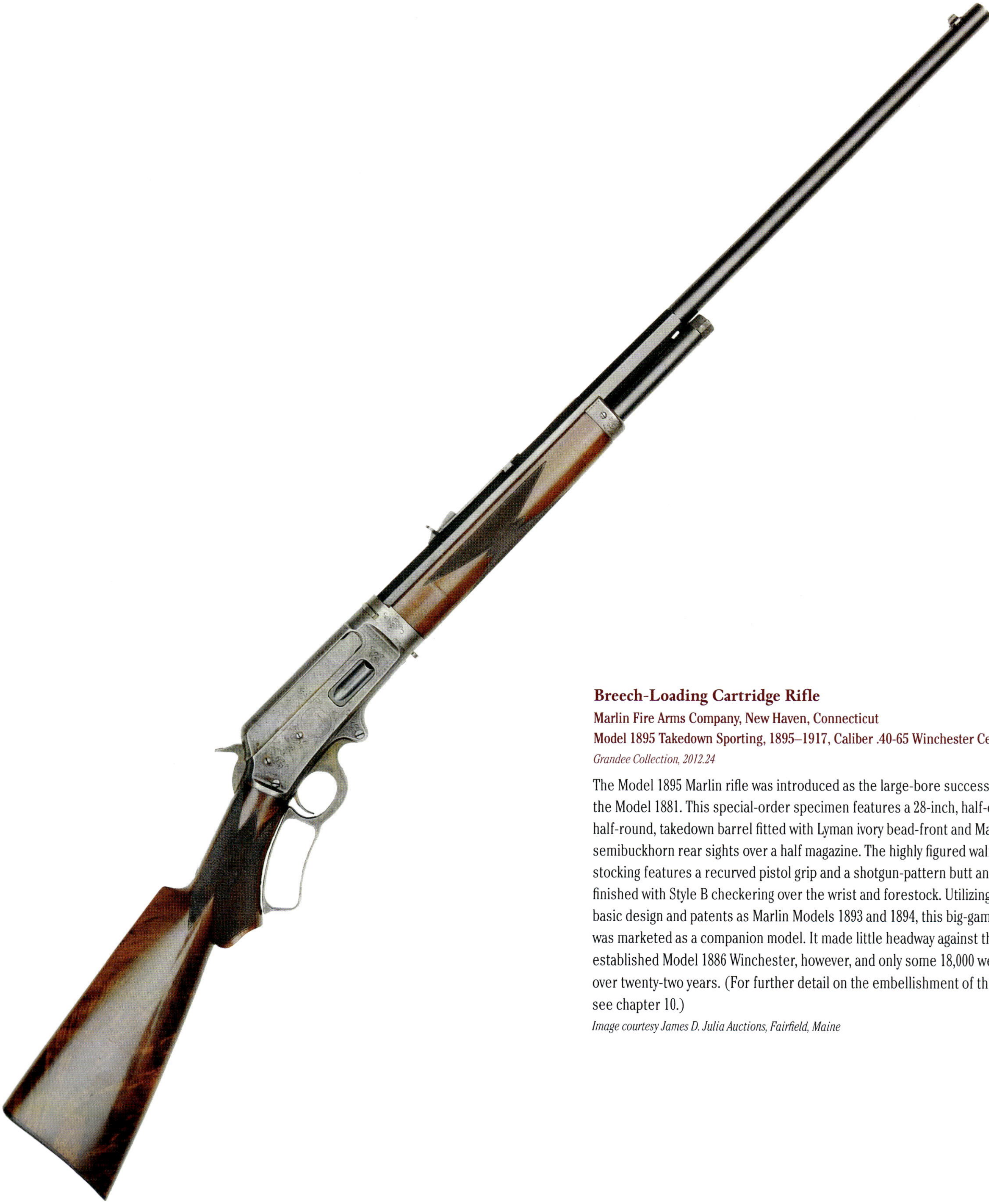

Breech-Loading Cartridge Rifle

Marlin Fire Arms Company, New Haven, Connecticut

Model 1895 Takedown Sporting, 1895–1917, Caliber .40-65 Winchester Centerfire

Grandee Collection, 2012.24

The Model 1895 Marlin rifle was introduced as the large-bore successor to the Model 1881. This special-order specimen features a 28-inch, half-octagon/half-round, takedown barrel fitted with Lyman ivory bead-front and Marlin semibuckhorn rear sights over a half magazine. The highly figured walnut stocking features a recurved pistol grip and a shotgun-pattern butt and is finished with Style B checkering over the wrist and forestock. Utilizing the same basic design and patents as Marlin Models 1893 and 1894, this big-game rifle was marketed as a companion model. It made little headway against the already established Model 1886 Winchester, however, and only some 18,000 were made over twenty-two years. (For further detail on the embellishment of this piece, see chapter 10.)

Image courtesy James D. Julia Auctions, Fairfield, Maine

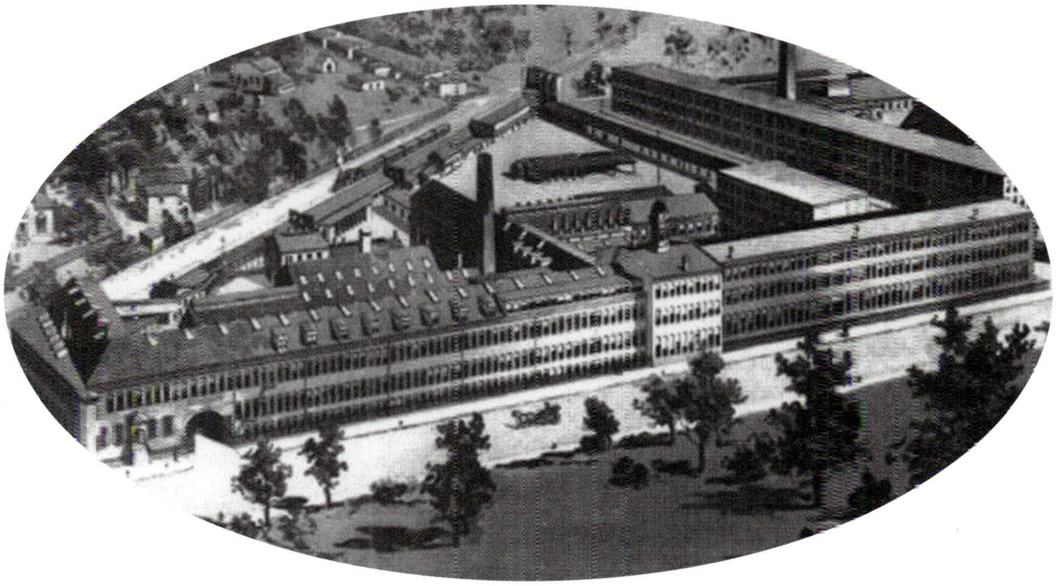

Chapter 8

WINCHESTER FIREARMS

NEITHER A FIREARMS INVENTOR nor one particularly familiar with arms manufacture, Oliver Winchester (1810–1880) nonetheless was an astute businessman with an eye for opportunity and the ability to recognize a product with potential marketability. Already a success in the garment trade, in 1855 Winchester made a shrewd investment in the Volcanic Repeating Arms Company, then producing iron-frame, lever-action magazine repeaters at Norwich, Connecticut. Within two years, Winchester had taken complete control of the firm's assets, machinery, and patents, which he then restructured under his management as the New Haven Arms Company in the latter town. Within a decade, he transformed this struggling business into the Winchester Repeating Arms Company, a promising enterprise that, by 1900, would dominate the commercial longarm market with a succession of repeating weapons that won widespread acclaim in the frontier West and around the world.[1]

The Winchester factory, lithograph print, circa 1895.
Image courtesy private collection

The Winchester company built its success on a series of exclusive, predecessor patent designs for lever-actuated repeating arms having a tubular magazine beneath the barrel holding caseless, or self-consuming, projectiles. Encompassing the patents of Walter Hunt (1847/49), Lewis Jennings (1849), Horace Smith (1851), and Horace Smith and Daniel B. Wesson (1854), the Volcanic arm proved a commercial failure—like its predecessors—because of the quite weak and unreliable performance of its ammunition. In all, some 3,350 pistols and 1,000 longarms of the brass-framed Volcanic design saw production between 1855 and 1860.[2] By the latter year, the New Haven Arms Company's superintendent, Benjamin Tyler Henry, had modified and scaled up the Volcanic's toggle-link, breech-locking mechanism to accommodate a new, self-contained metallic cartridge of his own design—the .44 Henry Flat rimfire. The fifteen-shot Henry rifle, in iron (rare) and brass-framed variations, introduced the first practical lever-action repeating rifle with real firepower—rivaled only by the seven-shot Spencer repeater, which was not, however, a self-cocking arm like the Henry. Approximately 13,500 Henry rifles were manufactured between 1862 and 1866, and these fast-firing arms gained some considerable notoriety during the Civil War and in the West. Frontiersman Luther "Yellowstone" Kelly, for example, acquired and used such a rifle on the upper Missouri in 1868. "With the Henry . . . I killed many a buffalo, as well as other game," Kelly later recalled, "and it stood me in good hand when I was forced to defend myself in encounters with hostile Indians."[3]

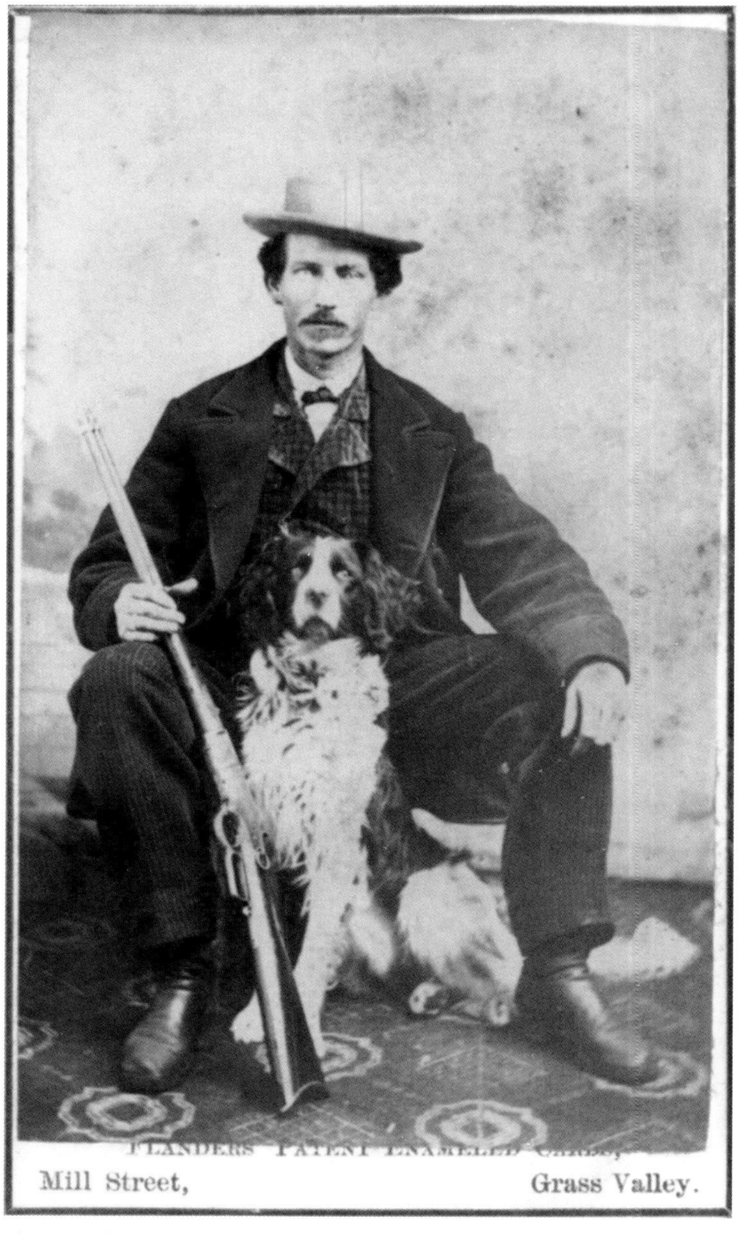

Photographer unknown. "Flanders Patent Enameled Cards." [Untitled.] Carte de visite, Grass Valley, California, circa 1870. Straddling his devoted spaniel, this western gent displays his favored firearm, a Winchester Model 1866 saddle-ring carbine.

Dickinson Research Center, National Cowboy & Western Heritage Museum, RC2007.030.2

Sales of the Henry rifle provided Winchester with sufficient solvency to reorganize his business in 1867 as the Winchester Repeating Arms Company,[4] as well as to bring a much-improved arm to market. Though a dependable weapon, the Henry rifle presented design drawbacks in its rather awkward front-loading system and in the long, open slot along the magazine's underside, which allowed dirt and debris to foul the cartridges and the action. New factory superintendent Nelson King overcame these problems by devising a separate, closed magazine tube joined with a convenient cartridge-loading port, or aperture, in the rifle's frame. A wooden forestock further protected the magazine tube, while the loading port was closed with a spring-loaded gate. Now a truly functional repeating rifle, the "Improved Henry"—or Model 1866 Winchester as it later was known—enjoyed widespread appreciation in the West. Dazzled by its brass frame and its potent firepower, admiring Plains Indians dubbed it the "Yellow Boy." Employing the same cartridge and toggle-link mechanism as the Henry, in excess of 170,000 Model 1866 rifles, carbines, and muskets saw production between their inception in 1867 and their discontinuance in 1898.[5]

With the Model 1873 Winchester, the New Haven enterprise commenced its dominance of the domestic

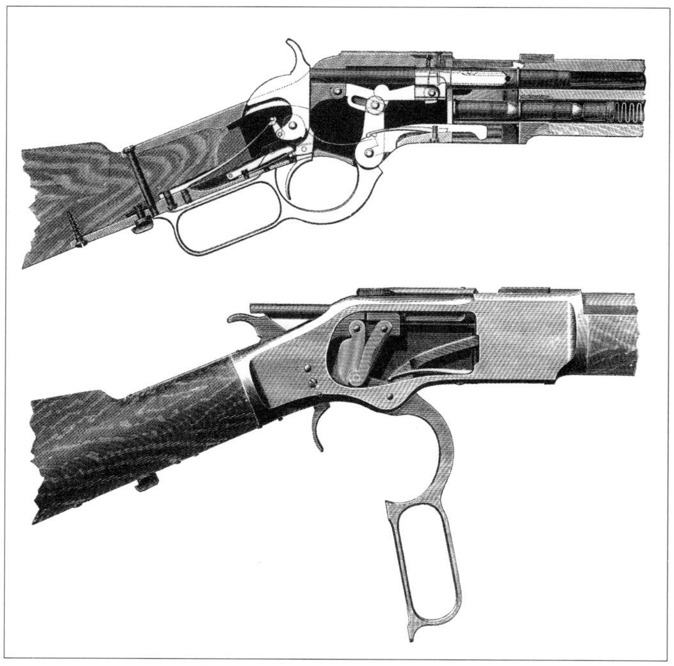

Cutaway drawings illustrating the internal workings of the Winchester Model 1873 repeating rifle and its long-serving, toggle-link, bolt-locking mechanism—the same system used in the Volcanic, Henry, 1866, and 1876 model arms.

Drawings from the Winchester Repeating Arms Company's 1899 sales catalog

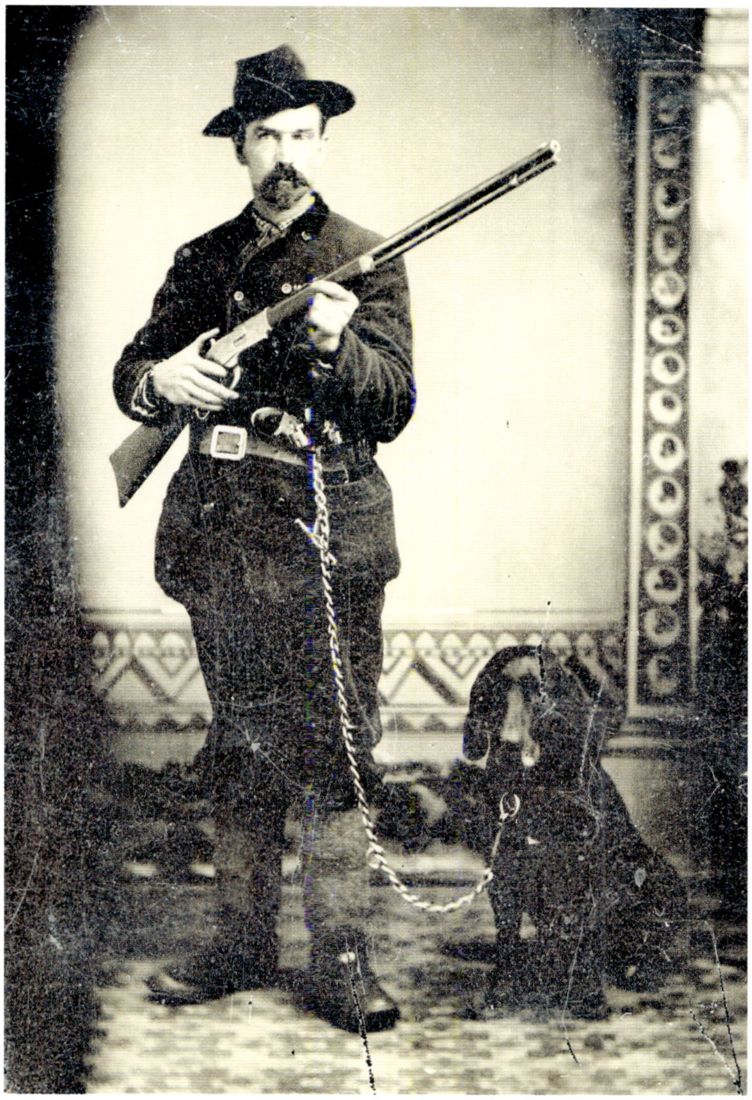

Photographer unknown. [Untitled.] Sixth-plate tintype, America, circa 1880–1885. Standing at the ready and accompanied by his rather mournful hound, this huntsman poses with a standard-grade Winchester Model 1873 lever-action, repeating sporting rifle having a case-hardened frame and an adjustable buckhorn rear sight. Two indistinguishable, pocket-sized cartridge revolvers are thrust into the fellow's belt.
Dickinson Research Center, National Cowboy & Western Heritage Museum, 2003.034

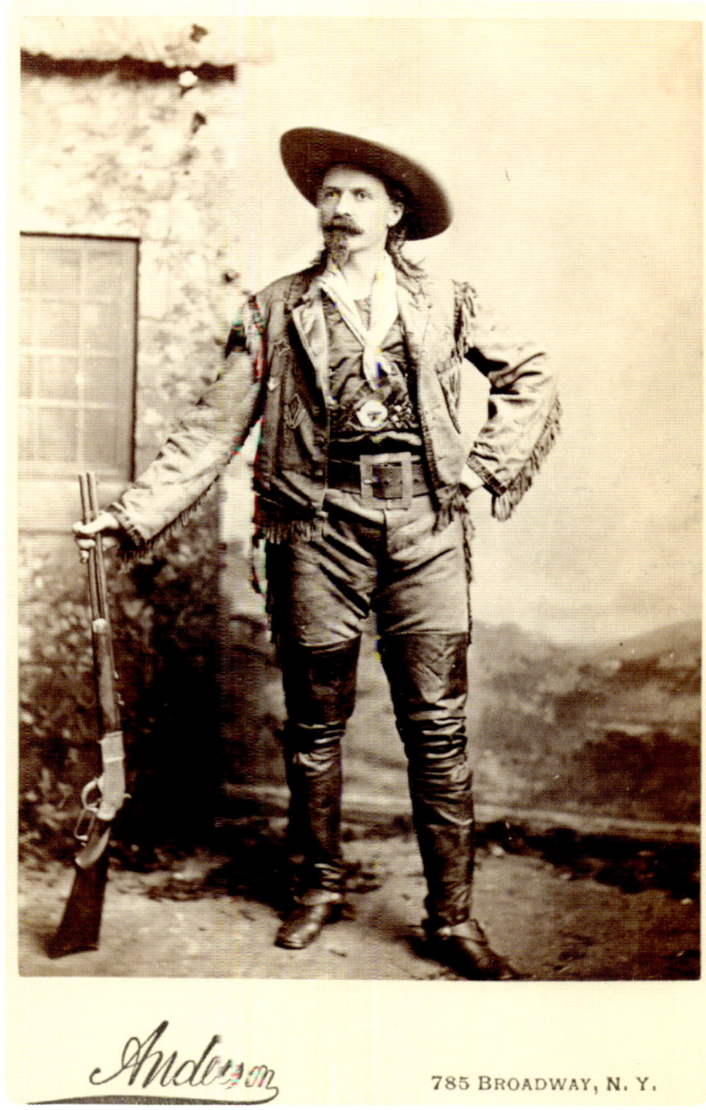

Anderson, photographer. [*William F. Cody.*] Cabinet card, New York, New York, circa 1882. Replete in theatrical garb, a yet-to-be-world-famous Buffalo Bill strikes a pose with one of his several Model 1873 Winchester rifles—a weapon he declared "the boss" among frontier arms (here, a deluxe variation featuring checkered, pistol-grip stocking).
Dickinson Research Center, National Cowboy & Western Heritage Museum, 2003.160

repeating-longarm market. Incorporating a lighter and stronger receiver of iron and firing a somewhat more potent .44-40 centerfire cartridge, the new rifle won enthusiastic devotees throughout the West. Prominent among them was William F. "Buffalo Bill" Cody, who wrote the firm in 1875: "I have been using and have thoroughly tested your latest improved rifle. Allow me to say that I have tried and used nearly every kind of a gun made in the United States, and for general hunting, or Indian fighting, I pronounce your improved Winchester the boss."[6] In rifle, carbine, and musket configurations, the Model 1873 was the first Winchester arm available in different chamberings (.32-20, .38-40, and .44-40 centerfire and .22 Short and Long rimfire) and the first available with a variety of options and special-order features in barrel configurations, sights, triggers, stock patterns, and finishes—all of which broadened its remarkably long-lasting appeal. In excess of 720,000 Model 1873s had been made when production ceased in 1919—long after the arm had been rendered technically obsolescent by subsequent Winchester models.[7]

Responding to entreaties for a more powerful repeater, Winchester introduced the Model 1876 at the

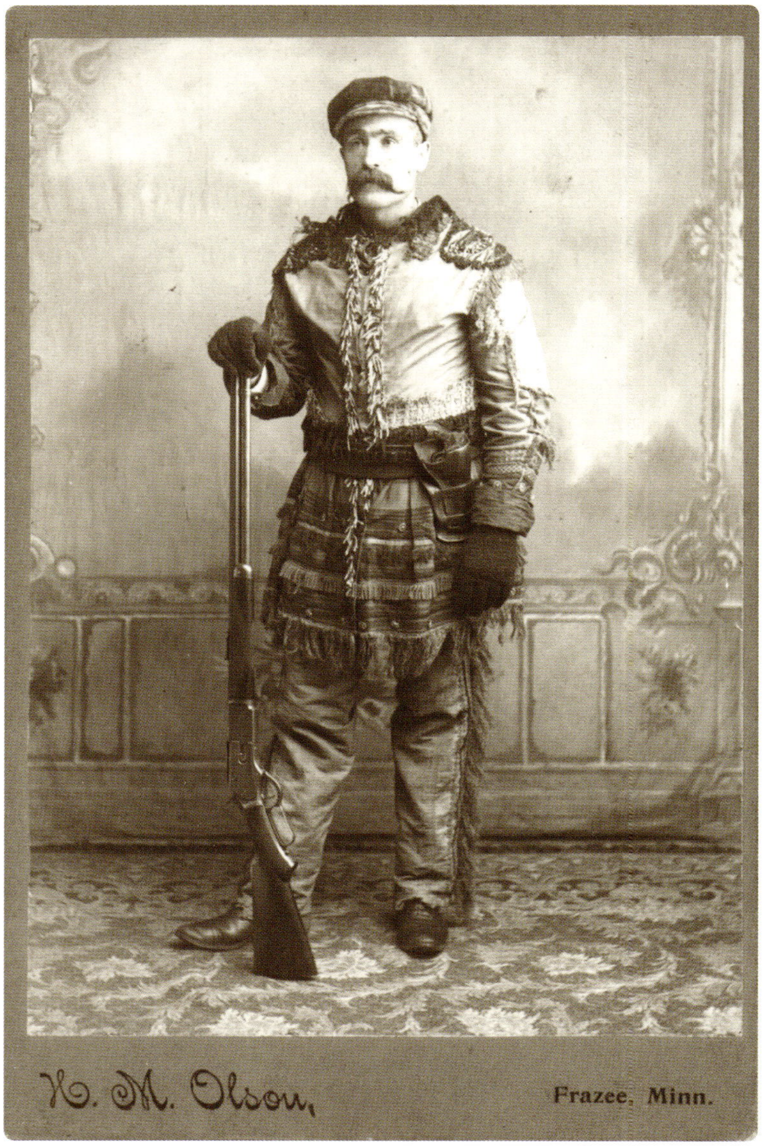

H. M. Olson, photographer. [Untitled.] Cabinet card, Frazee, Minnesota, circa 1885–1890. Posed in a Native-inspired hunting outfit, this evidently well-to-do sportsman models with a large-frame Winchester Model 1876 repeating magazine rifle. A deluxe, special-order arm, the piece features checkered, fancy walnut stocking with a pistol grip and a shotgun-pattern butt.
Dickinson Research Center, National Cowboy & Western Heritage Museum, RC2006.006.1

nation's Centennial Celebration in Philadelphia, where it was recognized as "the best magazine rifle for sporting purposes yet produced."[8] The last repeater to utilize the toggle-link, breech-locking mechanism descended from Smith & Wesson's Volcanic design, the Model 1876 presented a robust, scaled-up version of the Model 1873 well able to accommodate powerful long-cased centerfire cartridges (.40-60, .45-60, .45-75, and .50-95 Express). With this weapon, which commenced commercial production in 1877, Winchester won the appreciation of many western sport hunters—among them the future president Theodore Roosevelt, who declared in 1885, "The [Model 1876] Winchester, stocked and sighted to suit myself, is by all odds the best weapon I ever had. . . . [It] is deadly, accurate, and handy as any, stands very rough usage, and is unapproachable for the rapidity of its fire and the facility with which it is loaded."[9] From their initial sales to their discontinuance in 1897, Winchester produced nearly 64,000 Model 1876 rifles, carbines, and muskets. Much like the Model 1873, these already had been superseded some years before by more mechanically advanced Winchester (and competitor) models.[10]

In 1879 John M. Browning, then a virtually unknown gunsmith located in Ogden, Utah Territory, designed a self-cocking, lever-actuated, falling-block, single-shot rifle so eminently practical that Browning Brothers manufactured only about 550 rifles before Winchester bought the patent and manufacturing rights in 1883 and introduced it as the Model 1885. (Thus began a sixteen-year association in which Winchester acquired some forty Browning designs resulting in nine production models, of which four were highly popular, lever-action repeating rifles.)[11] Like the Sharps and Remington single-shots, the Model 1885 Winchester provided almost unlimited variation in barrel lengths and weights, sight and trigger types, stock configurations, and custom features. Roughly sixty different cartridge chamberings were available, including at least thirty appropriate for big-game hunters (ranging from .38-55 to .50-110 WCF). Although a latecomer to western sportsmen, the Model 1885 Winchester won the praises of many, including grizzly bear hunter William Wright. He avowed that his .45-100-caliber rifle "killed at both ends," but he went on, "I liked it better than any other I have ever carried. Personally, I could depend on this old rifle for a sure three shots in twelve seconds, by holding two spare cartridges between the fingers of my right hand.”[12] Manufactured in high-wall and low-wall (and thick-wall and thin-wall) receiver patterns depending on cartridge size, nearly 140,000 Model 1885 rifles, carbines, and muskets realized production before their discontinuance in 1920.[13]

Winchester's commercial dominance in repeating rifles was ensured with its next lever-action—the famed Model 1886. The brainchild of prolific gun-designer John M. Browning, this rifle incorporated an ingenious vertical-bolt-locking mechanism that was remarkably smooth in operation and stronger than any high-powered cartridge then developed. Thousands chambered for the popular .45-70 Government cartridge served enthusiastic hunters in the West, and those desiring a faster or more potent round could choose among Winchester's .40-82, .45-90, or .50-110 Express loadings. Western sportsman W. A. Perry lauded the .50-110 Express version as “[l]ight, strong, and rapid of manipulation, terrific in killing power; there is no animal on this continent that can escape from a cool, nervy man armed with one of these superb weapons.”[14] An abiding favorite, nearly 160,000 Model 1886 rifles, carbines, and muskets were produced between 1886 and 1935.[15]

Seeking to replace the slow-selling Model 1873 lever-action .22 rifle (and compete with Colt's .22 Lightning

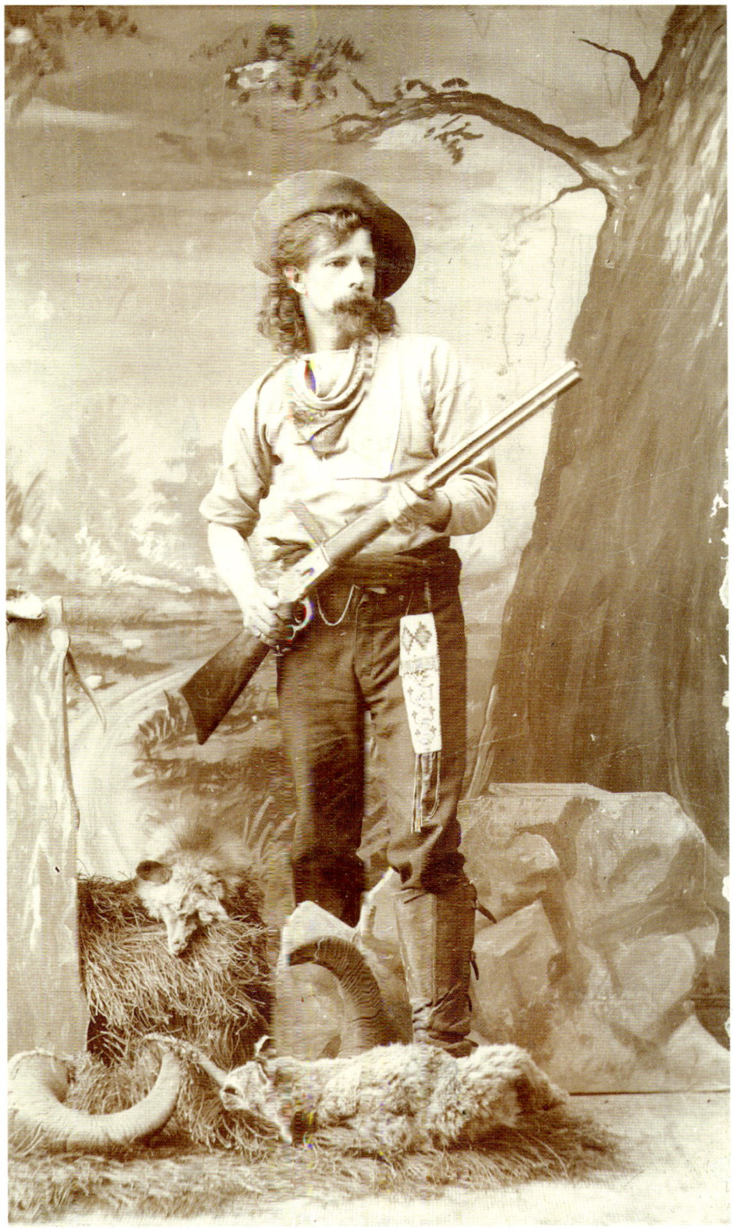

Photographer unknown. [Untitled.] Sepia print, America, circa 1895–1900. With one sleeve up and one sleeve down, this latter-day frontier hunter strikes a ready pose with his standard-grade Winchester Model 1886 lever-action, repeating magazine rifle. A skinning knife is thrust in his sash, while a beaded knife sheath—a likely studio prop—hangs empty.
Dickinson Research Center, National Cowboy & Western Heritage Museum, RC2007.061

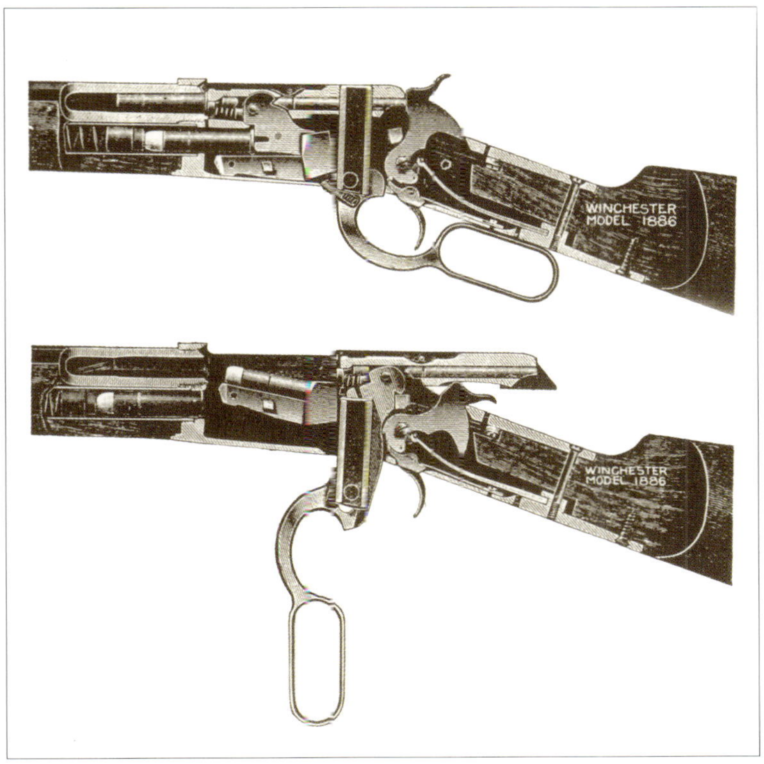

Cutaway drawings illustrating the internal workings of the Winchester Model 1886 repeating rifle and its John M. Browning–designed, lever-actuated, vertical-bolt-locking mechanism.
Drawings from the Winchester Repeating Arms Company's 1899 sales catalog

repeater), Winchester next introduced the Browning-designed Model 1890 slide-action magazine rifle. Chambered separately for .22 Short, Long, Long Rifle, or the special .22 Winchester Rim Fire cartridge, the Model 1890 achieved substantial sales, particularly with target galleries, arcades, and carnivals. When Model 1890 manufacture ceased in 1932 at nearly 850,000 rifles, it had outstripped Colt's Lightning .22 production by almost ten to one.[16] Like the rival Marlin firm, Winchester also sought to upgrade its lever-action line with improved designs. With the Model 1892, the company offered an improved successor to the classic Model 1873—a Browning-designed, short-frame action for pistol-sized centerfire cartridges (.25-20, .32-20, .38-40, and the ever-popular .44-40). Virtually a scaled-down version of the Model 1886 Winchester, the Model 1892 enjoyed substantial sales both at home and abroad, and it became a staple longarm in many Hollywood Western movies. Just over 1 million Model 1892 rifles and carbines (and a mere handful of muskets) saw manufacture from their inception to their cessation in 1941.[17]

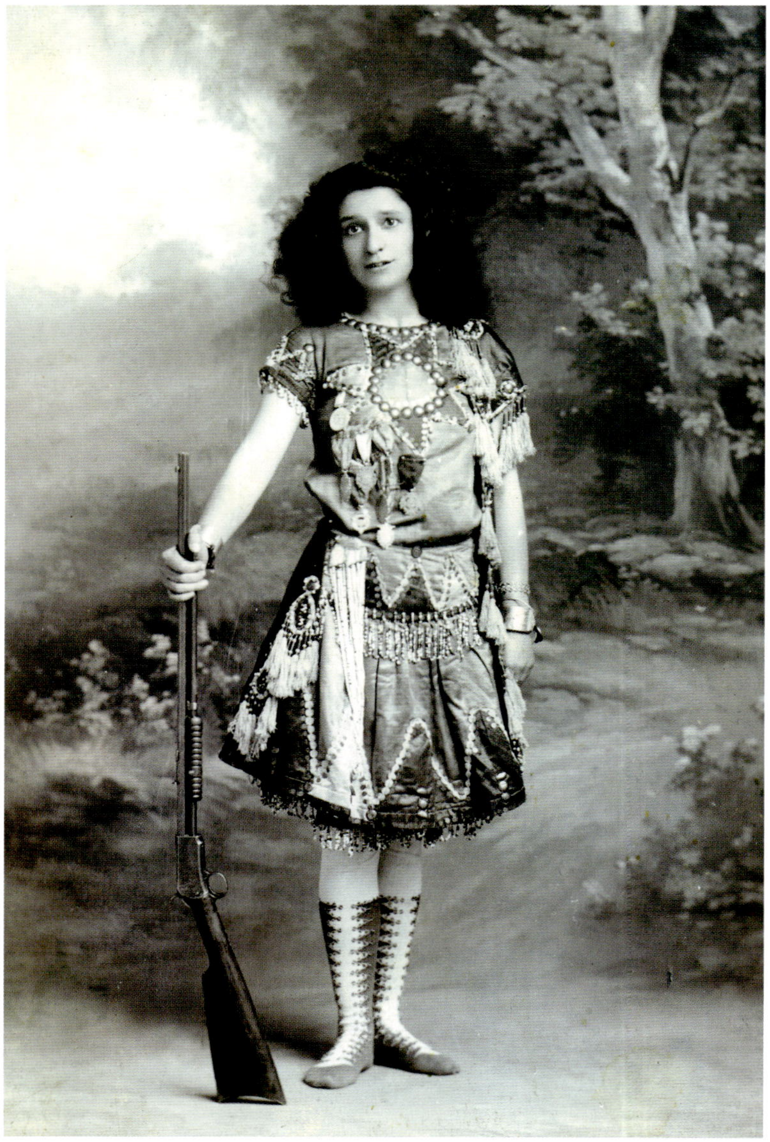

Durbin, photographer. [*Bonita—Champion Lady Rifle Shot.*] England (?), 1900–1905. Posed in a theatrical costume draped with a half-dozen shooting medals, a wistful-looking Bonita holds forth her Winchester Model 1890 slide-action, repeating magazine rifle. Many thousands of these fast-shooting and accurate rifles served lesser marksmen in galleries and arcades around the world for forty years.
Image courtesy private collection

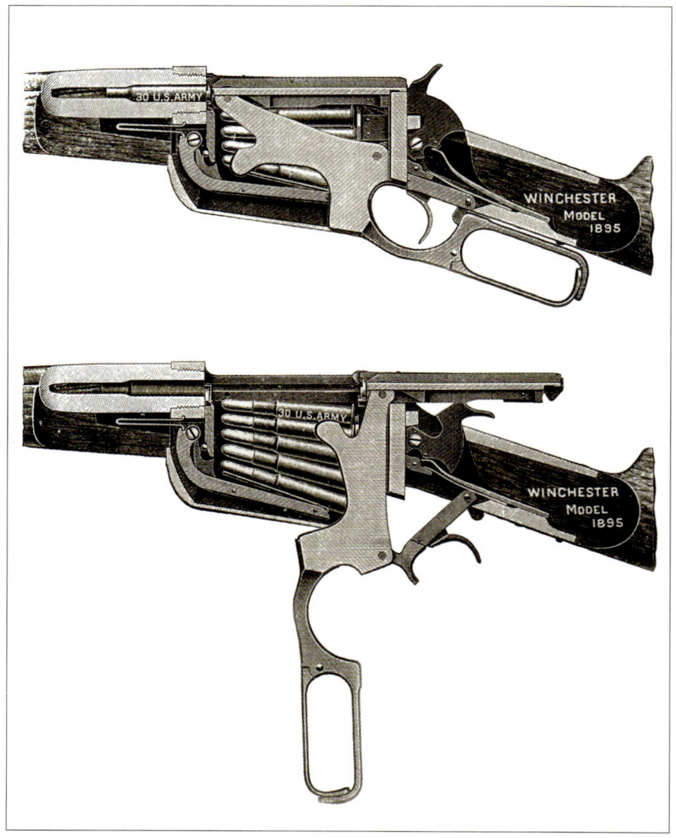

Cutaway drawings illustrating the internal workings of the Winchester Model 1895 repeating rifle and its John M. Browning–designed, lever-actuated, box magazine system.
Drawings from the Winchester Repeating Arms Company's 1899 sales catalog

The 1890s brought further technical improvements to American sporting firearms with the development of fast-burning, high-pressure smokeless powders, which, in turn, initiated the trend toward high-velocity, smaller-caliber ammunition. The Browning-designed Model 1894 Winchester lever-action became the first commercial arm adapted for a smokeless cartridge—the famed .30-30 or .30 WCF. Also chambered in calibers .25-35, .32-40, .32 Winchester Special, and .38-55, this classic American deer rifle ultimately became the best-selling repeating sporting rifle ever made. Model 1894 rifle and carbine variants surpassed 6 million units at their initial cessation in 2006.[18] For larger, dangerous game, sportsmen had access to yet another Browning-designed Winchester, the Model 1895. This rifle introduced the first high-velocity lever-action with a box magazine beneath the receiver in lieu of the long-familiar tubular magazine under the barrel. Ultimately available in nineteen commercial and military calibers ranging from the .30-40 Krag to the powerful .405 WCF, Model 1895 rifles, carbines, and muskets realized a total production of some 425,000 units (many of them of military pattern) between 1896 and 1931.[19]

Not unlike Colt's, the Winchester firm operated as a center of excellence in American arms manufacturing during the second half of the nineteenth century. By 1900 the company had introduced thirteen distinct model arms (including shotguns) and manufactured nearly 1,700,000 guns over thirty-four years. Winchester continued as a leader in longarm manufacture throughout much of the twentieth century. It began an economic decline in 1980 that saw the trademark produced under license by the U.S. Repeating Arms Company and finally culminated in the closure of the New Haven factory in 2006. Thereafter, certain classic Winchester models have seen production under licensed subsidiaries of Fabrique Nationale d'Herstal (FN) of Belgium.[20]

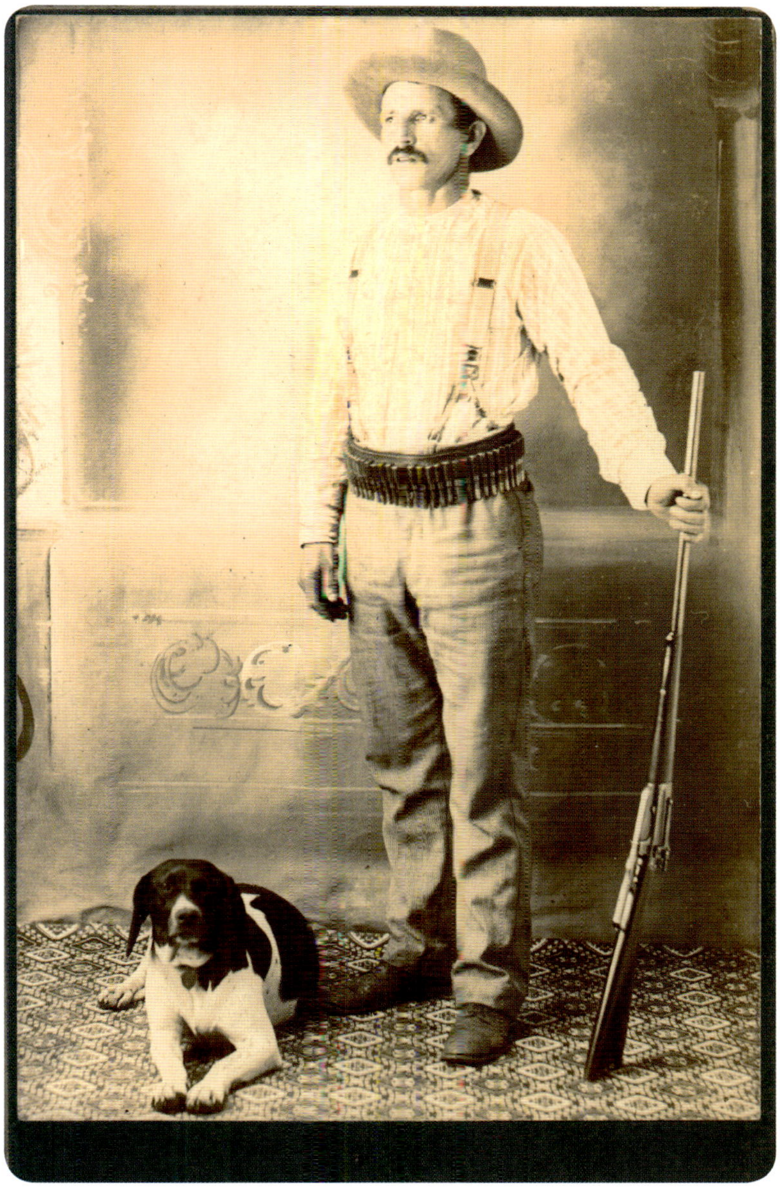

Photographer unknown. [Untitled.] Cabinet card, America, circa 1900–1905. This plainly dressed sportsman models with his hunting dog and a cartridge belt full of ammunition for his standard-grade Winchester Model 1895 lever-action repeating magazine rifle. A John Browning design, this model utilized a box magazine in lieu of the familiar tubular magazine beneath the barrel.
Dickinson Research Center, National Cowboy & Western Heritage Museum, 2004.230.02

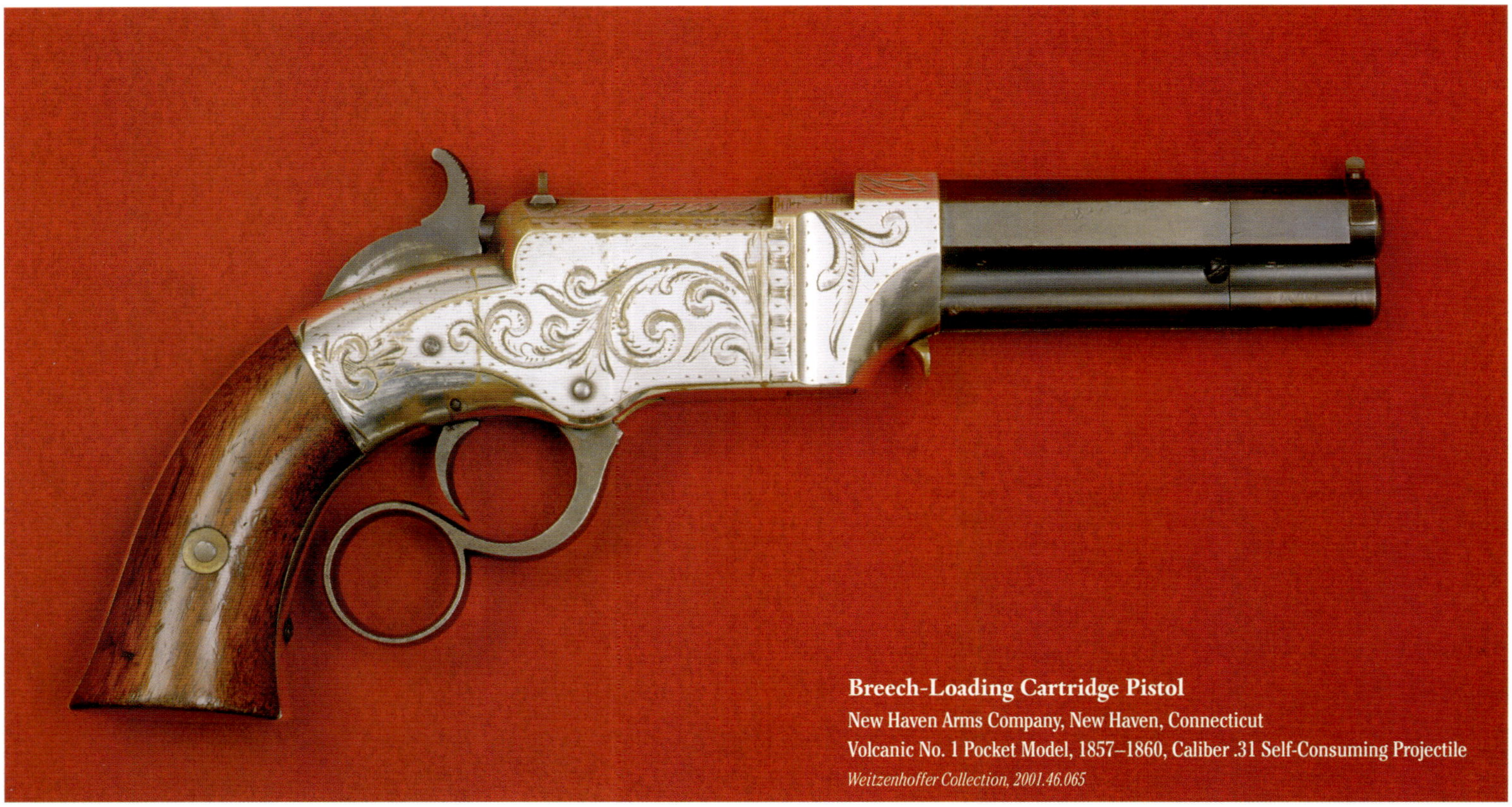

Breech-Loading Cartridge Pistol
New Haven Arms Company, New Haven, Connecticut
Volcanic No. 1 Pocket Model, 1857–1860, Caliber .31 Self-Consuming Projectile
Weitzenhoffer Collection, 2001.46.065

A predecessor of the Winchester repeating rifle, this handgun incorporates a lever-actuated, toggle-link breech mechanism joined with an integral tubular magazine beneath the barrel. The pistol fired special self-consuming (caseless) ammunition consisting of a hollow, conical bullet that held its own powder charge and a combustible primer. Accommodating only six shots, the pocket-sized Volcanic pistol offered no advantages over contemporary revolvers, such as Colt's 1849 Pocket Model. This specimen exhibits standard engraving treatment—a sparse border motif with a modest, flowing scroll pattern on a clear field. An 1859 ad in *The Democratic Age* stated that the retail cost for such a piece was then $13.50—only $1.50 more than an unadorned example.

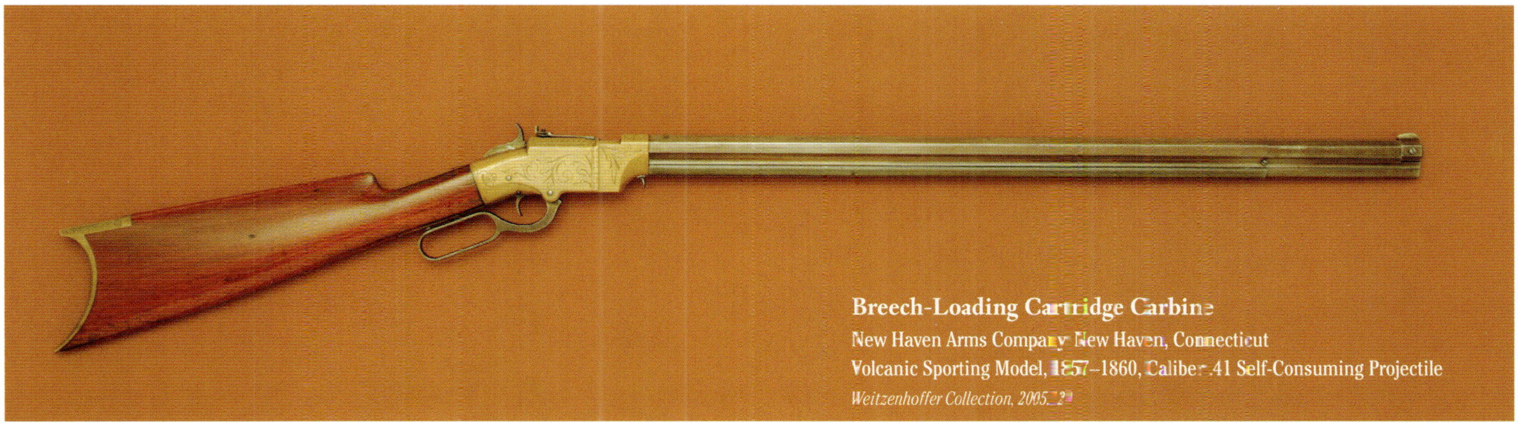

Breech-Loading Cartridge Carbine
New Haven Arms Company, New Haven, Connecticut
Volcanic Sporting Model, 1857–1860, Caliber .41 Self-Consuming Projectile
Weitzenhoffer Collection, 2005.[illegible]

Far more practical than the pistol, the Volcanic carbine allowed the shooter to better maintain the point of aim while operating the lever-action mechanism between shots. Though the weapon used comparatively underpowered ammunition, its twenty-round magazine offered impressive firepower in its day. In October 1858, *Frank Leslie's Illustrated Weekly* wrote of the then-novel arm, "It is placed beyond all competition by the rapidity of its execution. [Twenty] shots can be fired in less than one minute—a marvelous rapidity, in which it far outdoes the best revolving firearms yet produced." About 1,000 of these carbines were manufactured in three standard barrel lengths—here, the very scarce 25-inch variant, of which fewer than twenty are known. Foreshadowed in the Volcanic's toggle-link breech system, the merits of lever-action, repeating magazine arms would be amply demonstrated in the line of Winchester rifles that sprang from the design commencing in 1866.

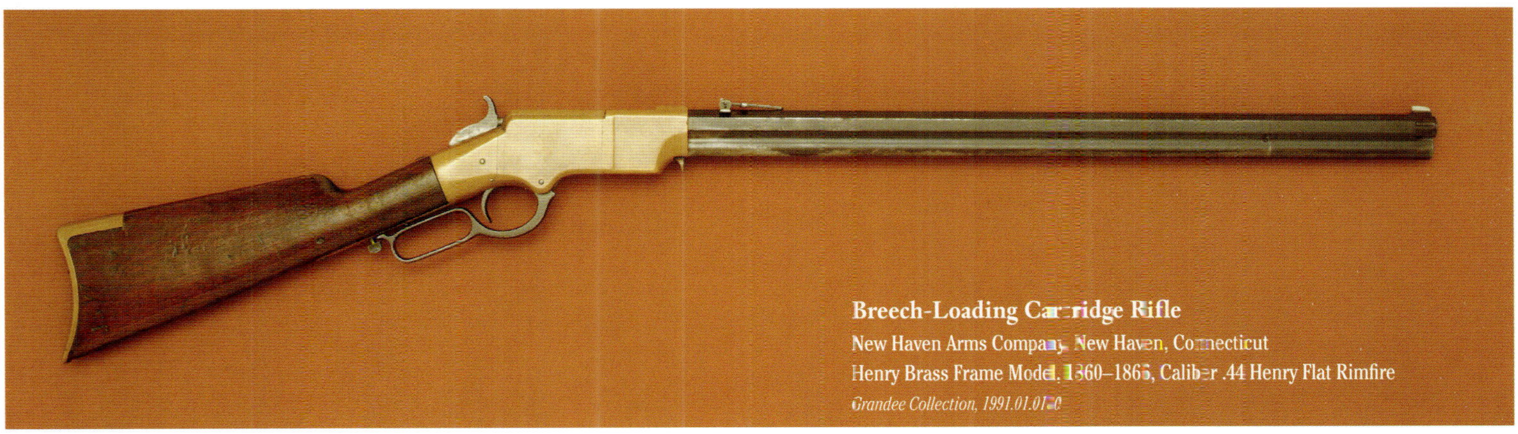

Breech-Loading Cartridge Rifle
New Haven Arms Company, New Haven, Connecticut
Henry Brass Frame Model, 1860–1866, Caliber .44 Henry Flat Rimfire
Grandee Collection, 1991.01.0[illegible]

The immediate forerunner of the famed line of Winchester lever-action repeaters, the Henry design ultimately proved commercially superior to its contemporary competitor, the Spencer repeating rifle. Discharging up to sixteen rounds in rapid succession, the Henry rifle provided the greatest firepower of the early 1860s, winning many enthusiastic converts in the frontier West. The arm's principal failing lay in its slotted, tubular magazine, which allowed dirt to foul the cartridges and toggle-link action. Machined integral with the iron barrel, the magazine also made the fast-firing gun somewhat muzzle heavy and unwieldy, particularly when fully loaded with cartridges. These objections would be virtually eliminated with the first Winchester—the Model 1866.

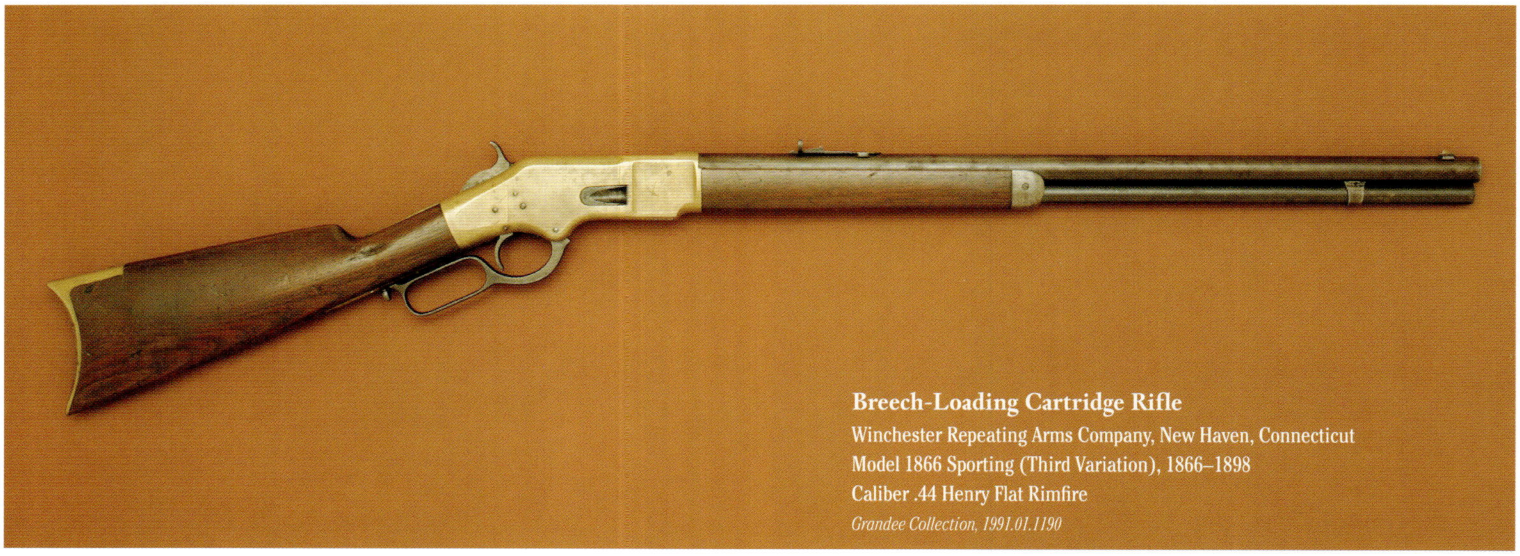

Breech-Loading Cartridge Rifle
Winchester Repeating Arms Company, New Haven, Connecticut
Model 1866 Sporting (Third Variation), 1866–1898
Caliber .44 Henry Flat Rimfire
Grandee Collection, 1991.01.1190

A successor to the Henry rifle, the Model 1866 Winchester also featured a brass receiver, and it employed the same, rather weak toggle-link, bolt-locking system. The rifle did introduce a much-improved magazine design with a convenient loading gate for inserting cartridges in the right side of the receiver and a new wooden forestock. Like its predecessor, the Model 1866 repeating rifle provided the western frontiersman with abundant firepower, but its comparatively underpowered ammunition was all but inadequate for big-game species such as elk, buffalo, and grizzly bear.

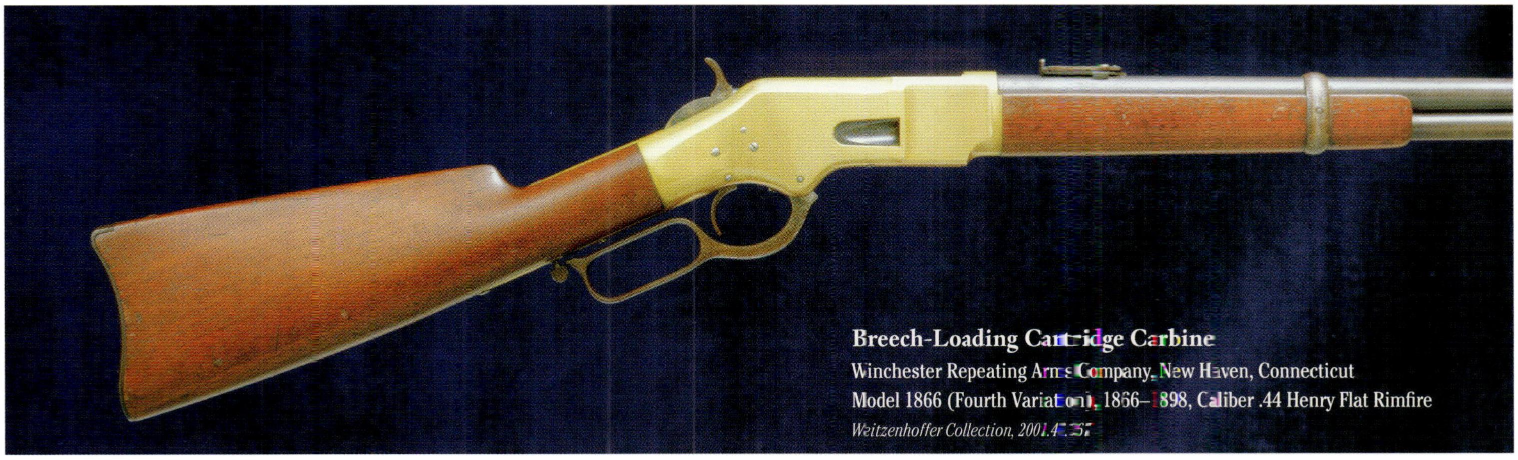

Breech-Loading Cartridge Carbine
Winchester Repeating Arms Company, New Haven, Connecticut
Model 1866 (Fourth Variation), 1866–1898, Caliber .44 Henry Flat Rimfire
Weitzenhoffer Collection, 2001.4.57

Formulated by Winchester engineer Nelson King, the Model 1866 marked a great technological improvement over the Volcanic and Henry repeaters, as it featured a fully closed magazine tube beneath the barrel and a convenient loading port on the right side of the frame. These alterations made it the first truly practical arm of its type. The model proved extremely popular in the West, establishing Winchester as a leader among domestic longarm manufacturers for the remainder of the nineteenth century and beyond. It also marked the beginning of very elaborately embellished Winchester arms, as can be seen in the Model 1866 rifles presented in chapter 10.

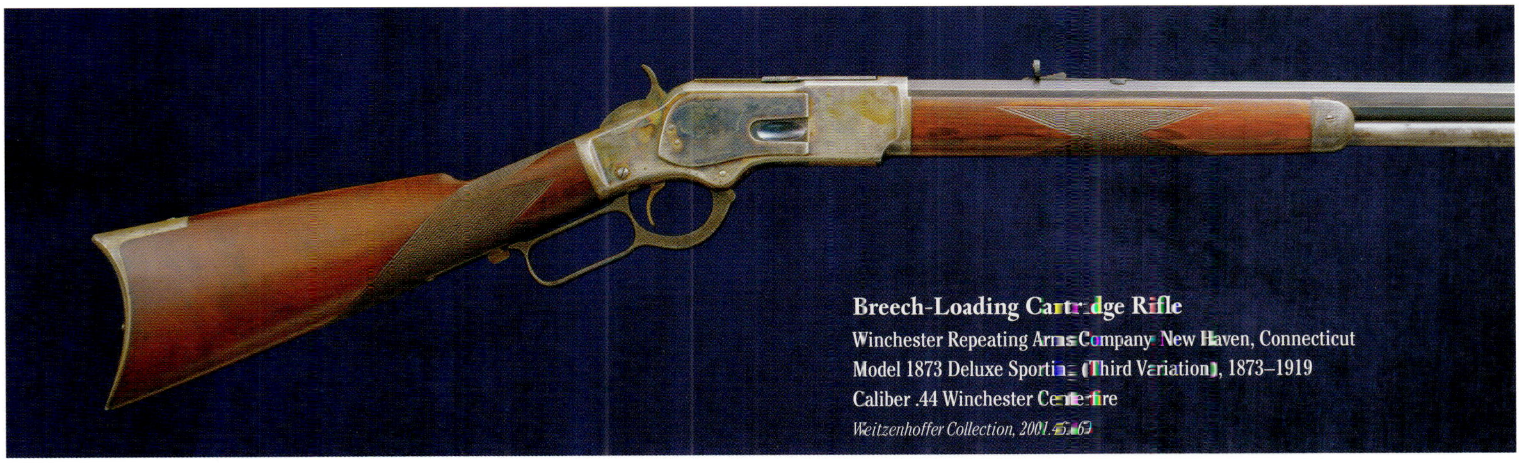

Breech-Loading Cartridge Rifle
Winchester Repeating Arms Company, New Haven, Connecticut
Model 1873 Deluxe Sporting (Third Variation), 1873–1919
Caliber .44 Winchester Centerfire
Weitzenhoffer Collection, 2001.45.69

Improving on the Model 1866 Winchester with a stronger frame of forged and milled iron and a more potent centerfire cartridge, the Model 1873 garnered immediate acclaim in the marketplace. Famous western scout and performer W. F. "Buffalo Bill" Cody proclaimed it "the boss" among repeating longarms, and its popularity became so widespread that Winchester ultimately declared it "The Gun that Won the West" (historically a title equally deserved by the Model 1866). Made in 1883, this specimen presents standard features except for its color-case-hardened frame and its checkered and varnished, select walnut stocking (deluxe extras that added $1 and $5, respectively, to the $27 retail cost). The original jointed cleaning rod remains in the buttstock.

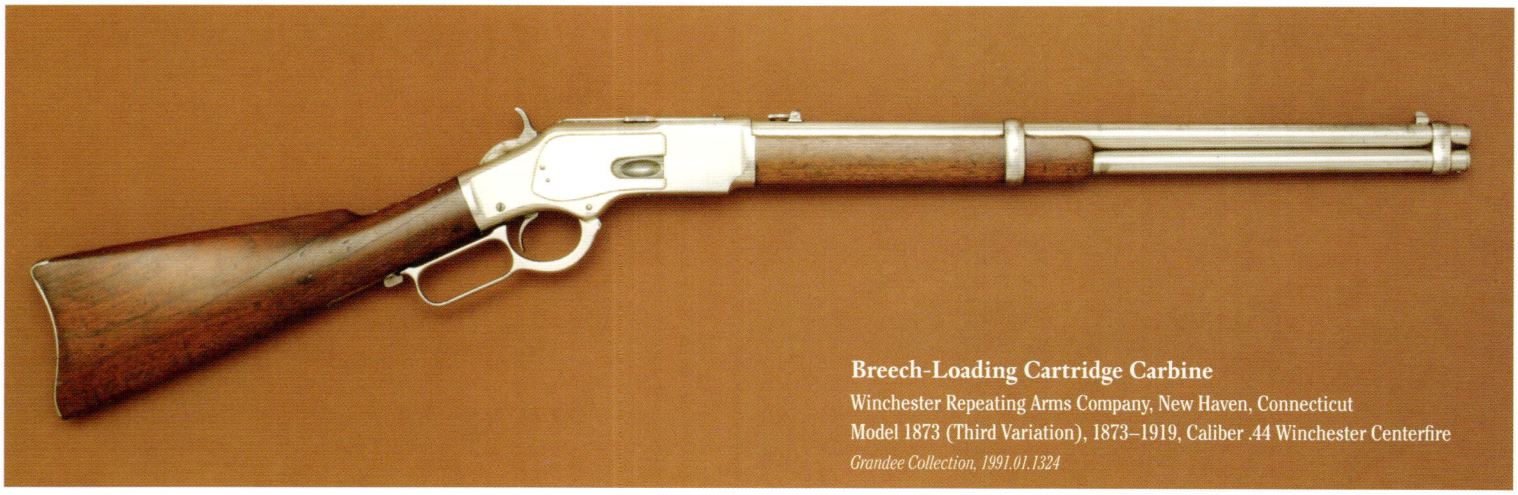

Breech-Loading Cartridge Carbine
Winchester Repeating Arms Company, New Haven, Connecticut
Model 1873 (Third Variation), 1873–1919, Caliber .44 Winchester Centerfire
Grandee Collection, 1991.01.1324

Standard with a 20-inch round barrel and twelve-shot magazine, the Model 1873 Winchester carbine proved an ideal repeating shoulder arm for use on horseback. Like the Model 1866 carbine (and all succeeding Winchesters of carbine pattern), this specimen has a saddle ring mounted on the left side of the receiver with which to secure the weapon in carriage. Carbines constituted about 36 percent of all Model 1873 production, but special-order features were relatively unusual for the type. Otherwise a standard example, this piece boasts full nickel plating, an embellishment that added $5 to the $24 retail cost. This is a rare extra; less than 0.10 percent of all Model 1873 production received full or partial plating in nickel.

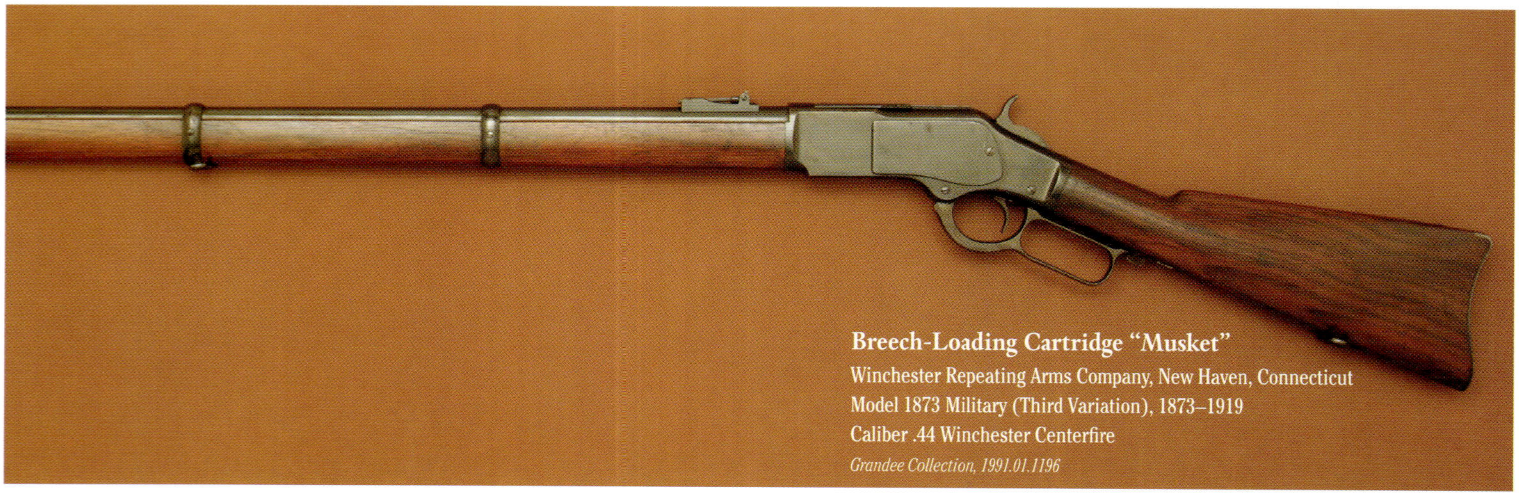

Breech-Loading Cartridge "Musket"
Winchester Repeating Arms Company, New Haven, Connecticut
Model 1873 Military (Third Variation), 1873–1919
Caliber .44 Winchester Centerfire
Grandee Collection, 1991.01.1196

Seeking military sales from the outset, the Winchester firm made "musket"-pattern repeaters in the lever-action Models 1866, 1873, 1876, 1886, 1892, and 1895. Termed an "Army Rifle" at the factory, these firearms were characterized by extended forestocks secured with one, two, or three barrel bands; sling swivels; ladder-type military sights; and recurved buttplates of carbine pattern. Most were supplied with triangular socket bayonets. Originally priced at $45, this standard Model 1873 musket incorporates a 30-inch round barrel with a 27-inch tubular magazine holding seventeen cartridges. Although the U.S. Army never adopted the arm, substantial numbers were sold to the Spanish government and other nations in Central America and the Middle East.

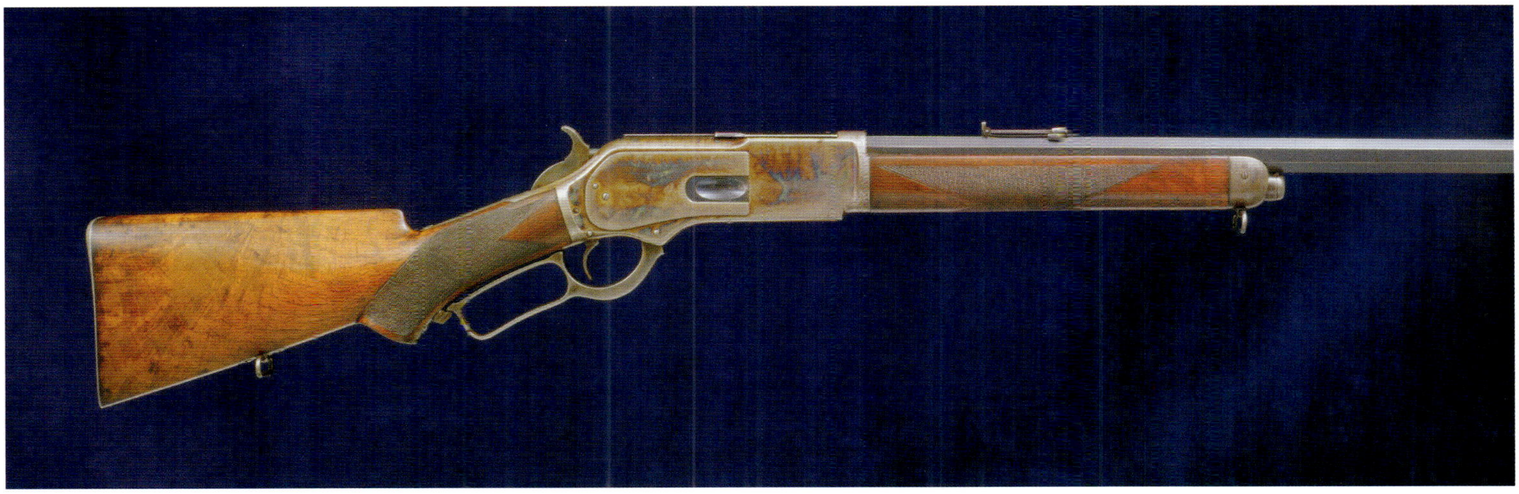

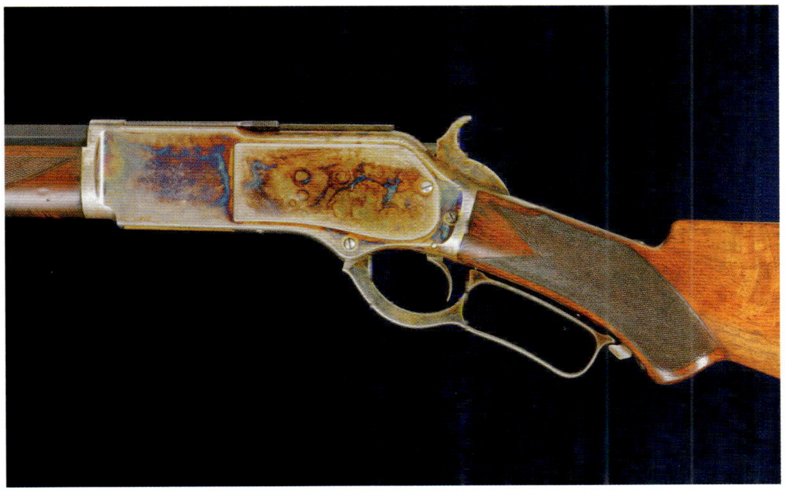

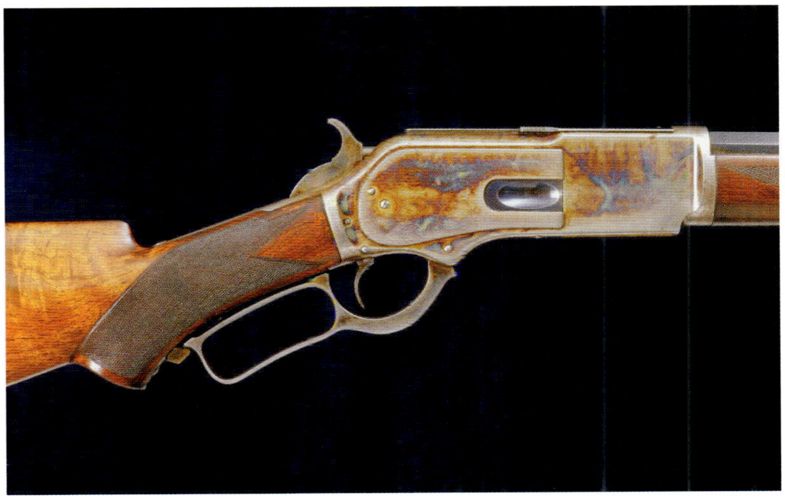

Breech-Loading Cartridge Rifle

Winchester Repeating Arms Company, New Haven, Connecticut
Model 1876 Deluxe Sporting (Late Second Variation), 1877–1897
Caliber .45-60 Winchester Centerfire
Weitzenhoffer Collection, 2002.3[illegible]

The Model 1876 Winchester, often termed the "Centennial" model for its year of introduction, presented a scaled-up version of the earlier Model 1873. It was designed to accommodate the long-cased, high-powered cartridges popular among sport hunters in the frontier West. Theodore Roosevelt considered it "the best gun for any [large] game to be found in the United States, for it is deadly, accurate, and handy as any . . . and unapproachable for the rapidity of its fire and the facility with which it is loaded." This special-order example features a full-octagon barrel with standard sights; a half, button-type magazine; sling swivels; and deluxe, figured walnut stocking having a shotgun-pattern butt and graceful pistol grip with Style H checkering over the wrist and forestock.

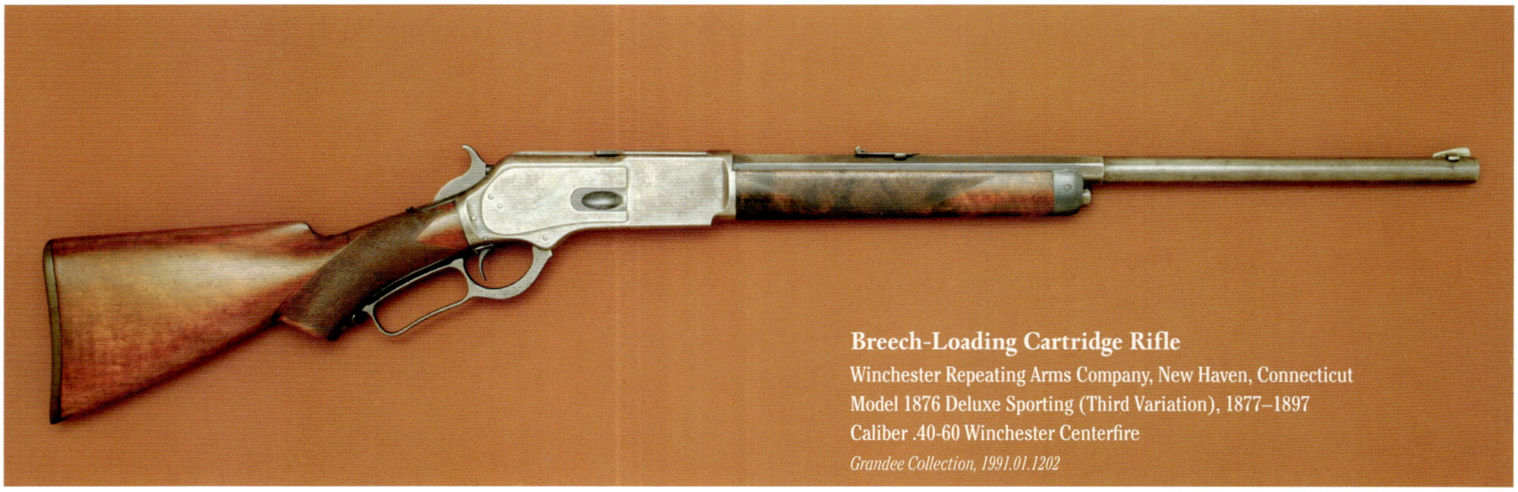

Breech-Loading Cartridge Rifle
Winchester Repeating Arms Company, New Haven, Connecticut
Model 1876 Deluxe Sporting (Third Variation), 1877–1897
Caliber .40-60 Winchester Centerfire
Grandee Collection, 1991.01.1202

Rifles with special-order and deluxe appointments were highly favored by sport hunters in the West—perhaps most notable among them future President Theodore Roosevelt. Much in keeping with Roosevelt's known taste in Winchesters, the special-order features of this specimen include half-octagon/half-round barrel, half-magazine, three-leaf rear sight, single set trigger, and select checkered walnut stocking with pistol grip and shotgun-pattern butt. (At some time in the past, the knife-blade front sight on this rifle was reinstalled backwards.)

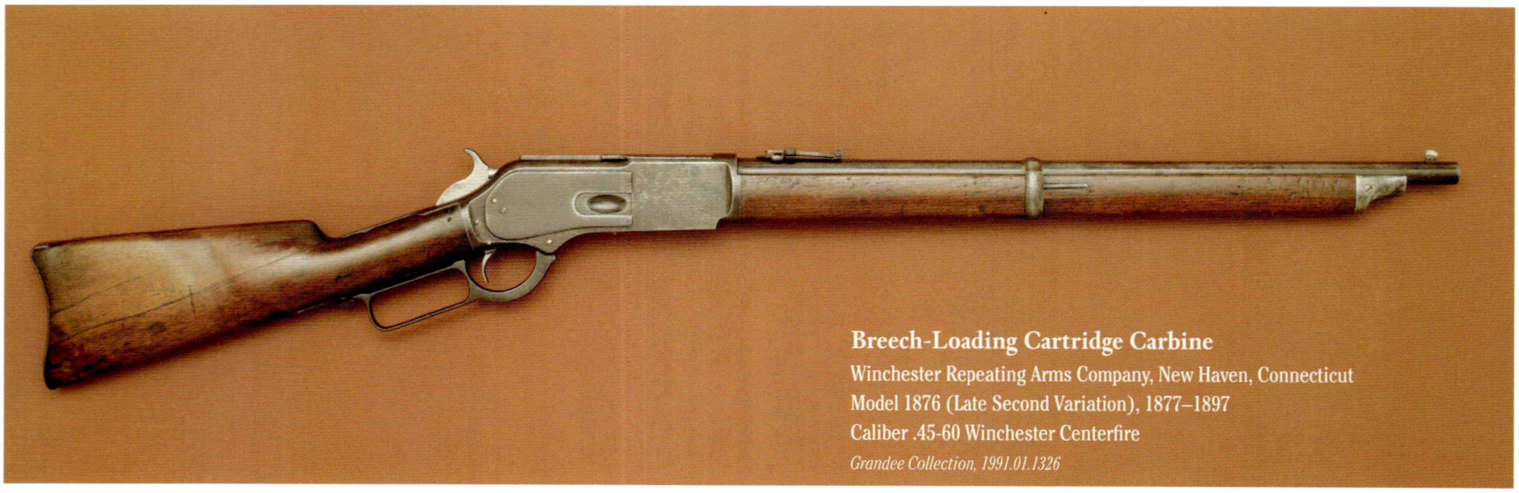

Breech-Loading Cartridge Carbine
Winchester Repeating Arms Company, New Haven, Connecticut
Model 1876 (Late Second Variation), 1877–1897
Caliber .45-60 Winchester Centerfire
Grandee Collection, 1991.01.1326

Actually appearing like a short "musket," the Model 1876 carbine differed from earlier and later Winchester carbine models in having an extended, 18-inch forestock that completely shrouded the tubular magazine beneath the 22-inch round barrel. Like other models, this carbine featured a saddle ring fixed on the left frame and a recurved, carbine or military-style buttplate. As noted elsewhere, the Model 1876 enjoyed wide acclaim in the American West. The carbine pattern, in particular, won the favor of the Northwest Mounted Police of Canada, which ordered some 850 of the weapons in caliber .45-75 WCF between 1878 and 1885—and kept them in service until nearly World War I.

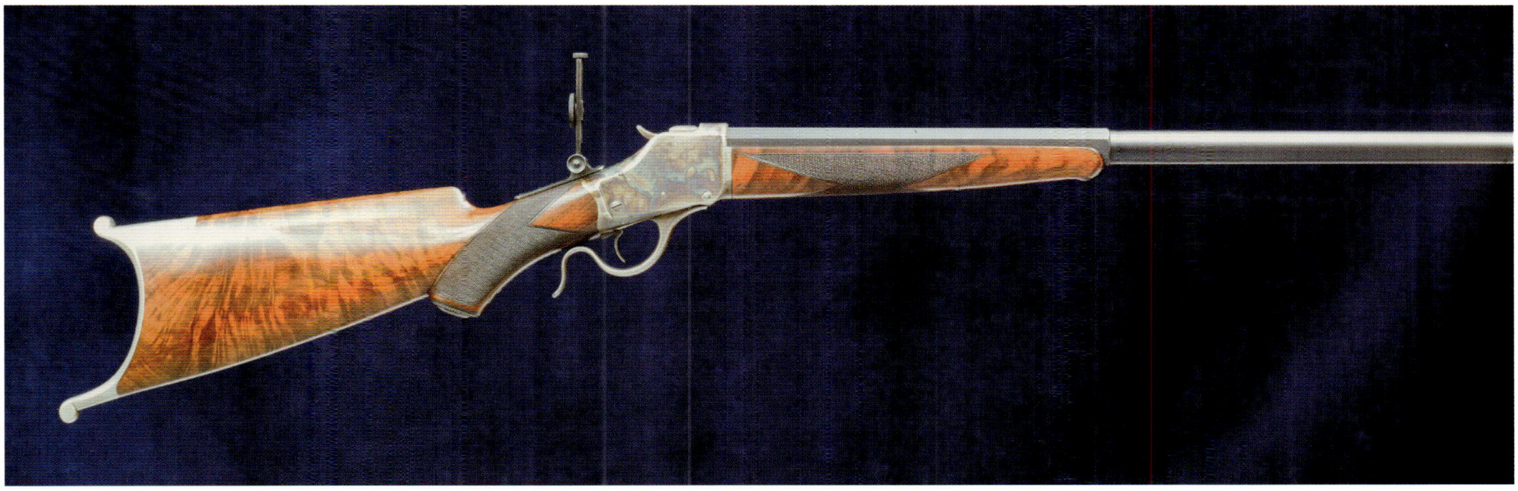

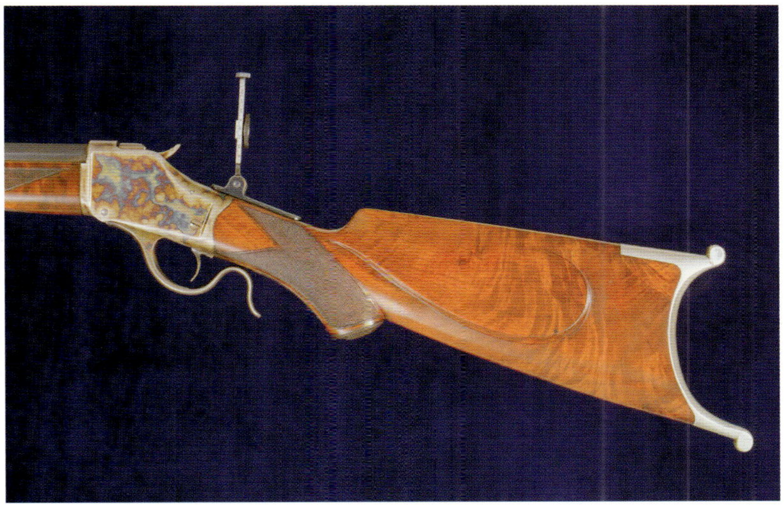

Breech-Loading Cartridge Rifle

Winchester Repeating Arms Company, New Haven, Connecticut
Model 1885 Deluxe High-Wall Target, 1885–1920
Caliber .32-40 Winchester Centerfire
Weitzenhoffer Collection, 2001.07.3

Originally conceived by Utah Territory arms designer John M. Browning, the Model 1885 Winchester became the most popular single-shot rifle of the late nineteenth and early twentieth centuries. Produced in two frame sizes (low-wall and high-wall) the pattern was available in a broad range of calibers, barrel weights, and sight configurations for hunting and target shooting. Made in 1898, this target variation features an adjustable, hooded wind-gauge front sight in combination with a midrange vernier peep sight mounted at the wrist. Additional special-order and deluxe appointments include a half-round/half-octagon barrel; select, crotch-grain walnut stocking checkered in Style H with a pistol grip and cheekrest; and a nickel-plated, Swiss-pattern buttplate.

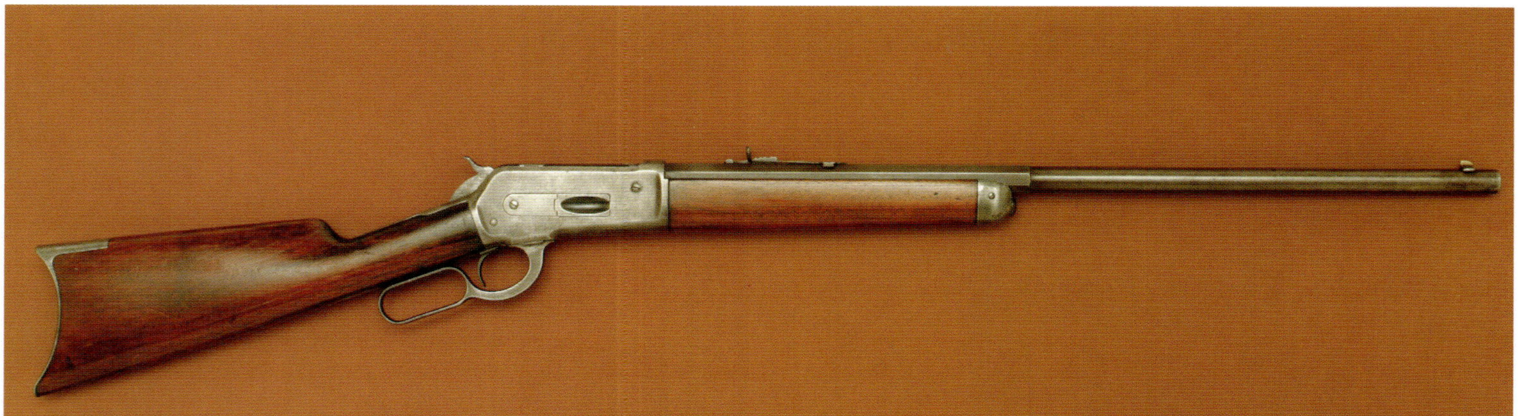

Breech-Loading Cartridge Rifle
Winchester Repeating Arms Company, New Haven, Connecticut
Model 1886 Sporting, 1886–1935, Caliber .38-56 Winchester Centerfire
Museum Purchase, 1999.20.6

A marked departure from Winchester's traditional toggle-link, breech-locking mechanism, the Model 1886 Winchester rifle, designed by westerner John M. Browning, incorporated an exceedingly strong action using vertical locking bolts to secure the breech. Capable of handling powerful centerfire cartridges such as the .50-110 Express and the popular .45-70 Government rounds, this repeater soon supplanted the Model 1876 Winchester and rapidly diminished the popularity of single-shot rifles such as the Sharps and Remington. Except for its half-octagon/half-round barrel configuration and its half, button-type magazine, this is a standard-production rifle.

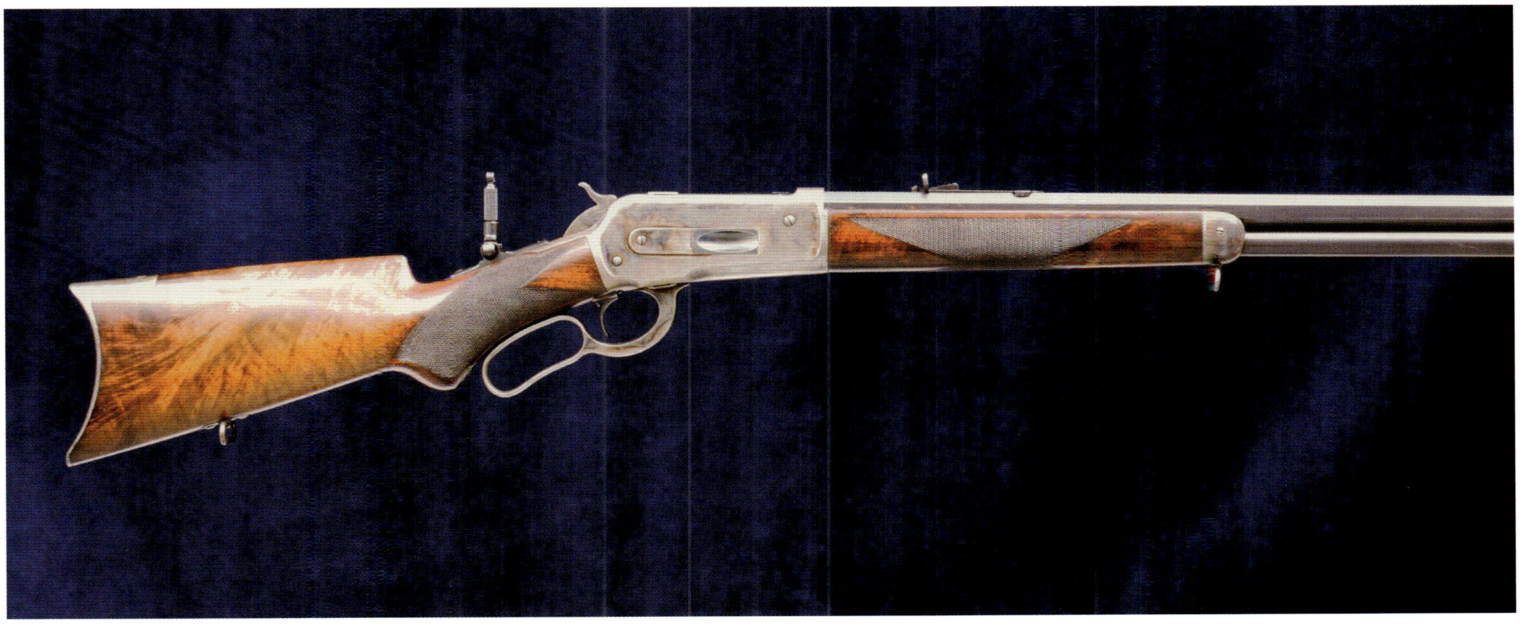

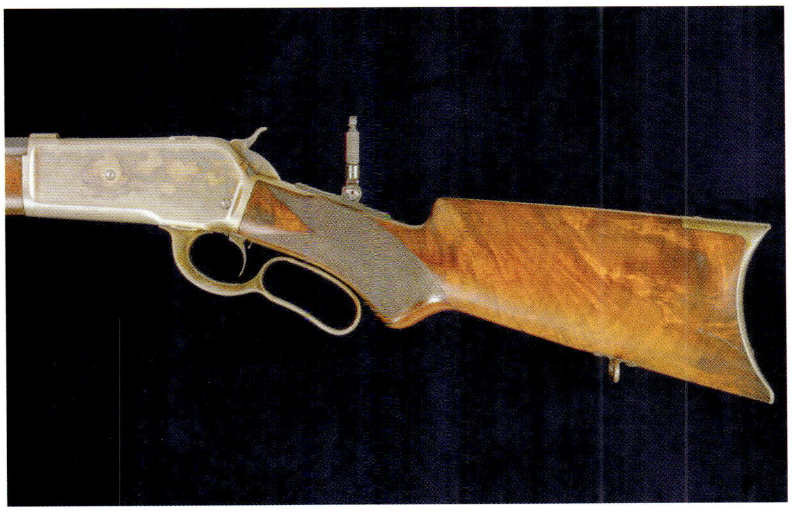

Breech-Loading Cartridge Rifle

Winchester Repeating Arms Company, New Haven, Connecticut
Model 1886 Deluxe Sporting, 1886–1935, Caliber .40-82 Winchester Centerfire
Weitzenhoffer Collection, 2001.-[illegible]

Conceived by renowned arms designer John M. Browning, the Model 1886 Winchester utilized an ingenious breech-locking system that easily accommodated the most powerful, large-caliber ammunition of the day. The model rapidly became a favorite among sportsmen such as Theodore Roosevelt for big-game hunting in the West. Reflecting something of the gentleman-sportsman's taste in sporting arms, this deluxe specimen incorporates several special-order features beyond the norm of standard production, including a Lyman adjustable peep sight mounted at the wrist, sling swivels mounted on the buttstock and forestock cap, a single set trigger, and select, flame-grained walnut stocking with an ebony-inlaid pistol grip and hand-checkering in factory style H.

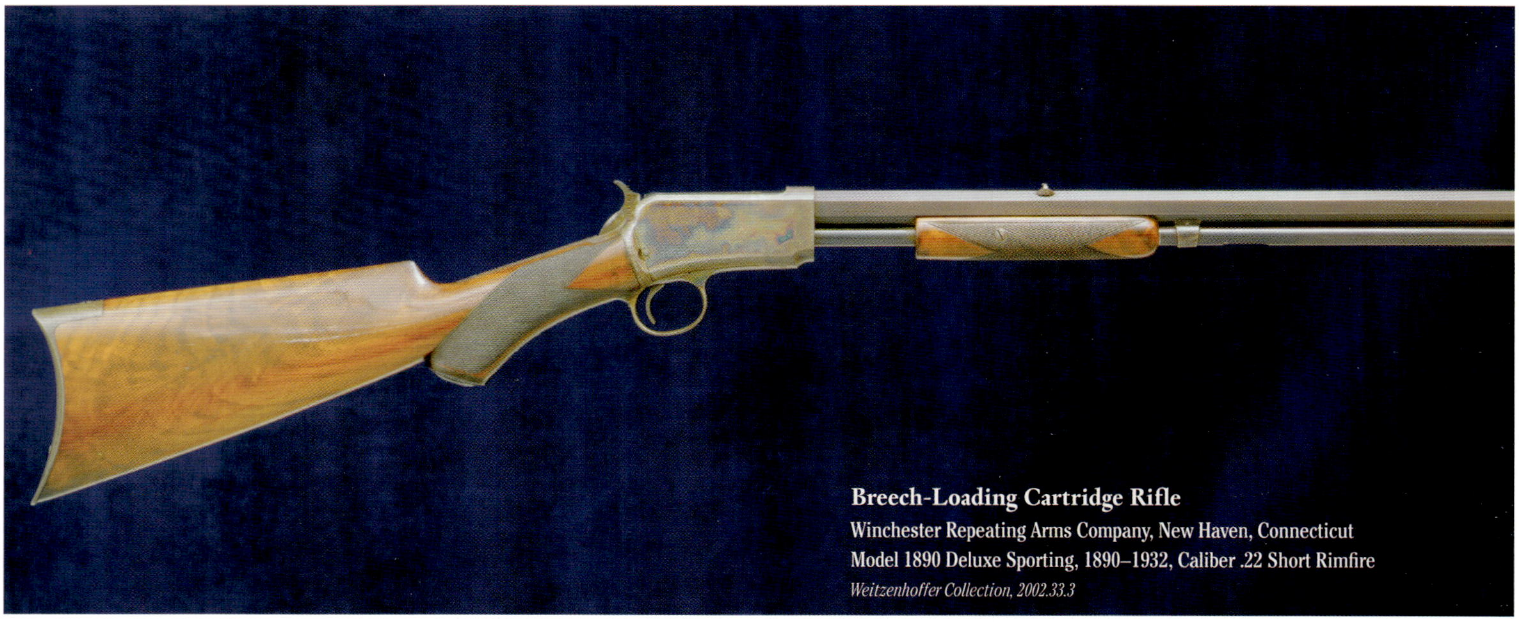

Breech-Loading Cartridge Rifle
Winchester Repeating Arms Company, New Haven, Connecticut
Model 1890 Deluxe Sporting, 1890–1932, Caliber .22 Short Rimfire
Weitzenhoffer Collection, 2002.33.3

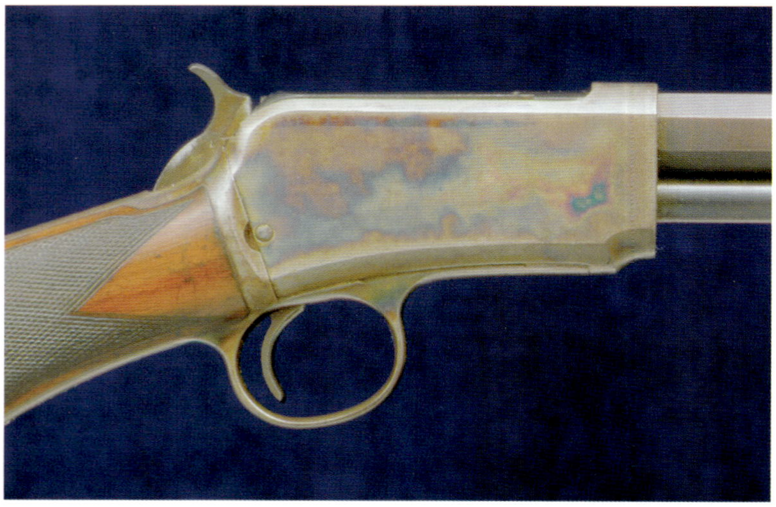

Another practical and long-lived design demonstrating the genius of John M. Browning, the Model 1890 was the Winchester firm's first slide-action, or pump-action, magazine rifle. The gun was introduced to compete with Colt's Lightning pump-action .22 brought out in 1887. Its simpler and superior design rapidly supplanted the Colt—from 1890 to 1900, Winchester produced about 99,000 Model 1890s against Colt's approximately 31,000 Lightning .22s. Retaining strong case coloring on its receiver, this deluxe rifle features a select, figured walnut buttstock with a special-order pistol grip and hand-checkering over the wrist and forepiece.

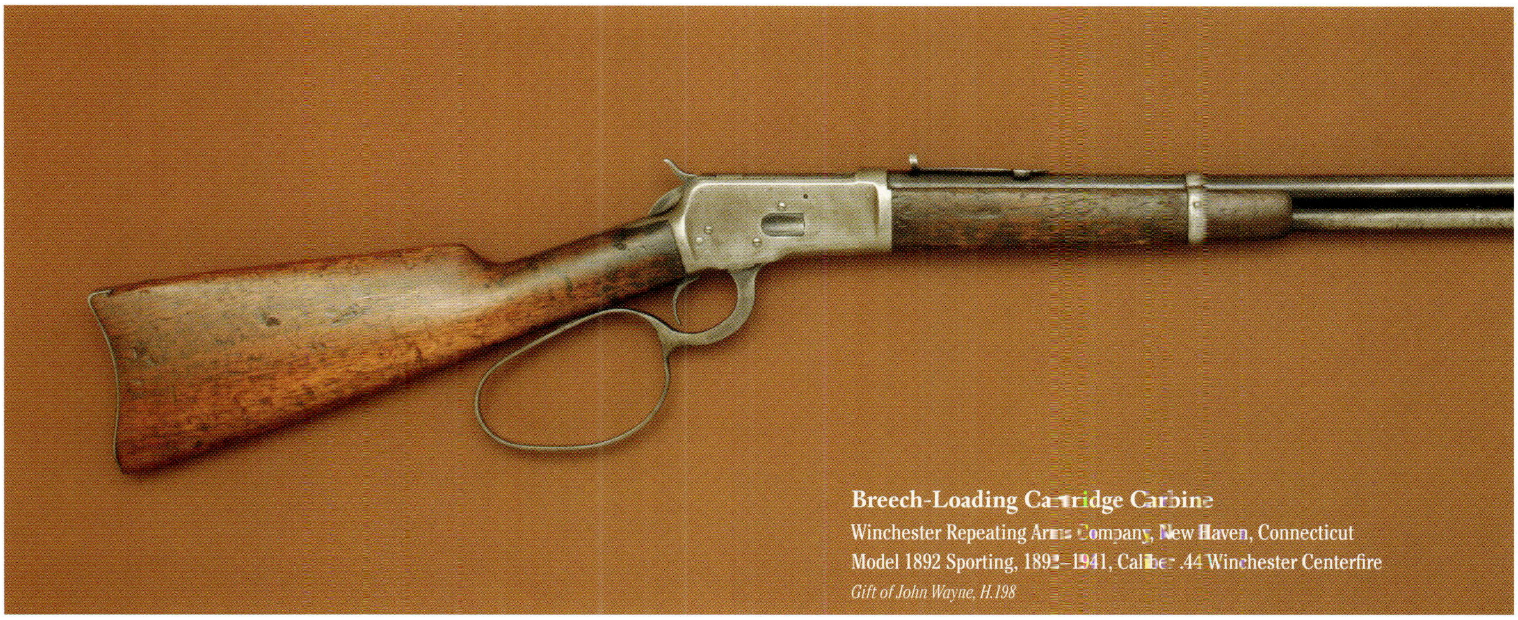

Breech-Loading Cartridge Carbine
Winchester Repeating Arms Company, New Haven, Connecticut
Model 1892 Sporting, 1892–1941, Caliber .44 Winchester Centerfire
Gift of John Wayne, H.198

Introduced as a functional improvement of the Model 1873 Winchester, the Model 1892 basically was a scaled-down version of the John M. Browning–designed Model 1886 utilizing his patented vertical-locking-bolt action for pistol-sized cartridges. The Model 1892 was produced and marketed in rifle, musket, and carbine configurations, and the latter pattern gained widespread favor not only in North America but also in South America and Australia. This well-worn Model '92 carbine—with its signature, modified lever loop of enlarged ovoid shape—represents something of a popular culture icon, as it frequently featured with actor John Wayne in Western movies during the 1950s, 1960s, and early 1970s.

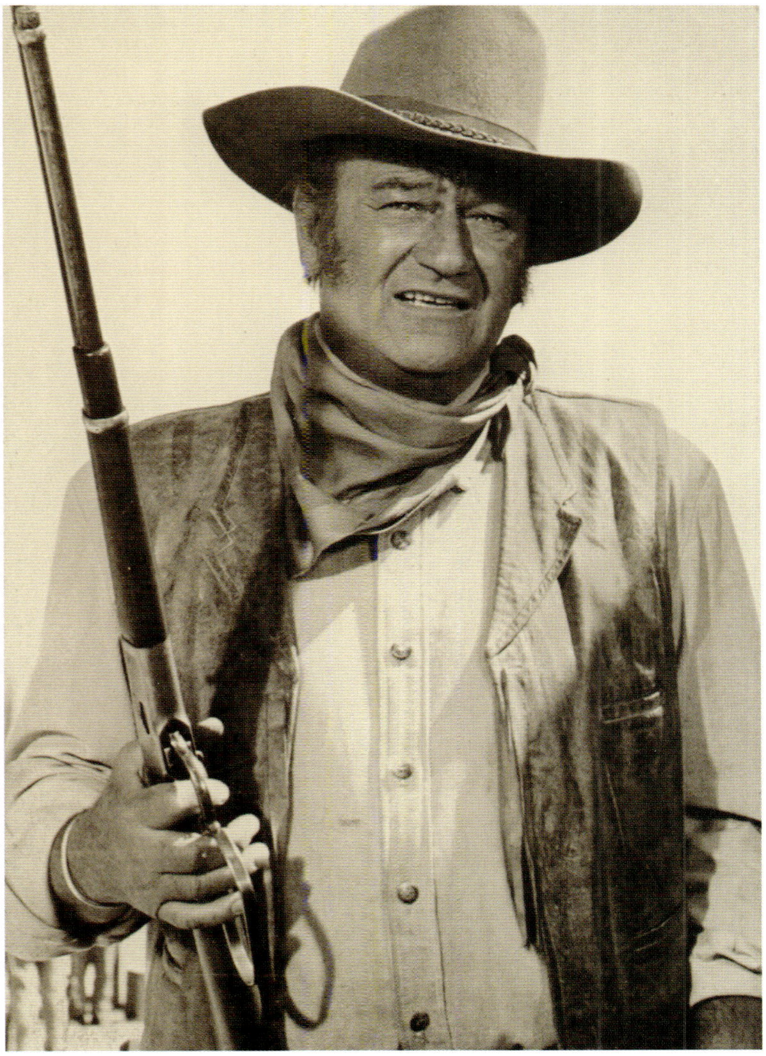

Ludlow Sales, publisher. [*John Wayne*.] Sepia postcard, New York, circa 1975. This still image from the movie *Rio Lobo* (1970) portrays iconic western actor John Wayne holding one of his equally iconic, large-loop, Model 1892 Winchester carbines—one of two, nearly identical examples held in the museum's collections.

Glenn D. Shirley Western Americana Collection, Dickinson Research Center, National Cowboy & Western Heritage Museum, RC2006.067.2.03261

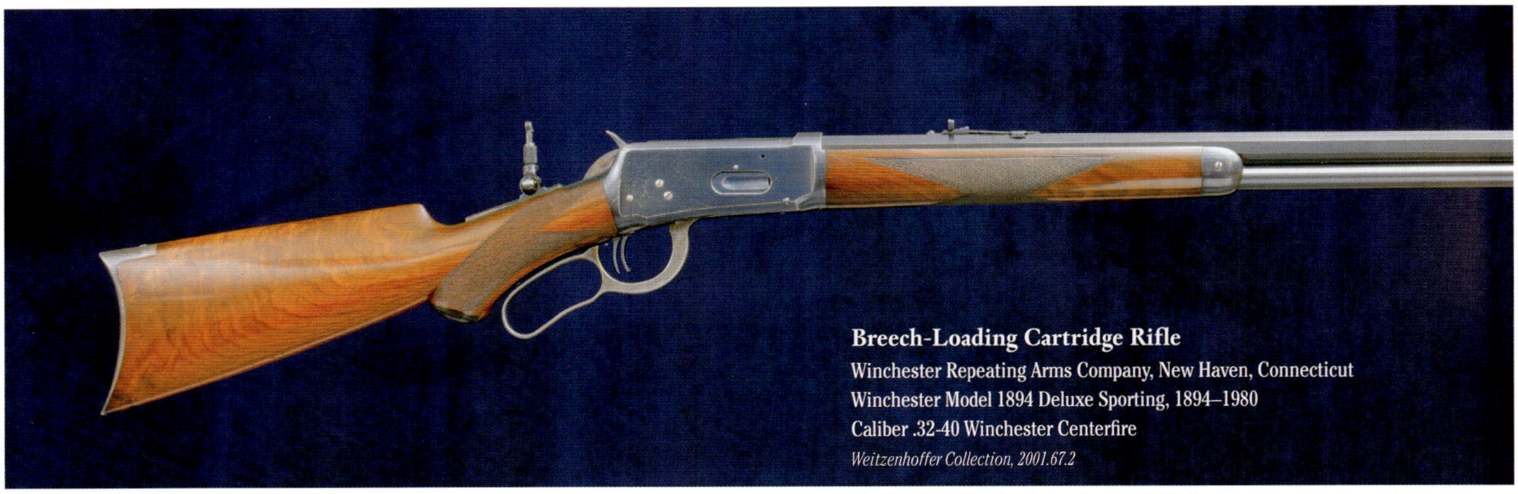

Breech-Loading Cartridge Rifle
Winchester Repeating Arms Company, New Haven, Connecticut
Winchester Model 1894 Deluxe Sporting, 1894–1980
Caliber .32-40 Winchester Centerfire
Weitzenhoffer Collection, 2001.67.2

The Model 1894, like the majority of Winchester arms introduced in the late nineteenth century, sprang from the prolific mind of John M. Browning. Designed for modern smokeless-powder ammunition, it proved to be by far the most practical and popular lever-action repeating rifle ever created. (Well over six million Model 1894s have been manufactured, and the pattern remains in limited production today.) This deluxe example boasts choice, figured walnut stocking with the pistol grip and forestock checkered in Style H. Other special features include a Marbles ramp-style rear sight, a Lyman tang-mounted peep sight, and double set triggers.

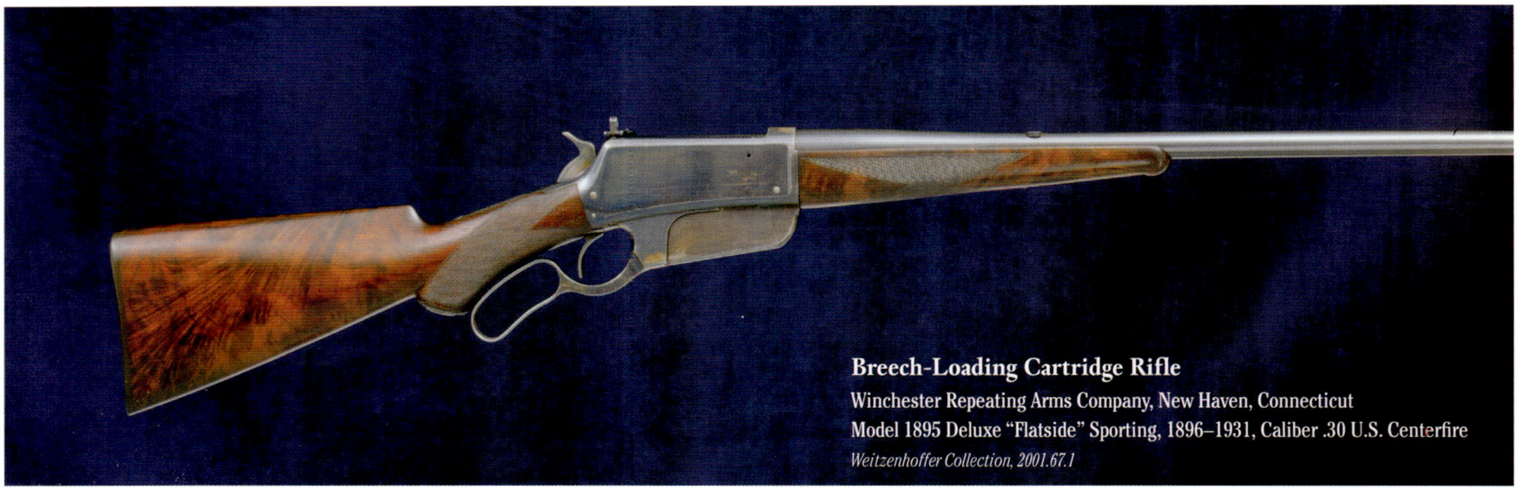

Breech-Loading Cartridge Rifle
Winchester Repeating Arms Company, New Haven, Connecticut
Model 1895 Deluxe "Flatside" Sporting, 1896–1931, Caliber .30 U.S. Centerfire
Weitzenhoffer Collection, 2001.67.1

The Model 1895 Winchester was the last large-bore, lever-action repeating rifle to incorporate the celebrated design skills of John M. Browning. A marked departure from earlier patterns, Model 1895s featured an innovative box magazine located under the receiver—instead of the familiar tubular magazine beneath the barrel. Because the model chambered a variety of high-powered smokeless cartridges, it, like the earlier Model 1886 Winchester, became a favorite of many big-game hunters. Produced in 1896, this specimen has an early, "flatside" (or unfluted) frame design—a rare and desirable variation among today's arms connoisseurs. It also features a rapid-taper barrel; a receiver-mounted, Lyman adjustable peep sight; and Style H–checkered, pistol grip buttstock and forestock of nicely figured, crotch-grain walnut.

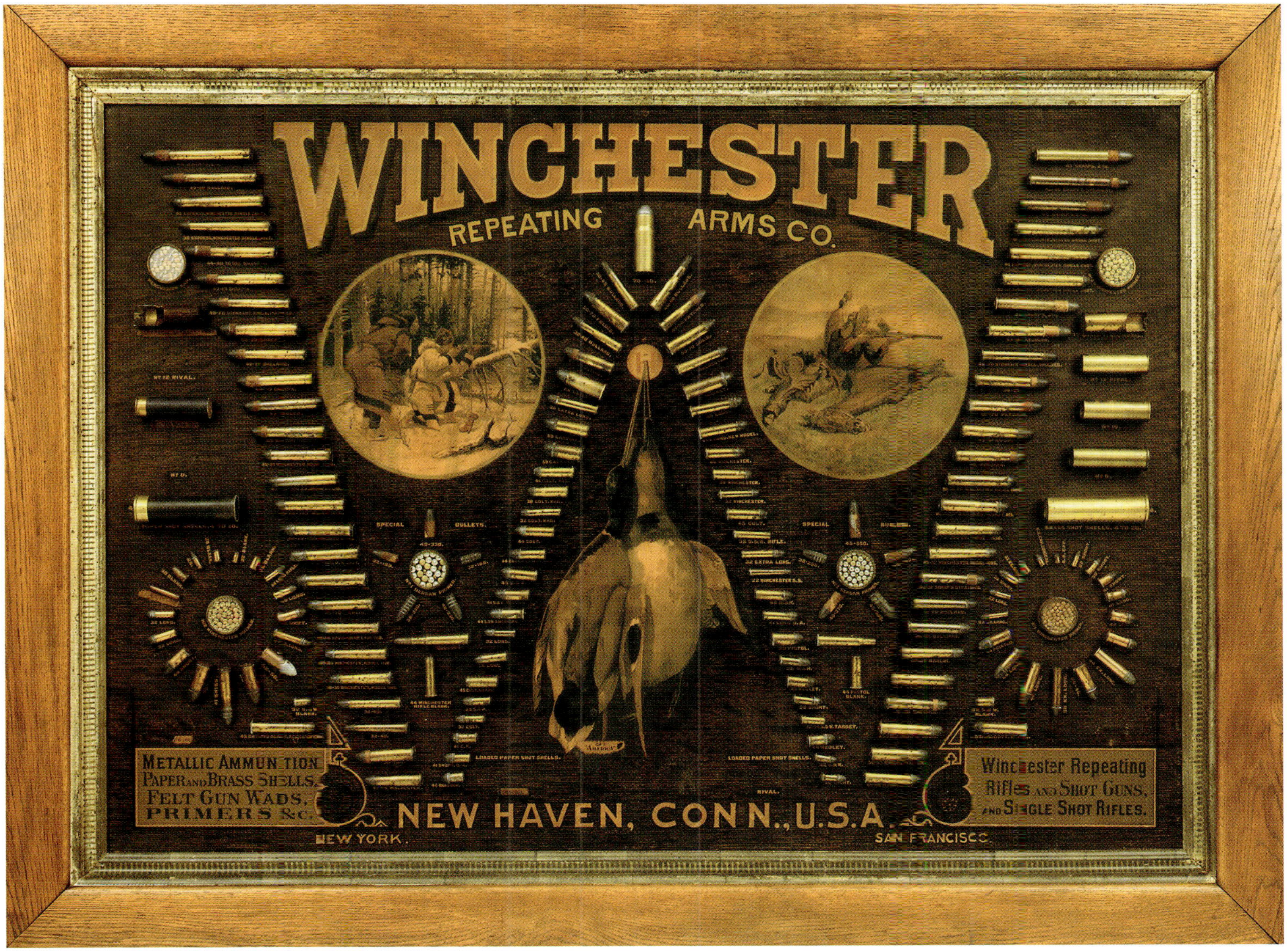

Promotional Cartridge Board

Winchester Repeating Arms Company, New Haven, Connecticut
"Single W" Pattern, 1890–1895

Museum Purchase, 1997.11

One in a series distributed by Winchester between 1874 and 1897 to advertise its line of sporting ammunition, such promotional vehicles appeared in sporting goods emporiums, gun shops, and hardware stores in which Winchester firearms also were sold. Usually referred to as the "Single W" or "Big W" board, this 1890-pattern presentation held the current panoply of 142 cartridges and shotshells with assorted bullets and primers. The board is decorated with hunting vignettes attributed to famed western artist Frederic Remington, while on the reverse is printed "*Presented With Compliments / of the / Winchester Repeating Arms Co. / Please Hang in a Prominent Place.*"

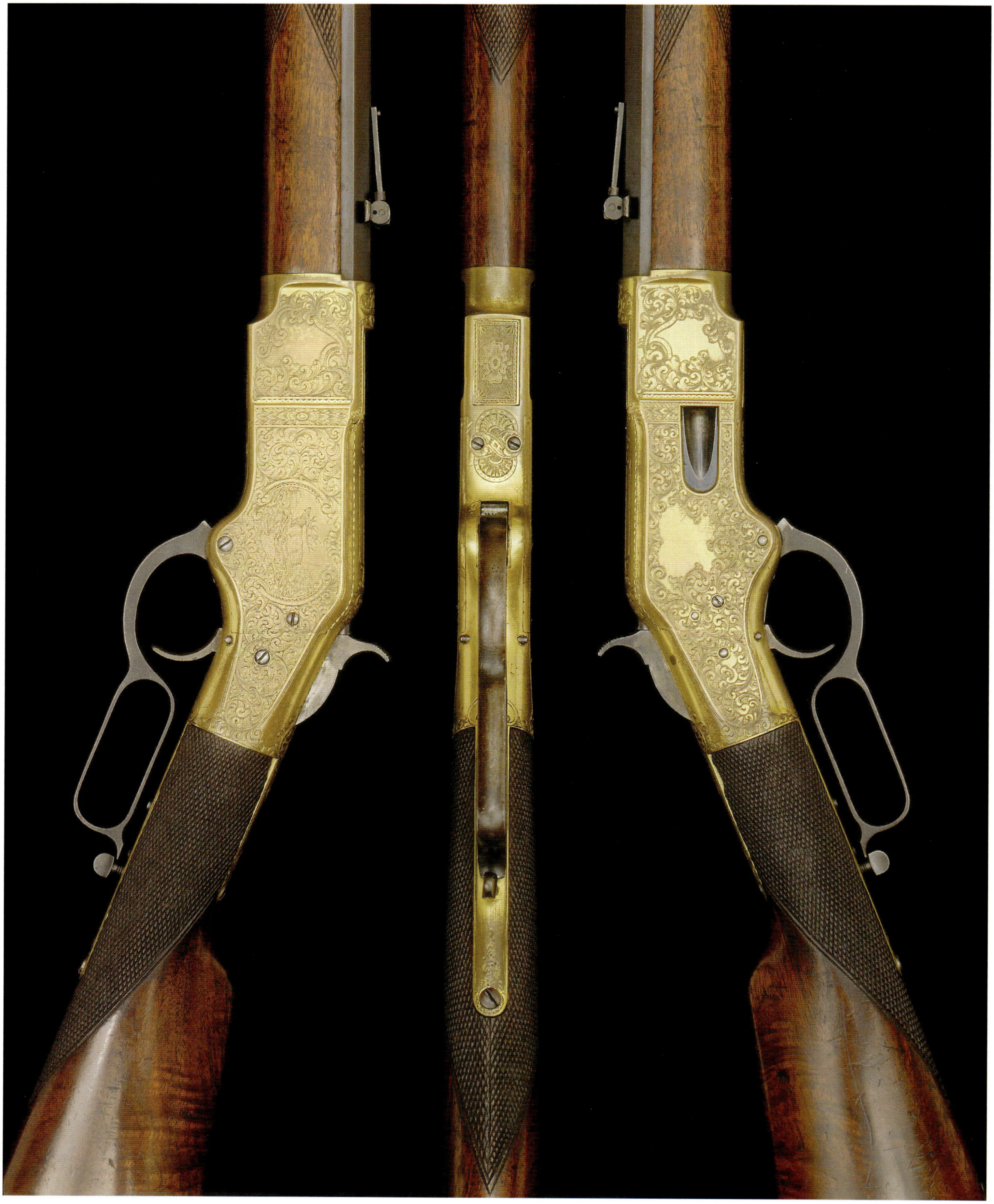

Deluxe Model 1866 Winchester rifle engraved by Louis Daniel Nimschke, 1868. For more information, see page 195.

Part III

From Functional Sculpture to Steel Canvas

Aesthetics in American Arms

Weapons have been an integral element of human culture from time immemorial. Such implements, whether intended for striking or piercing at arm's length or for projecting a deadly missile from afar, have undergone constant improvement in design and function since man first applied conscious intention to their creation. And in intentionally seeking improved function in his weaponry over the centuries, man has often, perhaps sometimes inadvertently, created implements possessing surprising elegance. Consider, for example, the recurved bow, the swept-hilt rapier, or the Kentucky long rifle—all weapons decidedly purposeful in intended function, yet all presenting an inherent grace of form, line, and proportion. Apart from their deadly fitness for the purposes for which they were conceived, each of these arms exhibits an appealing harmony of elements—an architecture—that, even in silhouette alone, may be appreciated as quite elegant in an aesthetic sense.

The further application of decoration, or embellishment, to an elegant arm usually raises it to an even higher aesthetic plane, setting it quite apart from the norm of the merely functional, albeit aesthetically pleasing, weapon. The application of decoration may run the gamut from mediocre to superb, but when practiced by a master artisan, it invariably elevates the object to the realm of decorative, sometimes even fine, art. Like the best of nineteenth-century American silver or furniture, so too with the best of domestic firearms—the refined and tasteful application of decorative enhancements oftentimes resulted in an appealing objet d'art.

The decoration of firearms via carving, checkering, chiseling, engraving, and inlaying extends back at least 450 years to the courtly weapons of the European aristocracy.[1] For several centuries such adornment remained the exclusive province of the titled and privileged (as did the weapons themselves), but in late eighteenth- and nineteenth-century America, where arms were ubiquitous and class distinctions comparatively moot, a finely embellished weapon could be acquired by anyone willing to bear the additional cost of the artisan's skilled touch. Over the second half of the nineteenth century, the unfolding democratization and increasing affluence of American society, the proliferation of machine-made arms, the establishment of a notable cadre of talented German-emigrant engravers, and the gradual development of a particularly American style of firearms engraving culminated in a remarkable golden age of domestic arms decoration. By the middle 1870s, certainly, Americans of varying social stature and economic means could share in the pleasure and pride of possessing a weapon that was both dependable and handsomely decorated.[2] By the 1890s, the finest decorated American arms easily vied with those of Europe.

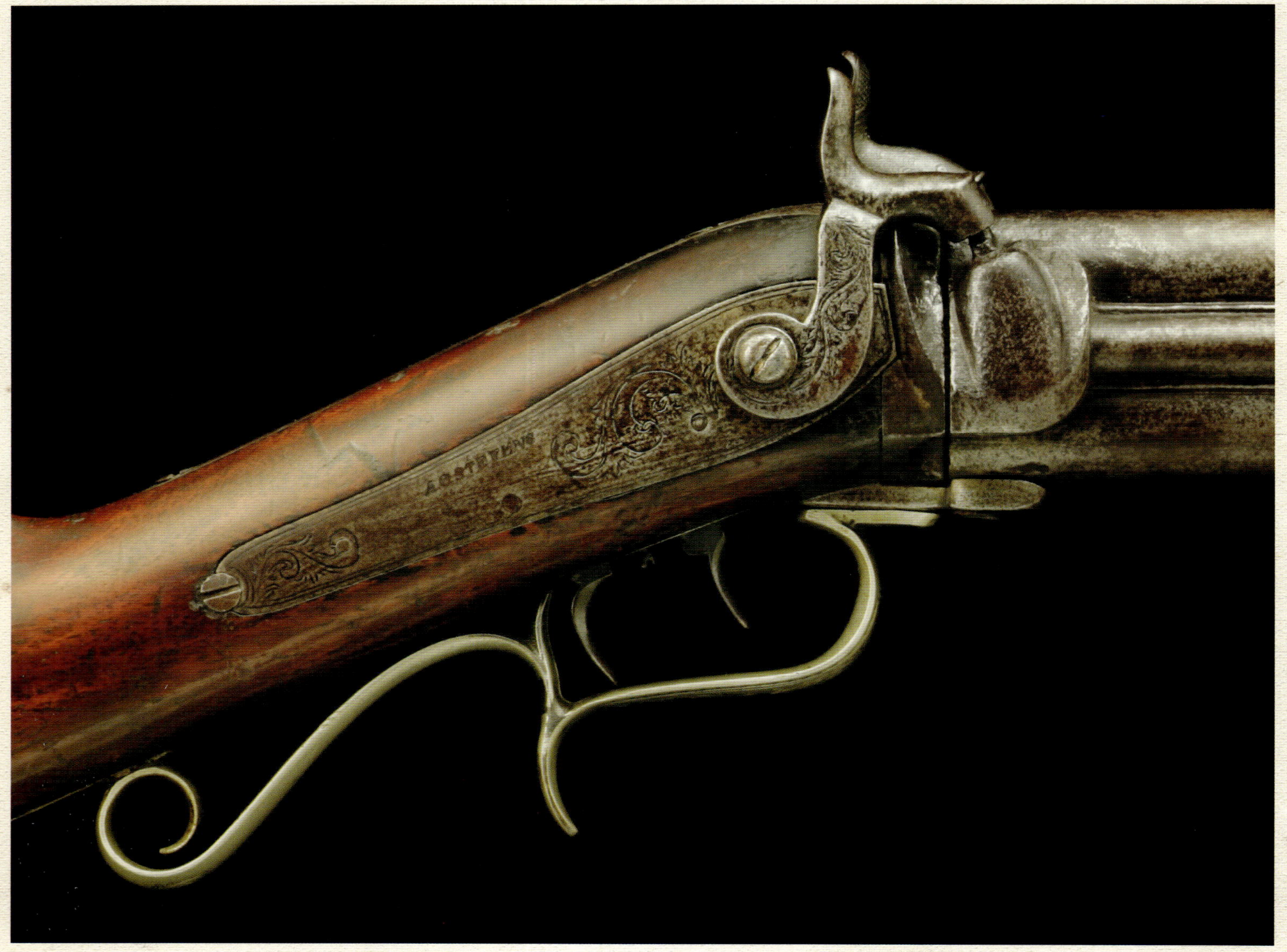

(*Detail*) Muzzle-Loading Percussion Rifle-Shotgun, A. C. Stevens, Hudson, New York, 1850–1860.

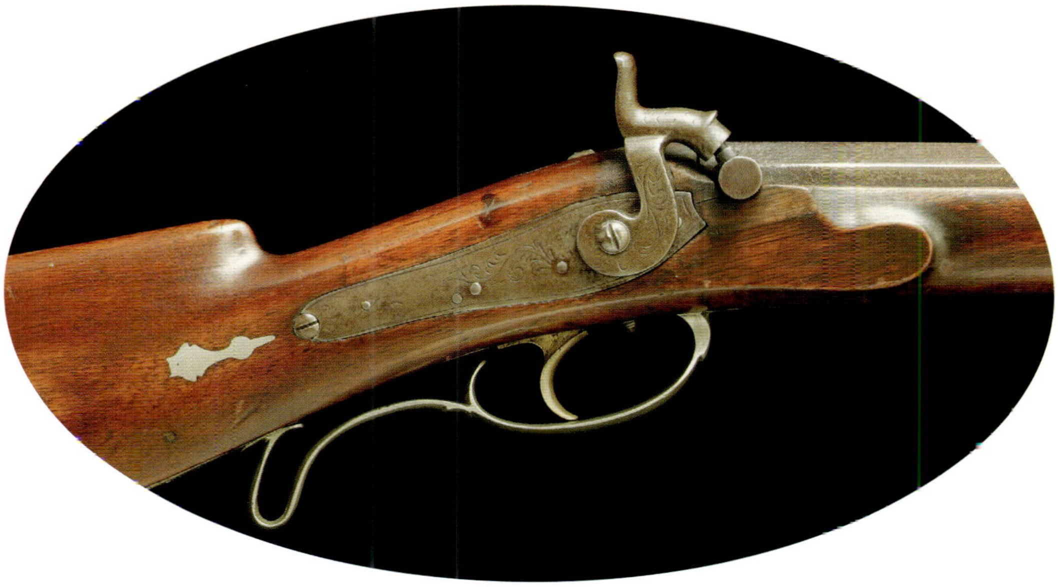

Chapter 9

MELDING FUNCTION AND FORM

IN THE REALM OF American arms, the Kentucky rifle perhaps best represents what scholar and critic John Kouwenhoven termed "vernacular" expression—those un-self-conscious forms or objects created by common people in an isolated environment that embody naturally satisfying patterns. Fashioned to meet specific needs in insular conditions, such objects typically manifest a simplicity not only eminently functional but also aesthetically pleasing. Although functionality certainly dictated the design evolution of the Kentucky rifle (as discussed in chapter 1), the vernacular expression of Kentucky riflesmiths also typically brought forth an attractive product. Even wholly unembellished examples of these distinctly American firearms exhibit an "intuitive sense of appropriate form" that approaches the sculptural in the harmony of their various elements and parts.[1]

Not a few arms historians and connoisseurs consider the Kentucky rifle a benchmark example of American vernacular expression in which technology and art are harmonized in the functional object. Firearms authority Merrill Lindsay, for example, has observed that

> Kentucky rifles derive their beauty from their severe adherence to functional line and proportion. The proportion to each other of the elements of the gun: the stock, the throat [wrist], the barrel length and fore end, as well as the amount of drop of the stock, . . . the straight, convex or concave profile lines of the stock, the thickness of the stock at the butt and the throat, . . . and finally the shape and extent of the cheek piece are summed up in one word. That word is *architecture*. . . . In the brief history of the European colonies in America, the Kentucky rifle may well be considered the outstanding indigenous art form.[2]

Likewise, in discussing the arm's patchbox, the dean of Kentucky rifle connoisseurs, Joe Kindig, Jr., has noted that this initially utilitarian feature "began

(*Detail*) Muzzle-Loading Percussion Kentucky-Pattern Sporting Rifle, Henry Deringer, Jr., Philadelphia, Pennsylvania, 1840–1850
Grandee Collection, 1991.01.1180

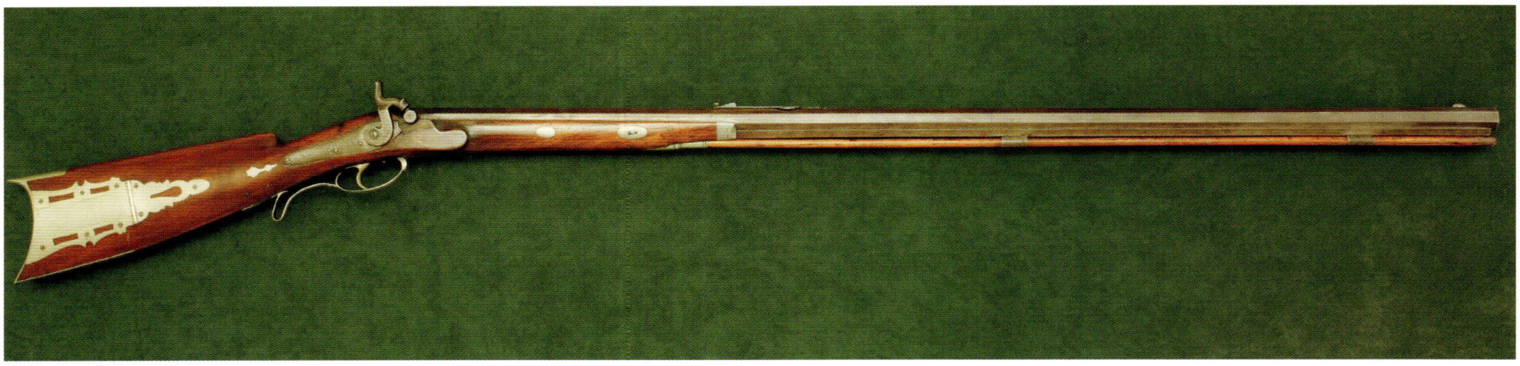

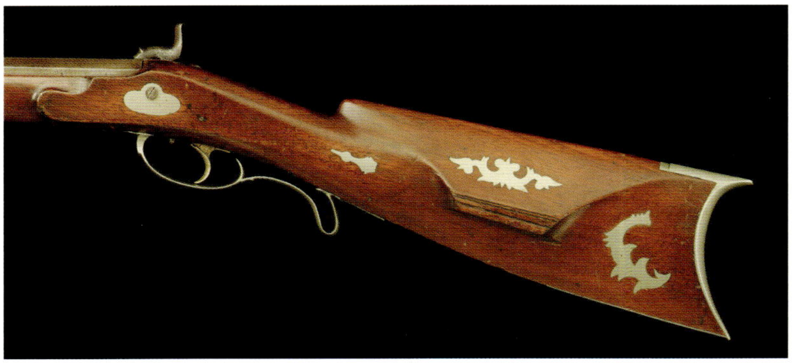

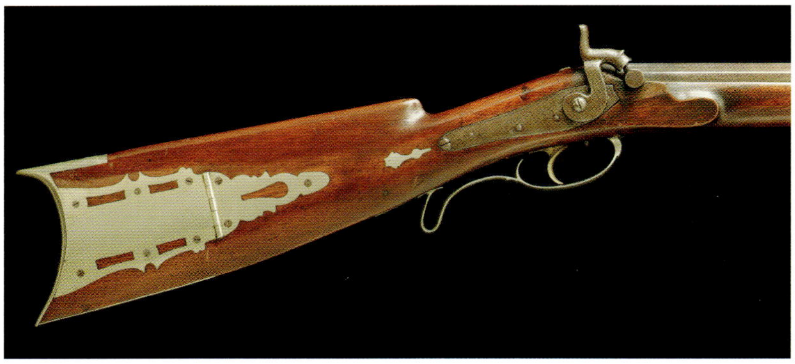

Muzzle-Loading Percussion Kentucky-Pattern Sporting Rifle
Henry Deringer, Jr., Philadelphia, Pennsylvania, 1840–1850
Caliber .48 Rifled Bore
Grandee Collection, 1991.01.1180

Perpetuating the graceful form and line of the Kentucky rifle tradition, this handsome, long-barreled, half-stock sporting rifle would have been prized by any well-to-do gentleman of the antebellum period. Well designed and crafted, the piece exhibits a tasteful application of embellishment, particularly in its finely proportioned, four-piece patchbox having nicely pierced-and-scalloped side panels and top piece. This and the other furniture and decorative elements of German silver, both functional and ornamental, exhibit a very precise fitting and inlay of metal to wood that is quite comparable to that found in the "best-quality" English sporting rifles and shotguns of the era. The meticulous artisanry lavished on this piece, clearly a special-order rifle, ranked near the apex of Deringer craftsmanship in this period.

to acquire a beauty through rococo design which made it the outstanding element of the Kentucky rifle and a purely American innovation worthy of being considered a work of art."[3] Perceived either in its various elements or as a whole, the Kentucky long rifle—even of middling quality—presents an inherently graceful expression.

The aesthetic merit of the Kentucky rifle lay not only in "appropriate form" but also in the craftsman-gunsmith's sensibility and skill in fitting and finishing—the "niceness" of detail with which the firearm was completed. At the hands of a master craftsman, the precise—gapless—fitting of all wood and metal elements; the superior polishing of all external surfaces; and the visually appealing application of harmonious color in barrel browning, lock case-hardening, and stock staining and polishing all contributed to aesthetic quality. (During the second half of the nineteenth century, major arms manufacturers such as Colt, Remington, Marlin, and Winchester achieved a remarkable vibrancy in color-case-hardening the frames and receivers of their arms, an aesthetic detail still much appreciated by contemporary connoisseurs.) It is unfortunate that few Kentucky rifles today exhibit their original fitting and finish, the majority having suffered two hundred years of wood shrinkage and surface wear. Then as today, however, the significant attributes of precision fitting and superior finish were understood and appreciated, contributing directly to a firearm's artistic quality and separating the truly fine from the simply serviceable.

Though seemingly anachronistic, machine-made arms produced under the American System also embodied an aesthetic merit oftentimes approaching the elegant. American sculptor and essayist Horatio Greenough

(1805–1852) was perhaps the first to publicly espouse the now-clichéd dictum that form follows function. Writing of American manufactures prior to the Civil War, he propounded the concept that the "unflinching adaptation of forms to function" resulted in a "majesty of the essential, without the trappings of pretension." This idea, which frequently fostered aesthetically pleasing design, applied not only to the clipper ships and trotting wagons that Greenough so admired but to many antebellum firearms as well. "It is neither the presence nor the absence of this or that part or shape or color that wins our eye," Greenough proclaimed in *Travels, Observations, and Experience of a Yankee Stonecutter* (1852); "it is the consistency and harmony of the parts juxtaposed, the subordination of details to masses, and of the masses to the whole."[4]

Applied to firearms, functionality dictated not only that a new or novel breech-loading or repeating mechanism operate properly but that the resulting weapon also handle, aim, and shoot well. The latter considerations often resulted in arms of pleasing—even attractive—form and line. Colt, for example, in seeking to reduce the weight and improve the balance and "pointability" of his various revolvers during the late 1850s, created a series of New Model arms notable for their pleasing proportion and sculptural silhouette. Starting with the rather chunky architecture of the existing Holster, Belt, and Pocket models, the heavy, slab-sided barrel-lug areas were completely recontoured, thinned, and rounded via milling machine. As well, the large-hinged loading levers were replaced with sleek rack-and-pinion designs of much more graceful contour. Finally, the rear diameter of the cylinders of the largest (Army) and smallest (Police) models were reduced in diameter (rebated), and flutes were milled between the chambers. (Cylinder fluting soon was discontinued on the large-frame, Army model to maintain structural integrity; it would reappear with the adoption of metallic-cartridge ammunition.) These thoughtfully designed improvements and alterations created a new generation of attractive Colt revolvers whose architecture today is admired by many connoisseurs and collectors as the ultimate in streamlined arms design in the percussion era. As one firearms historian recently observed, Colt's New Model revolvers of the 1860s "would not have been out of place had they been made during the Art Deco period."[5]

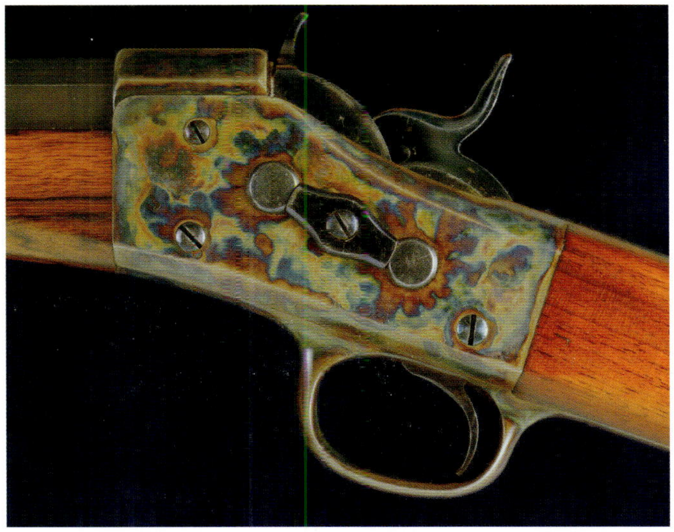

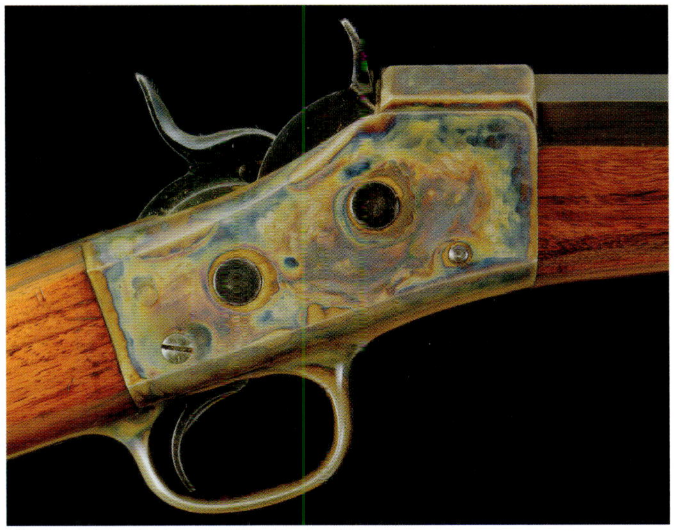

Breech-Loading Cartridge Rifle
E. Remington & Sons, Ilion, New York
No. 1 "Rolling Block" Sporting, 1868–1888, Caliber .38 Extra Long Centerfire
Weitzenhoffer Collection, 2001.676

In addition to the precise fitting of a firearm's metal and wood components, surface finish and coloration play a crucial role in a gun's aesthetic merit. Beyond the familiar, chemical- and heat-induced browning or blueing of iron and steel elements, the color case hardening of metal parts was the common means of surface hardening and corrosion protection in the nineteenth-century gun industry. The process involved packing and heating components in a sealed box with bone, charcoal, leather, or other carbonizing materials. The end result was an often attractive array of mottled and vibrant surface colors—which today hold special appeal among firearm connoisseurs. Though most century-old specimens are found with a faded or silvered finish, the receiver of this rifle still exhibits coloration as brilliant as when assembled. Gradations of surviving case coloring can be seen on other firearms throughout this work.

Aesthetic merit arising from functional design was hardly exclusive to Colt's firearms. Many domestic single-shot and repeating rifles, such as those produced by Sharps, Remington, Marlin, and Winchester, also present a pleasing clarity of outline in silhouette. In most of these, the

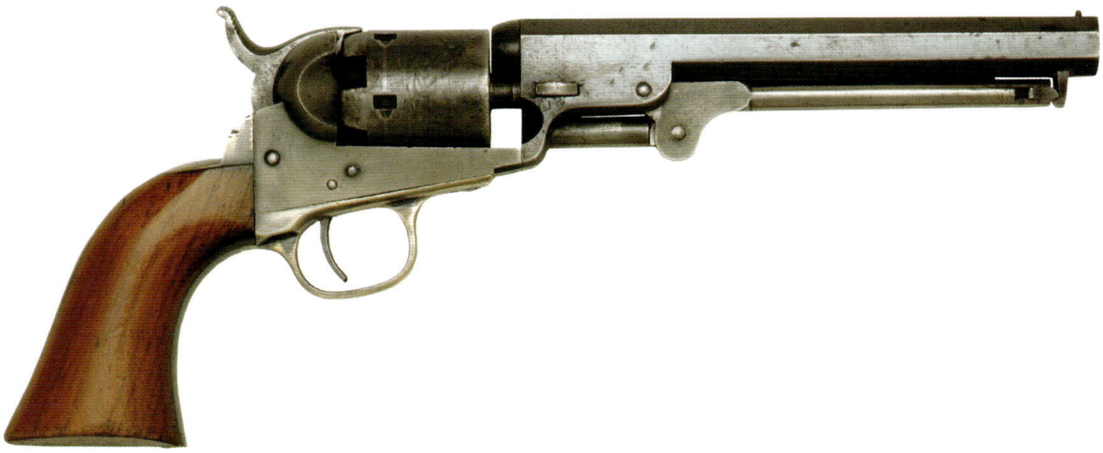

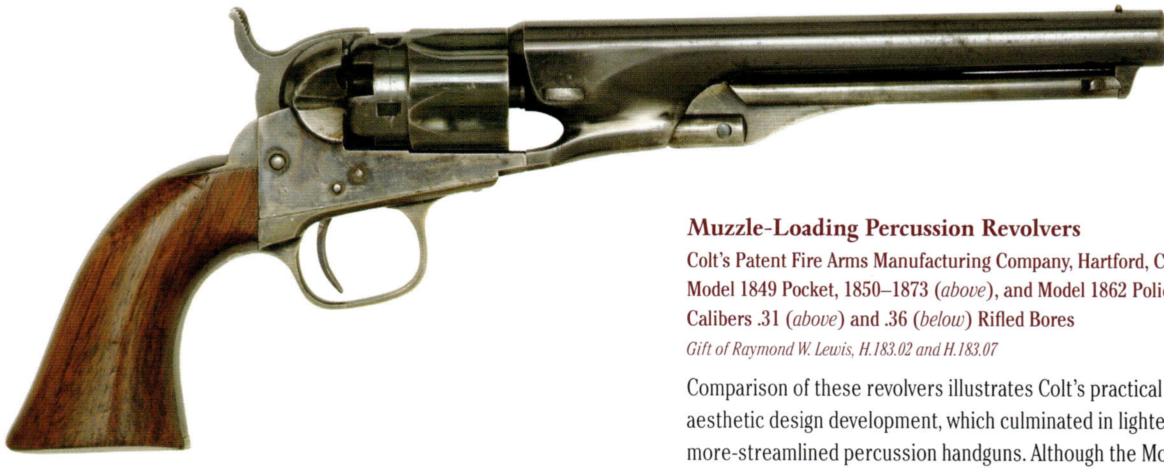

Muzzle-Loading Percussion Revolvers
Colt's Patent Fire Arms Manufacturing Company, Hartford, Connecticut
Model 1849 Pocket, 1850–1873 (*above*), and Model 1862 Police, 1861–1873 (*below*)
Calibers .31 (*above*) and .36 (*below*) Rifled Bores
Gift of Raymond W. Lewis, H.183.02 and H.183.07

Comparison of these revolvers illustrates Colt's practical as well as its aesthetic design development, which culminated in lighter, handier, and more-streamlined percussion handguns. Although the Model 1849 Pocket revolver remained in production throughout the percussion era, it clearly was superseded by the more elegant Model 1862 Police. The latter revolver's sophisticated machining of the recontoured barrel-lug area, its more functional and more graceful loading lever assembly, and its rebated and half-fluted cylinder embodies a superb example of "functional sculpture" in nineteenth-century firearms manufacture. In company with Colt's Model 1860 Army and Model 1861 Navy revolvers (illustrated in chapter 3), the New Model Police is esteemed by many arms connoisseurs as the pinnacle of handgun design in the percussion era.

advancing and receding plains of the barrel, receiver, and stock exhibit a smooth, flowing, and continuous expansion of mass from muzzle to butt that is altogether harmonious. Although some elements of the action, such as the hammer, lockplate, frame, trigger guard, or finger lever, may break the overall longitudinal flow somewhat, their potential interruption often is modified by their recurved design, or by surface rounding, scalloping, rebating, or beveling via artful milling and profiling. Additionally, the presence of available "special-order" elements—such as half-octagon/half-round and/or rapid-taper barrels, half magazines, and pistol-grip and schnabel-pattern stocking—oftentimes contributes directly to a rifle's ultimate aesthetic merit in regard to form and line.[6]

In sum, not a few nineteenth-century American firearms embody aesthetically pleasing qualities quite apart from any applied embellishment. Many reflect the characteristics noted by Henry Sharp in *Modern Sporting Gunnery* (1906): "Given a plain gun . . . innocent of the smallest ornament, it, nevertheless, should have a beauty of outline and proportion, for lacking which it would be a mere lump of iron and wood. In short, a gun without engraving that still appeals to the connoisseur—in order to stand his critical appraisement—must be of the finest order in design, construction and

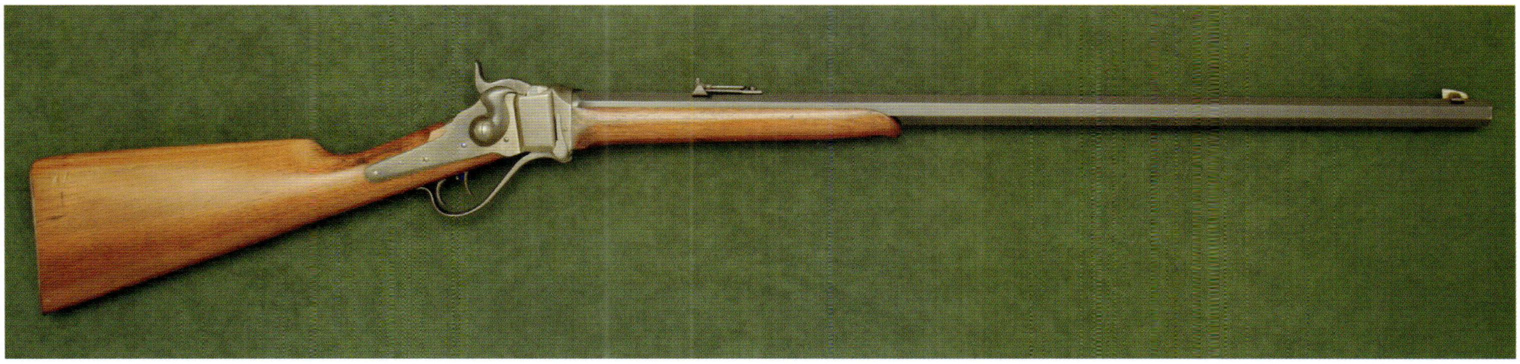

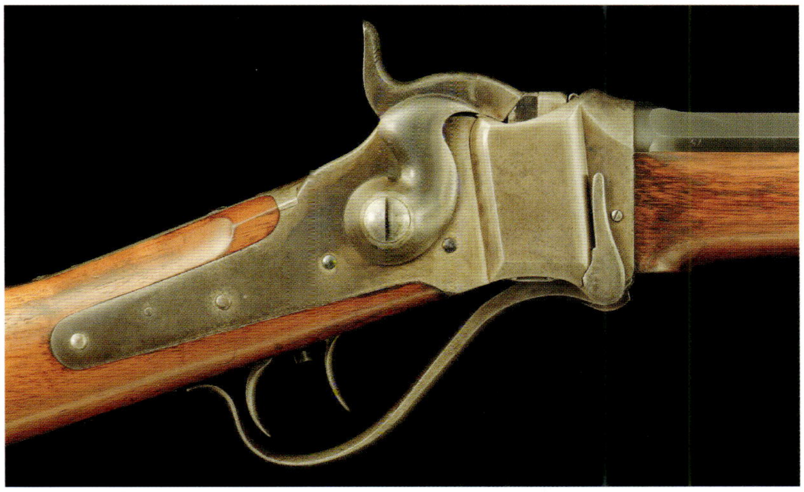

Breech-Loading Cartridge Rifle
Sharps Rifle Company, Bridgeport, Connecticut
Model 1874 "Old Reliable" Sporting, 1876–1881
Caliber .44-77 Sharps Necked Centerfire
Grandee Collection, 1991.01.1933

In addition to being a highly functional tool, this single-shot rifle presents pleasing aesthetic merit in form and line. Here, seemingly small details have some influence. For example, the transition from barrel to receiver is nicely smoothed by the curving scallop and chamfer of both elements in the breech area. Likewise, though it somewhat disrupts the flow of line, the hammer is pleasingly recurved and rounded, while the lockplate descends from the hammer in gentle recurves to meet the stock. (This particular rifle, recorded with a 30-inch octagonal barrel, double set triggers, and knife-blade front sight, shipped to Sharps dealer William Clemens at New Braunfels, Texas, in 1876.)

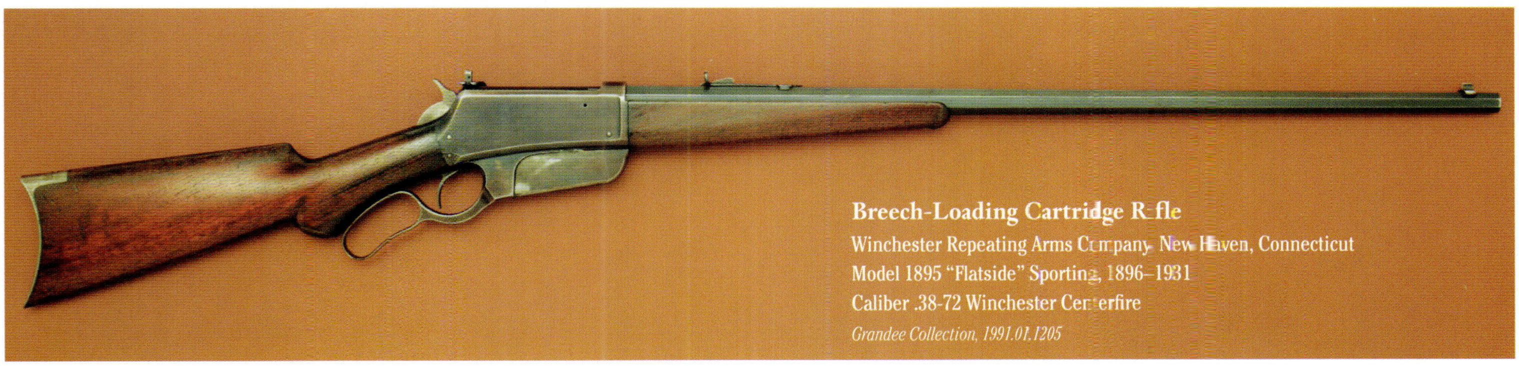

Breech-Loading Cartridge Rifle
Winchester Repeating Arms Company, New Haven, Connecticut
Model 1895 "Flatside" Sporting, 1896–1931
Caliber .38-72 Winchester Centerfire
Grandee Collection, 1991.01.1205

Even mechanically complex repeating firearms, such as this Model 1895 Winchester rifle, quite often present pleasing form and line. Here, various design elements contribute to a streamlined weapon that expands gracefully from muzzle to butt. These elements include a rapid-taper, full-octagon barrel; a schnabel-pattern forestock tip; a box magazine with a rounded lower-front line; a downward-curving pistol-grip stock and lever, and a traditional crescent-pattern buttplate that harkens back to the classic Kentucky rifle of a century earlier. This specimen also has the early, "flatside"-pattern receiver, a pattern soon supplanted by the standard-production fluted receiver (as illustrated at the end of the following chapter).

contour."[7] Whether of bench-craft or machine production, weapons such as the Kentucky rifle, the various Colt New Model revolvers, the Model 1874 Sharps sporting rifle, and the several Winchester repeating rifles all possess an intrinsic beauty of form and line. Though each, certainly, was conceived and produced in response to utilitarian and market considerations, they present in their differing designs a harmony of line, proportion, balance, and fitness that is pleasing, if not attractive, to the eye.[8] Such firearms, each possessing intrinsically fine handling characteristics and appealing visual design, well merit the connoisseur's valid recognition as "functional sculpture."

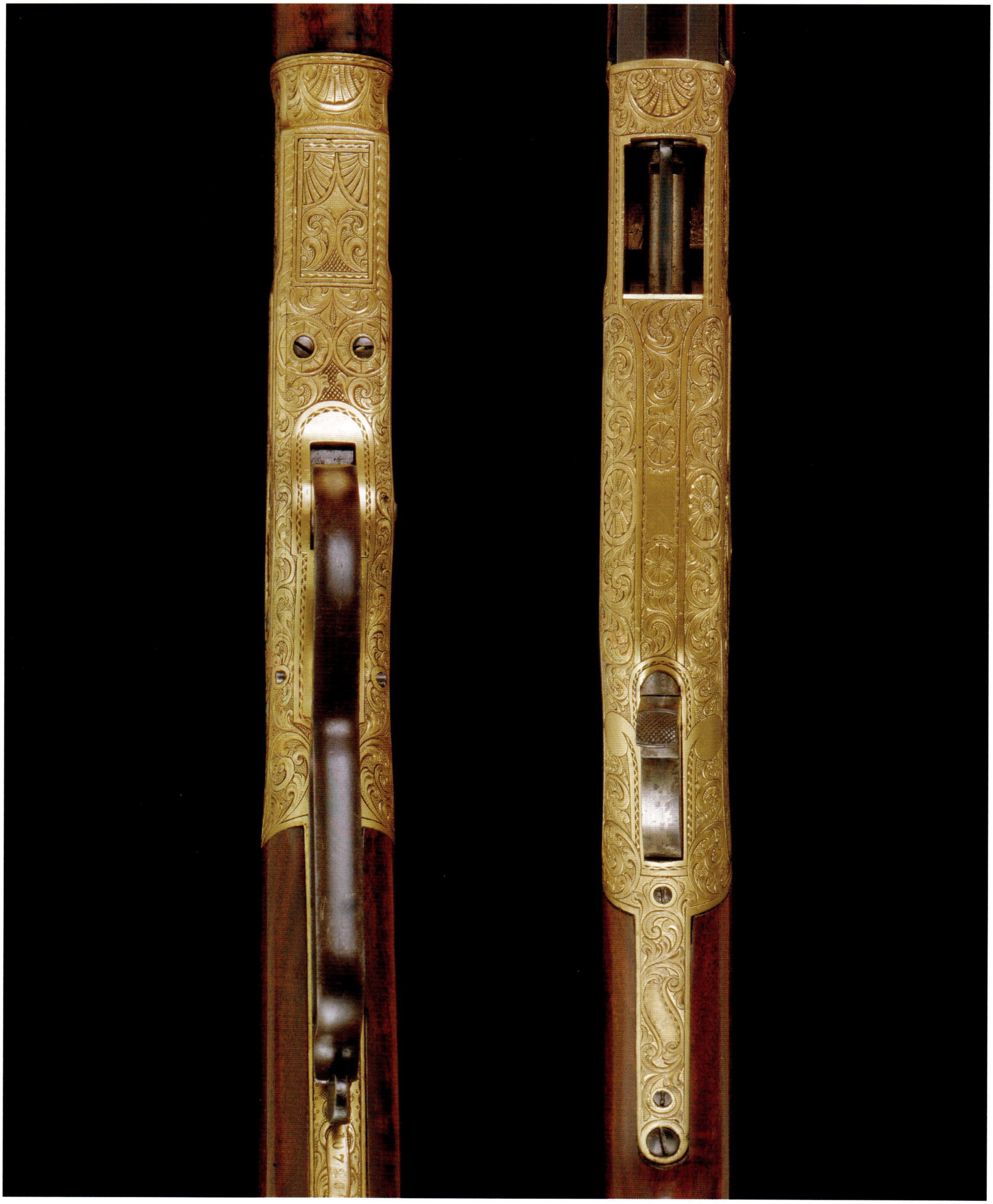

Deluxe Model 1866 Winchester rifle engraved by Conrad F. Ulrich. For more information, see page 196.

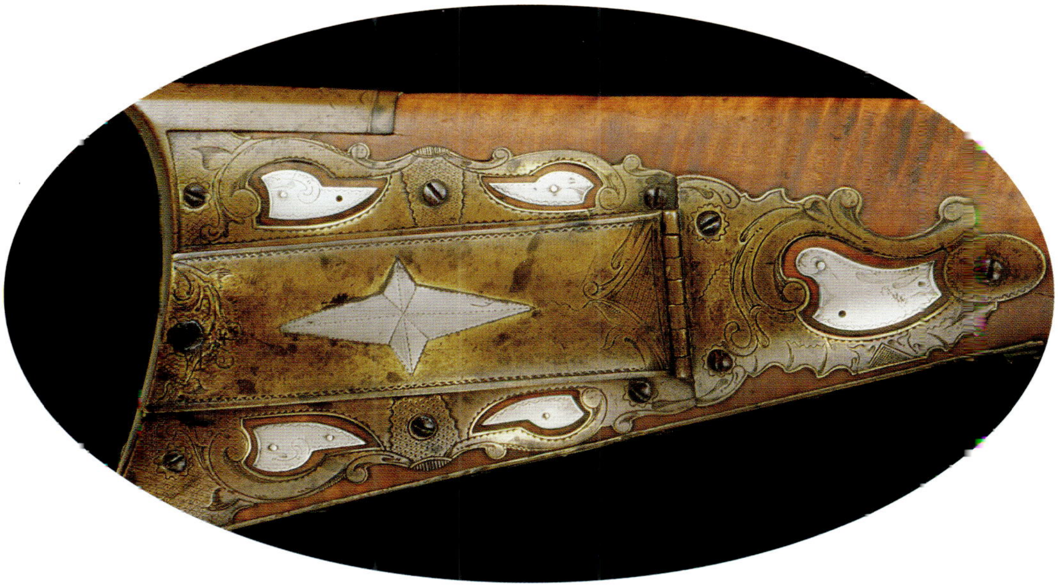

Chapter 10

EMBELLISHING THE UTILITARIAN

AT THE END OF THE eighteenth century, the Pennsylvania, or Kentucky rifle had attained a remarkable degree of functional and aesthetic perfection. The classic, or "Golden Age," rifles of the period 1785–1815 exhibited truly harmonious, rococo-style, C- and S-scroll, shell motif, and acanthus-leaf embellishment over their graceful relief-carved stocks and engraved brass or silver patchboxes. The finest of such "backwoods" stock carving and edge molding sometimes rivaled the best of carved Chippendale-style woodwork from the contemporary furniture-making shops of Philadelphia—then a leading center of American craftsmanship. The engraving on Kentucky patchboxes and lesser furniture elements, however, typically was limited to rather simple floral, shell, and C-scroll motifs, presenting a less sophisticated, sometimes primitive style. Silver or German silver figures inlet on the cheekpieces of the stocks might be lightly engraved, but these usually reflected rather naive religious talismans, folk-art symbols, or various renditions of the Federal-period American eagle. It should be noted that, except perhaps for the latter motif, these applied embellishments—unlike the rifles themselves—did not represent indigenous, "vernacular" expressions but instead reflected adoptions and adaptations of traditional stylistic influences brought from Europe.[1] (Though the museum holds no "Golden Age" Kentucky specimens, the Philip Creamer transitional rifle that is illustrated here does exhibit some classic Kentucky ornamentations in its stock molding and incised-line work and, most strongly, in its patchbox design and engraving.)

Both in form and in ornament, the aesthetic character of the Kentucky rifle commenced a gradual decline from the mid-1820s into the 1850s. With shorter barrels of lesser caliber, stocks also became shorter and thinner (sometimes almost slab sided), often compromising the graceful flow of line so notable in earlier

(*Detail*) Muzzle-Loading Percussion Kentucky-Pattern Rifle
Philip Creamer, St. Clair County, Illinois, circa 1820–1825.
Gift of William W. and James R. Rodgers in memory of Bill and Marguerite Rodgers, Blackwell, Oklahoma, 2001.56.1

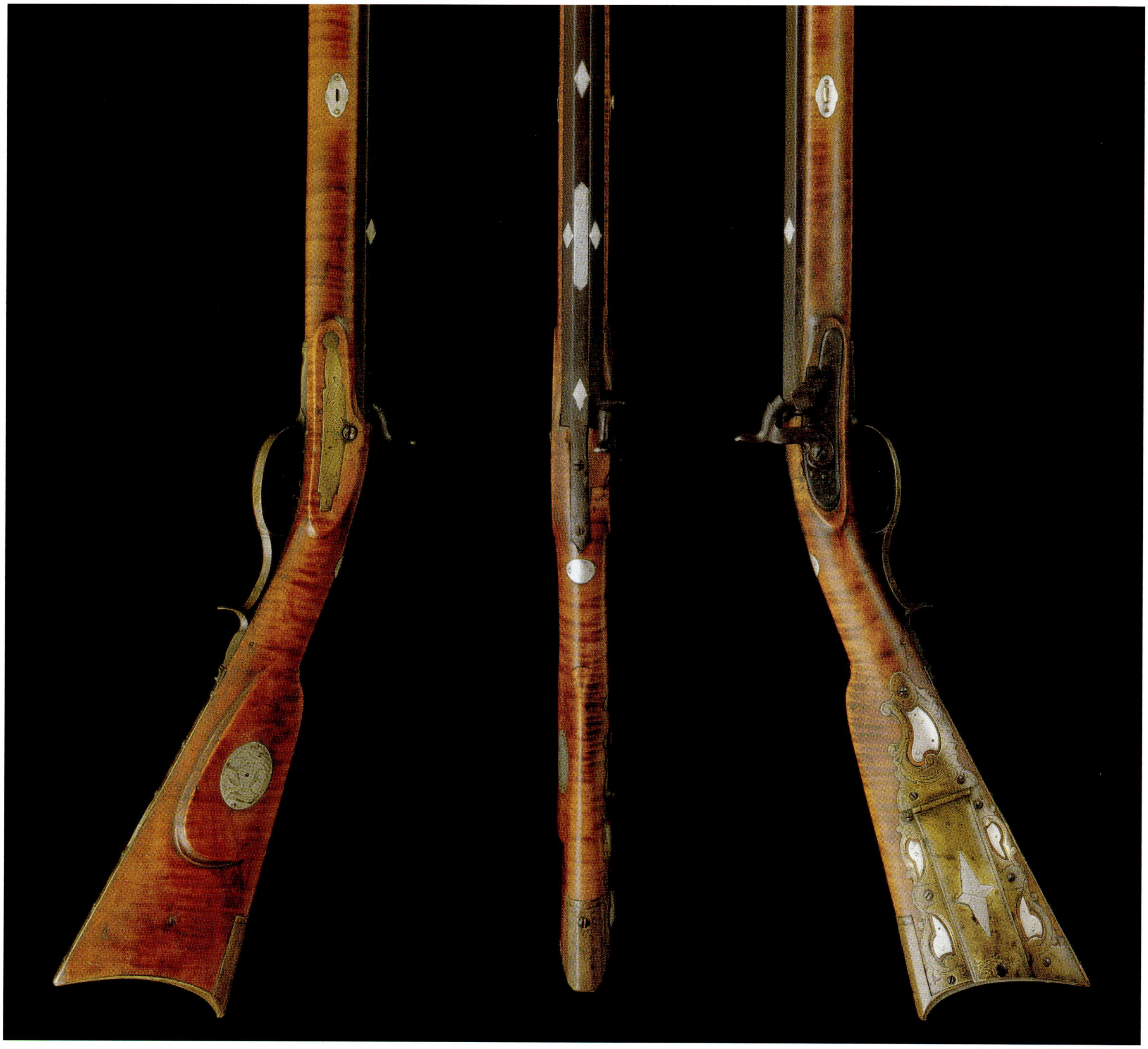

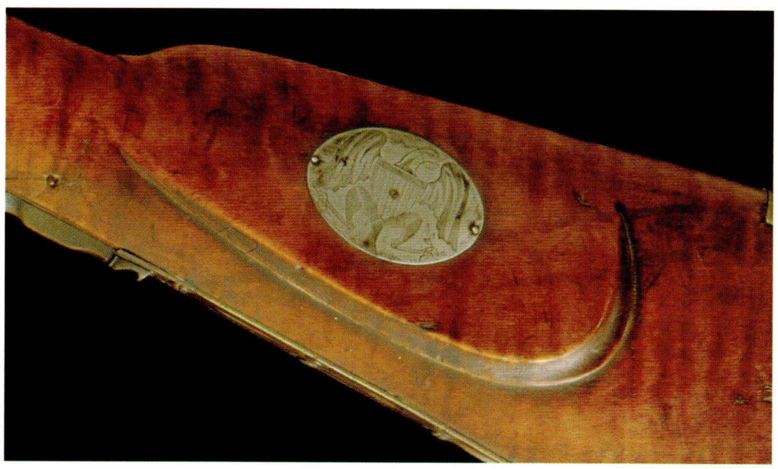

Muzzle-Loading Percussion Kentucky-Pattern Rifle

Philip Creamer, St. Clair County, Illinois, circa 1820–1825, Caliber .55 Rifled Bore

Gift of William W. and James R. Rodgers in memory of Bill and Marguerite Rodgers, Blackwell, Oklahoma, 2001.56.1

This transitional Kentucky-pattern rifle retains vestiges of classic, "Golden Age" stock carving in its decorative, incised-and-molded woodwork tracing the ramrod slot and in its raised-carved and edge-molded cheekpiece. The four-piece patchbox of pierced brass features an inlaid four-point compass or star of silver on the cover, while the piercings on the side and head pieces are further inlet with shaped pieces of silver. The patchbox periphery is patterned in the classic rococo style, and all elements also are embellished with foliate arabesque, crosshatched, and border engraving. Opposite, the silver oval in the cheekpiece is engraved with a classic, Federal-period American eagle. The rifle's full-octagon barrel is inlaid with silver diamonds or lozenges; the largest carries the maker's name in engraved script on the top flat.

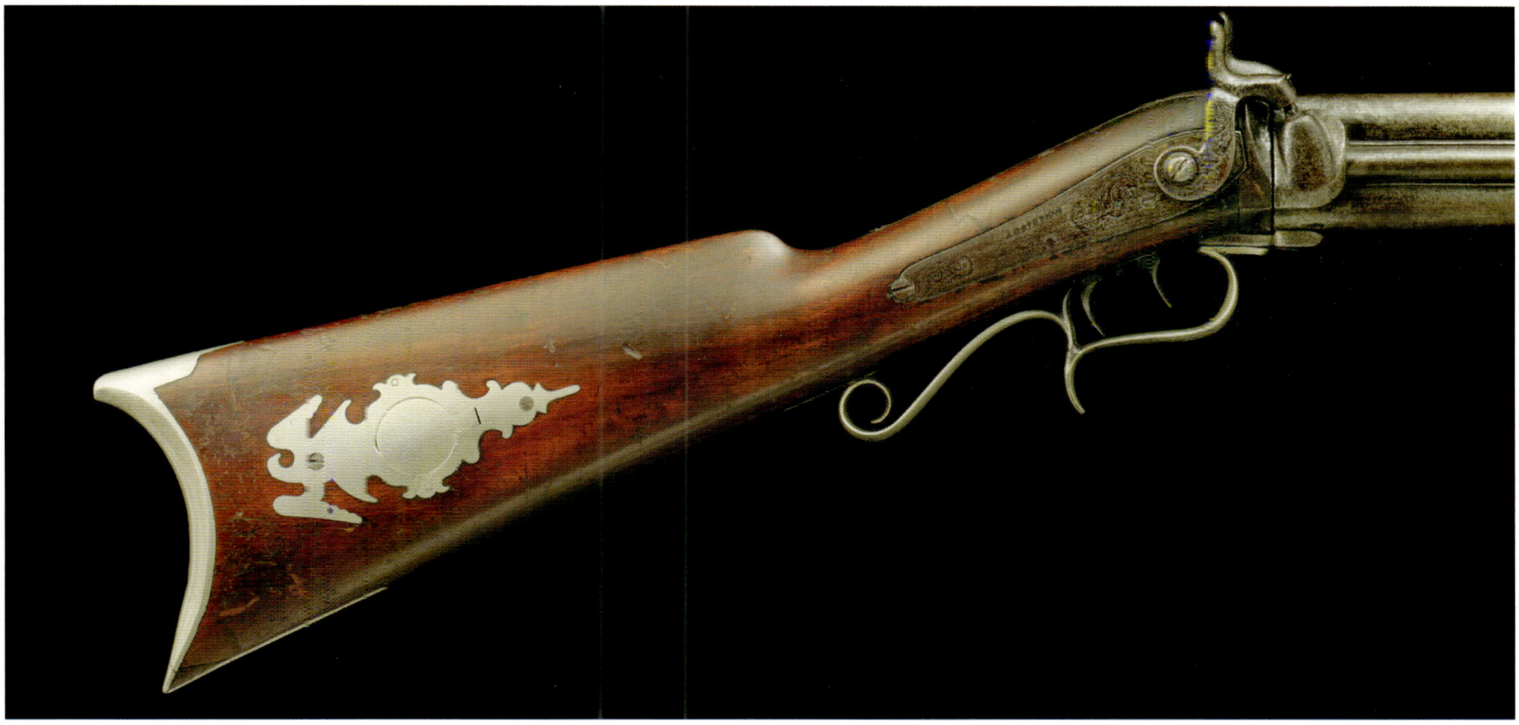

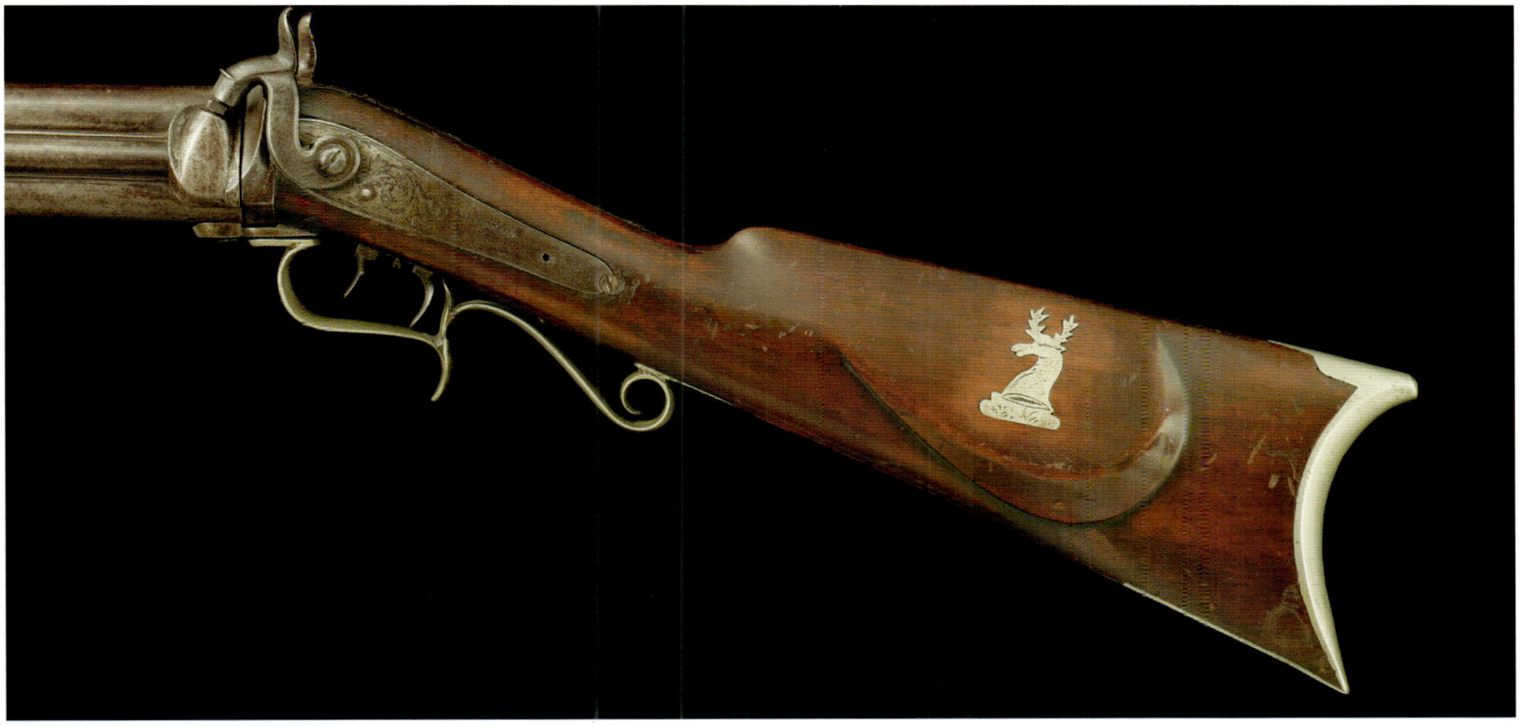

Muzzle-Loading Percussion Rifle-Shotgun (detail)

A. C. Stevens, Hudson, New York, 1850–186[illegible], Caliber .45 over 12 Gauge

Grandee Collection, 1991.01.2320

Illustrated in full in chapter 1, the buttstock of this combination rifle-shotgun displays typical ornamentation using figural inlays of German silver. Adding a touch of elegance, the right side features a decorative circular capbox surmounted with a rendition of the then-popular Federal eagle motif and finished below with a pendant finial. (The entire composition in general outline echoes the design of some domestic wall clocks and mirrors of the antebellum period.) The left side boasts a lightly engraved, stylized stag's head inlet in the oval-pattern cheekpiece, perhaps suggestive of the owner's hunting proclivities.

specimens. Stocking of plainer, unfigured maple (though often artificially striped) became more common. Classic rococo-style embellishment in stock carving faded away by 1820, to be replaced with minimal—if any—incised work. The ornately shaped and engraved, rococo-pattern patchbox, which personified the height of Kentucky rifle decoration, gradually disappeared with the widespread adoption of percussion ignition—to be replaced by the 1840s with smaller and plainer receptacles of round or oval pattern. Perhaps in compensation, later half-stocked Kentucky rifles often boasted a profusion of decorative brass or German silver inlays in various geometric and figurative motifs (among the latter, standing or reclining deer appear quite frequently). A variety of eastern and midwestern "sporting" rifles (that is, neither of strictly Kentucky nor Plains pattern) made during the antebellum decades also incorporated similar inlaid embellishments in both animal and patriotic motifs. Nearly all of this indigenous decoration remained the province of the local, self-taught gunsmith-engraver. The more exacting and exuberant work of the professional engraver would come to the fore in concert with the rise and adoption of machine production in the middle decades of the nineteenth century.[2]

By the 1840s, certainly, the engraving of a firearm's various metal elements had become the accepted hallmark of deluxe embellishment both in Europe and in America. The influence of the American System on arms production, which often resulted in lowered unit cost, allowed several domestic manufacturers to provide hand-engraved guns as stock items. Such decoration usually took the form of simple scroll or arabesque-pattern engraving on a clear field. Several pepperbox pistol manufacturers, for instance, marketed their machine-made weapons with a standard scroll decoration on their frames and/or nipple shields in the 1830s and 1840s. Samuel Colt, too, provided similar, rather minimal embellishment on some of his Paterson-made pistols, typically in the form of an open vine-style scroll. By the 1850s, an increasing number of skilled German-emigrant engravers brought their Old World design influence and metal-cutting talents to American arms decoration. Over succeeding decades, they developed a distinctly "American style" of engraving characterized by a lush and relatively bold, smoothly flowing scrollwork set off on a stippled, cross-hatched, punch-dot, or frosted ground that provided an illusion of depth. Various animal heads and figures at

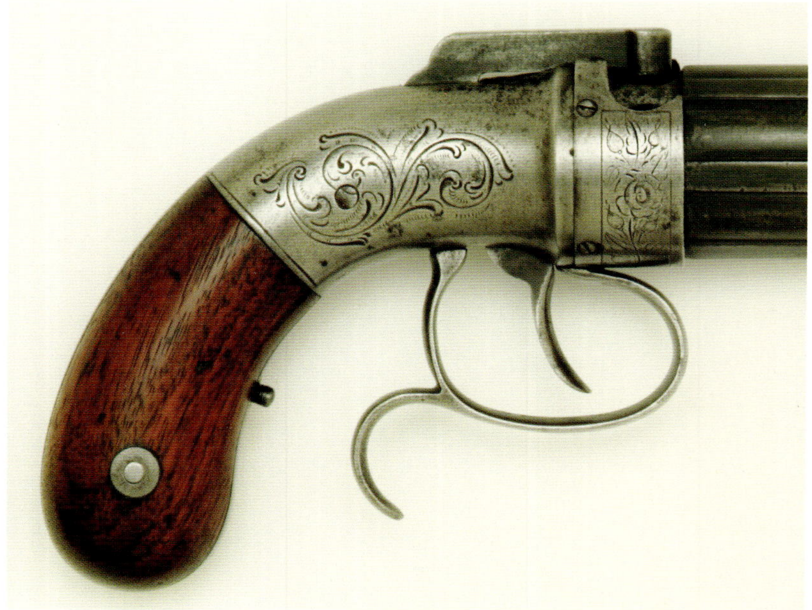

Muzzle-Loading Percussion Pepperbox Pistol (detail)
Allen & Thurber, Worcester, Massachusetts, 1847–1854, Caliber .36 Smooth Bore
Grandee Collection, 1991.01.3184

Illustrated and discussed in full in chapter 2, this repeating pepperbox pistol has a round-bodied frame that displays the typical "standard" engraving as provided on mass-manufactured American arms in the immediate antebellum decades. Rendered on a plain field, the foliate scrollwork is rather robust and fairly detailed, having deep-cut curved accent lines to lend form to the composition. A lighter floral engraving decorates the nipple shield below the hammer. Typically, such ornament then fell within the retail cost of the product, whereas later, more sophisticated work represented additional—sometimes considerable—cost to the purchaser.

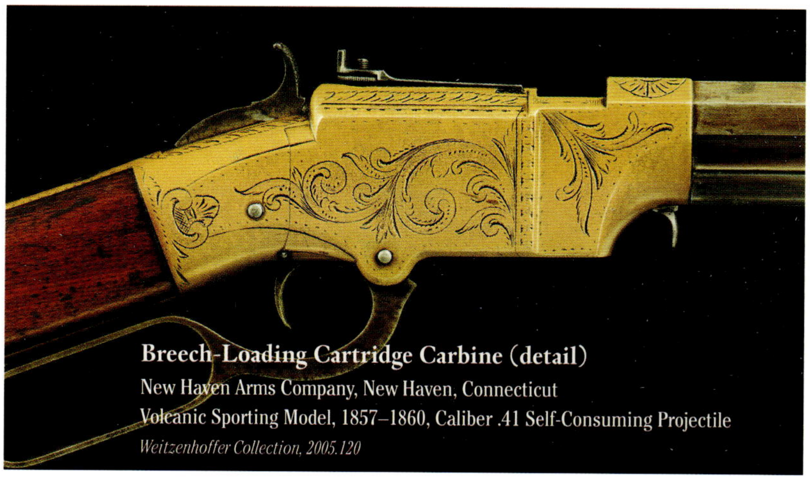

Breech-Loading Cartridge Carbine (detail)
New Haven Arms Company, New Haven, Connecticut
Volcanic Sporting Model, 1857–1860, Caliber .41 Self-Consuming Projectile
Weitzenhoffer Collection, 2005.120

Shown in full and discussed in chapter 8, this Volcanic carbine illustrates the gradual evolution of hand-cut engraving on American arms during the mid-nineteenth century. Here, the aesthetic treatment moves beyond the cursory, open-arabesque patterns of earlier single-shot and pepperbox handgun frames to include considerable border decoration over the top of the receiver and buttplate, as well as a more flamboyant foliate-scroll pattern covering most of the receiver sides. Though still rendered on a plain field, the execution presents some illusion of depth in its deeper cuts, feathered tendril ends, and chip-cut veining or shading. This was the standard engraving pattern seen on Volcanics, which proved more common in sales than did plain specimens. The embellishment possibly was rendered by contract engraver George Davis of Windsor, Connecticut.

first were artfully worked into the scrollwork and the scroll finials, but certainly by 1870 engravers had commenced rendering game animals and vignettes of the chase in full—either on plain open fields or in panels defined by ornamental borderlines or wide "ribbon" borders.[3]

As with the majority of Kentucky rifle embellishment, such decorative work rarely was an indigenous, or "vernacular," expression. Rather, it drew on, and elaborated, a long tradition of accepted European forms, many of which could be referenced in various pattern books (such as the series published by Gustave Ernst of Zella, Thuringia, Germany, 1839–53) that presented various styles and motifs of engraving for gun decorators. Although some students of American gun engraving have asserted a Middle Eastern origin for most of its stylistic attributes, such is certainly not the case. Even a cursory review of Owen Jones's classic, *The Grammar of Ornament* (1856), clearly demonstrates that the decorative elements found in firearms engraving draw on a multitude of cultural sources generated over several millennia. Persian ornamentation did indeed introduce floral and animal motifs, but these—along with Grecian wave and key borders, Roman acanthus leaf and scroll motifs, and Gothic-inspired arched borders and floral scroll and animal head finials—were all subsumed in the often-sumptuous expressions of Renaissance decorative artwork. There, the proportional arrangement and sheer exuberance of ornamentation reached a zenith that later influenced French and German interpretations in the seventeenth and eighteenth centuries and, in turn, impacted English and American styles during the nineteenth century.[4]

American firearms manufacturers began promoting their products at domestic trade fairs and European exhibitions in the 1840s and 1850s. In addition to displaying their most advanced designs, the principal firms also presented specimens of the most lavish character, employing the applied arts of engraving and inlaying in metal and carving and incising fine woods, ivory, and pearl, to heighten the appeal and prestige of their products. (The plating of a gun's metallic parts with silver, gold, and—somewhat later—nickel also constituted an important element in firearms decoration.) Colt and Winchester overshadowed their competitors in the provision of deluxe and embellished pieces at these special exhibitions. At the great Philadelphia Centennial Exhibition in 1876, for instance, these two manufacturers displayed some three hundred and two hundred arms, respectively, dozens of which were extravagantly ornamented. As with their dominance in production, sales, and brand recognition, both of these companies made and marketed many more highly finished arms for the buying public than did their rivals. The number of decorated handguns created by Colt over the second half of the nineteenth century dwarfed the output of Remington and Smith & Wesson combined, while Winchester far overshadowed Remington, Sharps, and Marlin in the creation of embellished longarms.[5]

Among mid-nineteenth-century American arms makers, none excelled Samuel Colt as a consummate promoter and salesman. Ever the marketing innovator, Colt introduced "engraved" panoramic scenes on the cylinders of his various repeating weapons as standard adornment at his Paterson factory during the later 1830s, and he continued the practice throughout the percussion era on all models having unfluted cylinders. Less expensive—and far less time-consuming—than traditional hand-engraving, this unusual industrial process involved the machine application of detailed, often-quite-attractive scenes created by master banknote engraver Waterman L. Ormsby (1809–1883). Ormsby also invented and patented the extraordinary roll-die engraving device, or "grammagraph," which made repeated impressions from a singular die. A dozen different tableaus presented artistic interpretations of Colt's firearms in dramatic use, and two of them portrayed historic incidents involving the Republic of Texas (the Colonel's first appreciative market): the naval victory against Mexico off Campeche, and Ranger captain John C. Hays's later-famed "Big Fight" against marauding Comanche Indians. The decoration proved sufficiently appealing that, around 1852, the Colonel also distributed a handsome broadside advertisement portraying three of the most artful and popular of Ormsby's tableaus. Several rival handgun firms, such as the Bacon Manufacturing Company, the Manhattan Fire Arms Manufacturing Company, and the Metropolitan Arms Company, followed Colt's lead with roll-engraved cylinder scenes during the 1850s and 1860s. Others, such as the Massachusetts Arms Company, embellished their revolver cylinders via decorative acid etching rather than die impression.[6]

Beginning in the early 1850s, Colt also initiated the regular practice of providing custom, hand-cut engraving

Colt Cylinder Scenes

A selection of cylinder scenes impressed on Colt percussion and conversion firearms

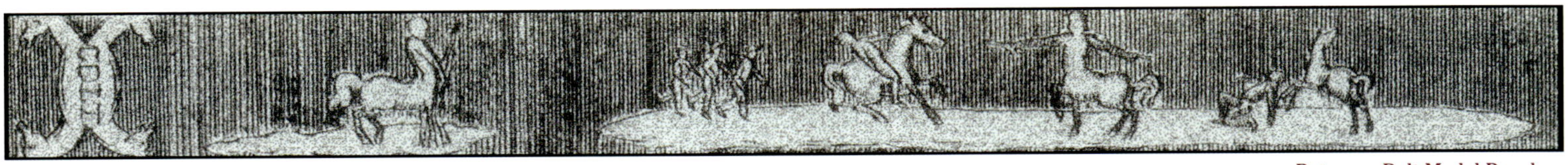

Paterson Belt Model Revolver

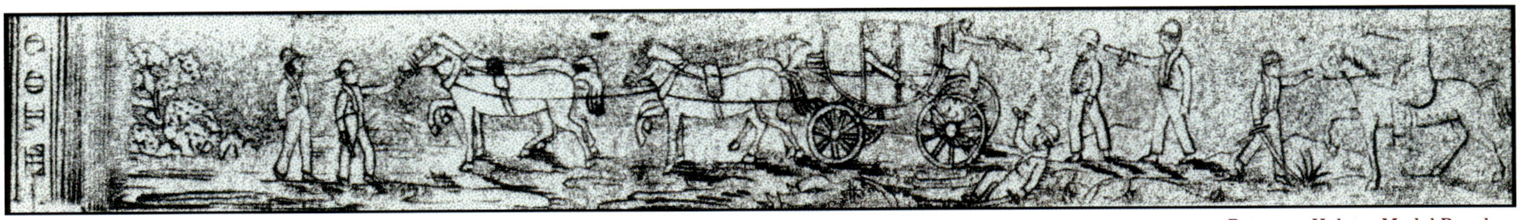

Paterson Holster Model Revolver

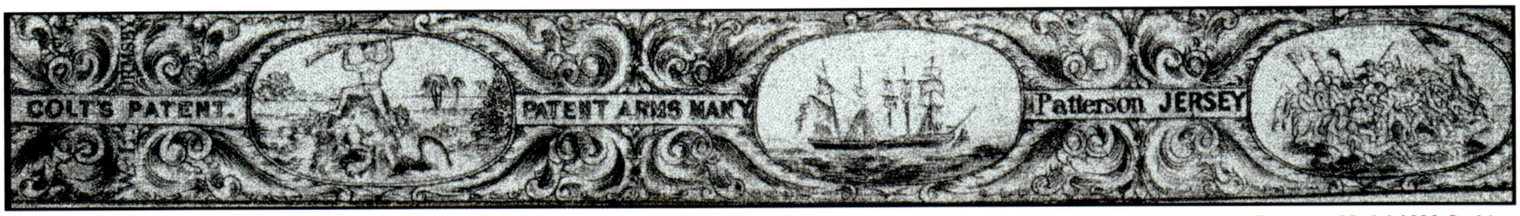

Paterson Model 1839 Carbine

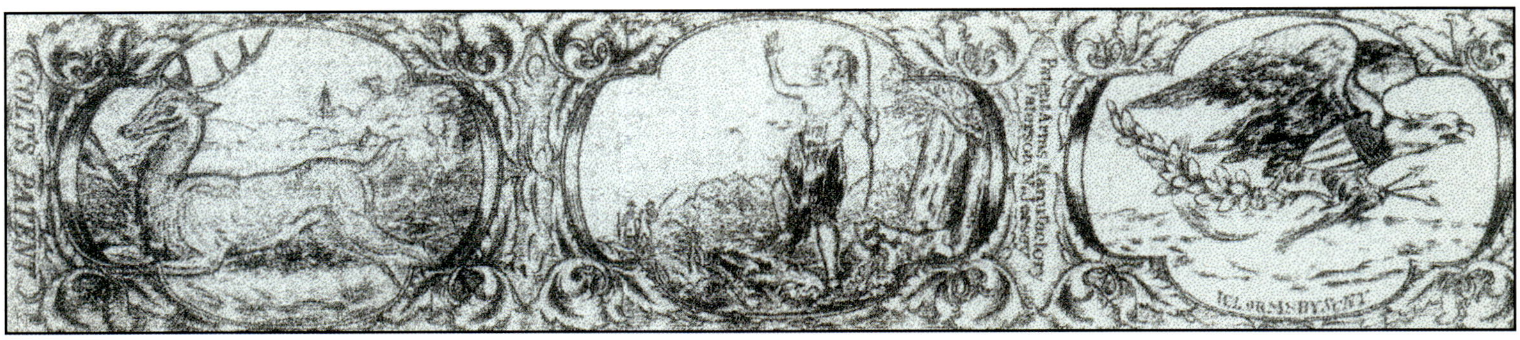

Paterson Model 1839 Shotgun

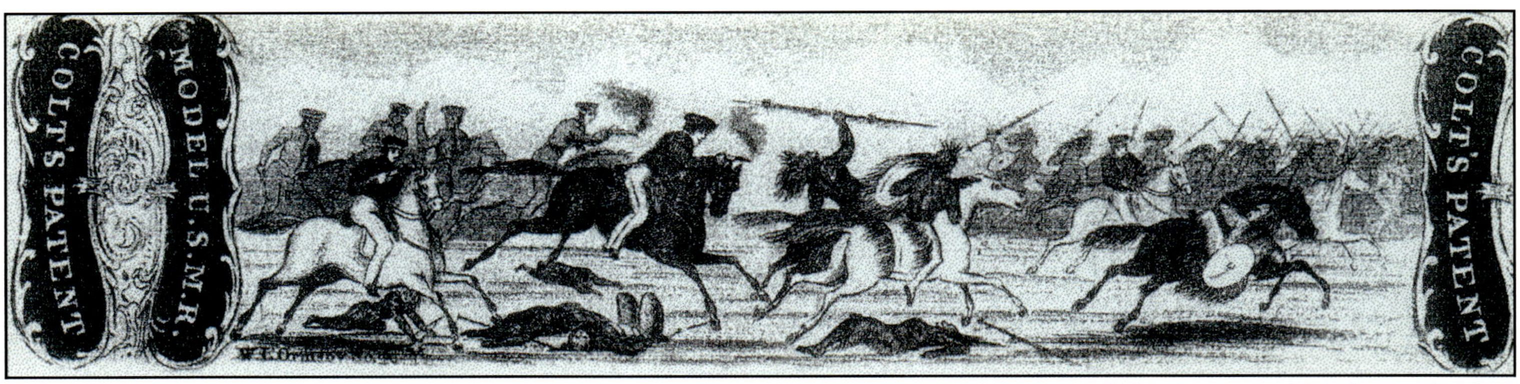

Walker, Dragoon, and Baby Dragoon Revolvers

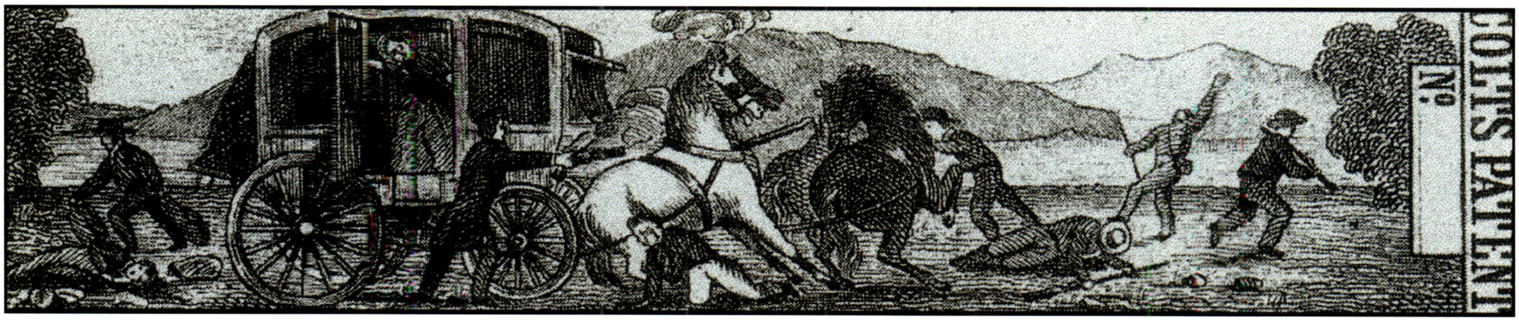

Model 1849 Pocket, 1855 Root, and 1862 [1865] Pocket Navy Revolvers

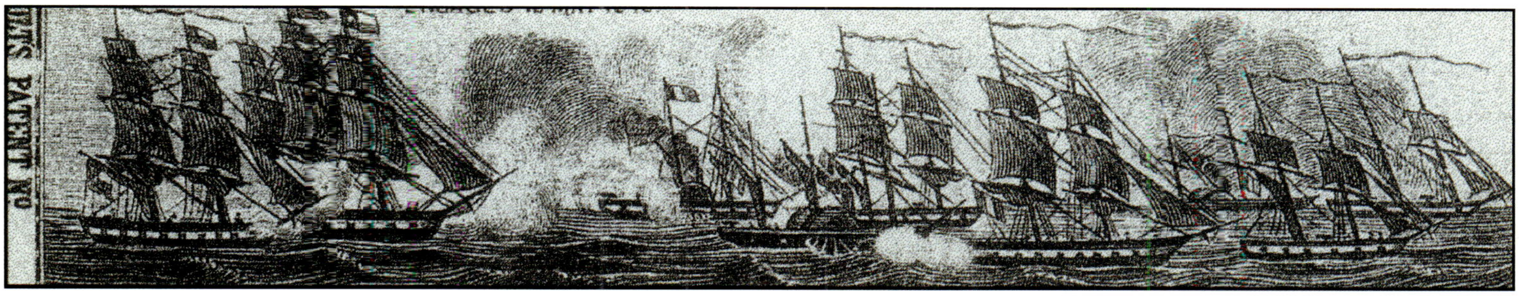

Model 1851 Navy, 1860 Army, and 1861 Navy Revolvers

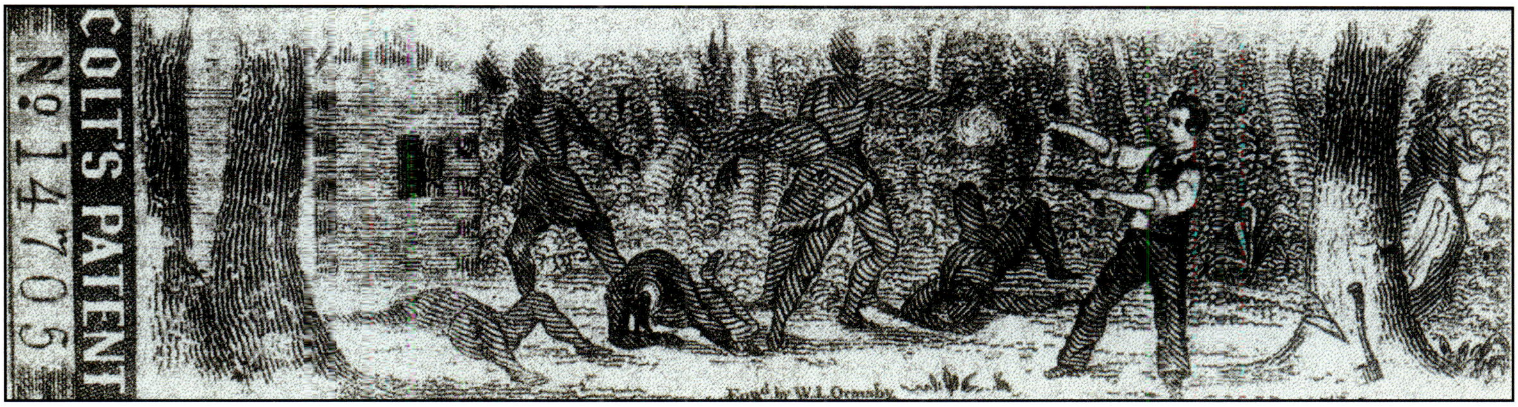

Model 1855 Root Revolver with Octagonal Barrel

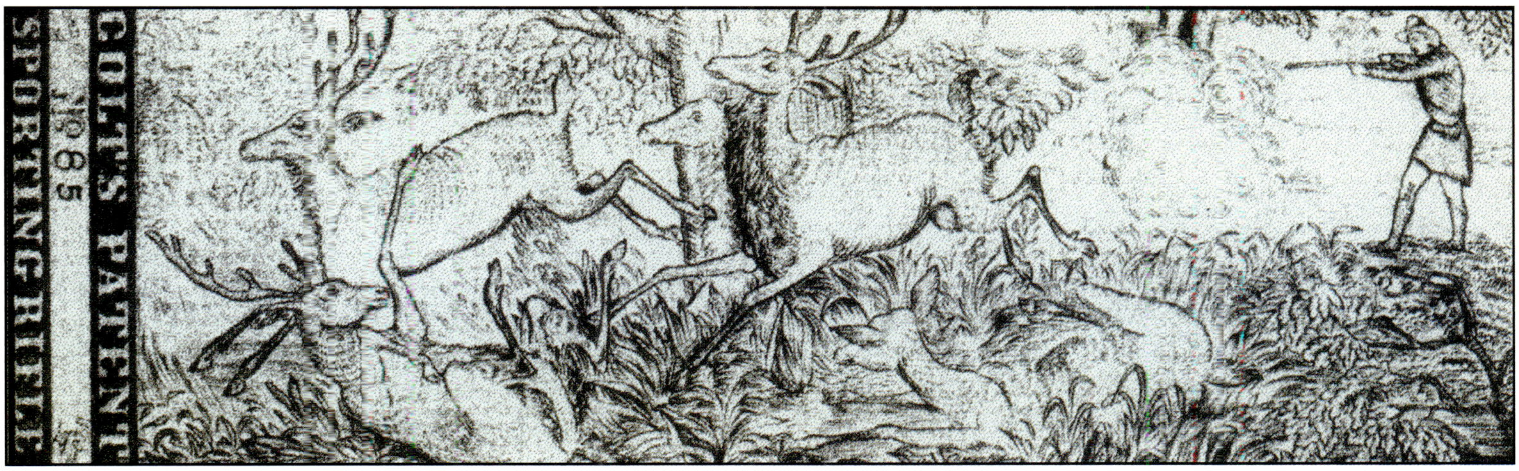

First Model 1855 Root Sporting Rifle

Colt cylinder-scene images reproduced from The Book of Colt Firearms *by R. L. Wilson with the permission of Mr. S. P. Fjestad, publisher, Blue Book Publications, Incorporated*

for his more discriminating clientele, transforming the utilitarian firearm into a lavishly finished objet d'art. His early presentations of deluxe firearms to distinguished and influential personalities ultimately made Colt firearms symbols of celebrity throughout the Victorian age, and they created an aesthetic standard emulated by other domestic arms makers. For their finest work, virtually all of the major manufacturers relied on highly skilled, German-emigrant engravers, die cutters, and carvers who labored as freelancers, as contractors, or as in-house employees. Leading firms such as Colt, Smith & Wesson, Winchester, and Marlin utilized familial dynasties of skilled engravers and carvers—notably, among the Young and Ulrich clans—to execute their "highly finished" arms. These master artisans added a luxurious aesthetic to the mechanical intricacy and machine-made precision of late nineteenth-century American firearms.

Among freelance artisans in the second half of the nineteenth century, none proved more prominent—or prolific—than Louis Daniel Nimschke (1832–1904) of New York City. Over a span of fifty years, he embellished weapons for more firearms manufacturers, individual gun makers, dealers, and singular clients than any other American engraver, perhaps ornamenting as many as five thousand pieces—or an average of one hundred a year! In varying number, all the famed marques of the era, including Colt, Remington, Smith & Wesson, Sharps, Marlin, and Winchester, benefited by his artistic attention. Nimschke was known for his well-designed and superbly executed American scrollwork, but he was equally adept at tighter and more-intricate English scrollwork and at the rendition of in-scroll animal heads, crests, monograms, as well as larger panel scenes of big game, the chase, or classical and patriotic motifs. Assiduously copied by numerous—still anonymous—local contemporaries, engraving after Nimschke's distinctive American-scroll pattern today is often generically referred to as the "New York" style. Unlike most of his copyists, however, Nimschke actually signed (or more often initialed) an appreciable number of his works.[7]

Colonel Colt's stable of in-house and contract artisans included the renowned Gustave Young (1827–1895), who certainly ranks among America's premier nineteenth-century firearms engravers. As an indisputable master of precision, intertwining scrollwork and gold inlay, Young rendered some of the finest ornamentation of the time. His animal-head, in-scroll finials became a notable trademark of his hand, and he produced superlative gold-inlaid handguns for both Colt and Smith & Wesson. Young did most of his work for Colt's (circa 1854–69) and subsequently for Smith & Wesson (circa 1869–93). Though the great majority of his artistry was lavished on handguns, he is known to have embellished numerous sporting longarms on contract—in particular, for Sharps. His artistic legacy lay not only in embellished metal but also in the training of sons Oscar (1854–1912) and Eugene (1863–1924) in the trade, both of whom decorated arms for both Colt and Smith & Wesson in the late nineteenth and early twentieth centuries.[8]

Gustave Young was followed as the head of Colt's engraving department by yet another talented German emigrant, Cuno A. Helfricht (1851–1927). Apprenticed to his father, and further trained in Berlin, he started with the company in 1875 and soon became its principal contractor. Helfricht introduced several new engraving styles and patterns, including figural panel scenes on a few select handguns. At the height of gun engraving's popularity, he and his several assistants embellished literally hundreds of Colt firearms in standard patterns every year over the last two decades of the nineteenth century. Thereafter (perhaps with the closing of the frontier West), the demand for ornamented firearms commenced a gradual decline. By 1910, decoration—except for standard work on shotguns—was applied only on special order. Notable early twentieth-century engravers for Colt's, either freelance or in-house, included Austrian-born Rudolph J. Kornbrath (1882–1946) and Wilbur A. Glahn (1888–1951), the latter the descendant of yet another dynasty of German-emigrant engravers in America.[9]

As a general observation, many more handguns than longarms received embellishment during the golden age of engraving in America (circa 1850–1900). Although shotguns made by the major manufacturers typically came to market with at least a cursory, standard border-and-scroll decoration, few rifles received any embellishment except upon special order. Engraved Remington and Sharps single-shot rifles, for example, appeared only rarely, and they are today notably scarce.[10] Among large-frame repeating rifles, however, copious hand-engraving proved comparatively more common, and this was particularly true of the Marlin and Winchester lines, whose receivers presented an ample canvas for the artisan's touch. By the 1880s, these firms provided a rather extensive repertoire of big-game vignettes and chase tableaus amid varying patterns of fine scrollwork. And both of these firms shared in the talents of yet another German-emigrant dynasty of skilled engravers—the Ulrichs.

Conrad Ulrich (1815–189_), a German-emigrant machinist and gun assembler laboring at Colt's, fathered a dynasty of engravers that ornamented the finest Winchester and Marlin rifles for two succeeding generations. Three of his sons, Conrad F. Ulrich (1844–1925), Herman L. Ulrich (1846–1937), and John Ulrich (1850–1924), all of whom apprenticed at Colt's during the early to middle 1860s, became the most talented practitioners of engraved classical motifs, big-game vignettes, and hunting tableaus during the later decades of the nineteenth century. On the relatively easy-to-cut, brass-framed Model 1866 Winchester rifle, each of the Ulrich brothers did elaborate, relief-engraved panel scenes of superlative quality. On the harder steel frames of later Winchester and Marlin rifles, the brothers sometimes resorted to flush gold inlay in producing quite-animated game figures. Conrad F. Ulrich began contract work for Winchester in the early 1870s. Though most recognized for his decoration of Winchesters, he also engraved Ballard rifles and ultimately became a regular employee of the Marlin Fire Arms Company (circa 1895–1907). Equally if not more talented, Herman L. Ulrich did engraving at different times for Colt, Winchester, and the American Bank Note Company. John Ulrich, in contrast, labored his entire career as an engraver and stock carver for Winchester (circa 1870–1907), and a greater number of signed examples of his work have survived.[11]

The work of the three Ulrich brothers was quite similar in quality and execution, and by the mid-1890s they had established a rich repertoire of standard game figures and compositions suitable for "off-the-shelf" embellishment at both Winchester and Marlin. In fact, their notable skills—and the continuing appreciation of embellished sporting arms in late nineteenth-century America—were celebrated by the Marlin and Winchester firms with the publication of unusual marketing catalogs in 1896 and 1897, respectively. *Special High Grade Marlin Repeaters* and *Winchester . . . Highly Finished Arms* showcased the Ulrichs' preeminent talents—albeit without naming them—through detailed illustrations and descriptions of the standard gradations of flush and relief engraving, inlaying, and stock carving and checkering then available.[12]

The Winchester catalog, for example, offered ten gradations of engraving and inlaying ranging in price from $5 to $250 and eight patterns of stock carving and checkering priced from $5 to $50. The publication declared with reasonable pride, "[We] have unsurpassed facilities for producing fancy finished guns of all prices and descriptions. Inlaying in gold, silver or platinum, gold or silver plating, engraving, carving or fancy checking, is done in the most artistic manner by the company's own employes [*sic*]. Stocks of fancy woods can be supplied, if desired, at special prices." (These included bird's-eye maple and English or Circassian walnut.)[13] Not unlike the guns themselves, these illustrated catalogs—along with L. D. Nimschke's personal pattern book, published in facsimile in 1965—provide the contemporary connoisseur, collector, and student with an invaluable pictorial legacy documenting the aesthetic achievement of domestic arms decorators during their most dynamic and prolific era.

Nineteenth-century American engraving embodied a generally fancy ornamentation that reflected the Victorian era's infatuation with the intricate and the extravagant. It presented not only an evolving variety of lavish scrollwork patterns but also a rich diversity of fantasy and animal figures coupled with sumptuous borders, panels, frames, and cartouches. When well-conceived and masterfully rendered, such embellishment actually complemented the firearm's architecture, allowing the gun's inherent lines to flow more freely and thus smooth the visual transition from one element to another. For the most part, twentieth- and twenty-first-century American firearms engravers have perpetuated these stylistic precedents in scroll and figural applications. Both the Marlin and Winchester companies, for example, continued to offer most of the same styles and gradations of the Ulrich brothers' engraving repertoire well into the twentieth century. Likewise, in the Colt firm's continuing decoration of the venerable Single Action Army revolver, most of its more recent, Custom Shop artisans have largely reiterated the classic American-scroll motifs developed by Cuno Helfricht more than a century ago.[14]

In the following portfolio of engraved firearms, the presentation follows that adopted in part 2—leading off with Colt and followed by Remington, Smith & Wesson, Sharps, Marlin, and Winchester. As well, a few other decorated firearms of nineteenth-century domestic origin are included where deemed most appropriate.

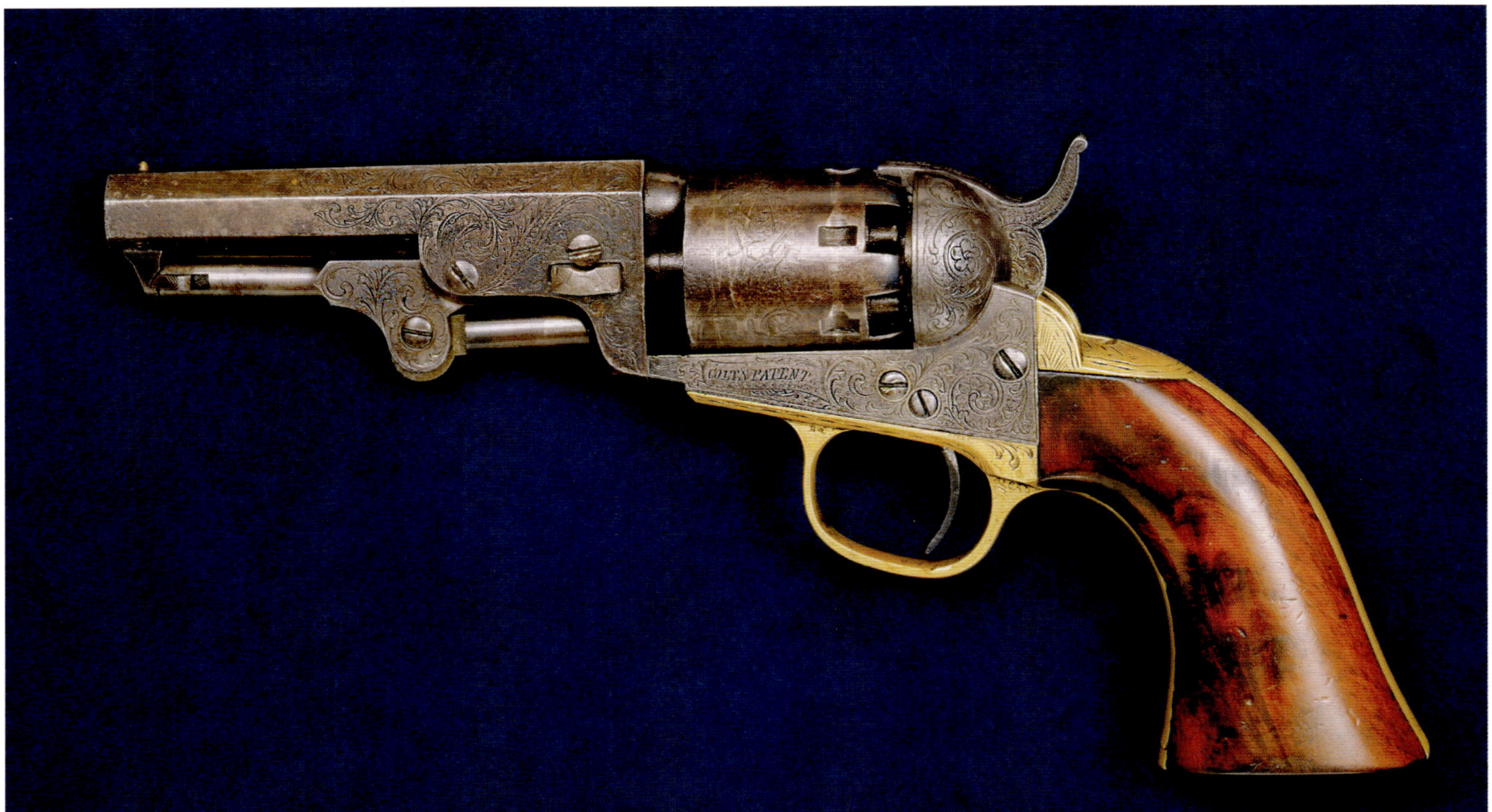

Muzzle-Loading Percussion Revolver
Colt's Patent Fire Arms Manufacturing Company, Hartford, Connecticut
Model 1849 Pocket, 1850–1873, Caliber .31 Rifled Bore
Weitzenhoffer Collection, 2001.46.009

Manufactured in 1865, this special-order piece is hand-engraved and fitted with select walnut grips. Embellished by an artisan in Gustave Young's shop, the revolver displays typical coverage in later, vine-style scrollwork on the hammer, gripstraps, trigger guard, frame, barrel, and loading lever. The rather elegant scroll pattern lacked the punched-dot (or stippled) background seen on earlier vine and doughnut-scroll designs, but it afforded a cleaner and more economical treatment that reduced application time by about 25 percent. Like other specimens engraved in this style, a riband or ribbon motif appears on both sides of the barrel lug; these sometimes carried patriotic inscriptions such as "UNION" or "LIBERTY." Like all other Model 1849 Pocket revolvers, this example also is ornamented with a roll-died, stagecoach holdup cylinder scene.

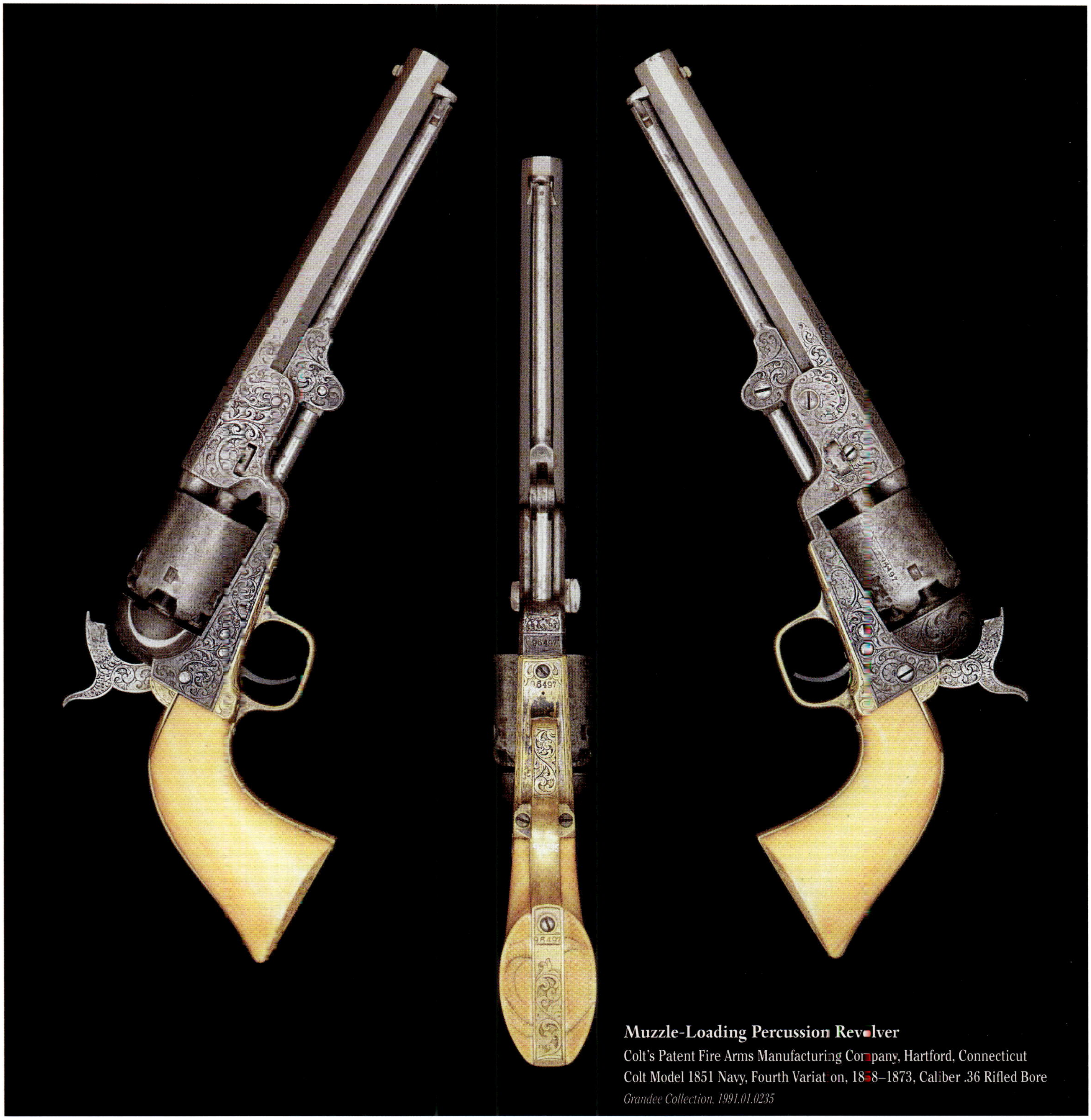

Muzzle-Loading Percussion Revolver
Colt's Patent Fire Arms Manufacturing Company, Hartford, Connecticut
Colt Model 1851 Navy, Fourth Variation, 1858–1873, Caliber .36 Rifled Bore
Grandee Collection, 1991.01.0235

This piece exhibits the sumptuous ornamentation available on American arms during the latter half of the nineteenth century. Made by Colt's in 1860, the revolver probably shipped "in-the-white" to a custom dealer like Schuyler, Hartley & Graham of New York, who in turn ordered it embellished by master engraver Louis Daniel Nimschke. The revolver displays fine foliate-scroll engraving over a punched-dot ground on the loading lever, barrel lug, frame, hammer, gripstraps, and trigger guard. Nimschke garnished the hammer with animal heads, hand-engraved "Colts / Patent" on the left frame, and rendered a whimsical beast on the left barrel lug. As a finale, the bottom of the special-order, elephant ivory grips were hand-checkered.

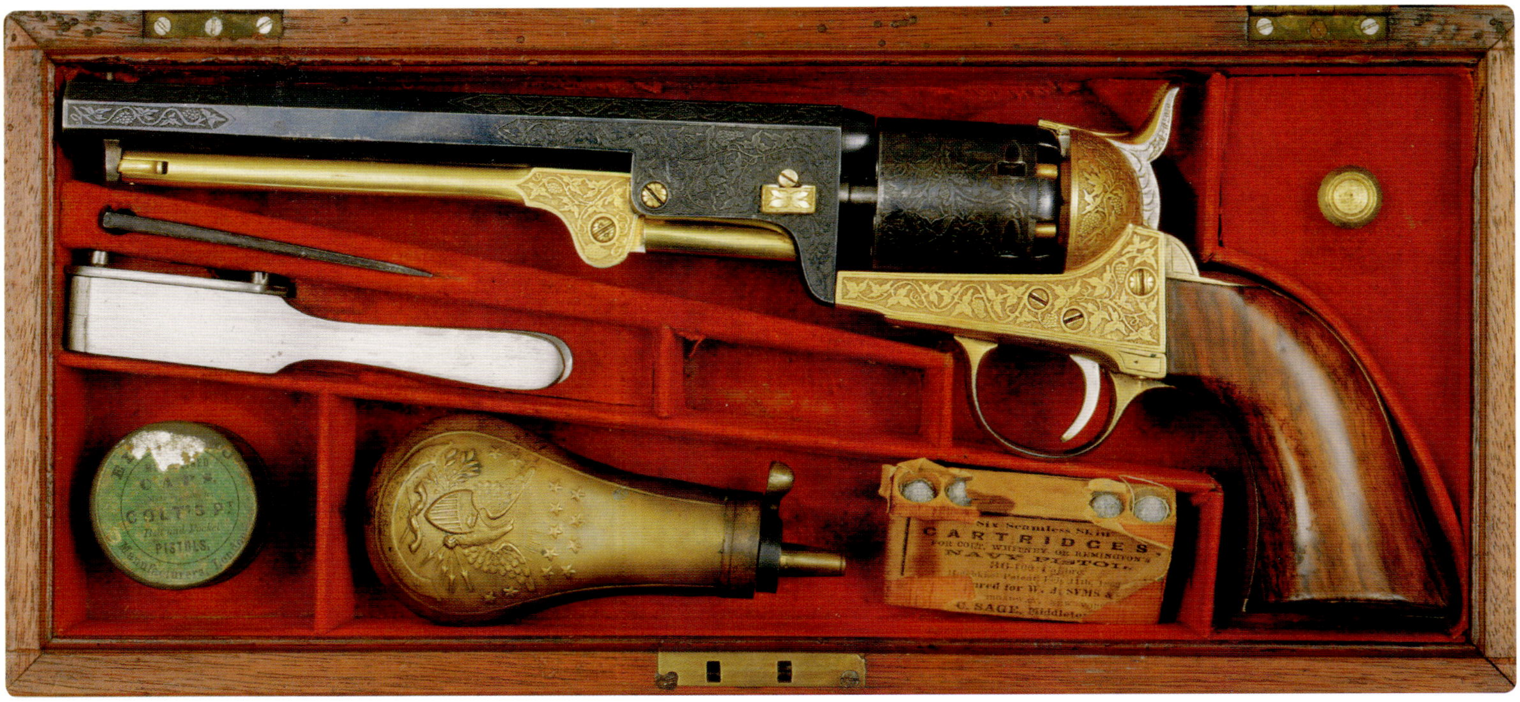

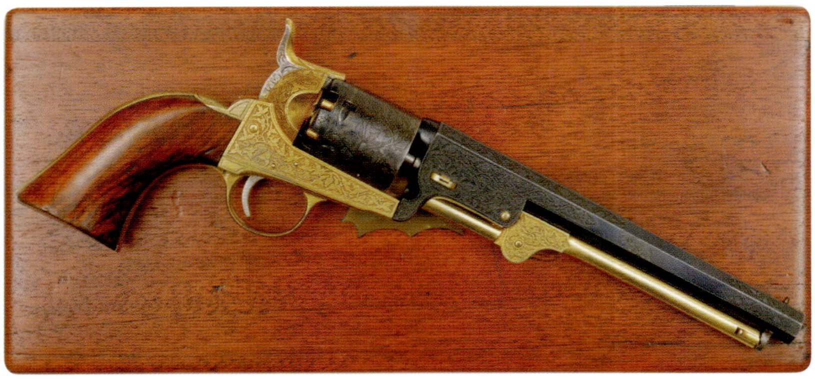

Muzzle-Loading Percussion Revolver

Attributed to Gilon, Liege, Belgium
Colt Brevete Model 1851 Navy, circa 1855–1860
Caliber .38 Rifled Bore
Weitzenhoffer Collection, 2001.46.015

Marked "COLT / BREVETE" on the top flat of the barrel, this well-made revolver was produced under a licensing arrangement with Colt's. This specimen, like most Belgian copies, is quite faithful to the original design and displays extraordinary embellishment. All of the pistol's metal surfaces exhibit elaborate acid etching in a delicate grapevine motif accentuated with leaves and clusters of grapes. The blued barrel and cylinder are strikingly set off by the gold-plated hammer, frame, gripstraps, loading lever, and wedge. Varnished grips of select, figured walnut complete the deluxe presentation. The piece resides in a compartmented Colt London casing fashioned of mahogany and lined in burgundy velvet. Associated accessories include a Colt powder flask, bullet mold, percussion cap tin, and packet of combustible cartridges.

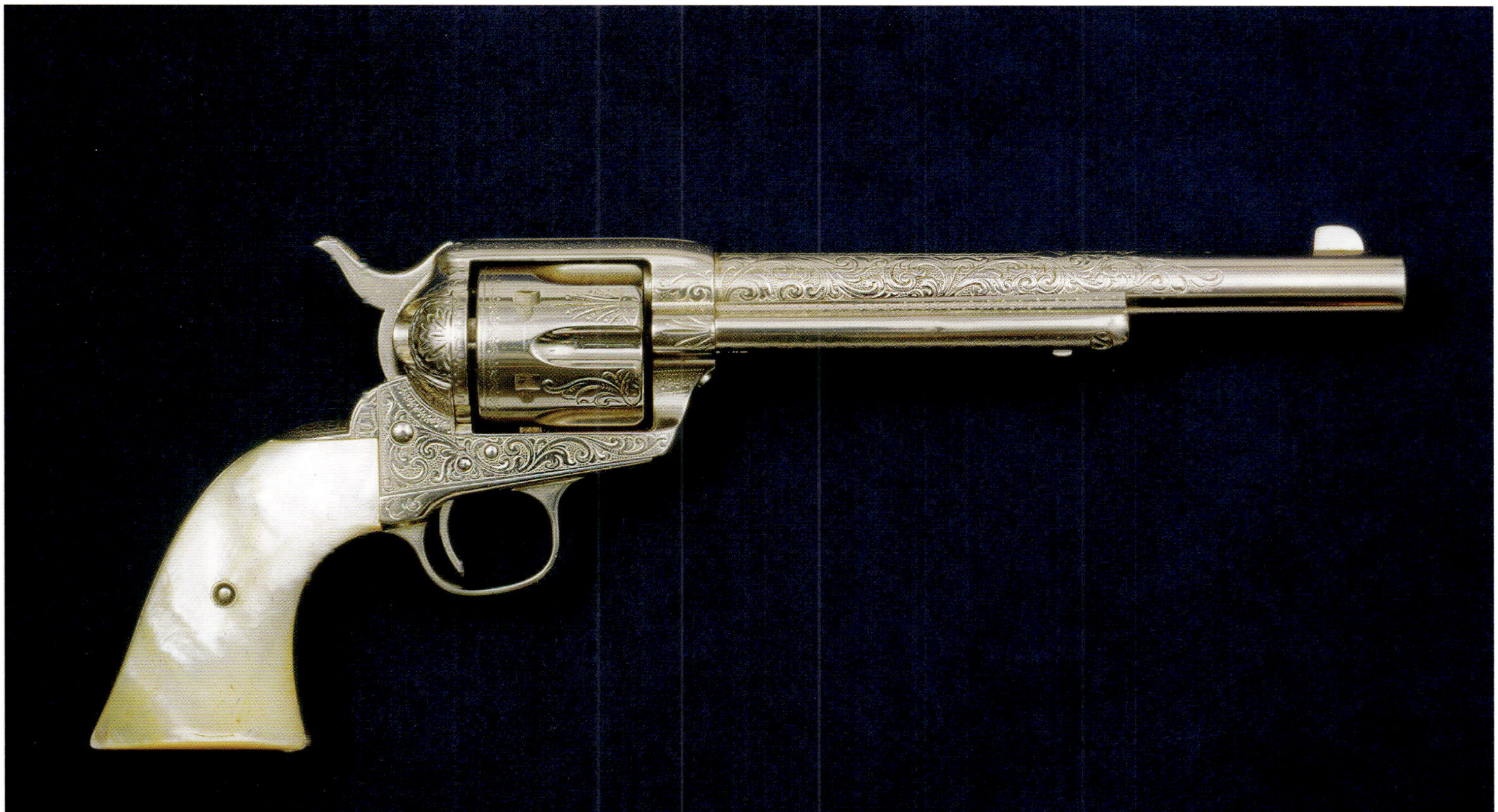

Breech-Loading Cartridge Revolver
Colt's Patent Fire Arms Manufacturing Company, Hartford, Connecticut
Model 1873 Single Action Army, 1872–1940, Caliber .44 Winchester Centerfire
Weitzenhoffer Collection, 2001.46.037

This superb Single Action Army revolver exhibits several custom features that place it well beyond the norm of regular production. Factory engraved in Cuno Helfricht's shop in 1891, the revolver displays broad coverage in an elegant foliate-scroll pattern set off with a punched-dot, or stippled, background. (Between 1872 and 1900, less than 1 percent of all Colt Single Action Army production was engraved by the factory.) The piece also is fitted with special-order two-piece mother-of-pearl grips fashioned in Helfricht's shop and is finished (probably refinished) in full nickel plating. Deluxe pistols such as this were prized by many westerners, from ranchman Theodore Roosevelt to showman "Buffalo Bill" Cody.

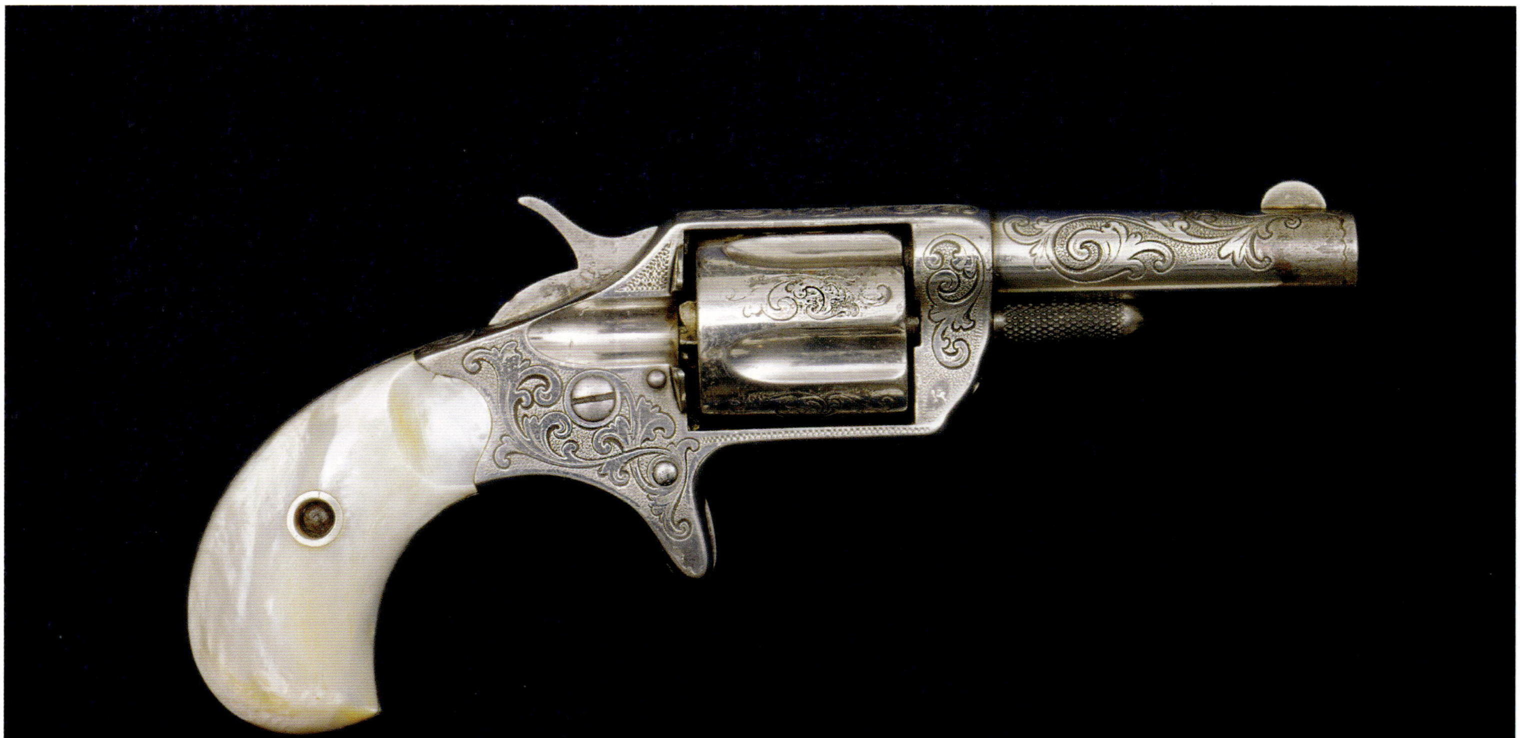

Breech-Loading Cartridge Revolver
Colt New Line Pocket Model, circa 1874–1876
Colt's Patent Fire Arms Manufacturing Co., Hartford, Connecticut
Caliber .30 Short or Long Rimfire
Weitzenhoffer Collection, 2001.046.046

Elegantly engraved, silver-plated, and fitted with mother-of-pearl grips, this special-order pistol passed through the workshop of master artisan Cuno Helfricht in 1875. The gun exhibits special foliate-scroll-style engraving typified by comparatively broad and flowing patterns that are deeply cut along selected edges and backed by a stippled ground, creating an almost three-dimensional look. The engraver's expertise is evident in the crisply cut and delicately shaded "crossover" seen in the scrollwork pattern just above the spur trigger. Coupled with its custom finish and grips, this revolver's extensive engraving suggests that it was appreciated by a man—or perhaps much more likely a woman—of considerable status, taste, and wealth.

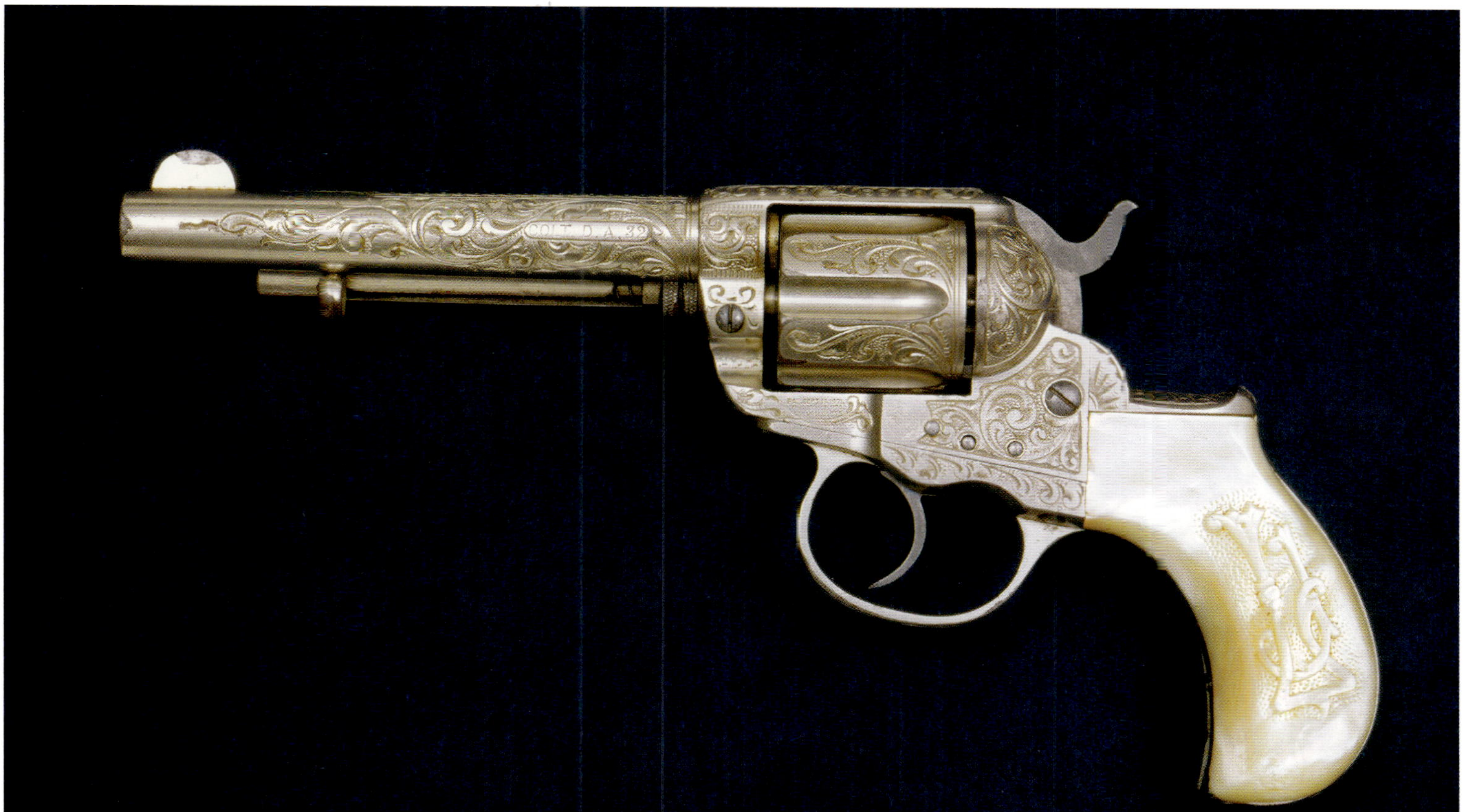

Breech-Loading Cartridge Revolver
Colt's Patent Fire Arms Manufacturing Company, Hartford, Connecticut
Model 1877 Lightning, 1877–1909, Caliber .32 Long Colt Centerfire
Weitzenhoffer Collection, 2001.46.050

Displaying the sumptuous ornamentation rendered by America's late nineteenth-century stock carvers and arms engravers, this revolver probably was embellished under contract for Colt's by the Cuno Helfricht shop in Hartford, although it could have been finished by a New York City artisan such as Louis D. Nimschke. The gun features profuse scroll engraving on a stippled, or punched-dot, ground covering the frame, cylinder, barrel, ejector rod housing, and backstrap. After engraving, the revolver received a full nickel-plated finish (except for screws, hammer, and trigger) and a pair of relief-carved and stippled mother-of-pearl grips with "L J" monograms. The revolver is desirable not only for its fine custom embellishment but also for its rare, nonstandard chambering in .32 Colt caliber.

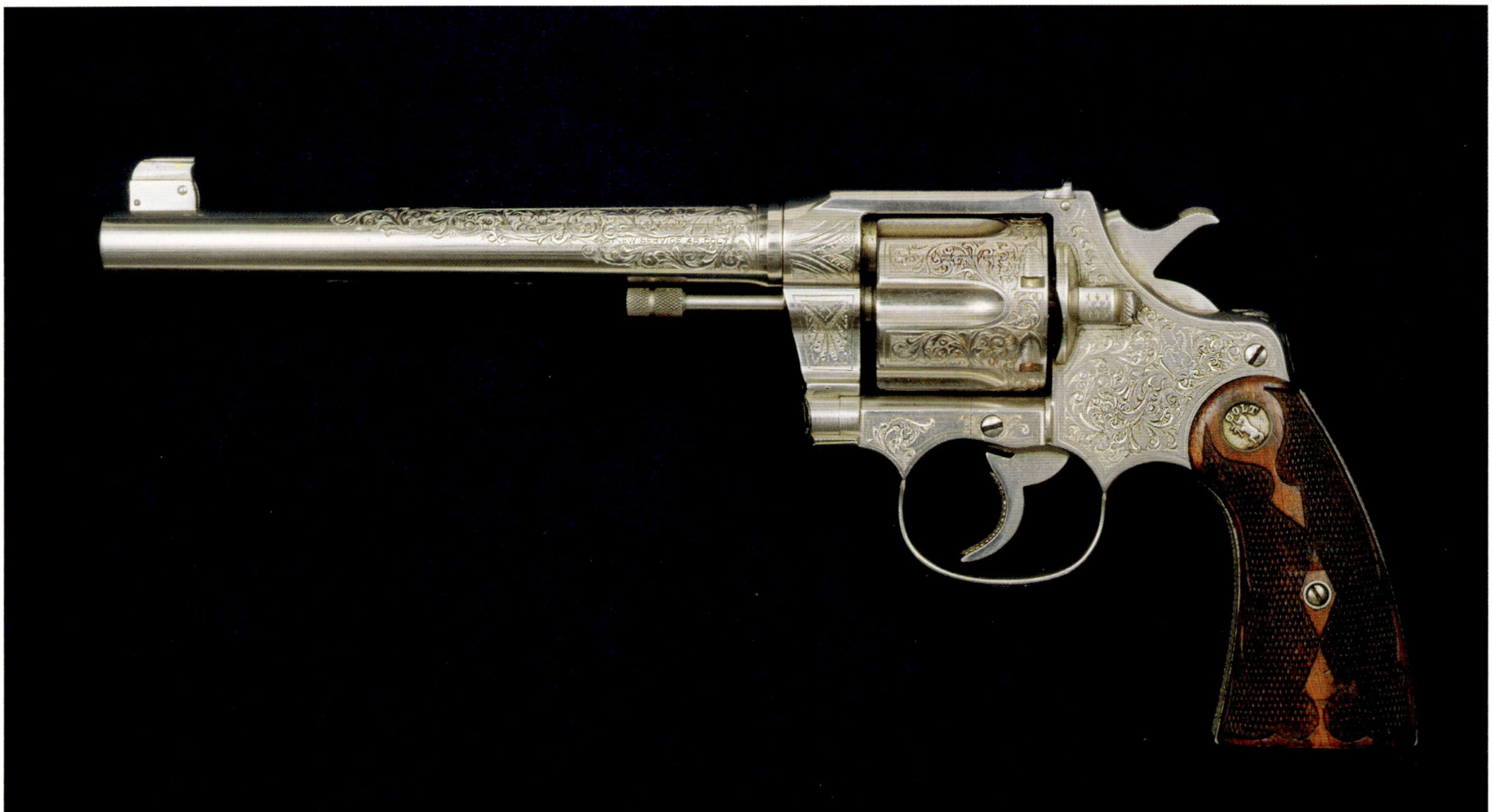

Breech-Loading Cartridge Revolver
Colt's Patent Fire Arms Manufacturing Company, Hartford, Connecticut
New Service Flattop Target Model, 1898–1944, Caliber .45 Long Colt Centerfire
Weitzenhoffer Collection, 2001.46.057

Although demand for embellished arms declined during the early twentieth century, Colt's still retained master engraver Wilbur A. Glahn, who succeeded Cuno Helfricht around 1920 and remained the company's leading artisan until 1950. This lavishly finished New Service revolver, however, is believed to highlight the talents of freelance and contract engraver Rudolph J. Kornbrath, who did work for Colt's in the 1920s and 1930s. Kornbrath's execution features an unusual, complex, and quite extensive floral-scroll motif over the barrel, frame, and cylinder. Typical of the target variant, this piece also includes hand-checkering on the backstrap, trigger guard strap, and trigger. The engraving is set off with a full nickel-plated finish and hand-checkered walnut grips inset with "Rampant Colt" medallions.

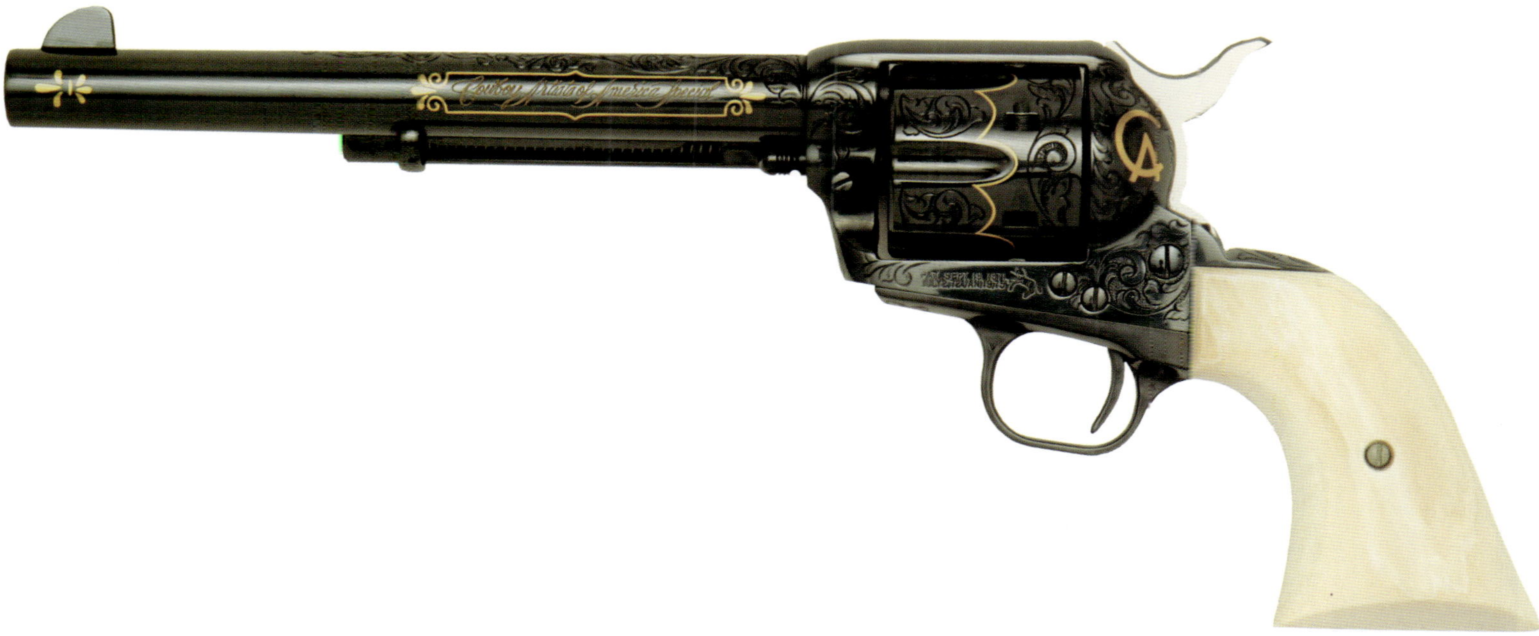

Breech-Loading Cartridge Revolver

Colt's Patent Fire Arms Manufacturing Company, Hartford, Connecticut
Model 1873 Single Action Army, Second Generation, 1956–1984
Caliber .44 Winchester Centerfire
Gift of David Lovell, 2000.35.34

Serial numbered 1TL, this deluxe revolver was embellished (and cased) in the late 1970s for well-known western artist Tom Lovell as one of the presentation "Cowboy Artists of America Special" series. Engraved and gold-inlaid by master artisan Alvin A. White for Colt's Custom Shop, the piece also boasts two-piece grips of rare elephant ivory. White's work encompasses profuse American-style scroll engraving on a punched-dot ground on the barrel, frame, cylinder, gripstraps, and trigger guard and includes a facsimile of the artist's signature engraved on the backstrap. Fourteen-carat gold inlay adorns the barrel, frame, and cylinder, with a gold-inlaid "CA" (Cowboy Artists) monogram on the recoil shield.

Remington Engraving

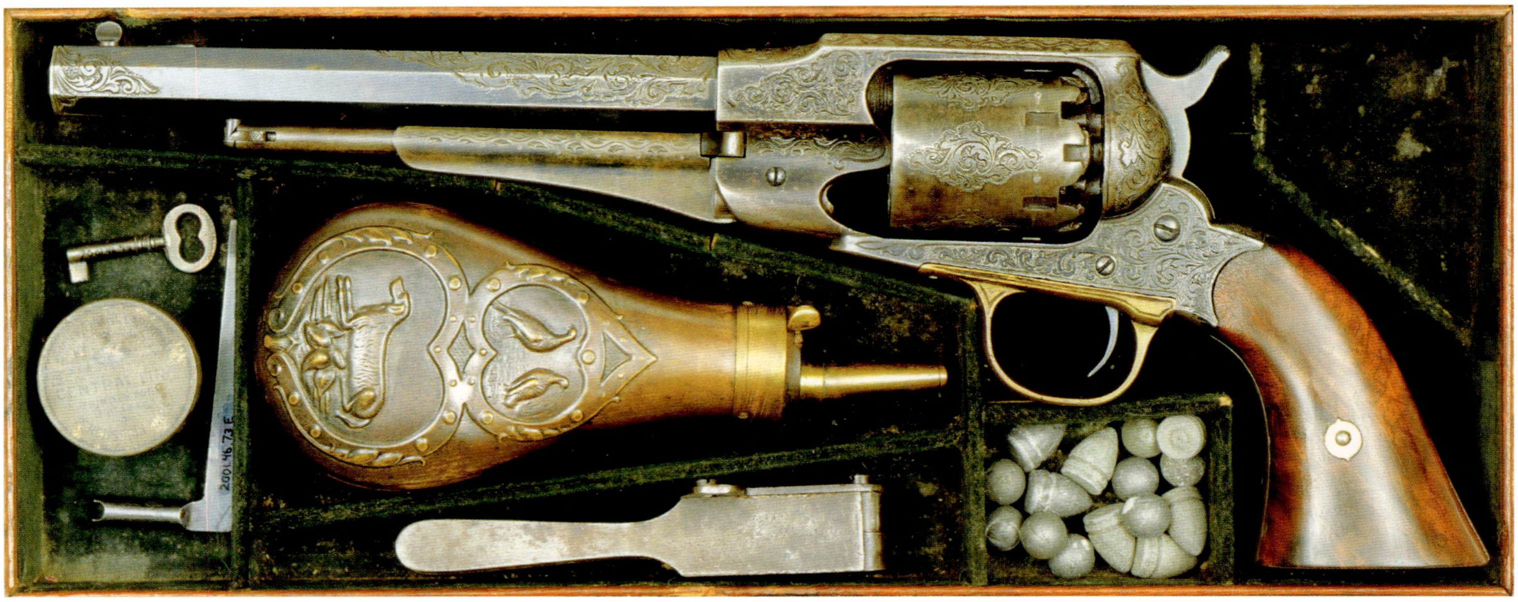

Muzzle-Loading Percussion Revolver
E. Remington & Sons, Ilion, New York
New Model Army, 1863–1875, Caliber .44 Rifled Bore
Weitzenhoffer Collection, 2001.46.073

Housed in a compartmented mahogany casing lined in dark green velvet, this New Model Army revolver, engraved and fitted with figured grips of walnut, presents the integration of machine production and custom embellishment that distinguished American firearms during the latter nineteenth century. The engraving, although fairly profuse and well executed, does not reflect factory styles of the period and probably represents aftermarket embellishment by a freelance artisan. The casing also holds a typical Remington powder flask embossed with game birds and a reclining dog, the correct two-cavity bullet mold, an L-shaped combination tool, a percussion cap tin, a case key, and an assortment of round and conical projectiles. Although seldom valued as highly by firearms enthusiasts, cased and engraved Remington handguns actually are much scarcer than Colts of similar garniture.

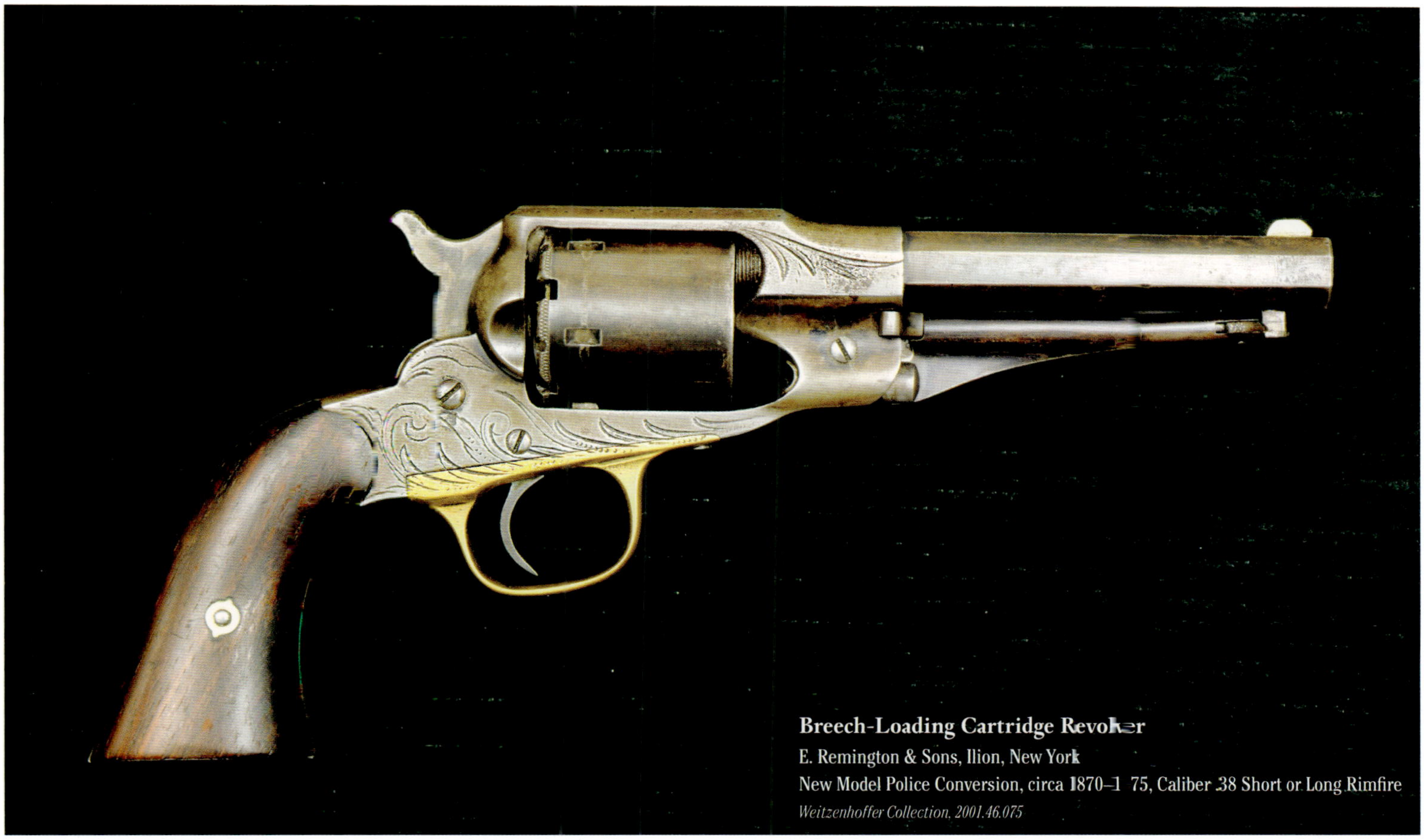

Breech-Loading Cartridge Revolver
E. Remington & Sons, Ilion, New York
New Model Police Conversion, circa 1870–1875, Caliber .38 Short or Long Rimfire
Weitzenhoffer Collection, 2001.46.075

Intended to compete with Colt's Police and Pocket Navy models, the Remington New Model Police revolver initially appeared as a percussion muzzle-loader in or around 1865. Much of its production, however, culminated either as metallic-cartridge conversions or as original cartridge breechloaders. Many of these were sold with interchangeable percussion and metallic-cartridge cylinders in an effort to serve both the old technology and the new. Factory engraved in a simplistic scroll-and-border treatment on a clear ground and mounted with rosewood grips, this example exceeds standard Remington production, but it hardly approaches the embellishment then available on special-order American firearms. This cursory work probably was done to dress up a then-obsolete firearm.

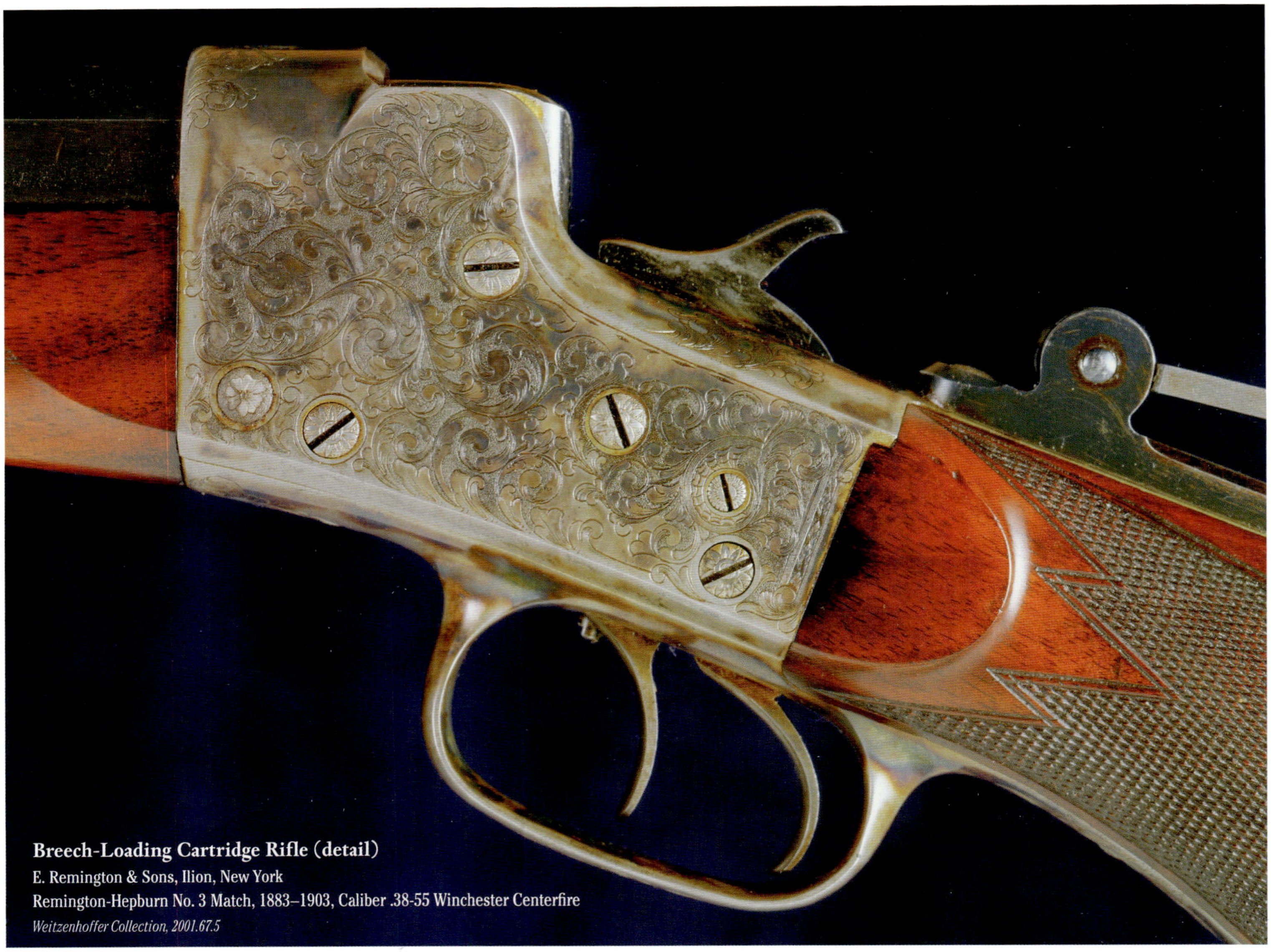

Breech-Loading Cartridge Rifle (detail)
E. Remington & Sons, Ilion, New York
Remington-Hepburn No. 3 Match, 1883–1903, Caliber .38-55 Winchester Centerfire
Weitzenhoffer Collection, 2001.67.5

Also illustrated and described in chapter 4, Remington's No. 3 Match rifle was marketed in two variations determined by stocking quality and sights. This higher-grade, "B Quality" specimen features checkered, deluxe walnut stocks with a pistol grip, cheekrest, and nickel-plated, Swiss-pattern buttplate and forestock tip. Specified sights include an adjustable, hooded wind-gauge front with spirit-level and a folding midrange vernier peep rear. Special-order extras include the half-round/half-octagon barrel and double set trigger. Beyond all of these factory refinements, the piece also exhibits exquisite English-style foliate-scroll engraving over the frame and including the screw heads. Though unsigned, this fine decorative work is most likely attributable to German-emigrant master engraver Louis Daniel Nimschke, who was fully capable of rendering such tight and meticulous embellishment.

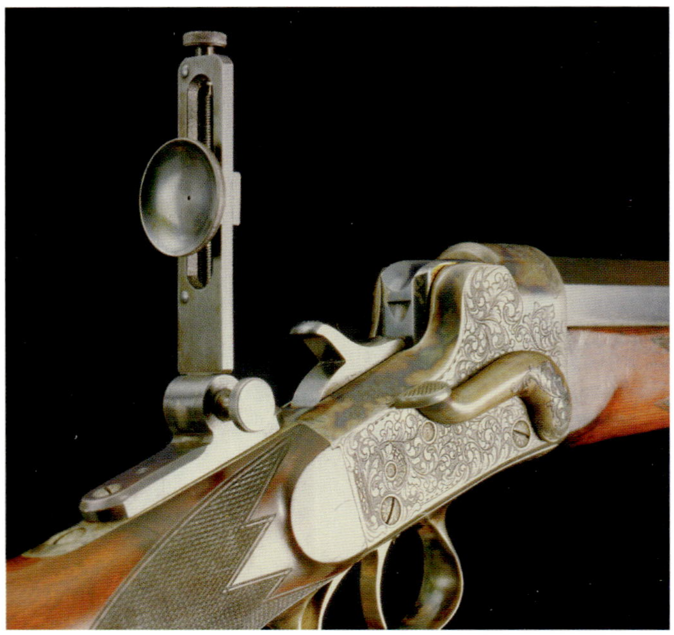

Smith & Wesson Engraving

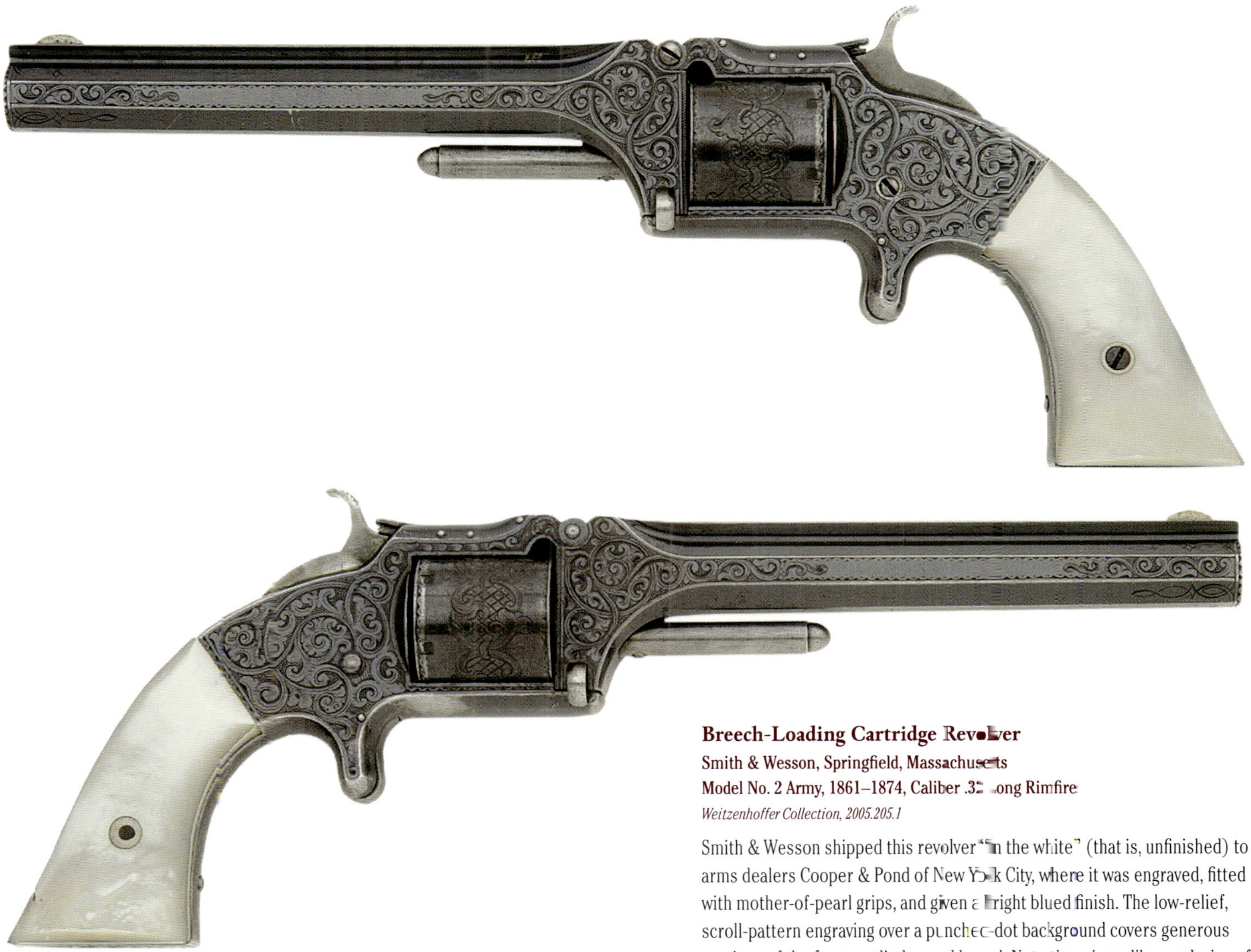

Breech-Loading Cartridge Revolver
Smith & Wesson, Springfield, Massachusetts
Model No. 2 Army, 1861–1874, Caliber .32 Long Rimfire
Weitzenhoffer Collection, 2005.205.1

Smith & Wesson shipped this revolver "in the white" (that is, unfinished) to arms dealers Cooper & Pond of New York City, where it was engraved, fitted with mother-of-pearl grips, and given a bright blued finish. The low-relief, scroll-pattern engraving over a punched-dot background covers generous portions of the frame, cylinder, and barrel. Note the mirror-like rendering of the scroll composition on each side of the frame and barrel, as well as the several, meticulous "crossovers" in the scrolling patterns. Though unsigned, the engraving reflects the talented and prolific hand of Louis Daniel Nimschke of New York City, who is known to have embellished many Smith & Wesson handguns for Cooper & Pond.

Images courtesy Amoskeag Auction Company, Manchester, New Hampshire

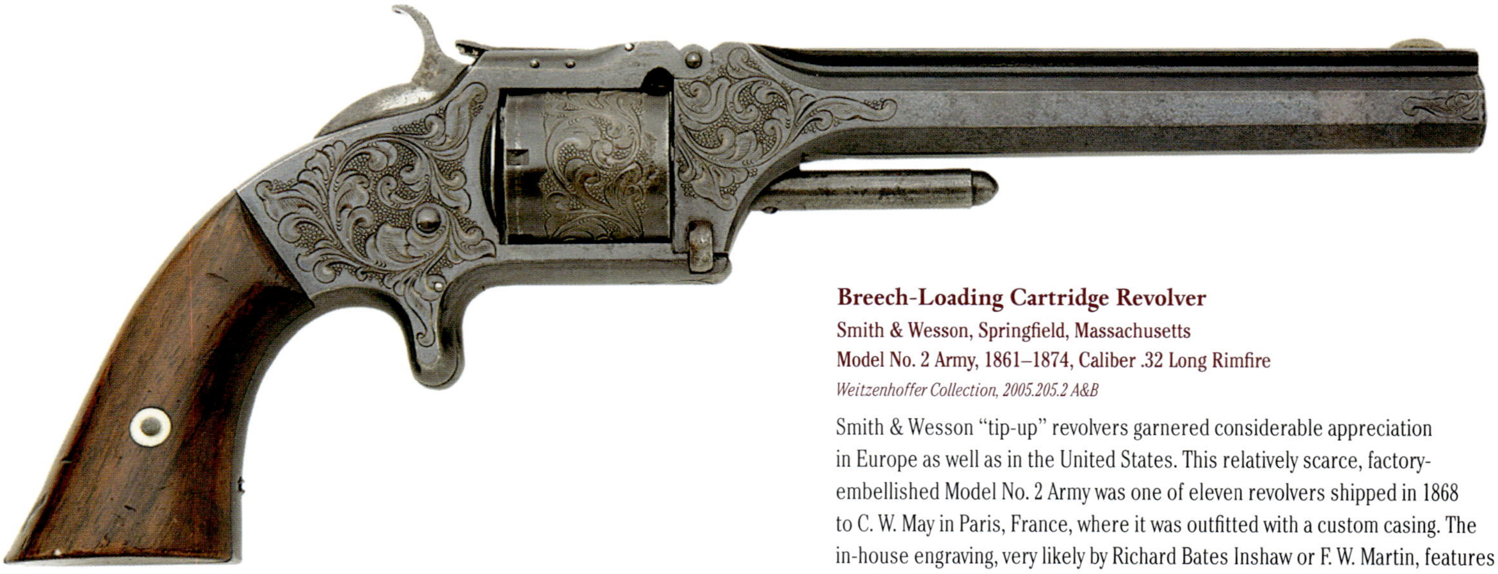

Breech-Loading Cartridge Revolver
Smith & Wesson, Springfield, Massachusetts
Model No. 2 Army, 1861–1874, Caliber .32 Long Rimfire
Weitzenhoffer Collection, 2005.205.2 A&B

Smith & Wesson "tip-up" revolvers garnered considerable appreciation in Europe as well as in the United States. This relatively scarce, factory-embellished Model No. 2 Army was one of eleven revolvers shipped in 1868 to C. W. May in Paris, France, where it was outfitted with a custom casing. The in-house engraving, very likely by Richard Bates Inshaw or F. W. Martin, features finely executed, flush acanthus-leaf scroll patterns over a stippled ground on the frame, cylinder, and barrel. The grip panels are varnished rosewood. The French-made casing has a beveled lid, contour-fit interior, and cartridge block for forty-one rounds of ammunition.
Images courtesy Amoskeag Auction Company, Manchester, New Hampshire

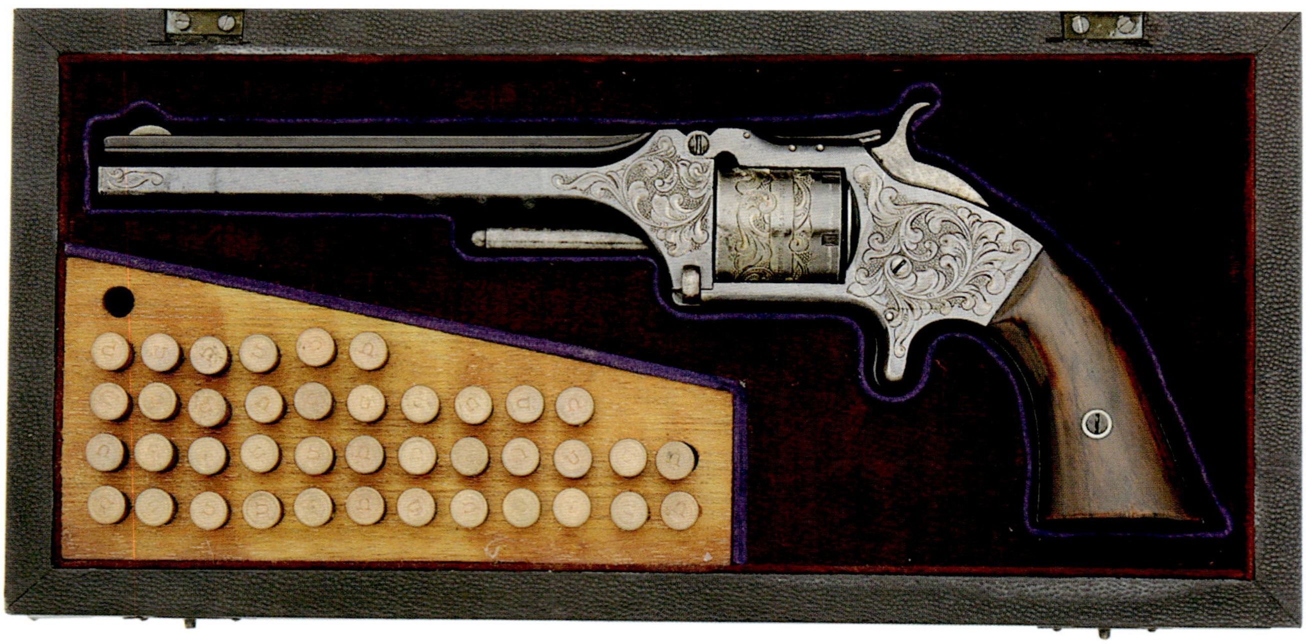

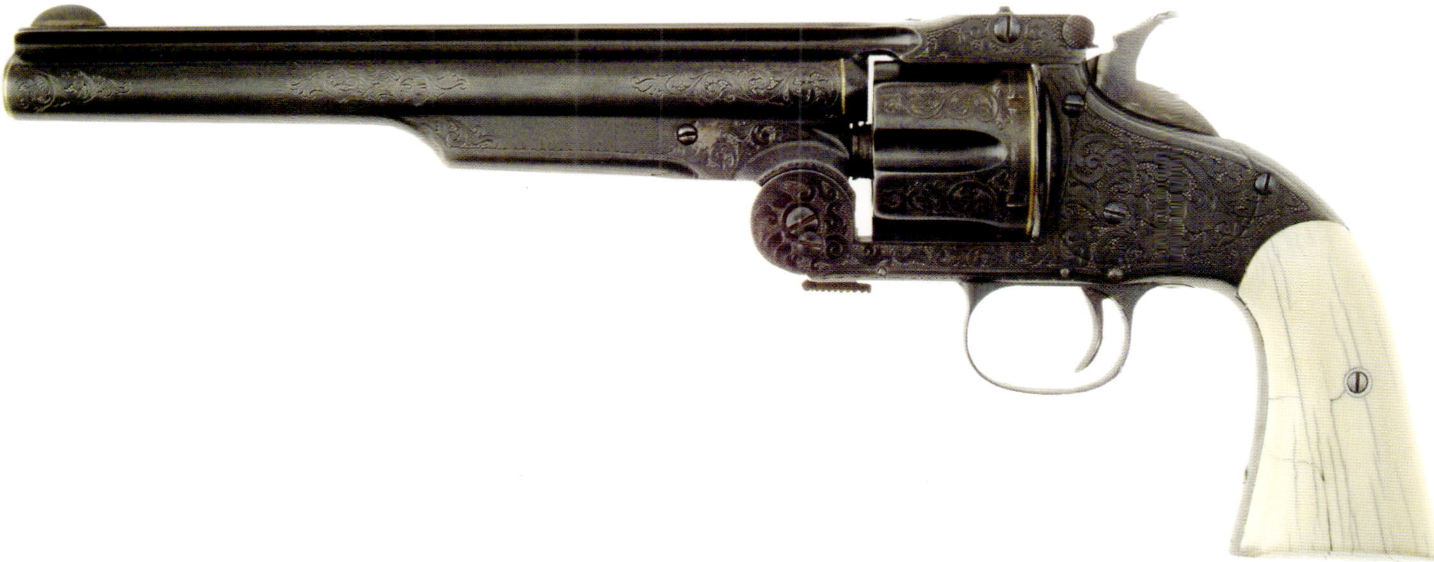

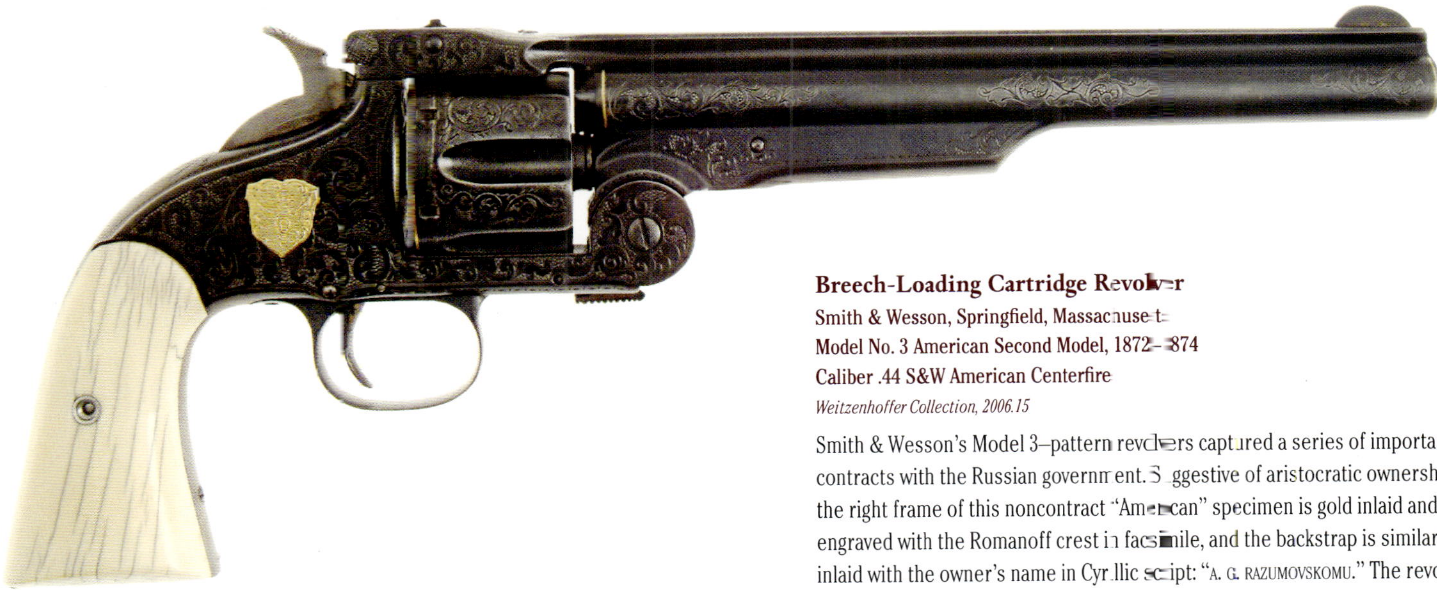

Breech-Loading Cartridge Revolver

Smith & Wesson, Springfield, Massachusetts

Model No. 3 American Second Model, 1872–1874

Caliber .44 S&W American Centerfire

Weitzenhoffer Collection, 2006.15

Smith & Wesson's Model 3–pattern revolvers captured a series of important contracts with the Russian government. Suggestive of aristocratic ownership, the right frame of this noncontract "American" specimen is gold inlaid and engraved with the Romanoff crest in facsimile, and the backstrap is similarly inlaid with the owner's name in Cyrillic script: "A. G. RAZUMOVSKOMU." The revolver is extensively engraved in the bold foliate arabesque-scroll style over punched-dot ground popularized by freelance artisan Louis D. Nimschke; it features gold-line inlays at the breech of the cylinder and the breech and muzzle of the barrel. The piece also boasts fitted grip panels of rare mastodon ivory.

Images courtesy James D. Julia Auctions, Fairfield, Maine

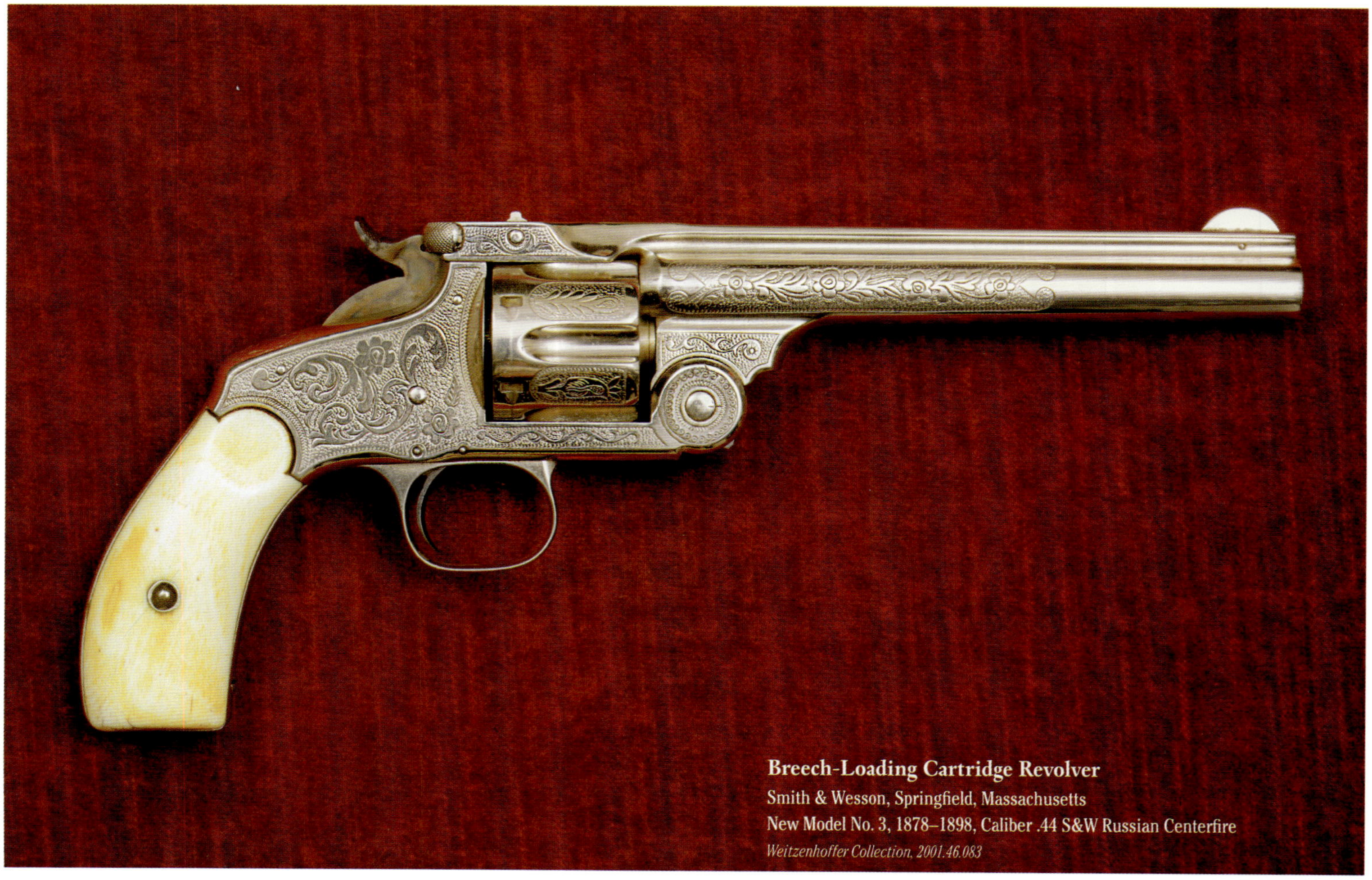

Breech-Loading Cartridge Revolver
Smith & Wesson, Springfield, Massachusetts
New Model No. 3, 1878–1898, Caliber .44 S&W Russian Centerfire
Weitzenhoffer Collection, 2001.46.083

Like other major American arms makers during the second half of the nineteenth century, Smith & Wesson provided special-order embellishments in the form of custom plating, fine engraving, and deluxe stocking. Factory engraving, however, proved relatively uncommon, and it commands commensurately greater attention and demand among collectors today. The consequence of such demand may be seen in this specimen, which serves as a caveat to the unwary arms enthusiast. Although the revolver's ivory stocking may be original, its engraving and nickel-plated finish are of much more recent date. The engraving, in particular, reflects neither the style nor the skill of late nineteenth-century artisans such as the Young brothers or L. D. Nimschke. Firearms exhibiting such aftermarket "enhancements" typically command little more appreciation than a plain specimen in like condition.

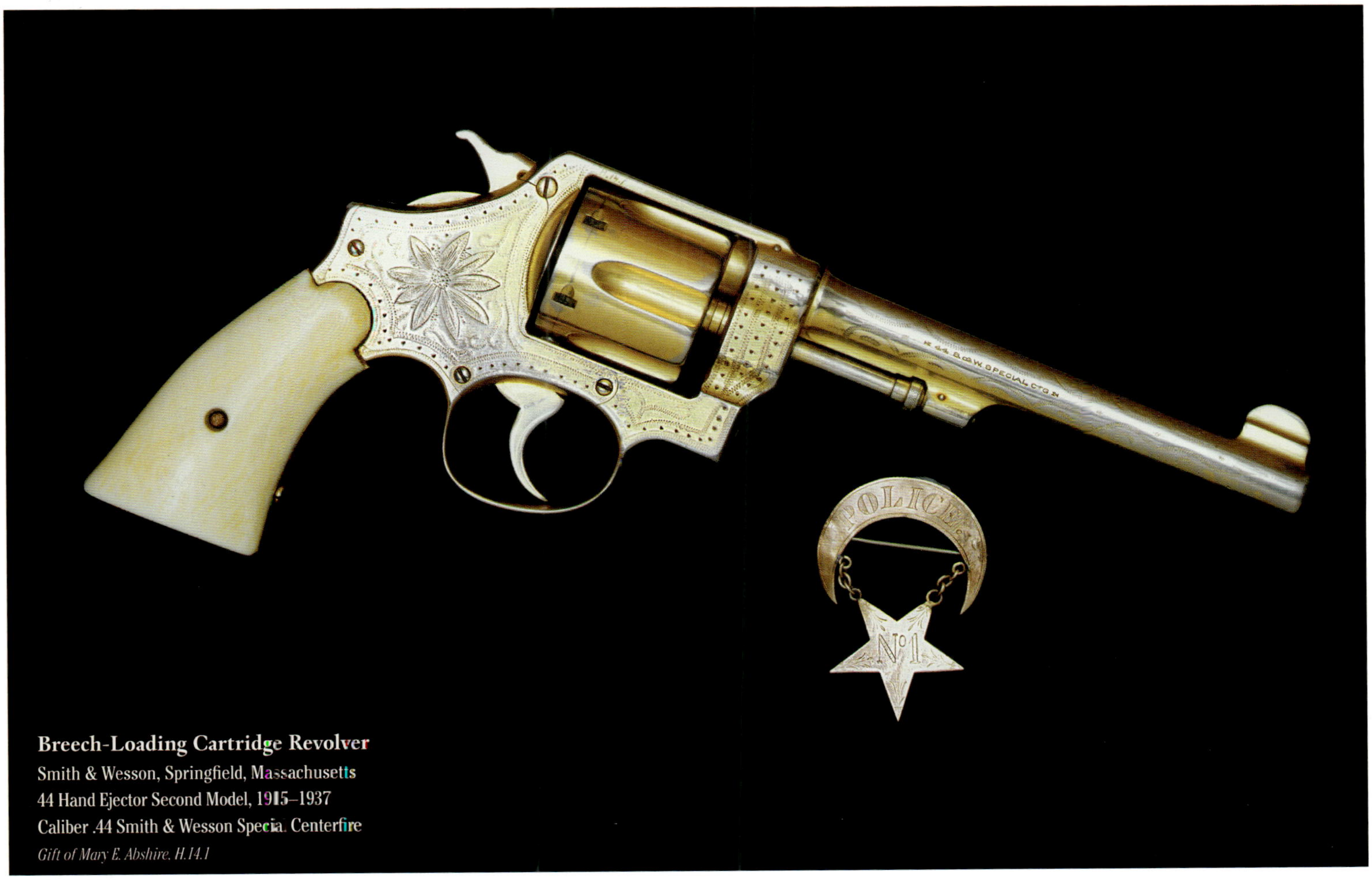

Breech-Loading Cartridge Revolver
Smith & Wesson, Springfield, Massachusetts
44 Hand Ejector Second Model, 1915–1937
Caliber .44 Smith & Wesson Special Centerfire
Gift of Mary E. Abshire. H.14.1

This Smith & Wesson revolver was one of two originally shipped in 1922 with standard finish and grips to pawnbrokers and jewelers Wolf & Klarr of Fort Worth, Texas, then also major gun dealers (considered the Hartley & Graham of the Southwest). Typical of the appearance favored by a jewelry house, the piece exhibits comparatively sparse and rather light "engraving," with a punched-dot and hairline border treatment and a cursory floral motif on the frame. The revolver was further adorned with a gold-washed finish and grips of elephant ivory. Finished identically on special order, both revolvers in the shipment were presented to famed frontier peace officer William "Bill" Tilghman by some admiring citizens of Oklahoma City, Oklahoma, where Tilghman had served as chief of police in the early 1910s.

Merwin Hulbert Engraving

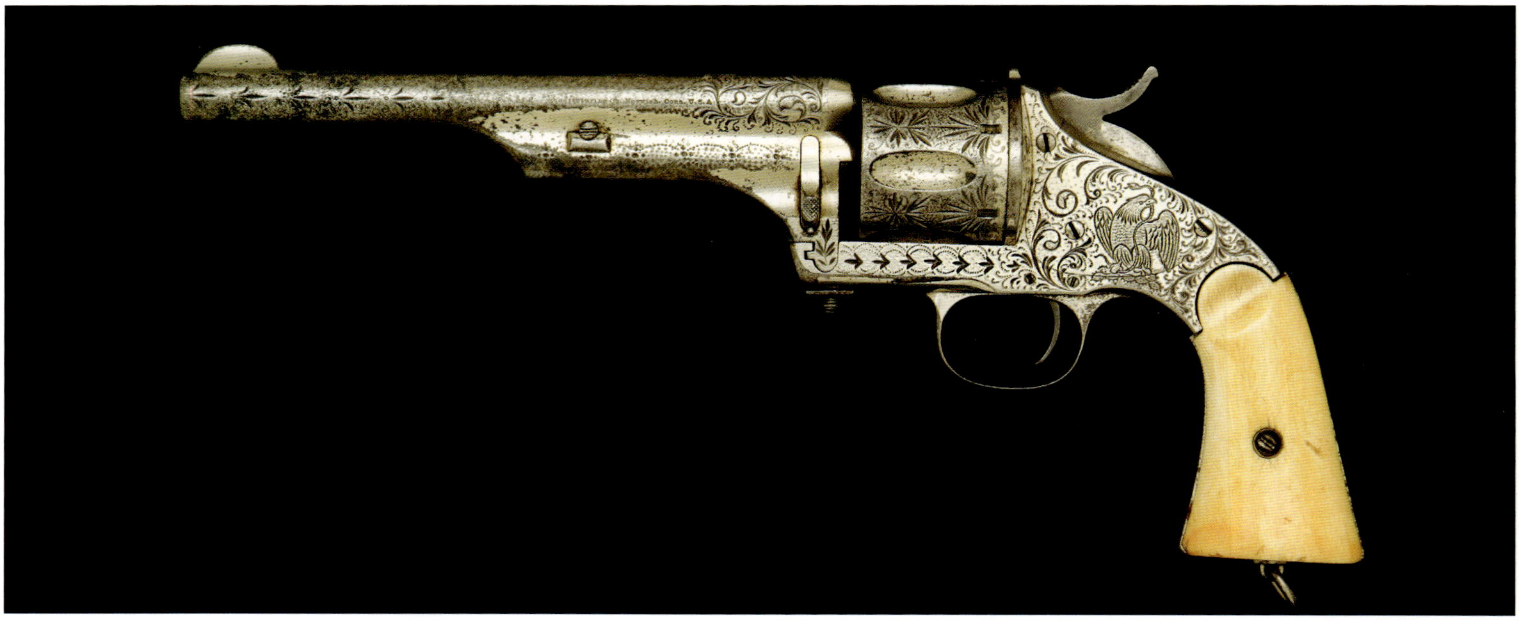

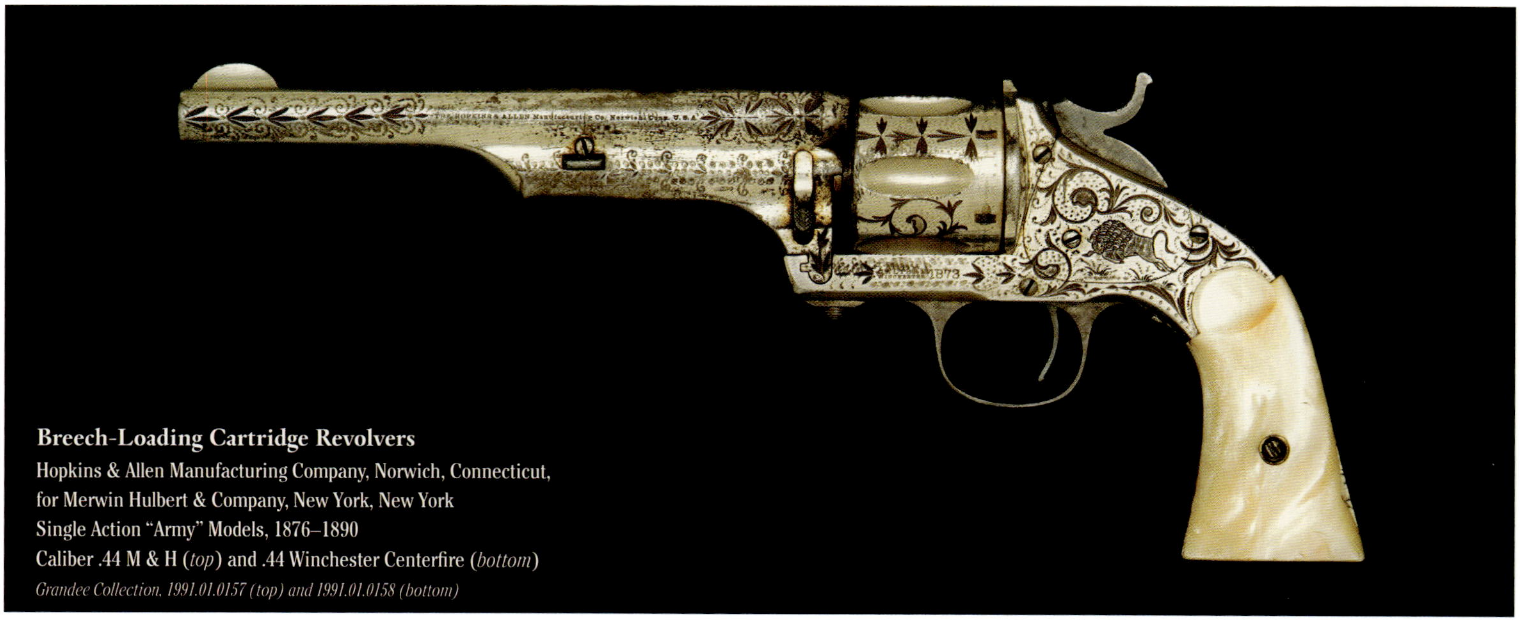

Breech-Loading Cartridge Revolvers
Hopkins & Allen Manufacturing Company, Norwich, Connecticut,
for Merwin Hulbert & Company, New York, New York
Single Action "Army" Models, 1876–1890
Caliber .44 M & H (*top*) and .44 Winchester Centerfire (*bottom*)
Grandee Collection, 1991.01.0157 (top) and 1991.01.0158 (bottom)

In addition to their rather sophisticated cartridge extraction mechanism (see chapter 5), Merwin Hulbert revolvers could be purchased with varying degrees of embellishment. These pieces are of the early, open-top pattern, having scooped rather than fluted cylinders. Both guns exhibit the very distinctive—yet comparatively simplistic and inexpensive—engraving style typical of the company: a broad, very cursory scroll pattern interspersed with single-cut curves and widely spaced punched dots. More precisely engraved vignettes often decorate the frames in open fields or panels. Here, the Mexican national motif enhances the upper piece, while a rampant buffalo graces that below. Nickel-plated finish was commonplace on these arms, but the grips of elephant ivory (top) and mother-of-pearl (bottom) represented (like the engraving) special-order embellishments of added cost.

Sharps Engraving

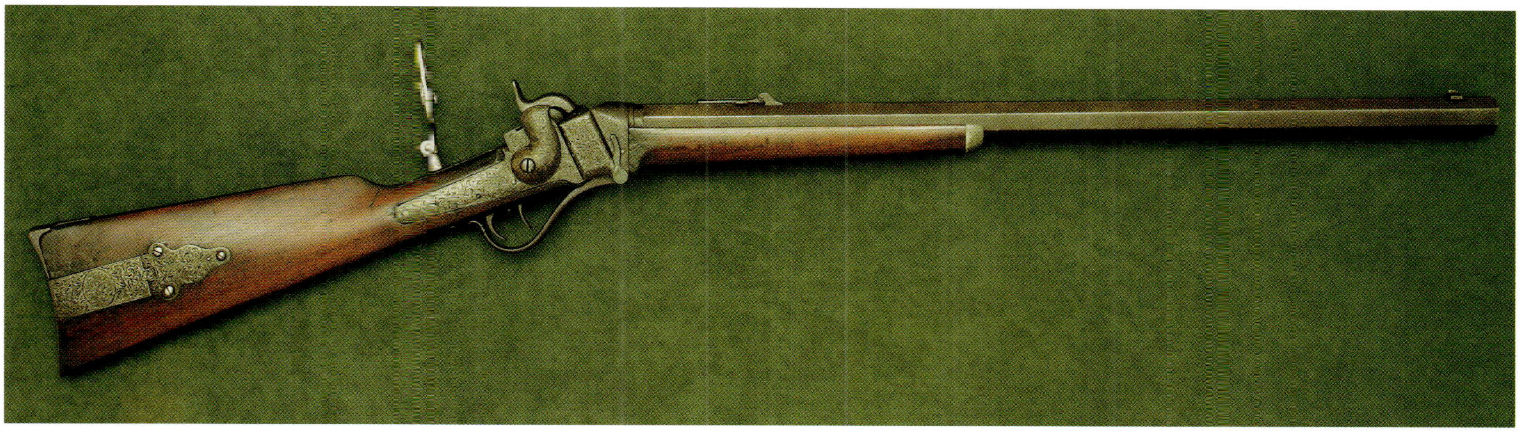

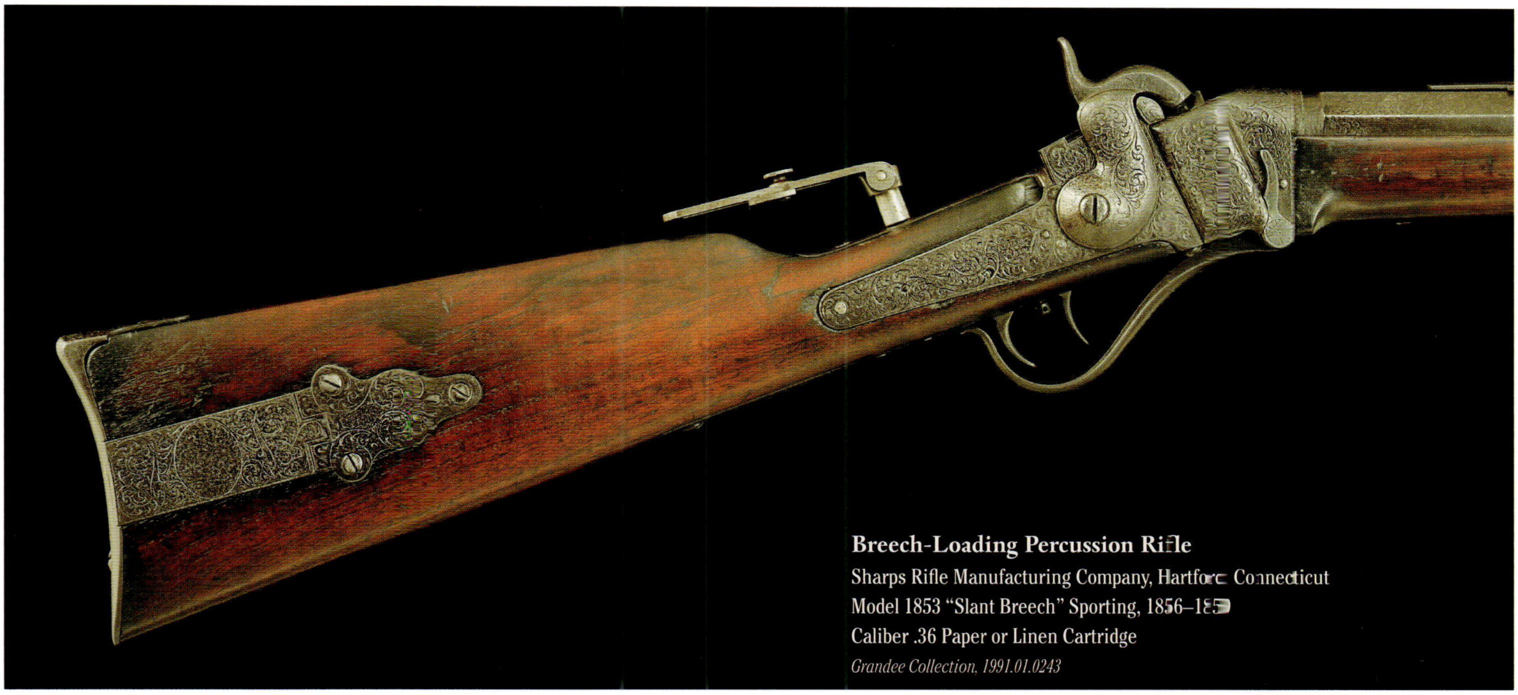

Breech-Loading Percussion Rifle
Sharps Rifle Manufacturing Company, Hartford, Connecticut
Model 1853 "Slant Breech" Sporting, 1856–185[illegible]
Caliber .36 Paper or Linen Cartridge
Grandee Collection, 1991.01.0243

A deluxe rifle for the wealthy sport hunter, this piece is engraved in a style somewhat akin to the "doughnut scroll" pattern over the patchbox, lockplate, hammer, receiver, and barrel breech. The decorative work, including an early oval panel scene on the patchbox cover, was recorded in the company's records as "extra fancy" and charged at a cost of $50. The scrollwork is reminiscent of that of freelance engraver Louis Daniel Nimschke, although it more likely was done by Gustave Young's Hartford shop, which took over Sharps' contract embellishment around 1852. Within the Sharps product line, the Model 1853 sporting rifle received such "factory" decoration more often than did any other model arm.

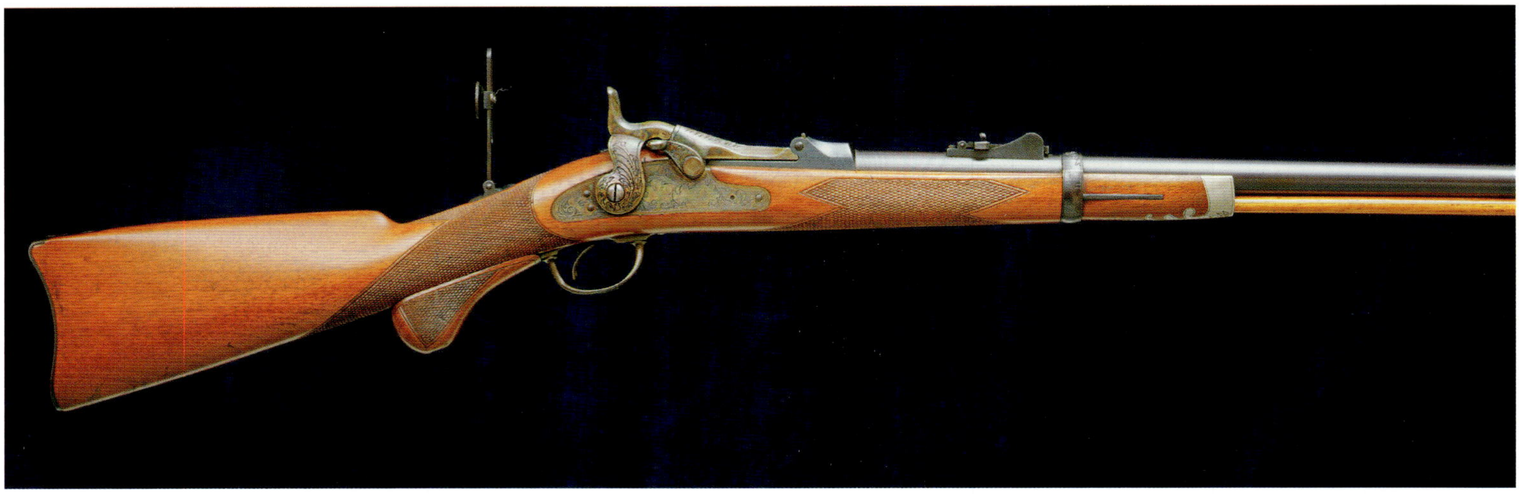

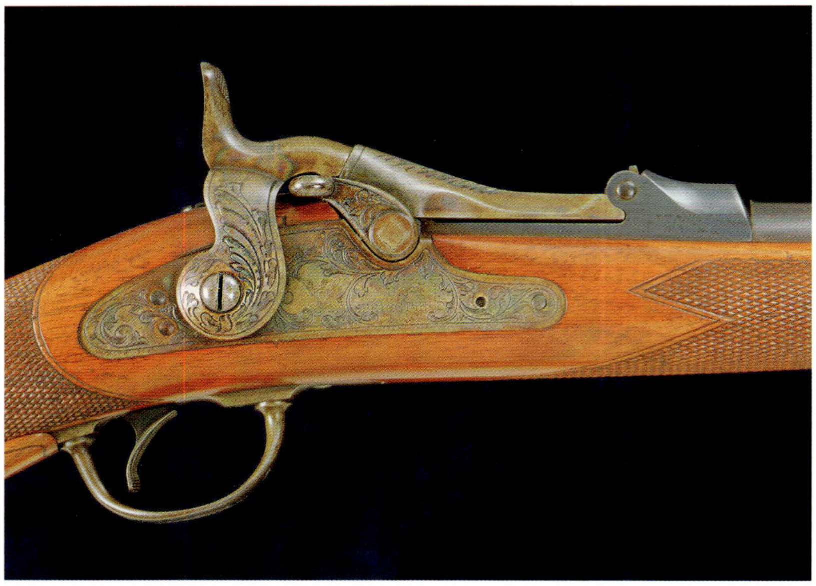

Breech-Loading Cartridge Rifle
United States Armory, Springfield, Massachusetts
U.S. Model 1875 Officer's Sporting, 1875–1885
Caliber .45-70 Government Centerfire
Weitzenhoffer Collection, 2001.46.091

Among the rarest and most desirable of all U.S. military firearms, this "Trapdoor" Officer's rifle epitomizes the old army adage that rank has its privileges. Only about 490 of these deluxe guns were made, each featuring a custom single set trigger, as well as special-order embellishments, stocking, and sights. Simple yet attractive foliate-scroll engraving—perhaps rendered by one-time Springfield employee F. W. Martin—decorates the buttplate, hammer, lockplate, breechblock, receiver, and trigger guard. The walnut stocking displays a delicately cast and engraved German silver forestock cap, as well as precise hand-checkering over the detachable pistol grip, wrist, and forestock. Special sights include folding Beach front and folding aperture rear adjustable for both windage and elevation.

Marlin Engraving

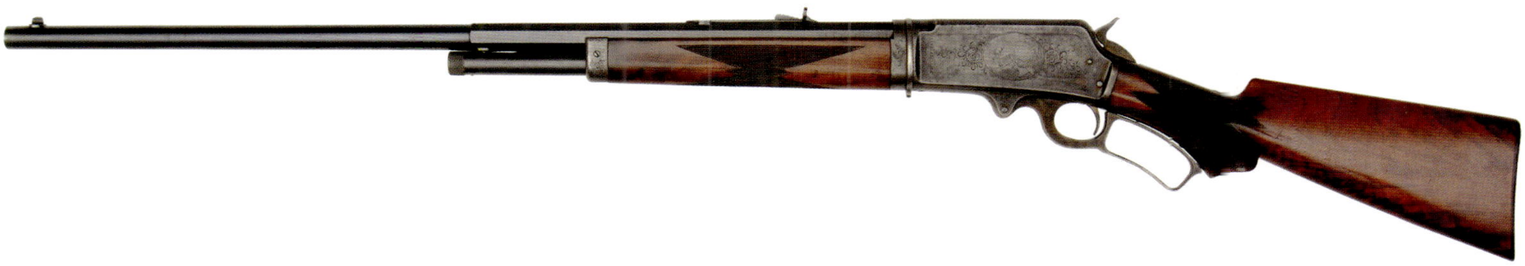

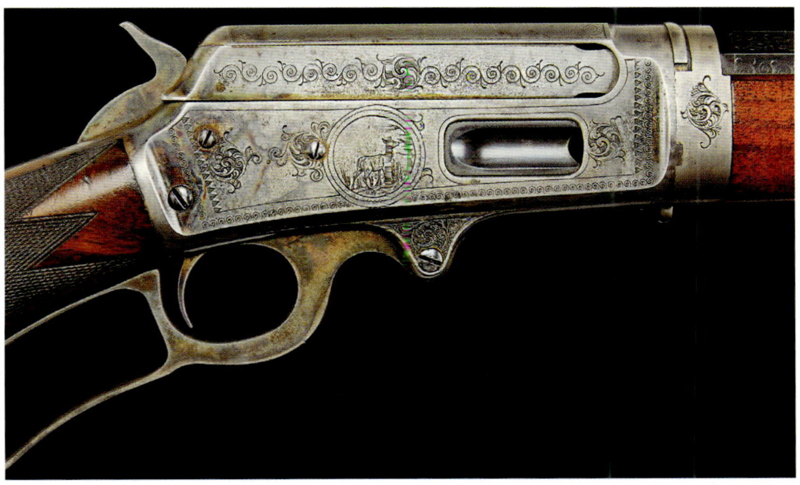

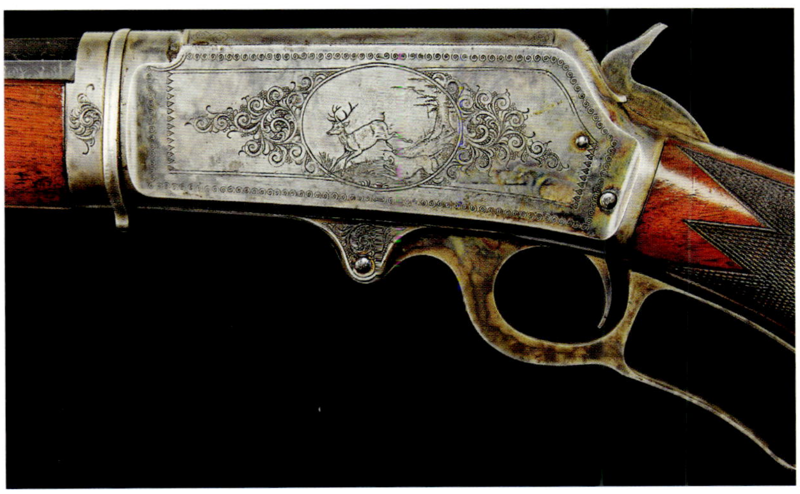

Breech-Loading Cartridge Rifle
Marlin Fire Arms Company, New Haven, Connecticut
Model 1895 Takedown Sporting, 1895–1917
Caliber .40-65 Winchester Centerfire
Grandee Collection, 2012.24

Discussed more generally in chapter 7, this deluxe, special-order Model 1895 Marlin repeater is finely engraved by master artisan Conrad F. Ulrich in Style No. 2. On the left side of the receiver is a fairly large oval tableau, or panel scene, of a running bull elk and hound in a rather detailed forest setting. On the right side appears a smaller, circular panel scene of a cow elk. Both of these big-game vignettes are flanked by fine scroll-and-arabesque engraving on constrained punched-dot fields. As well, the upper and lower tangs, the lever boss, the takedown ring, the forestock cap, and the barrel breech are engraved in matching style. This decoration is nicely complemented by highly figured, flame-and-shell-grained American walnut stocking that is hand-checkered in Style B over the wrist and forestock and finished with clear piano varnish.
Images courtesy James D. Julia Auctions, Fairfield, Maine

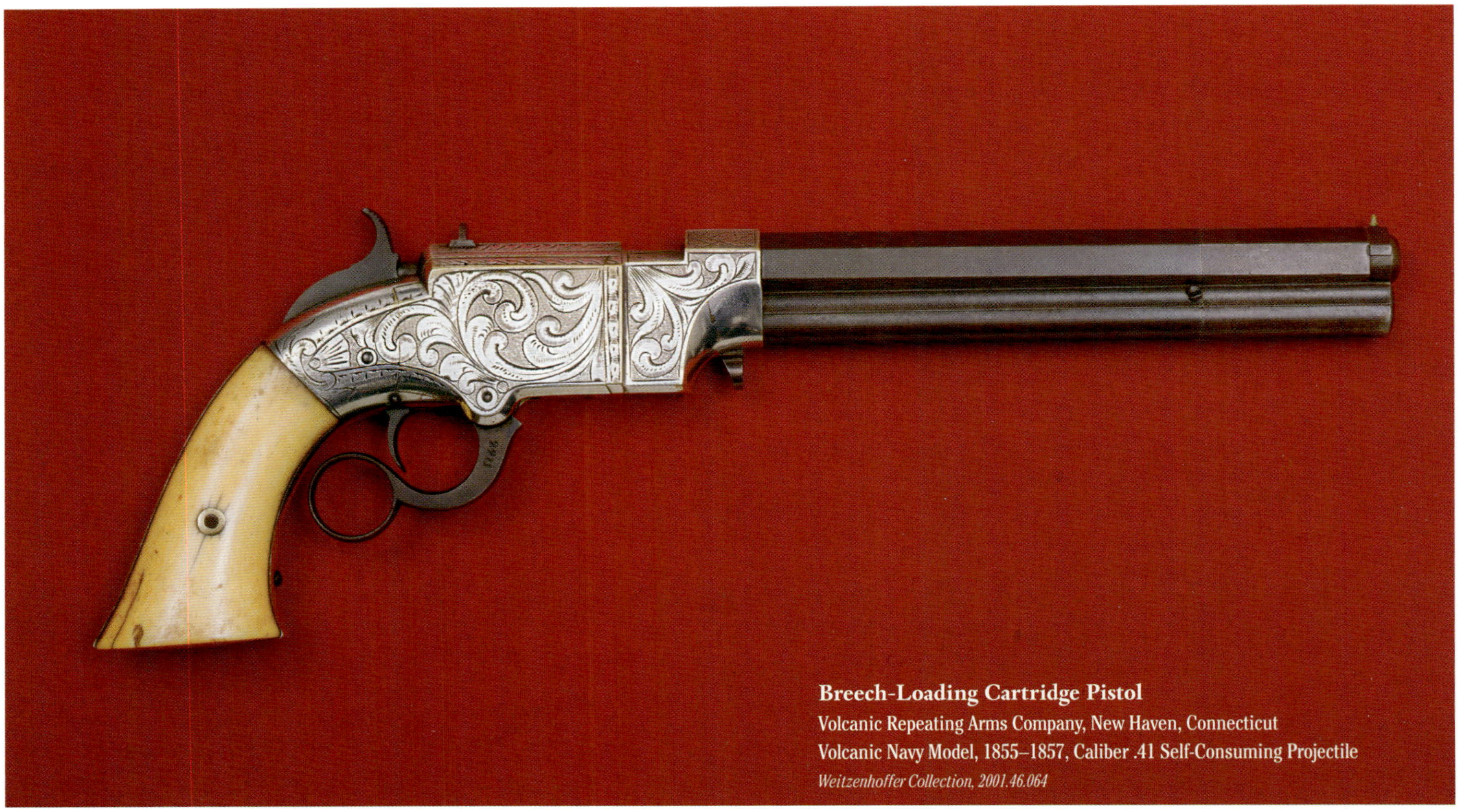

Breech-Loading Cartridge Pistol
Volcanic Repeating Arms Company, New Haven, Connecticut
Volcanic Navy Model, 1855–1857, Caliber .41 Self-Consuming Projectile
Weitzenhoffer Collection, 2001.46.064

A direct predecessor of the famed series of Winchester repeating rifles, this handgun incorporates a lever-actuated, toggle-link breech mechanism joined with an integral tubular magazine beneath the barrel. The pistol fired special self-consuming (caseless) ammunition consisting of a hollow, conical bullet containing its own powder charge and combustible primer. This very deluxe, ivory-stocked, engraved, and silver-plated pistol presents a far more extensive and lavish treatment than the company's standard embellishment (see chapter 8 and above). The broad and relatively open acanthus-style scrollwork is emphasized by a delicately crosshatched, or matte, background not seen in standard decoration. The engraving on this pistol, in fact, is virtually identical in coverage and style to that on the pair of deluxe pistols that belonged to the company's president, Oliver Winchester.

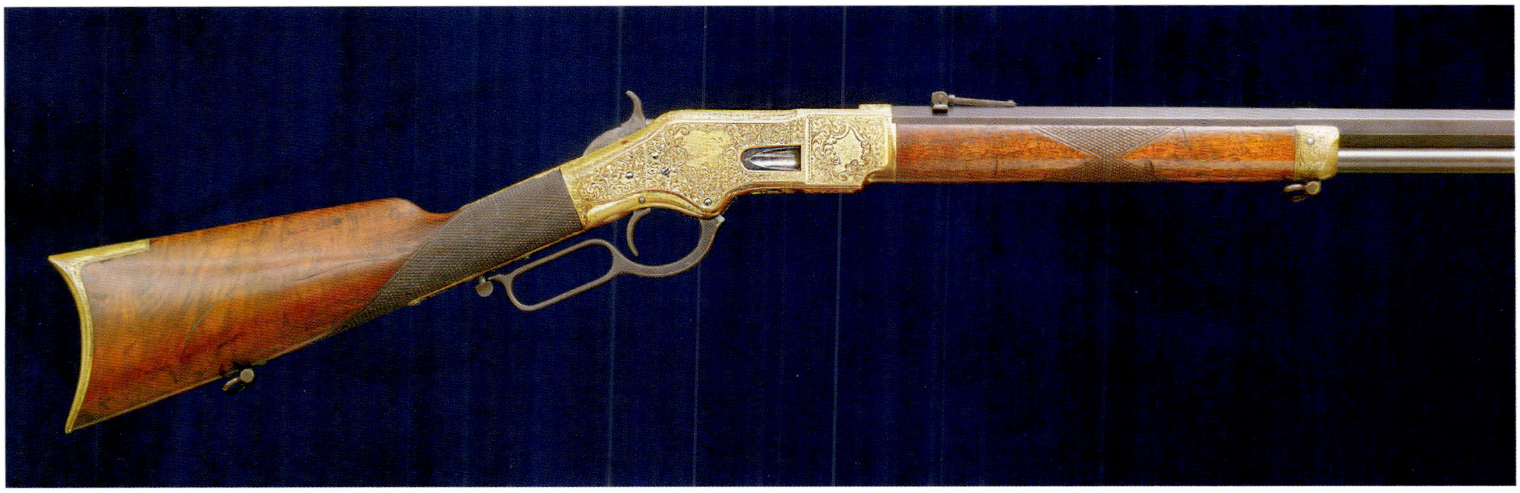

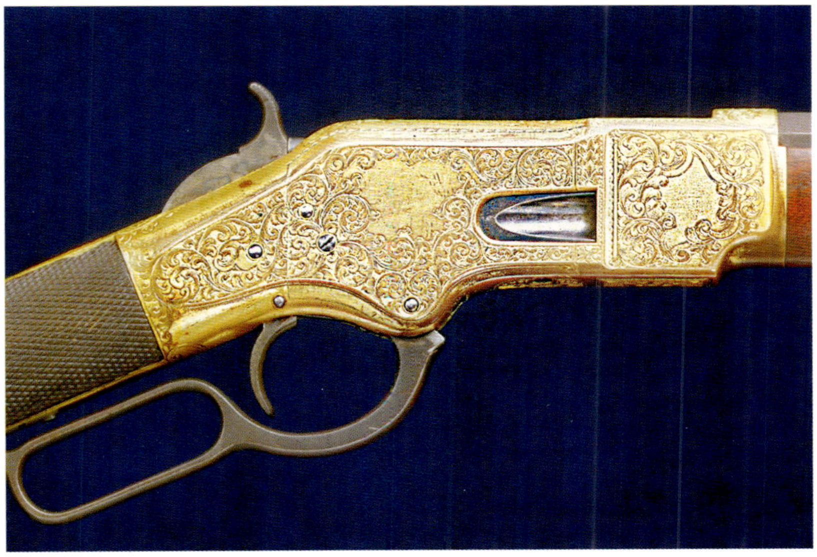

Breech-Loading Cartridge Rifle

Winchester Repeating Arms Company, New Haven, Connecticut
Model 1866 Sporting, 1866–1898, Caliber .44 Henry Flat Rimfire
Weitzenhoffer Collection, 2001.46.066

Although the Ulrich brothers (Conrad, Herman, and John) are the best-known engravers of deluxe nineteenth-century Winchesters, equally exquisite ornamentation came from the bench of freelance artisan Louis Daniel Nimschke of New York City. Among his finest work, the gold-plated receiver, forestock cap, and buttplate of this piece display crisp, relatively dense, yet flowing scrollwork over a punched-dot ground. The left frame bears a superbly rendered panel scene of a stag pursued by a hound, a motif used by Nimschke on several Model 1866 rifles. Unlike most artisans of his era, Nimschke frequently signed his work—here with the tiny initials "L.D.N." in a riband on the bottom of the frame and with "LDN" below the hound in the panel scene. Facsimile "pulls," or negative copy prints for this rifle appear in Nimschke's published scrapbook, *L. D. Nimschke Firearms Engraver* (1964), page 56.

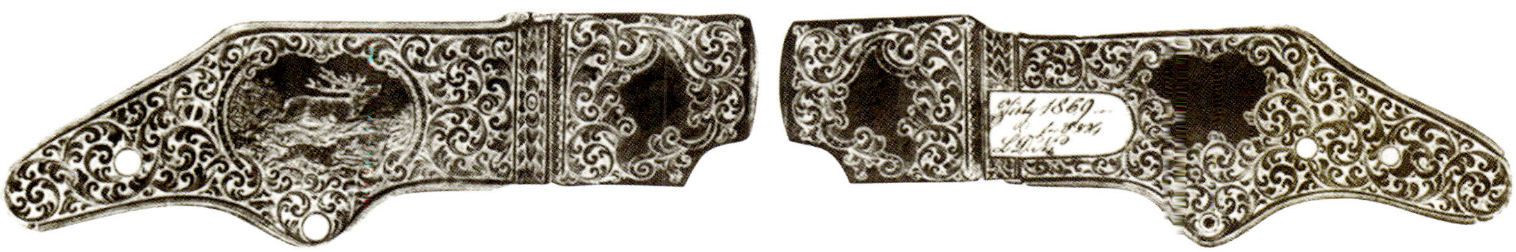

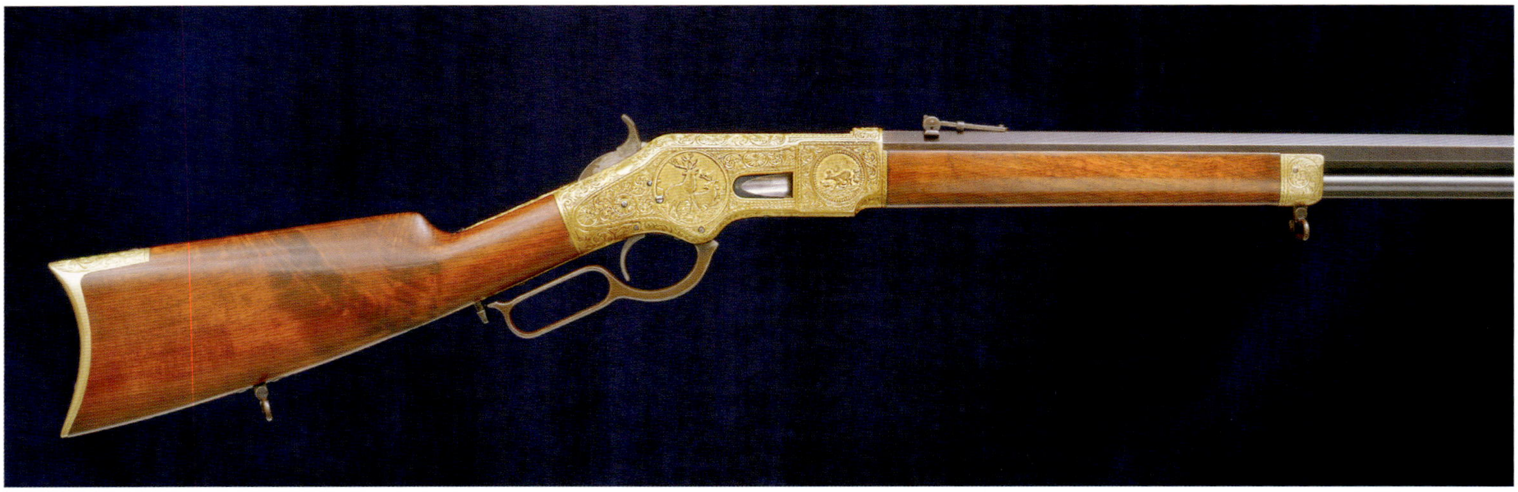

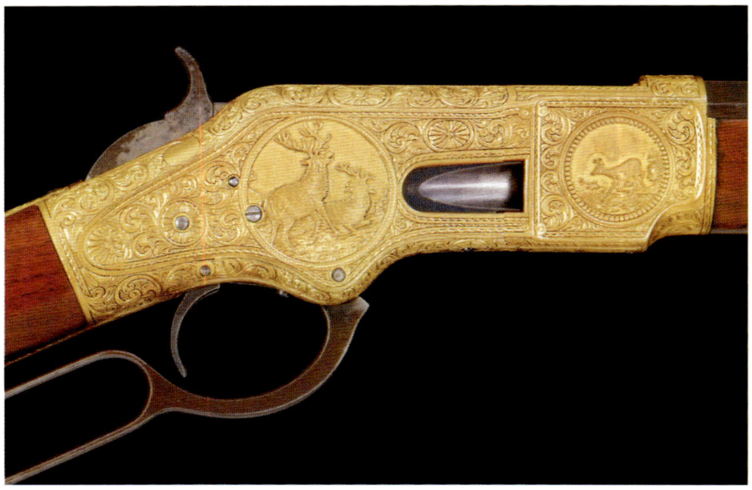

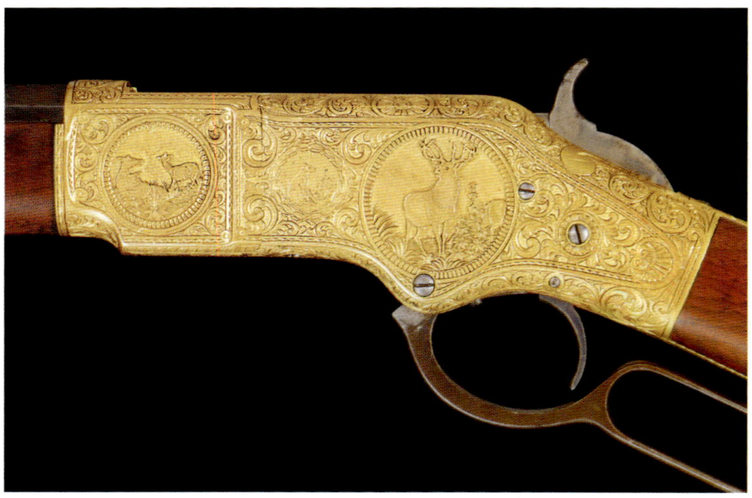

Breech-Loading Cartridge Rifle
Winchester Repeating Arms Company, New Haven, Connecticut
Model 1866 Sporting, 1866–1898, Caliber .44 Henry Flat Rimfire
Museum Purchase, 2001.35

A masterpiece in artistic style, quality, and execution, this Model 1866 showpiece was engraved by Conrad F. Ulrich late in 1875. A protégé of master engraver Gustave Young at Colt's, the elder Ulrich brother proved instrumental in establishing Winchester's reputation for exhibition-grade firearms. This exquisite example features superb floral scrollwork and shell motifs over a punched-dot, or stippled, ground on the buttplate, receiver, and forestock cap. The left frame presents two relief-engraved panel scenes of a bull elk and a doe with fawns, plus a small, wreath-enclosed rendering of a hound. The right frame displays two additional relief-cut panels depicting elk and antelope. The lower tang is twice stamped "C. F. Ulrich." This magnificent exhibition rifle is very likely one of those displayed by Winchester at the Philadelphia Centennial celebration in 1876.

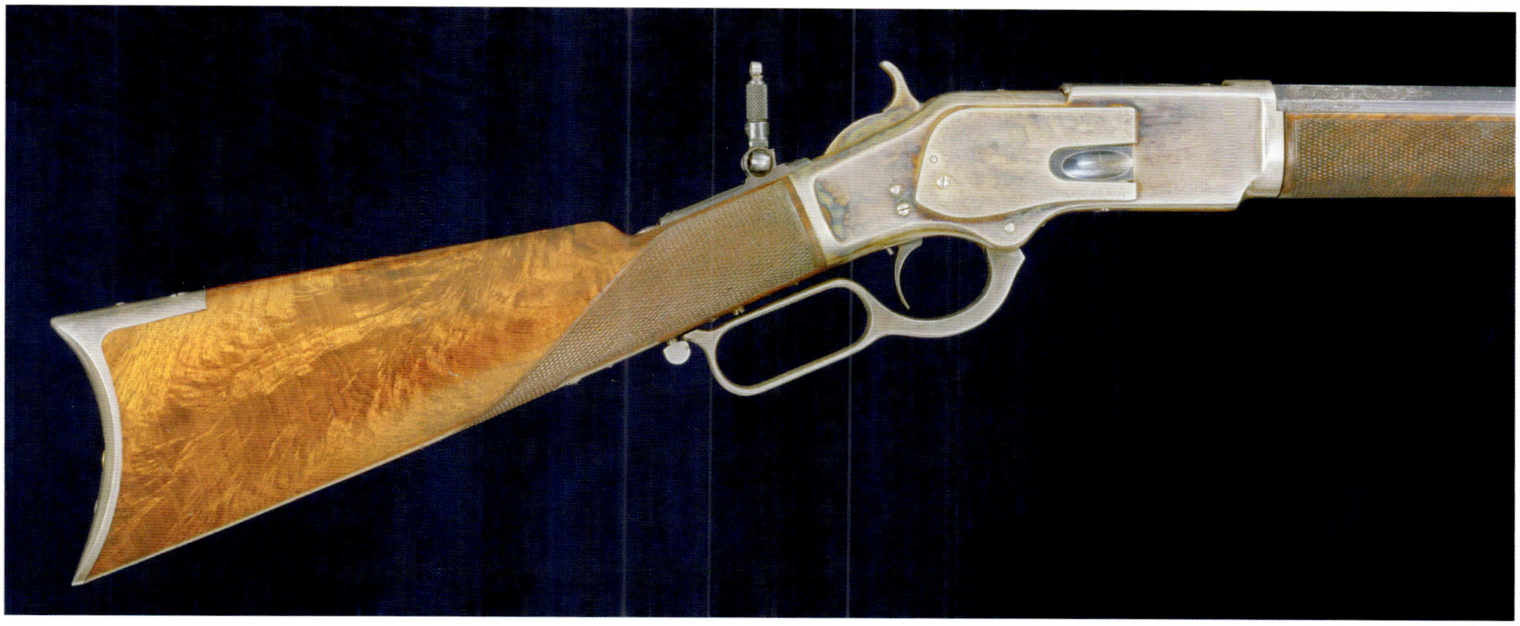

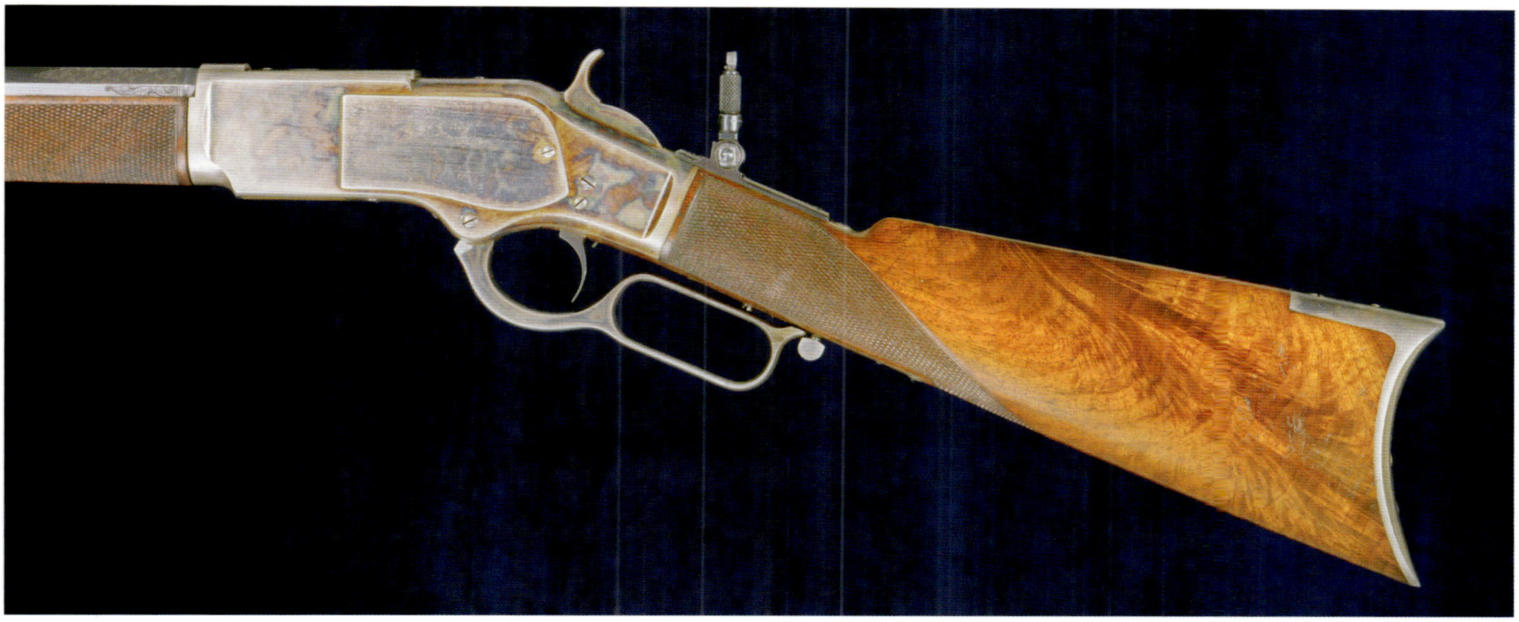

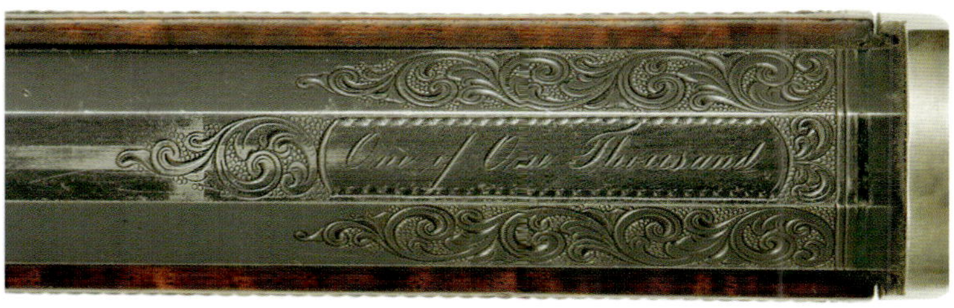

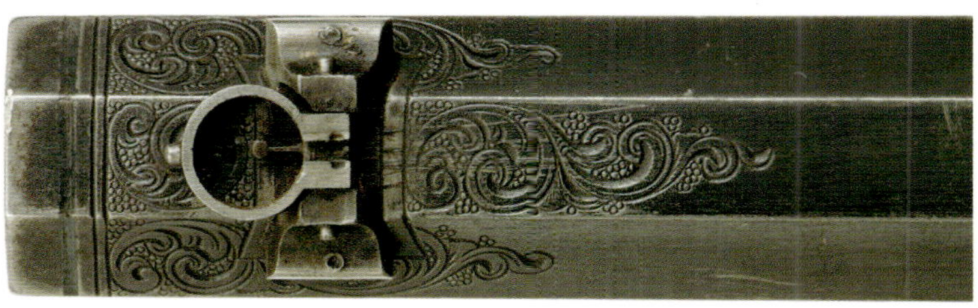

Breech-Loading Cartridge Rifle

Winchester Repeating Arms Company, New Haven, Connecticut
Model 1873 "One of One Thousand" Sporting (Late First Variation), 1873–1919
Caliber .44 Winchester Centerfire
Weitzenhoffer Collection, 2001.46.068

Among the ultimate Winchester rarities, this rifle is one of a very select group singled out for its superb accuracy and appropriately embellished by the factory. The firm's 1875 trade catalog stated: "All of those barrels . . . found to make targets of extra merit will be made up into guns with set-triggers and extra finish, marked 'one of a thousand,' and sold at $100." Made in 1876, the barrel of this piece features fine scroll engraving and inlaid silver bands at both the muzzle and the breech, with "One of One Thousand" hand-engraved on the top flat. (Only 133 of 720,600 Model 1873s were so marked.) The deluxe, XXXX-grade walnut stocking is neatly checkered at the wrist and forestock. The rifle features the prescribed single-set trigger and has Beach front, single-leaf folding rear, and adjustable peep sights.

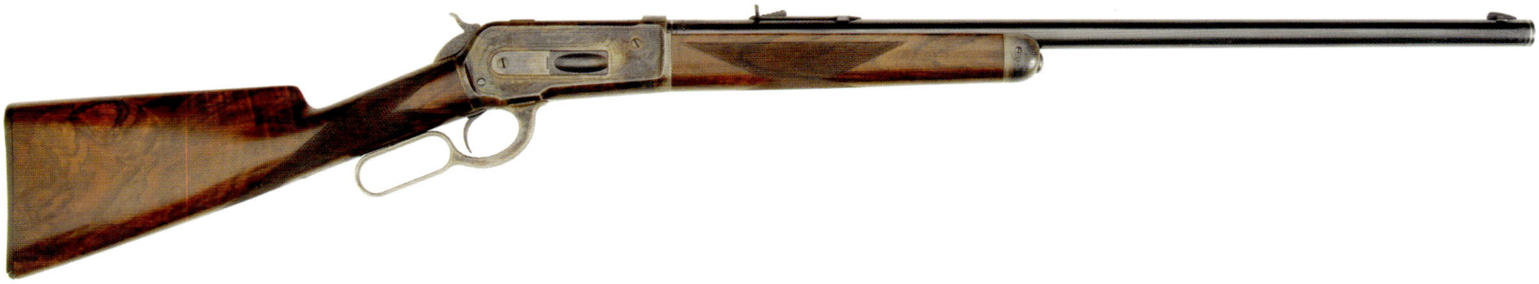

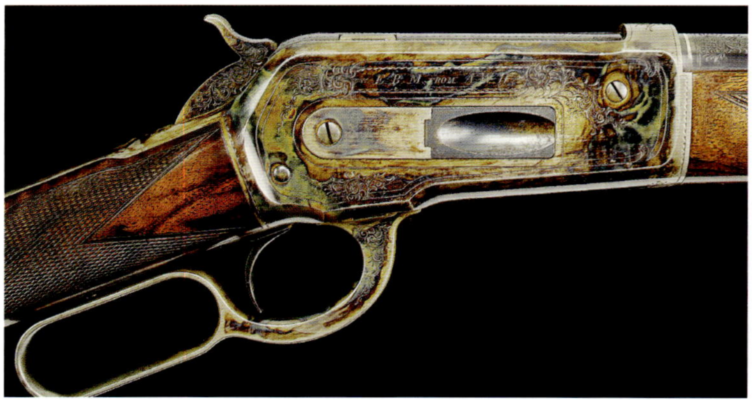

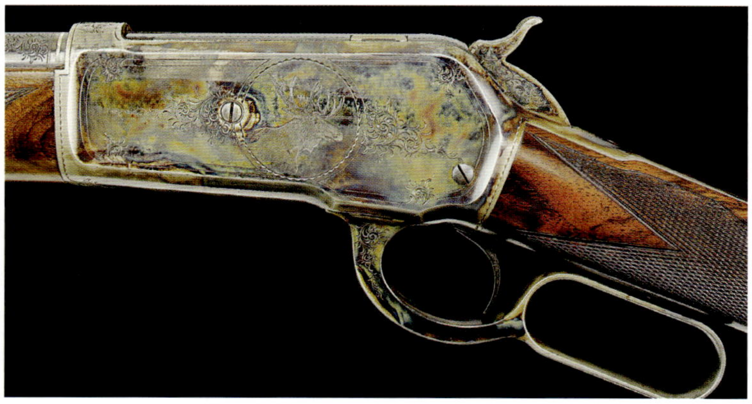

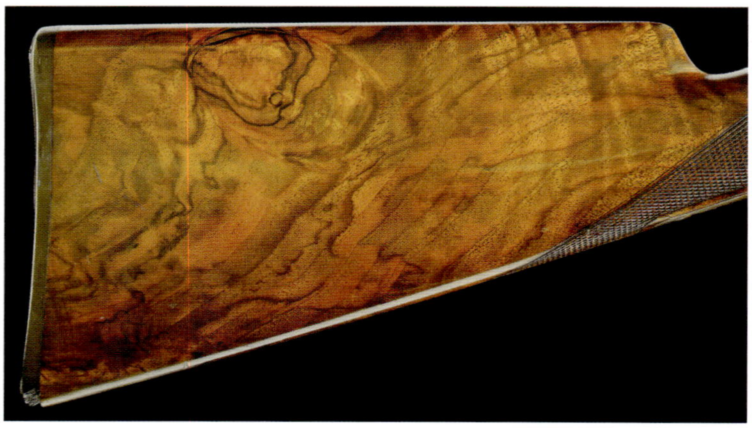

Breech-Loading Cartridge Rifle

Winchester Repeating Arms Company, New Haven, Connecticut
Model 1886 Sporting, 1886–1935, Caliber .45-90 Winchester Centerfire
Museum Purchase, 2012.23

This special-order, factory-engraved Model 1886 Winchester features a 26-inch round barrel with half, button-type magazine, Rocky Mountain front and semibuckhorn rear sights, and deluxe figured stocking with a shotgun-pattern butt. Signed by master artisan John Ulrich on the lower tang, the rifle is engraved in Style No. 10 with a No. 7 bull-moose vignette on the left side of the receiver—as later illustrated in Winchester's special 1897 *Highly Finished Arms* catalog. The right side also is engraved in identical arabesque scrollwork over a clear field and with a presentation panel above the loading gate. The sides of the lever, hammer, and forestock cap, as well as the top tang, bolt, and barrel, are engraved to match. The barrel also boasts inlaid platinum bands at the breech and muzzle. The select stocking of burled walnut features rare flat-point checkering in Style H over the wrist and forestock and is sealed with a hand-rubbed oil finish.

Images courtesy James D. Julia Auctions, Fairfield, Maine

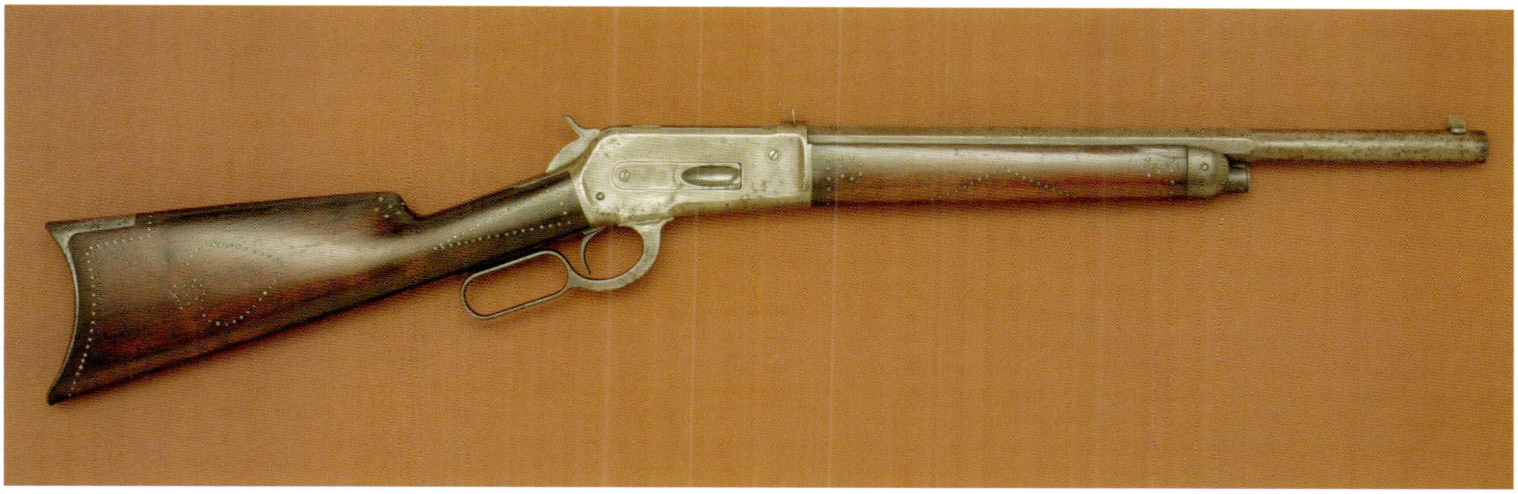

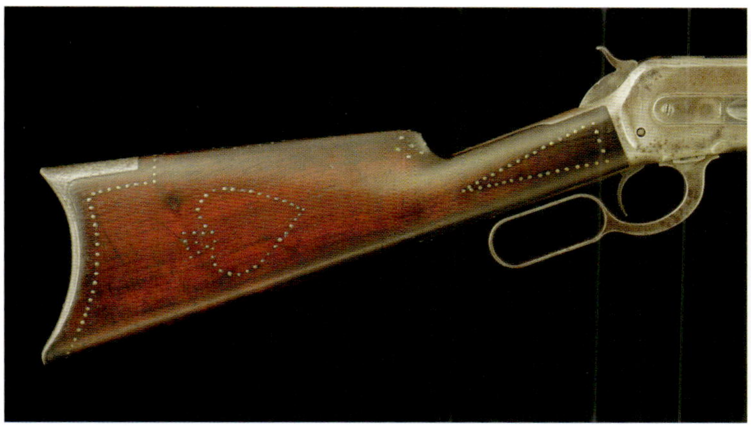

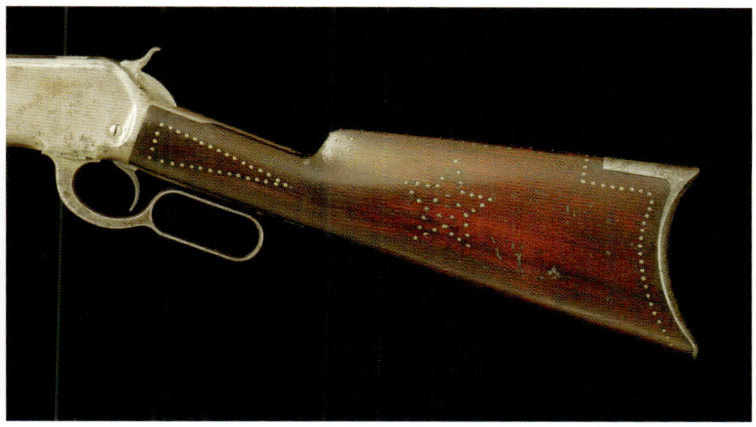

Breech-Loading Cartridge Rifle

Winchester Repeating Arms Company, New Haven, Connecticut

Model 1886 Sporting, 1886–1935, Caliber .45-70 Government Centerfire

Grandee Collection, 1991.01.1328

Evocative of cowboy use, this well-worn rifle displays aftermarket alterations and decorations far removed from the factory. Its once-full-octagon barrel has been shortened and turned down to half-octagon configuration, and the magazine tube has likewise been shortened. The walnut stock displays late nineteenth-century vernacular expression in its embellishment with a multitude of small brass nails. Meticulously laid out and precisely spaced, these present handsome spade and star motifs at right and left center, respectively, with additional applications tracing the upper and lower profiles of the stock. Though hardly the expression of a master artisan, such finely applied ornamentation certainly demanded time, talent, and close attention in execution.

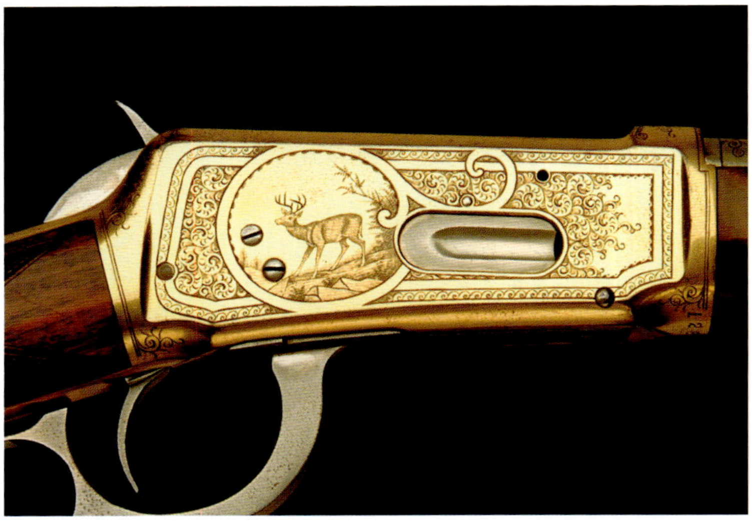

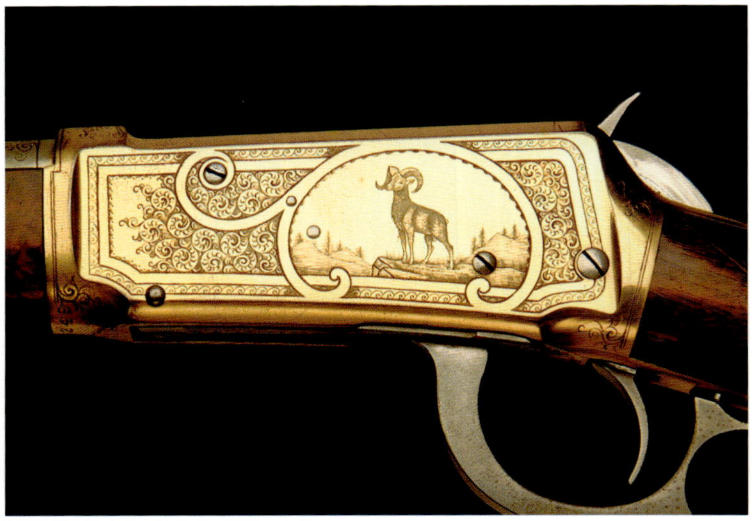

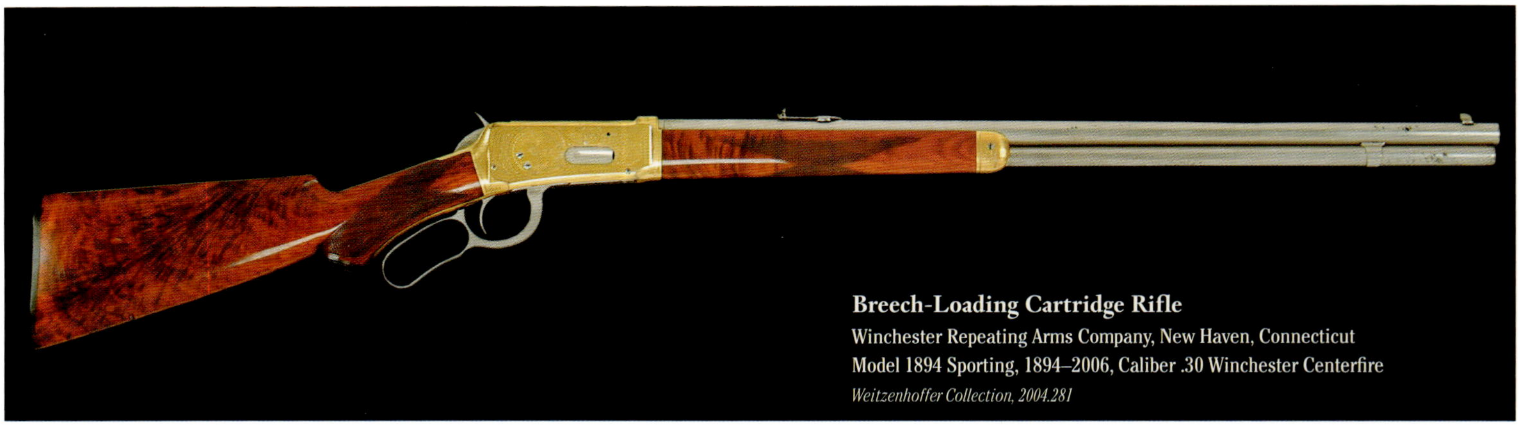

Breech-Loading Cartridge Rifle
Winchester Repeating Arms Company, New Haven, Connecticut
Model 1894 Sporting, 1894–2006, Caliber .30 Winchester Centerfire
Weitzenhoffer Collection, 2004.281

This Model 1894 "exhibition" rifle presents a fine example of the deluxe embellishment available in Winchester's 1897 *Highly Finished Arms* catalog. The highly figured, XXXX-grade American walnut stocking—with special-order pistol grip and shotgun-pattern butt—is checkered in Style H and finished with clear piano varnish. The receiver and forestock cap are engraved in Style 6 and gold-plated, while the remainder of the rifle is nickel-plated. Probably embellished by Winchester engraver John Ulrich, the receiver features vignettes of whitetail buck (right side) and bighorn sheep (left side) within border ribands centered amid tight arabesque, English-style scrollwork. In 1900, all this decorative detailing added some $70 to the standard retail cost of about $25. Factory records indicate that this deluxe gun remained with the company for nearly thirty years, charged to "sample consignment" or "exhibition showcase" twenty-seven times between 1904 and 1928, when it was finally sold into the market. Today the rifle presents clear evidence not only of the Winchester firm's aesthetic achievements a century ago but also something of its marketing strategies.

Images courtesy James D. Julia Auctions, Fairfield, Maine

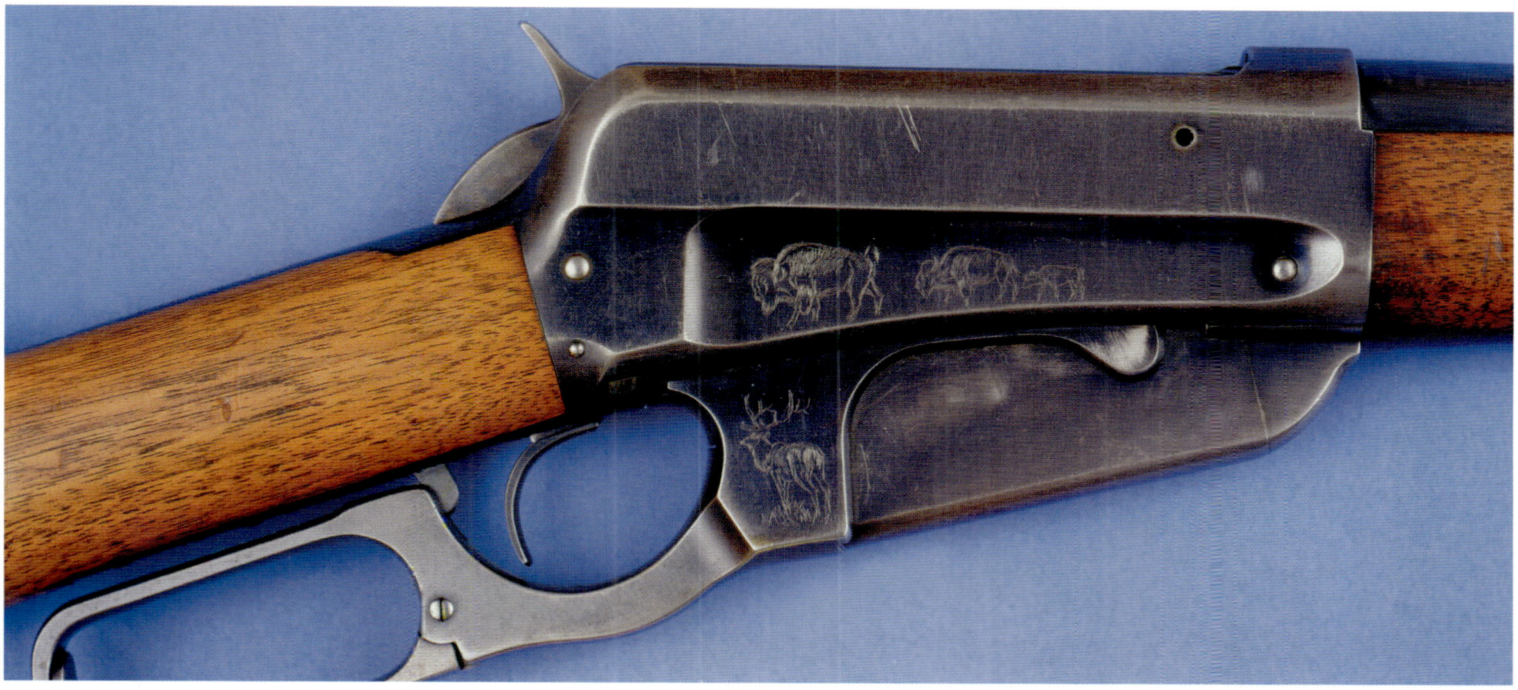

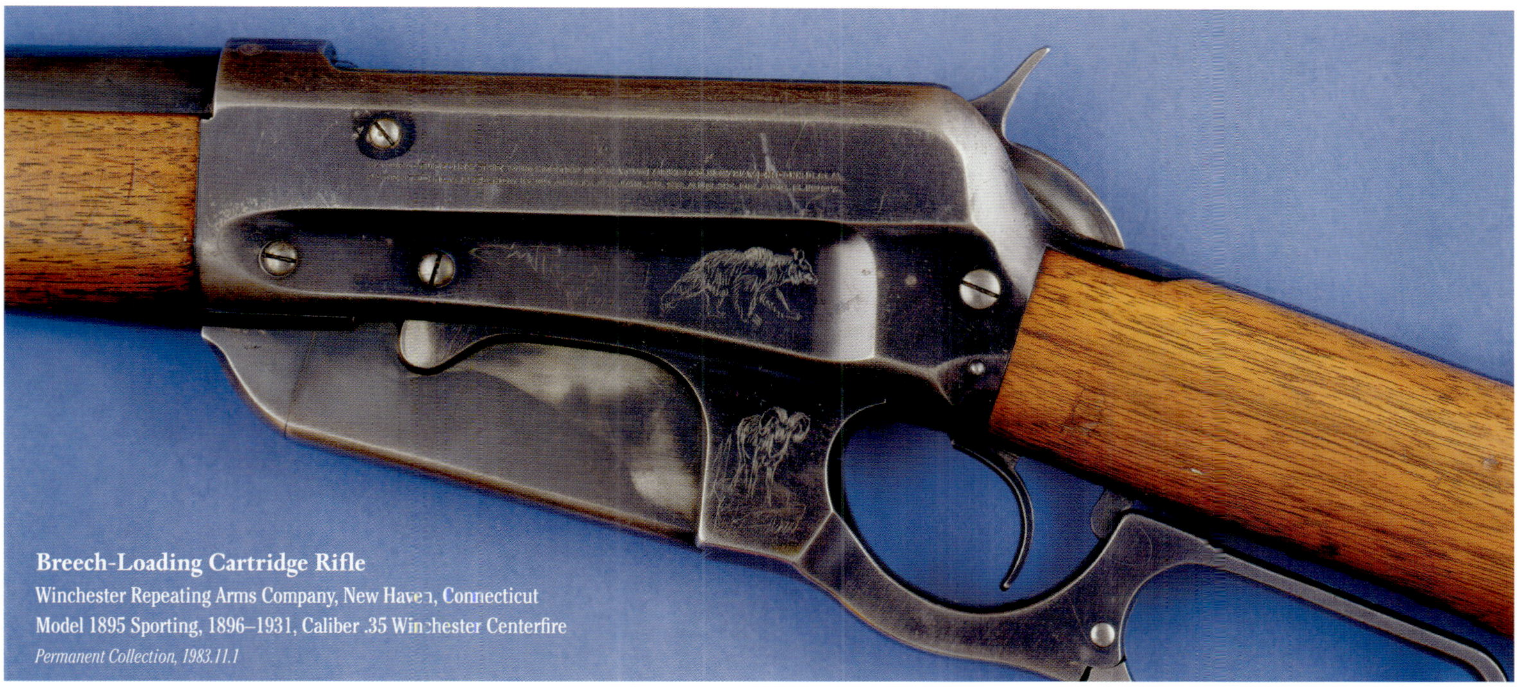

Breech-Loading Cartridge Rifle
Winchester Repeating Arms Company, New Haven, Connecticut
Model 1895 Sporting, 1896–1931, Caliber .35 Winchester Centerfire
Permanent Collection, 1983.11.1

Not all decorated sporting arms of the late nineteenth and early twentieth centuries received the attention of master artisans—a very select few were decorated instead by fine artists. This Model 1895 Winchester was lightly engraved with big-game animal motifs in 1913 by famed western artist Charles M. Russell for his hunting friend Frank Bird Linderman. The embellishments were rendered by Russell using a pocketknife by campfire light! On the right side are likenesses of a buffalo family and a bull elk, while on the left are a grizzly bear and a bighorn sheep along with Russell's "CMR" monogram. Linderman later recalled that the figures "seemed to walk out of the metal as though pricked into life by the blade." When finished, Russell observed, "Now the old gal will always have fresh meat in sight."

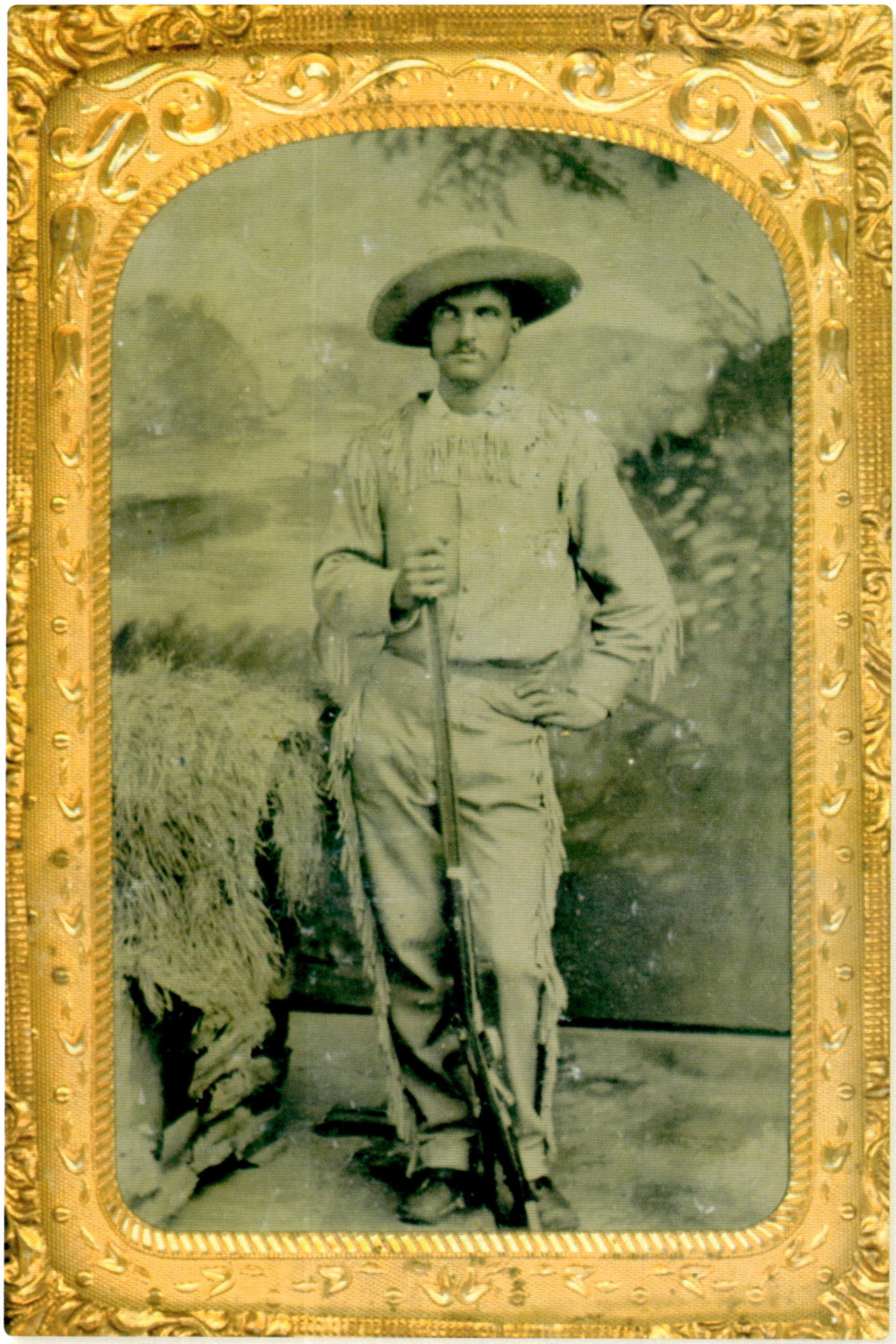

Photographer unknown. [Untitled]. Ninth-plate tintype, America, circa 1870. This buck-skinned frontier type holds a rather fancy, half-stock percussion sporting rifle highlighted with decorative German silver furniture that includes a finger-grip-pattern trigger guard and an inlet, hinged cap box.

Dickinson Research Center, National Cowboy & Western Heritage Museum, 2005.278.1

NOTES

PART I

From Craftsman to Industrial Mechanic: Evolution in American Arms

1. M. L. Brown, *Firearms in Colonial America: The Impact on History and Technology, 1492–1792* (Washington, D.C.: Smithsonian Institution Press, 1980), 149; Frank M. Sellers, *American Gunsmiths* (Highland Park, N.J.: Gun Room Press, 1983), 241. For historical background on the fundamental place and instrumental role of guns in early America, see Clayton E. Cramer, *Armed America: The Remarkable Story of How and Why Guns Became as American as Apple Pie* (Nashville, Tenn.: Nelson Current, 2006); and Jan E. Dizard, Robert Merrill Muth, and Stephen P. Andrews, Jr., eds., *Guns in America: A Reader* (New York: New York University Press, 1999).

2. For a history of the successive American frontiers on which firearms played a role, see Ray Allen Billington, *Westward Expansion: A History of the American Frontier*, 2nd ed. (New York: Macmillan, 1960). For a treatment of the transformation of manufacturing over the first half of the nineteenth century and the important place of arms making therein, see Brooke Hindle and Steven Lubar, *Engines of Change: The American Industrial Revolution, 1790–1860* (Washington, D.C.: Smithsonian Institution Press, 1986). Treatments of the larger cultural, political, social, and philosophical contexts in which America's industrial transformation occurred can be perused in Daniel J. Boorstin, *The Americans: The National Experience* (New York: Random House, 1965), esp. chaps. 4–5; Pauline Maier, Merritt Roe Smith, Alexander Keyssar, and Daniel J. Kevles, *Inventing America: A History of the United States* (New York: W. W. Norton, 2003), chaps. 2–18; Daniel Walker Howe, *What Hath God Wrought: The Transformation of America, 1815–1848* (Oxford: Oxford University Press, 2007); and David E. Nye, *America as Second Creation: Technology and Narratives of New Beginnings* (Cambridge, Mass.: MIT Press, 2003).

3. For technical and cultural overviews of firearms evolution, see W. W. Greener, *The Gun and Its Development*, 9th ed. (London: Cassell, 1910); and Robert Held, *The Age of Firearms: A Pictorial History* (New York: Harper and Brothers, 1957), esp. chaps. 10–12. A selective technical and evolutionary treatment specific to the design and function of American firearms can be found in David F. Butler, *United States Firearms. The First Century, 1776–1875* (New York: Winchester Press, 1973).

4. William Hosley, "Guns, Gun Culture, and the Peddling of Dreams," in Dizard et al., *Guns in America*, 49. For some notion of the volume and diversity of arms (both commercial and surplus military) shipped to the West in the later nineteenth century, see Herbert G. Houze, *Arming the West: Schuyler, Hartley & Graham's Arms Shipments to the American Frontier, 1868–1886* (Woonsocket, R.I.: Andrew Mowbray, 2008).

CHAPTER 1

Firearms of the Artisan Gunsmith

1. See, for example, Brown, *Firearms in Colonial America*, 22–23, 87, 91, 393; Merrill Lindsay, *The Kentucky Rifle* (New York: Arma Press and Historical Society of York County [Pa.], 1972), 3; Joe K. Kindig III, *Artistic Ingredients of the Longrifle* (York, Pa.: George Shumway, 1989), 54–55; and Michael S. Raber, Patrick M. Malone, Robert B. Gordon, and Carolyn C. Cooper, *Conservative Innovators and Military Small Arms: An Industrial History of the Springfield Armory, 1794–1968* (South Glastonbury, Conn.: Raber Associates, 1989), 238–41, 292–95. For a cultural treatment of traditional, European gun-making practices via the guild and putting-out systems, see Claude Gaier, *Four Centuries of Liege Gunmaking* (Liege, Belgium: Librairie Halbart, 1976).

2. The entire Roessor probate inventory is reproduced in Henry J. Kauffman, *The Pennsylvania-Kentucky Rifle* (New York: Bonanza Books, 1960), 147. For illustrated descriptions of the individual gunsmith's tools and their use circa 1750–1850, see Brown, *Firearms in Colonial America*, 244–53; and Ned H. Roberts, *The Muzzle-Loading Cap Lock Rifle* (Manchester, N.H.: Clarke Press, 1944), 317–29.

3. Brown, *Firearms in Colonial America*, 244; Richard C. Rattenbury, "J. C. Petmecky: Texas Gun and Spur Maker," *Man at Arms* 11, no. 5 (1989): 11.

4. For discussions of the gun-making process during the era of the singular craftsman-gunsmith, see Kauffman, *Pennsylvania-Kentucky Rifle*, 160–68; and Roberts, *Muzzle-Loading Cap Lock Rifle*, 10–45. To peruse a first-person memoir of rifle making by traditional craft methods, see that of apprentice Milton Warren of Wolf Hills, Virginia, circa 1850, in Capt. John G. W. Dillin, *The*

Kentucky Rifle (New York: Ludlum and Beebe, 1946), 29–33. For a most informative audiovisual presentation of eighteenth-century Kentucky rifle production, also see the 2009 DVD release of *The Gunsmith of Williamsburg* (Colonial Williamsburg Productions, 1969). A general appreciation of the individual gun-making business can be found in James B. Whisker, *The Gunsmith's Trade* (Lewiston, N.Y.: E. Mellon Press, 1992).

5. Kauffman, *Pennsylvania-Kentucky Rifle*, 155–60, 165–66.

6. Curtis L. Johnson, *Gunmakers of Illinois, 1683–1900*, vol. 1 (York, Pa.: George Shumway, 1994), 121–23.

7. Kauffman, *Pennsylvania-Kentucky Rifle*, 160; Louis A. Garavaglia and Charles G. Worman, *Firearms of the American West, 1803–1865* (Albuquerque: University of New Mexico Press, 1984), 34–37, 229–30; Maier et al., *Inventing America*, 297–99, 310–12. For an exhaustive directory of domestic arms makers and manufacturers from the colonial era to the late twentieth century, see Sellers, *American Gunsmiths*.

8. Stephen V. Grancsay, "The Craft of the Early American Gunsmith," *Metropolitan Museum of Art Bulletin*, new series, vol. 6, no. 2 (1947): 54; Lindsay, *Kentucky Rifle*, 3–5.

9. Grancsay, "Craft," 56–57; Greener, *Gun and Its Development*, 620–21. Hanger quoted in Brown, *Firearms in Colonial America*, 337.

10. Kauffman, *Pennsylvania-Kentucky Rifle*, 5–8, 12–20; Brown, *Firearms in Colonial America*, 261–62, 265–68; Lindsay, *Kentucky Rifle*, 5–51, passim.

11. Charles E. Hanson, Jr., *The Plains Rifle* (Harrisburg, Pa.: Stackpole, 1960), 153–54; Dillin, *Kentucky Rifle*, 125–27; Kauffman, *Pennsylvania-Kentucky Rifle*, 93–98; Garavaglia and Worman, *Firearms of the American West, 1803–1865*, 81–93.

12. Pike quoted in Garavaglia and Worman, *Firearms of the American West, 1803–1865*, 8.

13. Richard C. Rattenbury, "The Mountain Rifle: Fabled Arm of the Fur Trappers," *Persimmon Hill* 25, no. 4 (1997): 47; Garavaglia and Worman, *Firearms of the American West, 1803–1865*, 33–34.

14. Garavaglia and Worman, *Firearms of the American West, 1803–1865*, 34–37; Rattenbury, "Mountain Rifle," 47–48.

15. Garavaglia and Worman, *Firearms of the American West, 1803–1865*, 38–41; Hanson, *Plains Rifle*, 37–38.

16. Hamilton quoted in Garavaglia and Worman, *Firearms of the American West, 1803–1865*, 45.

17. Ibid., 44–45. For an in-depth treatment of the Hawken enterprise, its business history, and its larger-than-life stature in western history and myth, see Charles E. Hanson, Jr., *The Hawken Rifle: Its Place in History* (Chadron, Nebr.: Fur Press, 1979).

18. Rattenbury, "Mountain Rifle," 49.

19. Carl P. Russell, *Guns on the Early Frontiers* (Berkeley: University of California Press, 1957), 130–39; Hanson, *Plains Rifle*, 5–8, 74–78; Garavaglia and Worman, *Firearms of the American West, 1803–1865*, 34–37, 350–55.

20. Greener, *Gun and Its Development*, 112–16; Garavaglia and Worman, *Firearms of the American West, 1803–1865*, 41–43.

21. Roberts, *Muzzle-Loading Cap Lock Rifle*, 190–91. Gregg and Stewart quoted in Richard C. Rattenbury, *Hunting the American West: The Pursuit of Big Game for Life, Profit, and Sport, 1800–1900* (Missoula, Mont.: Boone and Crockett Club, 2008), 124.

22. Garavaglia and Worman, *Firearms of the American West, 1803–1865*, 54–58, 95–98.

23. Ibid., 11–12, 59–60, 237–40.

CHAPTER 2
Firearms of the American System

1. For a scattered discussion of the transition from product-oriented to process-oriented firearms manufacture, see Gene S. Cesari, "American Arms Making Machine Tool Development," Ph.D. dissertation, University of Pennsylvania, 1970. For treatments of the origins and ramifications of the American System, see Otto Mayr and Robert C. Post, eds., *Yankee Enterprise: The Rise of the American System of Manufactures* (Washington, D.C.: Smithsonian Institution Press, 1981); and David A. Hounshell, *From the American System to Mass Production, 1800–1932* (Baltimore, Md.: Johns Hopkins University Press, 1984). For cultural treatments of the transition from craft to machine production in American society generally, and among domestic arms and ammunition manufacturers particularly, see Maier et al., *Inventing America*, 390–410; Merritt Roe Smith, "From Craftsman to Mechanic: The Harpers Ferry Experience, 1798–1854," in *Technological Innovation and the Decorative Arts*, Winterthur Conference Report 1973 (Charlottesville: University Press of Virginia, 1974), 103–39; and James J. Farley, *Making Arms in the Machine Age: Philadelphia's Frankford Arsenal, 1816–1870* (University Park: Pennsylvania State University Press, 1994). For a treatment of other early products manufactured with serial, machine production under the American System (specifically, axes, typewriters, and pocket watches), see Donald R. Hoke, *Ingenious Yankees: The Rise of the American System of Manufactures in the Private Sector* (New York: Columbia University Press, 1990).

2. Hounshell, *From the American System*, 25–28; Maier et al., *Inventing America*, 322–23. For a discussion of the methods and machines used in achieving interchangeability and the centrality and impact of powered machine tools in domestic arms making, see Felicia Johnson Deyrup, *Arms Making in the Connecticut Valley: A Regional Study of the Economic Development of the Small Arms Industry, 1798–1870* (York, Pa.: George Shumway, 1970), 89–99, 146–59. For detail on the evolution, design, and function of domestic arms-making machinery circa 1815–1870, see Charles H. Fitch, "Report on the Manufactures of Interchangeable Mechanism," in *Tenth Census of the United States (1880)*, vol. 2 (Washington, D.C.: Government Printing Office, 1883), 617–45; and Cesari, "American Arms Making."

3. Whitney quoted in Boorstin, *Americans*, 33.

4. Edwin A. Battison, *Muskets to Mass Production: The Men and the Times That Shaped American Manufacturing* (Windsor, Vt.: American Precision Museum, 1976), 8–9; Hounshell, *From the American System*, 29–32; David Freeman Hawke, *Nuts and Bolts of the Past: A History of American Technology, 1776–1860*

(New York: Harper and Row, 1989), 101–102. For a detailed refutation of Eli Whitney's purported early use of the milling machine and his supposed central role in establishing the American System, see Edwin A. Battison, "Eli Whitney and the Milling Machine," *Smithsonian Journal of History* 1, no. 2 (1966): 9–11, 19–26; and Robert S. Woodbury, "The Legend of Eli Whitney and Interchangeable Parts," *Technology and Culture* 1, no. 2 (1960): 235–53. A discussion of Whitney's failings and his legitimate contributions to arms manufacture also can be perused in Cesari, "American Arms Making," 35–88, passim; and in Merritt Roe Smith, "Eli Whitney and the American System of Manufacturing," in *Technology in America: A History of Individuals and Ideas*, ed. Carroll W. Pursell, Jr., 2nd ed. (Cambridge, Mass.: MIT Press, 1990), 45–61.

5. North contract quoted in Joseph W. Roe, *English and American Tool Builders* (New Haven, Conn.: Yale University Press, 1916), 134.

6. Battison, *Muskets to Mass Production*, 10–11; Hounshell, *From the American System*, 29; Mayr and Post, *Yankee Enterprise*, 75–78. For the types, importance, and extensive use of gauges, see Raber et al., *Conservative Innovators*, 133–37; and Cesari, "American Arms Making," 259–61, 356–57. For the origin and importance of the milling machine in arms making, see Cesari, "American Arms Making," 140–44; Merritt Roe Smith, "John H. Hall, Simeon North, and the Milling Machine: The Nature of Innovation among Antebellum Arms Makers," *Technology and Culture* 14, no. 4 (1973): 574–76; and Robert S. Woodbury, *History of the Milling Machine* (Cambridge, Mass.: Technology Press, 1960). For an early, perhaps overly laudatory, appreciation of North's contributions, see S. N. D. North and Ralph North, *Simeon North: First Official Pistol Maker of the United States* (Concord, N.H.: Rumford Press, 1913).

7. Merritt Roe Smith, *Harpers Ferry Armory and the New Technology: The Challenge of Change* (Ithaca, N.Y.: Cornell University Press, 1977), 124–34; Hounshell, *From the American System*, 35–38; and Cesari, "American Arms Making," 240–47. For details on Blanchard's serial, machine-production method, see Carolyn C. Cooper, "'A Whole Battalion of Stockers': Thomas Blanchard's Production Line and Hand Labor at Springfield Armory," *Journal of the Society for Industrial Archeology* 14, no. 1 (1988): 37–56; Cesari, "American Arms Making," 338–41; and Raber et al., *Conservative Innovators*, 314–20.

8. Smith, *Harpers Ferry Armory*, 210–12, 225–40; Hounshell, *From the American System*, 39–43; Battison, *Muskets to Mass Production*, 12–13; Hawke, *Nuts and Bolts*, 109–13; and Raber et al., *Conservative Innovators*, 333–35. For a treatment of Hall's rifle design and its manufacture, see R. T. Huntington, *Hall's Breechloaders* (York, Pa.: George Shumway, 1972).

9. Raber et al., *Conservative Innovators*, 458–66; Maier et al., *Inventing America*, 323; Deyrup, *Arms Making*, 13–16; Cesari, "American Arms Making," 253–54; Mayr and Post, *Yankee Enterprise*, 128–29. An intriguing diagram illustrating the sharing and diffusion of manufacturing knowledge among northeastern firearm and machine-tool manufacturers, entitled "Genealogy of the New England Gun Makers," appears in Roe, *English and American Tool Builders*, 139. The lagging state of firearms manufacture in England and Europe is treated, respectively, in Russell I. Fries, "British Response to the American System: The Case of the Small-Arms Industry after 1850," *Technology and Culture* 16, no. 3 (1975): 379–82, 388–93; and Gaier, *Four Centuries of Liege Gunmaking*, 117–37.

10. Excerpts from the British 1855 *Report of the Committee on Machinery* quoted in Nathan Rosenburg, ed., *The American System of Manufactures* (Edinburgh: Edinburgh University Press, 1969), 129, 143. For a full machine inventory at the Sharps-Robbins & Lawrence plant in Hartford circa 1853, see Frank M. Sellers, *Sharps Firearms* (North Hollywood, Calif.: Beinfeld Publishing, 1978), 32. A similar, comparative inventory for the fledgling New Haven Arms plant circa 1857 appears in Herbert G. Houze, *Winchester Repeating Arms Company: Its History and Development from 1866 to 1981* (Iola, Wis.: Krause Publications, 1995), 10. The significant contributions of the Robbins & Lawrence company to the American System are outlined in Peter G. Smithurst, *The Guns and Gun-Making Machinery of Robbins and Lawrence* (Windsor, Vt.: American Precision Museum, 2007), 2–10; and Cesari, "American Arms Making," 143–59.

11. For a broad presentation of the model firearms made by the major American manufacturers during the nineteenth century, see Norm Flayderman, *Flayderman's Guide to Antique American Firearms . . . and Their Values*, 7th ed. (Iola, Wis.: Krause Publications, 1998), 41–278, passim.

12. Dimick quoted in Garavaglia and Worman, *Firearms of the American West, 1803–1865*, 243.

13. Sellers, *Sharps Firearms*, 2–35, passim; Cesari, "American Arms Making," 149–51; Flayderman, *Flayderman's Guide*, 165–69.

14. The arms-related inventions and patent arms of Dr. Edward Maynard are treated in George J. Layman, *A Guide to the Maynard Breechloader* (Union City, Tenn.: Pioneer Press, 1998), 3–46; and James J. Grant, *Single-Shot Rifles* (Highland Park, N.J.: Gun Room Press, 1982), 277–97. Also see Flayderman, *Flayderman's Guide*, 309, 508, 573–74; and Garavaglia and Worman, *Firearms of the American West, 1803–1865*, 253.

15. Burton quoted in Rattenbury, *Hunting the American West*, 129.

16. Clemens quoted in R. L. Wilson, *The Peacemakers: Arms and Adventure in the American West* (New York: Random House, 1992), 139.

17. Flayderman, *Flayderman's Guide*, 322–28; Garavaglia and Worman, *Firearms of the American West, 1803–1865*, 97–98. For an early but still valuable treatment of pepperbox-type weapons, see Lewis Winant, *Pepperbox Firearms* (New York: Greenburg, 1952).

18. Flayderman, *Flayderman's Guide*, 45–51. For greater detail on Allen-patent arms, see Harold R. Mouillesseaux, *Ethan Allen, Gunmaker: His Partners, Patents and Firearms* (Ottawa: Museum Restoration Service, 1973); and Paul Henry, *Ethan Allen and Allen & Wheelock: Their Guns and Their Legacy* (Lincoln, R.I.: Andrew Mowbray, 2006).

19. R. L. Wilson, *Colt, an American Legend: The Official History of Colt Firearms* (New York: Abbeville Press, 1985), 10–36, 40, 131; Rattenbury, "Colt Revolvers," in *Handbook of Texas Online*, *www.tshaonline.org/handbook/online*; Hawke, *Nuts and Bolts*, 142–47. For detailed treatments of Colt's Paterson-made arms and their production, see Philip R. Phillips and R. L. Wilson, *Paterson Colt Pistol Variations* (Dallas: Jackson Arms, 1979); and R. L. Wilson, *The Paterson Colt Book: The Early Evolution of Samuel Colt's Repeating Arms* (Palo Alto, Calif.: Strutz-LeVett Publishing, 2001). For the landmark patent suit, see Martin Rywell, *The Trial of Samuel Colt* (Harriman, Tenn.: Pioneer Press, 1953).

20. Hounshell, *From the American System*, 46–50. Also see Herbert G. Houze, *Samuel Colt: Arms, Art, and Invention* (New Haven, Conn.: Yale University Press in association with Wadsworth Atheneum Museum of Art, 2006), 171–80. The latter includes an interesting group of circa 1857 photographs of Colt-factory machine tenders at their special-purpose machinery. For detail on the several contributions of Elisha King Root to Colt's manufacturing success, see Paul Uselding, "Elisha K. Root, Forging, and the 'American System,'" *Technology and Culture* 15, no. 4 (1974): 563–68; Cesari, "American Arms Making," 196–99, 208–209; and Hoke, *Ingenious Yankees*, 168–81.

21. Excerpts from "A Day at the Armory" quoted in Charles T. Haven and Frank A. Belden, *A History of the Colt Revolver and Other Arms Made by Colt's Patent Fire Arms Manufacturing Company from 1836 to 1940* (New York: William Morrow, 1940), 354–56. For a fuller discussion of the Colt armory and its success in the refinement of the arms-making process, see Ellsworth S. Grant, *The Colt Armory: A History of Colt's Manufacturing Company* (Lincoln, R.I.: Andrew Mowbray, 1995); and Cesari, "American Arms Making," 212–22.

22. Deyrup, *Arms Making*, 158–59; Haven and Belden, *History of the Colt Revolver*, 358; Hounshell, *From the American System*, 48–49; Hawke, *Nuts and Bolts*, 154–55. For an appraisal of the technical and economic benefits—and shortcomings—of achieving true interchangeability (as opposed to parts uniformity), see Robert A. Howard, "Interchangeable Parts Reexamined: The Private Sector of the American Arms Industry on the Eve of the Civil War," *Technology and Culture* 19, no. 4 (1973): 633–36, 643–44.

23. Olmsted quoted in Garavaglia and Worman, *Firearms of the American West, 1803–1865*, 294–95.

24. Colt's antebellum arms production is thoroughly enumerated in Wilson, *Colt, an American Legend*, 362–64. In addition to Remington, Colt's domestic competitors and copyists during the percussion period included the Bacon Manufacturing Company, J. M. Cooper & Company, the Manhattan Fire Arms Manufacturing Company, W. W. Marston–Union Arms, the Metropolitan Arms Company, and the Whitneyville Armory (the Whitney Arms Company after 1863).

25. Treatment of the development and impact of the metallic cartridge on domestic arms design and performance can be found in Butler, *United States Firearms*, 101–28, 152–73, 208–47.

26. For background on the genesis of Smith & Wesson's Springfield partnership and the firm's acquisition of the Rollin White patent, see Roy G. Jinks, *History of Smith & Wesson: No Thing of Importance Will Come without Effort* (North Hollywood, Calif.: Beinfeld Publishing, 1977), 34–37.

PART II
From Colt to Winchester: A Portfolio of Famed American Arms Makers

1. Howard, "Interchangeable Parts Reexamined," 637–40. For much fuller context, see U.S. Secretary of the Interior, *Manufactures of the United States in 1860: Compiled from the Original Returns of the Eighth Census* (Washington, D.C.: Government Printing Office, 1865); and Deyrup, *Arms Making*, 120–26.

2. Roger Burlingame, *The March of the Iron Men: A Social History of Union through Invention* (New York: Charles Scribner's Sons, 1938), 190.

3. For a treatment of F. W. Freund's work, see F. J. Pablo Balentine, *Freund & Brother: Pioneer Gunmakers to the West* (Newport Beach, Calif.: Graphic Publishers, 1997). For an overview of Browning's influential career and many patents, see John Browning and Kurt Gentry, *John M. Browning, American Gunmaker* (New York: Doubleday, 1964); and Richard C. Rattenbury, *The Browning Connection: Patent Prototypes in the Winchester Museum* (Cody, Wyo.: Buffalo Bill Historical Center, 1982).

4. See, for example, Deyrup, *Arms Making*, 128–30.

5. Houze, *Samuel Colt*, 123–27; Harold F. Williamson, *Winchester, the Gun That Won the West* (Washington, D.C.: Combat Forces Press, 1952), 444–47. For a fully illustrated and detailed treatment of Colt's licensed arms, see Roy Marcot and Ron Paxton, *Colt Brevete Revolvers* (Tucson, Ariz.: Northwood Heritage Press, 2012). For an idea of the diversity and significance of U.S. patents in nineteenth-century American arms making, consult Virgil D. Stockbridge, *Digest of Patents Relating to Breech-Loading and Magazine Small Arms, 1836–1873* (Washington, D.C.: Government Printing Office, 1875).

CHAPTER 3
Colt Firearms

1. For thoroughly illustrated, overview treatments of Colt's history and product line, see Wilson, *Colt, an American Legend*; and Houze, *Samuel Colt*. Those interested in an in-depth treatment of Colt model production and variations should consult R. L. Wilson, *The Book of Colt Firearms*, 3rd ed. (Minneapolis, Minn.: Blue Book Publications, 2008).

2. Gregg quoted in Garavaglia and Worman, *Firearms of the American West, 1803–1865*, 100.

3. Rattenbury, "Colt Revolvers," in *Handbook of Texas Online*; Wilson, *Paterson Colt Book*, 67; Flayderman, *Flayderman's Guide*, 67–69.

4. Wilson, *Colt, an American Legend*, 28–58, 71–111, 363–65; Flayderman, *Flayderman's Guide*, 70–80. For further detail on the widely popular Model 1849 and 1851 Colt revolvers, see P. L. Shumaker, *Colt's Variations of the Old Model Pocket Pistol,*

1848–1872 (Beverly Hills, Calif.: Fadco Publishing, 1957); Robert M. Jordan, *Colt's Pocket '49: Its Evolution Including the Baby Dragoon and Wells Fargo* (Loma Mar, Calif.: Darrow M. Watt, 2000); and Nathan L. Swayze, *'51 Colt Navies* (Highland Park, N.J.: Gun Room Press, 1991). For more detail on the Model 1855 Root-pattern arms, see Gerald Keogh, *Samuel Colt's New Model Pocket Pistols* (Ogden, Utah: printed by author, 1964). For the development of the New Model series (Models 1860, 1861, and 1862), see Houze, *Samuel Colt*, 103–109, 161–65. For further detail on the redesigned New Model Police (1862) and Pocket Navy (1865) revolvers, consult Robert M. Lee and R. L. Wilson, *The Art of the Gun: Magnificent Colts,* vol. 1 (Sparks, Nev.: Yellowstone Press, 2012), 237–38; and John D. Breslin, William Q. Pirie, and David E. Price, *Variations of Colt's New Model Police and Pocket Breech Loading Pistols* (Lincoln, R.I.: Andrew Mowbray, 2002), esp. chapter 4.

5. Wilson, *Colt, an American Legend*, 59–69; Flayderman, *Flayderman's Guide*, 72–75. For treatments of Colt's short-lived London enterprise, see Joseph G. Rosa, *Colonel Colt London: The History of Colt's London Firearms, 1851–1857* (London: Arms and Armour Press, 1976); and Cesari, "American Arms Making," 200–208.

6. Wilson, *Colt, an American Legend*, 153–63; Flayderman, *Flayderman's Guide*, 80–83. For greater technical and illustrative detail on the various Colt (and other revolver) conversions, see, respectively, R. Bruce McDowell, *A Study of Colt Conversions and Other Percussion Pistols* (Iola, Wis.: Krause Publications, 1997); Breslin, Pirie, and Price, *Variations of Colt's New Model Police and Pocket Breech Loading Pistols*; and Dennis Adler, *Metallic Colt Conversions: The History of the Guns and Modern Reproductions* (Iola, Wis.: Krause Publications, 2002).

7. Wilson, *Colt, an American Legend*, 165–73; Flayderman, *Flayderman's Guide*, 87–91.

8. *The Field* correspondent quoted in John E. Parsons, *The Peacemaker and Its Rivals* (New York: William Morrow, 1950), 54.

9. Wilson, *Colt, an American Legend*, 175–97; Flayderman, *Flayderman's Guide*, 83–86. For an in-depth treatment of the Model 1873 Colt Single Action Army revolver, see John Kopec, Ron Graham, and Kenneth Moore, *A Study of the Colt Single Action Army Revolver* (La Puente, Calif.: printed by authors, 1976).

10. Wilson, *Colt, an American Legend*, 199–204, 211–15; Flayderman, *Flayderman's Guide*, 91–94.

11. Lower quoted in Rattenbury, *Hunting the American West*, 143. The origin and known production details of the Colt double rifle are treated in Fred P. L. Mills, *A Study of Colt's Double Rifles* (Greenfield, Mass.: Secretarial Phone-O-Service, 1953).

12. Wilson, *Colt, an American Legend*, 253–65; Flayderman, *Flayderman's Guide*, 101–104; Garavaglia and Worman, *Firearms of the American West, 1866–1894*, 213–14. Detailed treatments of the Colt-Burgess and Colt Lightning rifles, respectively, can be found in Samuel L. Maxwell, Sr., *The Colt-Burgess Magazine Rifle* (Dallas: Taylor: Publishing, 1985); and Ted Tivey, *The Colt Rifle, 1884–1902* (New South Wales, Australia: Clouston and Hall, 1984).

CHAPTER 4

Remington Firearms

1. Roy Marcot, *Remington: America's Oldest Gunmaker* (Peoria, Ill.: Primedia Publications, 1998), 25–28, 51–53; Flayderman, *Flayderman's Guide*, 135–36, 138–39. In addition to surveying Remington model production, Marcot's work provides the most thorough business and marketing history of the company to date. For a copiously illustrated, full-color presentation of virtually all Remington's model arms and numerous variants, see Howard Michael Madaus and Simeon Stoddard, comps., *The Guns of Remington: Historic Firearms Spanning Two Centuries* (Dayton, Ky.: Biplane Productions, 1997).

2. Marcot, *Remington*, 32–41, 57–58; Flayderman, *Flayderman's Guide*, 136–38. In-depth information on Remington's large-frame percussion revolvers, their extensive Civil War service, and their conversion to metallic cartridges can be found in Donald L. Ware, *Remington Army and Navy Revolvers, 1861–1888* (Albuquerque: University of New Mexico Press, 2007).

3. Marcot, *Remington*, 30–32, 53–55, 67; Flayderman, *Flayderman's Guide*, 140–42.

4. Marcot, *Remington*, 64–65; Flayderman, *Flayderman's Guide*, 142–43. For full detail regarding Smoot models and variants, see Harry J. Parker, Ora Lee Parker, and Joan S. Reisch, *A Study of Remington's Smoot Patent and Number Four Revolvers* (Santa Ana, Calif.: Armslore Press, 2003).

5. James quoted in Louis A. Garavaglia and Charles G. Worman, *Firearms of the American West, 1866–1894* (Albuquerque: University of New Mexico Press, 1985), 301.

6. Marcot, *Remington*, 67–68, 120–21; Flayderman, *Flayderman's Guide*, 143–44.

7. Marcot, *Remington*, 83–87, 98–101; Flayderman, *Flayderman's Guide*, 149–55. Greater detail on the Remington rolling-block rifle can be found in Roy M. Marcot, *Remington Rolling Block Rifles, Carbines and Shotguns, Sporting and Target Rifles* (Tucson, Ariz.: Northwood Heritage Press, 2009); and Grant, *Single-Shot Rifles*, 114–37.

8. Marcot, *Remington*, 72–75, 121–22; Flayderman, *Flayderman's Guide*, 160–61. Further detail on the Remington Hepburn rifle may be found in Tom Rowe, *Remington's No. 3 Hepburn Sporting and Target Rifles* (Woonsocket, R.I.: Mowbray, 2011); and Grant, *Single-Shot Rifles*, 139–44.

9. Marcot, *Remington*, 107–10, 136–50.

CHAPTER 5

Smith & Wesson Firearms

1. Jinks, *History of Smith & Wesson*, 1–13, 21–37. For a handy, but detailed, explanation of nineteenth-century Smith & Wesson models and variants, see Jim Supica and Richard Nahas, *Standard Catalog of Smith & Wesson* (Iola, Wis.: Gun Digest Books, 2006), 36–71. A fuller treatment of Smith & Wesson's history and production can be found in Robert J. Neal and Roy G. Jinks, *Smith and Wesson, 1857–1945* (New York: A. S. Barnes, 1975).

2. Clemens quoted in Garavaglia and Worman, *Firearms of the American West, 1803–1865*, 324.

3. Jinks, *History of Smith & Wesson*, 38–57; Flayderman, *Flayderman's Guide*, 181–86.

4. Jinks, *History of Smith & Wesson*, 62–96; Flayderman, *Flayderman's Guide*, 197–200. For a thorough technical treatment of all of the large-frame, "top-break" models, see David R. Chicoine, *Smith & Wesson: Sixguns of the Old West* (Lincoln, R.I.: Andrew Mowbray, 2004), 91–143. For background on Merwin Hulbert revolvers, see Art Phelps, *The Story of Merwin, Hulbert & Co. Firearms* (Santa Ana, Calif.: printed by author, 1992); and Flayderman, *Flayderman's Guide*, 391–92.

5. "Cimmarron" quoted in John E. Parsons, *Smith & Wesson Revolvers: The Pioneer Single Action Models* (William Morrow, 1957), 80, 82–83.

6. Garavaglia and Worman, *Firearms of the American West, 1866–1894*, 278–83.

7. Jinks, *History of Smith & Wesson*, 82–91; Flayderman, *Flayderman's Guide*, 203; Garavaglia and Worman, *Firearms of the American West, 1866–1894*, 341–42.

8. Jinks, *History of Smith & Wesson*, 121–34, 143–47, 159–61, 194–98, 260–61; Flayderman, *Flayderman's Guide*, 201.

CHAPTER 6
Sharps Firearms

1. Sellers, *Sharps Firearms*, 2–7. Sellers's book presents by far the most comprehensive reference on the historical and design progression of all Sharps firearms (and ammunition) available. The purported Hall-Sharps connection is questioned per advisement of historian Merritt Roe Smith, who found no indication in the original Harpers Ferry Armory payroll records of Sharps having worked at the Rifle Works or the main armory at any time (Smith, undated reader's report, author's files). For a brief historical and technical overview of the Sharps "falling-block" design, also consult Frank de Haas, "The Sharps Side-Hammer Rifle," *Gun Digest*, 22nd ed. (1968): 138–45.

2. Both the *Scientific American* article and the Burtis advertisement are quoted in Garavaglia and Worman, *Firearms of the American West, 1803–1865*, 240–41.

3. Sellers, *Sharps Firearms*, 13–22.

4. Ibid., 29–64, 91–95; Flayderman, *Flayderman's Guide*, 165–67; Garavaglia and Worman, *Firearms of the American West, 1803–1865*, 243–44.

5. Sellers, *Sharps Firearms*, 66, 75–86, 175–87; Flayderman, *Flayderman's Guide*, 167–69.

6. Bicknell quoted in Sellers, *Sharps Firearms*, 308.

7. Ibid., 211–38; Flayderman, *Flayderman's Guide*, 170–71.

8. Shields quoted in Rattenbury, *Hunting the American West*, 135.

9. Sellers, *Sharps Firearms*, 255–81; Grant, *Single-Shot Rifles*, 172–78. Approximate total production is gleaned from Flayderman, *Flayderman's Guide*, 164–74.

CHAPTER 7
Marlin Firearms

1. Lt. Col. William S. Brophy, *Marlin Firearms: A History of the Guns and the Company That Made Them* (Mechanicsburg, Pa.: Stackpole Books, 1989), 2–6, 100–124; Flayderman, *Flayderman's Guide*, 114–17. An informative and detailed treatment of Marlin arms also may be found in Bill West, *Marlin and Ballard—Firearms and History* (Whittier, Calif.: Stockton Trade Press, 1968).

2. Brophy, *Marlin Firearms*, 127–44; Flayderman, *Flayderman's Guide*, 123–31. *Colorado Chieftain* advertisement quoted in Garavaglia and Worman, *Firearms of the American West, 1866–1894*, 168. For detailed historical and technical analysis of the Ballard-patent arms, see Eldon G. Wolff, *Ballard Rifles in the Henry J. Nunnemacher Collection*, 2nd ed. (Milwaukee, Wis.: North American Press, 1961); John T. Dutcher, *Ballard: The Great American Single Shot Rifle* (Denver, Colo.: printed by author, 2002), esp. 85–278, passim; and Grant, *Single-Shot Rifles*, 1–49.

3. For details on the many extras and options available on Marlin's nineteenth-century rifles, see Brophy, *Marlin Firearms*, 517–20, 530–40, 552, 633–37, 660–61, 670–74.

4. Ibid., 145–61; Flayderman, *Flayderman's Guide*, 117. Broadwater, Hubbell & Company advertisement quoted in Garavaglia and Worman, *Firearms of the American West, 1866–1894*, 195.

5. Brophy, *Marlin Firearms*, 162–75; Flayderman, *Flayderman's Guide*, 117–18.

6. Brophy, *Marlin Firearms*, 176–205; Garavaglia and Worman, *Firearms of the American West, 1866–1894*, 218; Flayderman, *Flayderman's Guide*, 118–19.

7. Brophy, *Marlin Firearms*, 210–16, 223–33; Flayderman, *Flayderman's Guide*, 120–21.

8. Brophy, *Marlin Firearms*, 235–43; Flayderman, *Flayderman's Guide*, 121.

CHAPTER 8
Winchester Firearms

1. For a biographical profile of Oliver Winchester and something of his considerable business acumen, see Williamson, *Winchester*, 19–21, 77–80, 368–69.

2. Thorough historical and technical treatments of antecedent Winchester firearms (and the long-serving toggle-link breech mechanism) are presented in R. Bruce McDowell, *Evolution of the Winchester* (Tacoma, Wash.: Armory Publications, 1985); and Edmund E. Lewis and Stephen W. Rutter, *Volcanic Firearms: Predecessor to the Winchester Rifle* (Woonsocket, R.I.: Andrew Mowbray, 2011). For quick reference, also see R. L. Wilson, *Winchester, an American Legend: The Official History of Winchester Firearms and Ammunition* (New York: Random House, 1991), 3–10; and Flayderman, *Flayderman's Guide*, 260–63. For a full overview of all Winchester production, see George R. Watrous, Thomas E. Hall, James C. Rokhoff, Pete Kuhlhoff, and Duncan Barnes, *The History of Winchester Firearms, 1866–1980* (Tulsa, Okla.: Winchester Press, 1980).

3. Wilson, *Winchester, an American Legend*, 10–21; Flayderman, *Flayderman's Guide*, 264. Kelly quoted in Garavaglia and Worman, *Firearms of the American West, 1866–1894*, 122. For an exhaustive technical analysis of the Henry rifle, also see Les Quick, *The Story of Benjamin Tyler Henry and His Famed Repeating Rifle* (Santa Ana,

Calif.: Graphic Publishers, 2008), 43–135. Historical treatment of the Henry rifle appears in Wiley Sword, *The Historic Henry Rifle: Oliver F. Winchester's Famous Civil War Repeater* (Lincoln, R.I.: Andrew Mowbray, 2002).

4. For fine economic and product-design histories of the Winchester enterprise, see Williamson, *Winchester*; and Houze, *Winchester Repeating Arms Company*.

5. Wilson, *Winchester, an American Legend*, 22–39; Flayderman, *Flayderman's Guide*, 264–65; Garavaglia and Worman, *Firearms of the American West, 1866–1894*, 123–29. For historical color regarding the Model 1866, also see John E. Parsons, *The First Winchester: The Story of the 1866 Repeating Rifle* (New York: William Morrow, 1955), 61–79.

6. Cody quoted in Williamson, *Winchester*, 67.

7. Wilson, *Winchester, an American Legend*, 41–55; Flayderman, *Flayderman's Guide*, 265–66. For detailed treatment of Model 1873 specifications and variants, see James D. Gordon, *Winchester's New Model of 1873: A Tribute* (Grant, Colo.: printed by author, 1997). An appreciation of the 1873 Winchester's role in the West can be found in Garavaglia and Worman, *Firearms of the American West, 1866–1894*, 183–88. To view many examples of the myriad special-order variants and extras available on this and succeeding Winchester rifles, see George Madis, *The Winchester Book* (Dallas: Taylor Publishing, 1971).

8. Quoted in Herbert G. Houze, *The Winchester Model 1876 "Centennial" Rifle* (Lincoln, R.I.: Andrew Mowbray, 2001), 72. This work provides a detailed developmental and technical treatment of the Model 1876 Winchester rifle in its various calibers.

9. Roosevelt quoted in Rattenbury, *Hunting the American West*, 146.

10. Wilson, *Winchester, an American Legend*, 58–70; Flayderman, *Flayderman's Guide*, 266–67. For detail on cartridge development and ballistics for the Model 1873 and Model 1876, see David F. Butler, *Winchester '73 and '76: The First Repeating Centerfire Rifles* (New York: Winchester Press, 1970), 57–66, 81–89. For a treatment of many Winchester competitor arms during the latter nineteenth century, see Samuel L. Maxwell, Sr., *Lever Action Magazine Rifles Derived from the Patents of Andrew Burgess* (Dallas: Taylor Publishing, 1976).

11. For detail on the Browning-Winchester association, see Houze, *Winchester Repeating Arms Company*, 130–51; Williamson, *Winchester*, 95–98; and Rattenbury, *Browning Connection*.

12. Wright quoted in Rattenbury, *Hunting the American West*, 140.

13. Wilson, *Winchester, an American Legend*, 133–43; Flayderman, *Flayderman's Guide*, 272–74. For thorough treatment of Model 1885 technical specifications and variations, see John Campbell, *The Winchester Single-Shot: A History and Analysis* (Lincoln, R.I.: Andrew Mowbray, 1995), esp. chaps. 3–7 and 12.

14. Perry quoted in Rattenbury, *Hunting the American West*, 148.

15. Wilson, *Winchester, an American Legend*, 83–90; Flayderman, *Flayderman's Guide*, 267–68.

16. Wilson, *Winchester, an American Legend*, 144–47; Flayderman, *Flayderman's Guide*, 276. For in-depth treatment of Model 1890 variations and technical specifications, see Ned Schwing, *Winchester Slide-Action Rifles*, vols. 1 and 2 (Iola, Wis.: Krause Publications, 2004), esp. chaps. 2–4.

17. Wilson, *Winchester, an American Legend*, 90–96; Flayderman, *Flayderman's Guide*, 269.

18. Wilson, *Winchester, an American Legend*, 96–102; Flayderman, *Flayderman's Guide*, 270. For Model 1894 (and Model 94) technical and variant detail, see Robert C. Renneberg, *The Winchester Model 94: The First 100 Years* (Iola, Wis.: Krause Publications, 1991).

19. Wilson, *Winchester, an American Legend*, 104–11; Flayderman, *Flayderman's Guide*, 271–72.

20. For a cursory detailing of the company's checkered financial demise, see the Winchester citation in Wikipedia: http://www.wikipedia.org/wiki/Winchester_Repeating_Arms_Company.

PART III
From Functional Sculpture to Steel Canvas: Aesthetics in American Arms

1. Extensive pictorial treatments of highly embellished firearms from across the centuries can be reviewed in Merrill Lindsay, *One Hundred Great Guns: An Illustrated History of Firearms* (New York: Walker, 1967); and Claude Gaier and Pietro Sabatte, *The Art of Gun Engraving* (New York: Knickerbocker Press, 1999).

2. R. L. Wilson, *Winchester: The Golden Age of American Gunmaking and the Winchester 1 of 1000* (Cody, Wyo.: Buffalo Bill Historical Center, 1983), 19.

CHAPTER 9
Melding Function and Form

1. This discussion of the aesthetic character of vernacular expression in American manufactures draws on John A. Kouwenhoven, *The Arts in Modern American Civilization* (New York: W. W. Norton, 1967), 13–15, 27, 56.

2. Lindsay, *Kentucky Rifle*, 2.

3. Joe Kindig, Jr., *Thoughts on the Kentucky Rifle in Its Golden Age* (Wilmington, Del.: George N. Hyatt, 1960), 31.

4. Greenough quoted in Kouwenhoven, *Arts in Modern American Civilization*, 83–84. Also see Horatio Greenough, *Form and Function: Remarks on Art*, ed. Harold A. Small (Berkeley: University of California Press, 1947). A fuller understanding of the functional and aesthetic character of domestic machine-made products also may be drawn from John F. Kasson, *Civilizing the Machine: Technology and Republican Values in America, 1776–1900* (New York: Grossman, 1976), 142–55, although this book is focused on antebellum American machinery.

5. Houze, *Samuel Colt*, 20. For background on the principles and requirements of acceptable object design, see David Pye, *The Nature of Design* (London: Studio Vista and Van Nostrand Reinhold, 1972), 21–39.

6. Much of the foregoing discussion and evaluation of domestic firearms aesthetics draws on the principles of artifact study as presented in Thomas J. Schlereth, ed., *Material Culture Studies in*

America (Nashville, Tenn.: American Association for State and Local History, 1989). See, in particular, the essays of Charles F. Montgomery, "The Connoisseurship of Artifacts," 143–52; and E. McClung Fleming, "Artifact Study: A Proposed Model," 162–73.

7. Sharp quoted in Christopher Austyn, *Gun Engraving* (Long Beach, Calif.: Safari Press, 1998), 21.

8. R. L. Wilson, *Steel Canvas: The Art of American Arms* (New York: Random House, 1994), 26.

CHAPTER 10
Embellishing the Utilitarian

1. Kindig, *Thoughts on the Kentucky Rifle*, 31–34; Siro R. Toffolon, "The Kentucky Rifle and Its Ornamentations," *Kentucky Rifle Association Bulletin* 6, no. 2 (Winter 1980): 2–5; Grancsay, "Craft," 59; Kindig, *Artistic Ingredients of the Longrifle*, 72–101, passim. The first work cited provides by far the best illustrated treatment of the artistic expression of "Golden Age" Kentucky riflesmiths.

2. Wilson, *Steel Canvas*, 37, 49; Flayderman, *Flayderman's Guide*, 537, 539.

3. Wilson, *Steel Canvas*, 77–78, 92; Wilson, *Paterson Colt Book*, 235–51; R. L. Wilson, "Firearms Engraving in Nineteenth Century America," *Bulletin of the American Society of Arms Collectors* 28 (Fall 1973): 2, 6, 15. Also see Frederick Knight, *Scroll Ornaments of the Early Victorian Period* (New York: Dover Pictorial Archive Publications, 1978).

4. Herbert G. Houze, *Colt Factory Engravers of the Nineteenth Century: Understanding Their Careers and Identifying Their Work* (Woonsocket, R.I.: Andrew Mowbray, 2012), 159–65; Wilson, *Steel Canvas*, 88, 95–97; Fredric A. Harris, *Firearms Engraving as Decorative Art: The Origin of American Firearms Engraving Motifs in the Decorative Art of the Middle East* (Seattle, Wash.: Barbara R. Harris, 1989), 5–44, passim; Owen Jones, *The Grammar of Ornament* (New York: Portland House, 1986), 33–47, 78–84, 101–27. Also see Alexander Speltz, *The Styles of Ornament* (New York: Dover Publications, 1959); and C. Roger Bleile, *Hand Engraving Glossary*, http://engravingglossary.com.

5. Wilson, *Steel Canvas*, 147–51, 181–83, 217–19, 224, 230–36; Wilson, *Colt, an American Legend*, 136, 172; Wilson, *Winchester, an American Legend*, 58. Also see Houze, *Samuel Colt*, 110.

6. Arthur Tobias, *Colt Cylinder Scenes, 1847–1851* (Los Angeles, Calif.: printed by author, 2011), 1–11, 18–20, 48–51, 60–64, 76–79, 98–109; Houze, *Colt Factory Engravers*, 85–87; Rattenbury, "Colt Revolvers," in *Handbook of Texas Online*; Wilson, *Book of Colt Firearms*, 27–185, passim; R. L. Wilson, *The Colt Engraving Book*, 2 vols. (New York: Bannerman's, 2002–2003), 1:29; Flayderman, *Flayderman's Guide*, 109–10, 302–11.

7. R. L. Wilson, *L. D. Nimschke, Firearms Engraver* (Teaneck, N.J.: John J. Malloy, 1964), xv–xxxiii; Houze, *Colt Factory Engravers*, 78–84; Wilson, *Steel Canvas*, 80, 87–88.

8. Wilson, *Colt Engraving Book*, 1:65–67, 193–97, 383–84; Houze, *Colt Factory Engravers*, 133–47; Roy G. Jinks, *Artistry in Arms: The Guns of Smith & Wesson* (Springfield, Mass.: Smith & Wesson and Connecticut Valley Historical Museum, 1991), 61–75. For pictorial examples of Young's ascribed work on Colt percussion revolvers, see Houze, *Colt Factory Engravers*, 138–42; Jordan, *Colt's Pocket '49*, 198, 208–19; and Wilson, *Steel Canvas*, 123–32. For similar imagery of the Young family's finest work for Smith & Wesson, see Wilson, *Steel Canvas*, 192–97.

9. Wilson, *Colt Engraving Book*, 1:380–83, 2:529–36 and 549–55; Houze, *Colt Factory Engravers*, 52–63. For Kornbrath's work, see E. C. Prudhomme, *Gun Engraving Review* (Shreveport, La.: printed by author, 1961), 1–14. For the lives and works of numerous lesser-known Colt engravers, circa 1850–1900, see Houze, *Colt Factory Engravers*; and Lee and Wilson, *The Art of the Gun*, 2:493–519.

10. For pictorial examples of engraved Sharps and Remington rifles (the majority likely rendered by Gustave Young and L. D. Nimschke, respectively), see Sellers, *Sharps Firearms*, 48–53, 228, 267; and Marcot, *Remington Rolling Block Rifles*, 106–24.

11. R. L. Wilson, *Winchester Engraving* (Palm Springs, Calif.: Beinfeld Publishing, 1990), 97–100, 133–34; Wilson, *Steel Canvas*, 172. Also see Herbert G. Houze, "The Appearance of Evidence: A Brief Examination of the Life and Work of Herman Leslie Ulrich," *Armax—The Journal of the Cody Firearms Museum* 4, no. 2 (1993): 37, 48–49, 52–53, 64–68; and Houze, *Colt Factory Engravers*, 121–31. For the earlier, "factory" engraving style usually attributed to Samuel Hoggson on predecessor Henry patent rifles, see Quick, *Story of Benjamin Tyler Henry*, 149–55; and Wilson, *Winchester Engraving*, 29–32, 36–38.

12. Marlin catalog reproduced in Brophy, *Marlin Firearms*, 530–40; and Wilson, *Steel Canvas*, 226–27. Winchester catalog reproduced in full in Wilson, *Winchester, an American Legend*, 114–30; and Wilson, *Winchester Engraving*, 217–33.

13. Winchester catalog quoted in Wilson, *Winchester Engraving*, 218.

14. For the comparative work of twentieth-century (and later) domestic arms engravers, see Prudhomme, *Gun Engraving Review*; C. Roger Bleile, *American Engravers* (North Hollywood, Calif.: Beinfeld Publishing, 1980); C. Roger Bleile, *American Engravers—The 21st Century* (Minneapolis, Minn.: Blue Book Publications, 2011); and Tom Turpin, *Custom Firearms Engraving: The Techniques and Treasures of the World's Greatest Artists* (Iola, Wis.: Krause Publications, 1999).

BIBLIOGRAPHY

Adler, Dennis. *Metallic Colt Conversions: The History of the Guns and Modern Reproductions*. Iola, Wisc.: Krause, 2002.

Austyn, Christopher. *Gun Engraving*. Long Beach, Calif.: Safari Press, 1998.

Balentine, F. J. Pablo. *Freund & Brother: Pioneer Gunmakers to the West*. Newport Beach, Calif.: Graphic Publishers, 1997.

Ball, Robert W. D. *Remington Firearms: The Golden Age of Collecting*. Iola, Wisc.: Krause, 1995.

Battison, Edwin A. "Eli Whitney and the Milling Machine." *Smithsonian Journal of History* 1, no. 2 (1966): 9–34.

———. *Muskets to Mass Production: The Men and the Times That Shaped American Manufacturing*. Windsor, Vt.: American Precision Museum, 1976.

———. "A New Look at the 'Whitney' Milling Machine." *Technology and Culture* 14, no. 4 (1973): 592–98.

Billington, Ray Allen. *Westward Expansion: A History of the American Frontier*. 2nd ed. New York: Macmillan, 1960.

Bleile, C. Roger. *American Engravers*. North Hollywood, Calif.: Beinfeld, 1980.

———. *American Engravers—The 21st Century*. Minneapolis, Minn.: Blue Book, 2011.

Boddington, Craig, ed. *America: The Men and Their Guns That Made Her Great*. Los Angeles: Peterson, 1981.

Boorstin, Daniel J. *The Americans: The National Experience*. New York: Random House, 1965.

Breslin, John D., William Q. Pirie, and David E. Price. *Variations of Colt's New Model Police and Pocket Breech Loading Pistols*. Lincoln, R.I.: Andrew Mowbray, 2002.

Broehl, Wayne G. *Precision Valley: The Machine Tool Companies of Windsor, Vermont*. Englewood Cliffs, N.J.: Prentice Hall, 1959.

Brophy, Lt. Col. William S. *Marlin Firearms. A History of the Guns and the Company That Made Them*. Mechanicsburg, Pa.: Stackpole Books, 1989.

Brown, M. L. *Firearms in Colonial America: The Impact on History and Technology, 1492–1792*. Washington, D.C.: Smithsonian Institution Press, 1980.

Browning, John, and Kurt Gentry. *John M. Browning, American Gunmaker*. New York: Doubleday, 1964.

Burlingame, Roger. *The March of the Iron Men: A Social History of Union through Invention*. New York: Charles Scribner's Sons, 1938.

Butler, David F. *United States Firearms: The First Century, 1776–1875*. New York: Winchester Press, 1971.

———. *Winchester '73 and '76: The First Repeating Centerfire Rifles*. New York: Winchester Press, 1970.

Buttrick, John. "The Inside Contract System." *Journal of Economic History* 12, no. 2 (1952): 205–21.

Campbell, John. *The Winchester Single-Shot: A History and Analysis*. Lincoln, R.I.: Andrew Mowbray, 1995.

Cesari, Gene S. "American Arms Making Machine Tool Development." Ph.D. dissertation, University of Pennsylvania, 1970.

Chicoine, David R. *Smith & Wesson Sixguns of the Old West*. Lincoln, R.I.: Andrew Mowbray, 2004.

Colonial Williamsburg Productions. *The Gunsmith of Williamsburg*. 2009 DVD.

Cooper, Carolyn C. "'A Whole Battalion of Stockers': Thomas Blanchard's Production Line and Hand Labor at Springfield Armory." *Journal of the Society for Industrial Archeology* 14, no. 1 (1988): 37–56.

Cooper, Carolyn C., and Merrill Lindsay. *Eli Whitney and the Whitney Armory*. Whitneyville, Conn.: Eli Whitney Museum, 1980.

Cramer, Clayton E. *Armed America: The Remarkable Story of How and Why Guns Became as American as Apple Pie*. Nashville, Tenn.: Nelson Current, 2006.

Davis, Theodore R. "The Buffalo Range." *Harper's New Monthly Magazine* 38, no. 6 (1869): 147–63.

De Haas, Frank. "The Sharps Side-Hammer Rifle." *Gun Digest*, 22nd ed. (1968): 138–45.

Deyrup, Felicia Johnson. *Arms Making in the Connecticut Valley: A Regional Study of the Economic Development of the Small Arms Industry, 1798–1870*. York, Pa.: George Shumway, 1970.

Dillin, Capt. John G. W. *The Kentucky Rifle*. New York: Ludlum and Beebe, 1946.

Dizard, Jan E., Robert Merrill Muth, and Stephen P. Andrews, Jr., eds. *Guns in America: A Reader*. New York: New York University Press, 1999.

Durant, Stuart. *Ornament from the Industrial Revolution to Today*. Woodstock, N.Y.: Overlook Press, 1986.

Dutcher, John T. *Ballard: The Great American Single Shot Rifle*. Denver, Colo.: printed by author, 2002.

Farley, Capt. J. P. *Rules for the Inspection of Army Revolvers and Gatling Guns*. Springfield, Mass.: U.S. National Armory, 1875.

Farley, James J. *Making Arms in the Machine Age: Philadelphia's Frankford Arsenal, 1816–1870*. University Park: Pennsylvania State University Press, 1994.

Fisher, Marvin. *Workshops in the Wilderness: The European Response to American Industrialization, 1830–1860*. New York: Oxford University Press, 1967.

Fitch, Charles H. "Report on the Manufactures of Interchangeable Mechanism." In *Tenth Census of the United States (1880)*, vol. 2, pp. 617–45. Washington, D.C.: Government Printing Office, 1883.

Flayderman, Norm. *Flayderman's Guide to Antique American Firearms . . . and Their Values*. 7th ed. Iola, Wisc.: Krause, 1998.

Fries, Russell I. "British Response to the American System: The Case of the Small-Arms Industry after 1850." *Technology and Culture* 16, no. 3 (1975): 377–403.

———. "A Comparative Study of the British and American Arms Industries, 1790–1890." Ph.D. dissertation, Johns Hopkins University, 1972.

Gaier, Claude. *Four Centuries of Liege Gunmaking*. Liege, Belgium: Librairie Halbart, 1976.

Gaier, Claude, and Pietro Sabatte. *The Art of Gun Engraving*. New York: Knickerbocker Press, 1999.

Garavaglia, Louis A., and Charles G. Worman. *Firearms of the American West, 1803–1865*. Albuquerque: University of New Mexico Press, 1984.

———. *Firearms of the American West, 1866–1894*. Albuquerque: University of New Mexico Press, 1985.

Gordon, James D. *Winchester's New Model of 1873: A Tribute*. 2 vols. Grant, Colo.: printed by author, 1997.

Gordon, Robert B. "Material Evidence of Manufacturing Methods Used in Springfield Armory." *Journal of the Society for Industrial Archeology* 14, no. 1 (1988): 57–75.

Grancsay, Stephen V. "The Craft of the Early American Gunsmith." *Metropolitan Museum of Art Bulletin*, new series, vol. 6, no. 2 (1947): 54–61.

———. *Master French Gunsmiths' Designs*. New York: Winchester Press, 1970.

Grant, Ellsworth S. *The Colt Armory: A History of Colt's Manufacturing Company*. Lincoln, R.I.: Andrew Mowbray, 1995.

Grant, James J. *Single-Shot Rifles*. Highland Park, N.J.: Gun Room Press, 1982.

Greener, W. W. *The Gun and Its Development*. 9th ed. London: Cassell, 1910.

Greenough, Horatio. *Form and Function: Remarks on Art*. Edited by Harold A. Small. Berkeley: University of California Press, 1947. Greenough essays originally published as *Travels, Observations, and Experience of a Yankee Stonecutter*, 1852.

Gusler, Wallace B., and James D. Lavin. *Decorated Firearms, 1540–1870*. Williamsburg, Va.: Colonial Williamsburg Foundation, 1977.

Habakkuk, H. J. *American and British Technology in the Nineteenth Century: The Search for Labor-Saving Inventions*. Cambridge: Oxford University Press, 1962.

Hanson, Charles E., Jr. *The Hawken Rifle: Its Place in History*. Chadron, Nebr.: Fur Press, 1979.

———. *The Plains Rifle*. Harrisburg, Pa.: Stackpole, 1960.

Harris, Fredric A. *Firearms Engraving as Decorative Art: The Origin of American Firearms Engraving Motifs in the Decorative Art of the Middle East*. Seattle, Wash.: Barbara R. Harris, 1989.

Haven, Charles T., and Frank A. Belden. *A History of the Colt Revolver and Other Arms Made by Colt's Patent Fire Arms Manufacturing Company from 1836 to 1940*. New York: William Morrow, 1940.

Hawke, David Freeman. *Nuts and Bolts of the Past: A History of American Technology, 1776–1860*. Perennial Library edition. New York: Harper and Row, 1989.

Hayward, John F. *The Art of the Gunmaker*. 2 vols. London: Barrie and Rockliff, 1962–63.

Held, Robert. *The Age of Firearms: A Pictorial History*. New York: Harper and Brothers, 1957.

Henry, Paul. *Ethan Allen and Allen & Wheelock: Their Guns and Their Legacy*. Lincoln, R.I.: Andrew Mowbray, 2006.

Hindle, Brooke. *Emulation and Invention*. New York: New York University Press, 1981.

———. *Technology in Early America*. Chapel Hill: University of North Carolina Press, 1966.

Hindle, Brooke, and Steven Lubar. *Engines of Change: The American Industrial Revolution, 1790–1860*. Washington, D.C.: Smithsonian Institution Press, 1986.

Hoke, Donald R. *Ingenious Yankees: The Rise of the American System of Manufactures in the Private Sector*. New York: Columbia University Press, 1990.

Hornung, Clarence P. *Treasury of American Design*. 2 vols. New York: Harry N. Abrams, 1976.

Hosley, William. *Colt: The Making of an American Legend*. Amherst: University of Massachusetts Press, 1996.

———. "Guns, Gun Culture, and the Peddling of Dreams." In Dizard et al., *Guns in America*, 47–75.

Hounshell, David A. *From the American System to Mass Production, 1800–1932*. Baltimore, Md.: Johns Hopkins University Press, 1984.

Houze, Herbert G. "The Appearance of Evidence: A Brief Examination of the Life and Work of Herman Leslie Ulrich." *Armax—The Journal of the Cody Firearms Museum* 4, no. 2 (1993): 35–71.

———. *Arming the West: Schuyler, Hartley & Graham's Arms Shipments to the American Frontier, 1868–1886*. Woonsocket, R.I.: Andrew Mowbray, 2008.

———. *Colt Factory Engravers of the Nineteenth Century: Understanding Their Careers and Identifying Their Work*. Woonsocket, R.I.: Andrew Mowbray, 2012.

———. *Samuel Colt: Arms, Art, and Invention*. New Haven, Conn.: Yale University Press in association with Wadsworth Atheneum Museum of Art, 2006.

———. *The Winchester Model 1876 "Centennial" Rifle*. Lincoln, R.I.: Andrew Mowbray, 2001.
———. *Winchester Repeating Arms Company: Its History and Development from 1866 to 1981*. Iola, Wisc.: Krause, 1995.
Howard, Robert A. "Interchangeable Parts Reexamined: The Private Sector of the American Arms Industry on the Eve of the Civil War." *Technology and Culture* 19, no. 4 (1973): 633–49.
Howe, Daniel Walker. *What Hath God Wrought: The Transformation of America, 1815–1848*. Oxford: Oxford University Press, 2007.
Hubbard, Guy. "Development of Machine Tools in New England." *American Machinist*, vols. 59–61 (1923–25).
Huntington, R. T. *Hall's Breechloaders*. York, Pa.: George Shumway, 1972.
Jervis, Simon, et al. *Art and Design in Europe and America, 1800–1900*. New York: E. P. Dutton, 1987.
Jinks, Roy G. *Artistry in Arms: The Guns of Smith & Wesson*. Springfield, Mass.: Smith & Wesson and Connecticut Valley Historical Museum, 1991.
———. *History of Smith & Wesson: No Thing of Importance Will Come without Effort*. North Hollywood, Calif.: Beinfeld, 1977.
Johnson, Curtis L. *Gunmakers of Illinois, 1683–1900*. Vol. 1. York, Pa.: George Shumway, 1994.
Jones, Owen. *The Grammar of Ornament*. New York: Portland House, 1986. Reprint of original 1856 London edition.
Jordan, Robert M. *Colt's Pocket '49: Its Evolution Including the Baby Dragoon and Wells Fargo*. Loma Mar, Calif.: Darrow M. Watt, 2000.
Kasson, John F. *Civilizing the Machine: Technology and Republican Values in America, 1776–1900*. New York: Grossman, 1976.
Kauffman, Henry J. *The Pennsylvania-Kentucky Rifle*. New York: Bonanza Books, 1960.
Kendall, George Wilkins. *Narrative of the Texan Santa Fe Expedition*. 2 vols. New York: Harper and Brothers, 1844.
Kennedy, Monty. *Checkering and Carving of Gun Stocks*. Harrisburg, Pa.: Stackpole Books, 1962.
Kentucky Rifle Association. *Kentucky Rifles and Pistols, 1750–1850*. Delaware, Ohio: Golden Age Arms Co. and James R. Johnston, 1976.
Keogh, Gerald. *Samuel Colt's New Model Pocket Pistols*. Ogden, Utah: printed by author, 1964.
Kindig, Joe, Jr. *Thoughts on the Kentucky Rifle in Its Golden Age*. Wilmington, Del.: George N. Hyatt, 1960.
Kindig, Joe K., III. *Artistic Ingredients of the Longrifle*. York, Pa.: George Shumway, 1989.
Knight, Frederick. *Scroll Ornaments of the Early Victorian Period*. New York: Dover Pictorial Archive, 1978. Reprint of Knight's *Scroll Ornaments, Designed for the Use of Silversmiths, Chasers, Die-Sinkers, Modellers, &c. &c.* (London, n.d.).
Kopec, John, Ron Graham, and Kenneth Moore. *A Study of the Colt Single Action Army Revolver*. La Puente, Calif.: printed by authors, 1976.
Kouwenhoven, John A. *The Arts in Modern American Civilization*. New York: W. W. Norton, 1967. First published as *Made in America: The Arts in Modern Civilization*, 1948.
Layman, George J. *A Guide to the Maynard Breechloader*. Union City, Tenn.: Pioneer Press, 1998.
Lee, Robert M., and R. L. Wilson. *The Art of the Gun: Magnificent Colts*. 2 vols. Sparks, Nev.: Yellowstone Press, 2012.
———. *The Art of the Gun*. Vol. 4, *Winchester Firearms Masterpieces*. Sparks, Nev.: Yellowstone Press, 2002.
Lewis, Edmund E., and Stephen W. Rutter. *Volcanic Firearms: Predecessor to the Winchester Rifle*. Woonsocket, R.I.: Andrew Mowbray, 2011.
Lindsay, Merrill. *The Kentucky Rifle*. New York: Arma Press and Historical Society of York County [Pa.], 1972.
———. *One Hundred Great Guns: An Illustrated History of Firearms*. New York: Walker, 1967.
Madaus, Howard Michael, and Simeon Stoddard, comps. *The Guns of Remington: Historic Firearms Spanning Two Centuries*. Dayton, Ky.: Biplane Productions, 1997.
Madis, George. *The Winchester Book*. Dallas: Taylor, 1971.
———. *Winchester Dates of Manufacture*. Brownsboro, Tex.: Art and Reference House, 1981.
Maier, Pauline, Merritt Roe Smith, Alexander Keyssar, and Daniel J. Kevles. *Inventing America: A History of the United States*. New York: W. W. Norton, 2003
Marcot, Roy M. *Remington: America's Oldest Gunmaker*. Peoria, Ill.: Primedia, 1998.
———. *Remington Rolling Block Rifles, Carbines and Shotguns; Sporting and Target Rifles*. Tucson, Ariz.: Northwood Heritage Press, 2009.
———. *Spencer Repeating Firearms*. Irvine, Calif.: Northwood Heritage Press, 1983.
Marcot, Roy, and Ron Paxton. *Colt Brevete Revolvers*. Tucson, Ariz.: Northwood Heritage Press, 2012.
Marlin Fire Arms Company. *Marlin Repeating Rifles, Illustrated Price List*. Sales catalog. San Francisco: Philip B. Bekeart, 1905.
Maxwell, Samuel L., Sr. *The Colt-Burgess Magazine Rifle*. Dallas: Taylor, 1985.
———. *Lever Action Magazine Rifles Derived from the Patents of Andrew Burgess*. Dallas: Taylor, 1976.
Mayr, Otto, and Robert C. Post, eds. *Yankee Enterprise: The Rise of the American System of Manufactures*. Washington, D.C.: Smithsonian Institution Press, 1981.
McDowell, R. Bruce. *Evolution of the Winchester*. Tacoma, Wash.: Armory, 1985.
———. *A Study of Colt Conversions and Other Percussion Pistols*. Iola, Wisc.: Krause, 1997.
Meek, James B. *The Art of Engraving*. Montezuma, Iowa: Brownell's Inc., 1974.
Mills, Fred P. L. *A Study of Colt's Double Rifles*. Greenfield, Mass.: Secretarial Phone-O-Service, 1953.
Mouillesseaux, Harold R. *Ethan Allen, Gunmaker: His Partners, Patents and Firearms*. Ottawa: Museum Restoration Service, 1973.

Neal, Robert J., and Roy G. Jinks. *Smith and Wesson, 1857–1945.* New York: A. S. Barnes, 1975.
North, S. N. D., and Ralph H. North. *Simeon North: First Official Pistol Maker of the United States.* Concord, N.H.: Rumford Press, 1913.
Norton, Charles B. *American Inventions and Improvements in Breech-Loading Small Arms.* Springfield, Mass.: Chapin and Gould, 1880.
Nye, David E. *America as Second Creation: Technology and Narratives of New Beginnings.* Cambridge, Mass.: MIT Press, 2003.
Parker, Harry J., Ora Lee Parker, and Joan S. Reisch. *A Study of Remington's Smoot Patent and Number Four Revolvers.* Santa Ana, Calif.: Armslore Press, 2003.
Parsons, John E. *The First Winchester: The Story of the 1866 Repeating Rifle.* New York: William Morrow, 1955.
———. *The Peacemaker and Its Rivals.* New York: William Morrow, 1950.
———. *Smith & Wesson Revolvers: The Pioneer Single Action Models.* New York: William Morrow, 1957.
Peterson, Harold L., ed. *Encyclopedia of Firearms.* New York: E. P. Dutton, 1967.
———. *How to Tell If It's a Fake.* New York: Charles Scribner's Sons, 1978.
Phelps, Art. *The Story of Merwin, Hulbert & Co. Firearms.* Santa Ana, Calif.: printed by author, 1992.
Phillips, Philip R., and R. L. Wilson. *Paterson Colt Pistol Variations.* Dallas: Jackson Arms, 1979.
Prudhomme, E. C. *Gun Engraving Review.* Shreveport, La.: printed by author, 1961.
Pursell, Carroll W., Jr., ed. *Technology in America: A History of Individuals and Ideas.* 2nd ed. Cambridge, Mass.: MIT Press, 1990.
Pye, David. *The Nature of Design.* London: Studio Vista and Van Nostrand Reinhold, 1972.
Quick, Les. *The Story of Benjamin Tyler Henry and His Famed Repeating Rifle.* Santa Ana, Calif.: Graphic Publishers, 2008.
Raber, Michael S. "Conservative Innovators, Military Small Arms, and Industrial History at Springfield Armory." *Journal of the Society for Industrial Archeology* 14, no. 1 (1988): 1–21.
Raber, Michael S., Patrick M. Malone, Robert B. Gordon, and Carolyn C. Cooper. *Conservative Innovators and Military Small Arms: An Industrial History of the Springfield Armory, 1794–1968.* South Glastonbury, Conn.: Raber Associates, 1989.
———. *Forge of Innovation: An Industrial History of the Springfield Armory, 1794–1968.* Edited by Richard Colton. n.p.: Eastern National, 2008.
Rattenbury, Richard C. *The Browning Connection: Patent Prototypes in the Winchester Museum.* Cody, Wyo.: Buffalo Bill Historical Center, 1982.
———. "Exhibition Preview: Samuel Colt—Arms, Art and Invention." *Persimmon Hill* 35, no. 2 (2007): 42–48.
———. *Hunting the American West: The Pursuit of Big Game for Life, Profit, and Sport, 1800–1900.* Missoula, Mont.: Boone and Crockett Club, 2008.
———. "J. C. Petmecky: Texas Gun and Spur Maker." *Man at Arms* 11, no. 5 (1989): 10–15.
———. "Management of Firearms Collections: Identification and Classification." *History News* 36, no. 3 (1981): unpaged. American Association for State and Local History Technical Leaflet 136.
———. "The Mountain Rifle: Fabled Arm of the Fur Trappers." *Persimmon Hill* 25, no. 4 (1997): 46–49.
———. *Packing Iron: Gunleather of the Frontier West.* Millwood, N.Y.: ZON International, 1993.
———. "Shooting a Stand: Buffalo Rifles on the West Texas Frontier." *Man at Arms* 9, no. 6 (1987): 12–20.
———. "Tools of Triumph and Tragedy: Firearms in the American West." In *Visions of the West: Art and Artifacts from the Private Collections of J. P. Bryan, Torch Energy Advisors Incorporated and Others.* Salt Lake City, Utah: Gibbs-Smith, 1999.
Rattenbury, Richard C., and Thomas E. Hall. *Sights West: Selections from the Winchester Museum Collection.* Cody, Wyo.: Buffalo Bill Historical Center; Lincoln: University of Nebraska Press, 1981.
Renneberg, Robert C. *The Winchester Model 94: The First 100 Years.* Iola, Wisc.: Krause, 1991.
Roberts, Ned H. *The Muzzle-Loading Cap Lock Rifle.* Manchester, N.H.: Clarke Press, 1944.
Roe, Joseph W. *Connecticut Inventors.* New Haven, Conn.: Yale University Press, 1934.
———. *English and American Tool Builders.* New Haven, Conn.: Yale University Press, 1916.
Rosa, Joseph G. *Colonel Colt London: The History of Colt's London Firearms, 1851–1857.* London: Arms and Armour Press, 1976.
———. *Guns of the American West.* New York: Crown, 1985.
Rose, Alexander. *American Rifle: A Biography.* New York: Delacorte Press, 2008.
Rosebush, W. E. *American Firearms and the Changing Frontier.* Spokane: Eastern Washington State Historical Society, 1962.
Rosenburg, Nathan, ed. *The American System of Manufactures.* Edinburgh: Edinburgh University Press, 1969.
Rowe, Tom. *Remington's No. 3 Hepburn Sporting and Target Rifles.* Woonsocket, R.I.: Mowbray, 2011.
Russell, Carl P. *Firearms, Traps and Tools of the Mountain Men.* New York: Alfred A. Knopf, 1967.
———. *Guns on the Early Frontiers.* Berkeley: University of California Press, 1957.
Rywell, Martin. *The Trial of Samuel Colt.* Harriman, Tenn.: Pioneer Press, 1953.
Schlereth, Thomas J., ed. *Material Culture Studies in America.* Nashville, Tenn.: American Association for State and Local History, 1989.
Schwing, Ned. *Winchester Slide-Action Rifles.* Vols. 1 and 2. Iola, Wisc.: Krause, 2004.
Sellers, Frank M. *American Gunsmiths.* Highland Park, N.J.: Gun Room Press, 1983.
———. *Sharps Firearms.* North Hollywood, Calif.: Beinfeld, 1978.

——. *The William M. Locke Collection*. East Point, Ga.: Antique Armory, 1973.
Sellers, Frank M., and Samuel Smith. *American Percussion Revolvers*. Ottawa: Museum Restoration Service, 1971.
Serven, James E. *The Collecting of Guns*. Harrisburg, Pa.: Stackpole, 1964.
Sharpe, Philip B. *The Rifle in America*. New York: William Morrow, 1943.
Sharps Rifle Company. *The Sharps Rifle Company Armory*. Sales catalog. New York: D. Appleton, 1875.
Shumaker, P. L. *Colt's Variations of the Old Model Pocket Pistol, 1848–1872*. Beverly Hills, Calif.: Fadco, 1957. Revised edition, 1966.
Slotkin, Richard. *The Fatal Environment: The Myth of the Frontier in the Age of Industrialization, 1800–1890*. New York: Atheneum, 1985.
Smith, Merritt Roe. "Army Ordnance and the 'American System' of Manufacturing, 1815–1861." In *Military Enterprise and Technological Change*, edited by Merritt R. Smith. Cambridge, Mass.: MIT Press, 1985, pp. 39–86.
——. "Eli Whitney and the American System of Manufacturing." In *Technology in America: A History of Individuals and Ideas*, edited by Carroll W. Pursell, Jr. 2nd ed. Cambridge, Mass.: MIT Press, 1990, pp. 45–61.
——. "From Craftsman to Mechanic: The Harpers Ferry Experience, 1798–1854." In *Technological Innovation and the Decorative Arts*, 103–39. Winterthur Conference Report 1973. Charlottesville: University Press of Virginia, 1974.
——. *Harpers Ferry Armory and the New Technology: The Challenge of Change*. Ithaca, N.Y.: Cornell University Press, 1977.
——. "John H. Hall, Simeon North, and the Milling Machine: The Nature of Innovation among Antebellum Arms Makers." *Technology and Culture* 14, no. 4 (1973): 573–91.
——. "Museum Review: The American Precision Museum." *Technology and Culture* 15, no. 3 (1974): 413–37.
Smith, William P. "The American Small Arms Industry: 1870–1890." Ph.D. dissertation, Northwestern University, 1966.
Smithurst, Peter G. *The Guns and Gun-Making Machinery of Robbins and Lawrence*. Windsor, Vt.: American Precision Museum, 2007. Article originally published in the Royal Armories Yearbook, vol. 7 (2002): 64–76.
Speltz, Alexander. *The Styles of Ornament*. New York: Dover, 1959.
Stockbridge, Virgil D. *Digest of Patents Relating to Breech-Loading and Magazine Small Arms, 1836–1873*. Washington, D.C.: Government Printing Office, 1875.
Supica, Jim, and Richard Nahas. *Standard Catalog of Smith & Wesson*. Iola, Wisc.: Gun Digest Books, 2006.
Swayze, Nathan L. *'51 Colt Navies*. Highland Park, N.J.: Gun Room Press, 1991.
Sword, Wiley. *The Historic Henry Rifle: Oliver F. Winchester's Famous Civil War Repeater*. Lincoln, R.I.: Andrew Mowbray, 2002.
Tarassuk, Leonid. *Arms and Art: American and European Firearms Decorated by Raymond J. Wielgus*. Chicago: Art Institute of Chicago, 1988.
Tivey, Ted. *The Colt Rifle, 1884–1902*. New South Wales, Australia: Clouston and Hall, 1984.
Tobias, Arthur. *Colt Cylinder Scenes, 1847–1851*. Los Angeles, Calif.: printed by author, 2011.
Toffolon, Siro R. "The Kentucky Rifle and Its Ornamentations." *Kentucky Rifle Association Bulletin* 6, no. 2 (Winter 1980): 2–5.
Turpin, Tom. *Custom Firearms Engraving: The Techniques and Treasures of the World's Greatest Artists*. Iola, Wisc.: Krause, 1999.
Twain, Mark. *Roughing It*. Hartford, Conn.: American Publishing Company, 1872.
U.S. Secretary of the Interior. *Manufactures of the United States in 1860: Compiled from the Original Returns of the Eighth Census*. Washington, D.C.: Government Printing Office, 1865.
Uselding, Paul J. "Elisha K. Root, Forging, and the 'American System.'" *Technology and Culture* 15, no. 4 (1974): 543–68.
——. "Technical Progress at the Springfield Armory, 1820–1850." *Explorations in American History* 9 (1972): 291–316.
Ware, Donald L. *Remington Army and Navy Revolvers, 1861–1888*. Albuquerque: University of New Mexico Press, 2007.
Watrous, George R., Thomas E. Hall, James C. Rokhoff, Pete Kuhlhoff, and Duncan Barnes. *The History of Winchester Firearms, 1866–1980*. Tulsa, Okla.: Winchester Press, 1980.
West, Bill. *Marlin and Ballard—Firearms and History*. Whittier, Calif.: Stockton Trade Press, 1968.
Whisker, James B. *The Gunsmith's Trade*. Lewiston, N.Y.: E. Mellon Press, 1992.
Williamson, Harold F. *Winchester, the Gun That Won the West*. Washington, D.C.: Combat Forces Press, 1952.
Wilson, R. L. "American Arms Engravers, c. 1830–1981." *Man at Arms* 3, no. 6 (1989): 21–23.
——, ed. *Antique Arms Annual*. San Antonio, Tex.: S. P. Stevens, Leo Bradshaw, and the Texas Gun Collectors Association, 1971.
——. *Blue Book Pocket Guide for Colt Dates of Manufacture*. Minneapolis, Minn.: Blue Book, 2008.
——. *The Book of Colt Firearms*. 3rd ed. Minneapolis, Minn.: Blue Book, 2008.
——. *Colt, an American Legend: The Official History of Colt Firearms*. New York: Abbeville Press, 1985.
——. *The Colt Engraving Book*. 2 vols. New York: Bannerman's (Martin Lane), 2002–2003.
——. *Colt's Dates of Manufacture, 1837–1978*. Hamilton, Ohio: I.D.S.A. Books, 1983.
——. "Firearms Engraving in Nineteenth Century America." *Bulletin of the American Society of Arms Collectors* 28 (Fall 1973): 2–20.
——, ed. *L. D. Nimschke, Firearms Engraver*. Teaneck, N.J.: John J. Malloy, 1964.
——. "Masterpieces of the Gun Engraver's Art: An Analysis of Masterworks from 1850–1970." In *Guns & Ammo Annual 1971*, 248–55. Los Angeles: Petersen, 1971.
——. *The Paterson Colt Book: The Early Evolution of Samuel Colt's Repeating Arms*. Palo Alto, Calif.: Strutz-LeVett, 2001.

———. *The Peacemakers: Arms and Adventure in the American West*. New York: Random House, 1992.
———. *Steel Canvas: The Art of American Arms.* New York: Random House, 1994.
———. *Winchester, an American Legend: The Official History of Winchester Firearms and Ammunition.* New York: Random House, 1991.
———. *Winchester Engraving*. Palm Springs, Calif.: Beinfeld, 1990.
———. *Winchester: The Golden Age of American Gunmaking and the Winchester 1 of 1000*. Cody, Wyo.: Buffalo Bill Historical Center, 1983.
Winant, Lewis. *Pepperbox Firearms.* New York: Greenburg, 1952.
Winchester Repeating Arms Company. *Winchester Repeating Arms Co., Highly Finished Arms.* Sales catalog, reprints are known. Springfield, Mass.: Press of Springfield Printing and Binding, 1897.
———. *Winchester Repeating Arms Co., Repeating Rifles, Shot Guns and Ammunition of All Kinds . . . 1899*. Sales catalog, reprint. Prescott, Ariz.: Wolf, 1990.
———. *Winchester Repeating Fire Arms and Metallic Cartridges . . . 1875*. Sales catalog. New Haven, Conn.: Punderson and Crisand, 1875.
Wolff, Eldon G. *Ballard Rifles in the Henry J. Nunnemacher Collection*. 2nd ed. Milwaukee: North American Press, 1961.
Woodbury, Robert S. *History of the Milling Machine*. Cambridge, Mass.: Technology Press, 1960.
———. "The Legend of Eli Whitney and Interchangeable Parts." *Technology and Culture* 1, no. 2 (1960): 235–53.

INDEX

References to illustrations appear in italic type.